Nanomaterials for Electrochemical Sensing

This book reviews the fundamentals of electrochemical sensors, the preparation of electrodes, potential materials for sensing applications, and different analytical methods used for electrochemical sensing applications. It further covers the designing of various electrodes and electrode materials, instruments, sensing mechanisms, advanced nanomaterials for sensing, and so forth. The scalability and commercialization of electrochemical sensors and the challenges and prospects of electrochemical sensors are also described.

Key Features:

- Provides an overview of the advances in the application of nanomaterials in sensing,
- Covers basic fabrication techniques of electrodes as an important part of electrochemical sensors and analysis,
- Reviews the use and analysis of different types of nanomaterials and nanocomposites used for fabrication of working electrodes,
- Emphasizes carbon-based nanomaterials, 2D nanomaterials, and advanced nanocomposites comprising various matrix systems such as conducting polymers, and
- Explores electron transfer, redox behaviour, fabrication techniques, data interpretation, and advanced nanomaterials as working electrode materials.

This book is aimed at researchers and graduate students in nanomaterials, electrochemistry, chemical engineering, and materials science.

Emerging Materials and Technologies
Series Editor: Boris I. Kharissov

The *Emerging Materials and Technologies* series is devoted to highlighting publications centered on emerging advanced materials and novel technologies. Attention is paid to those newly discovered or applied materials with potential to solve pressing societal problems and improve quality of life, corresponding to environmental protection, medicine, communications, energy, transportation, advanced manufacturing, and related areas.

The series takes into account that, under present strong demands for energy, material, and cost savings, as well as heavy contamination problems and worldwide pandemic conditions, the area of emerging materials and related scalable technologies is a highly interdisciplinary field, with the need for researchers, professionals, and academics across the spectrum of engineering and technological disciplines. The main objective of this book series is to attract more attention to these materials and technologies and invite conversation among the international R&D community.

Advancements in Nanomaterials for Energy Conversion and Storage
Edited by Piyush Kumar Sonkar and Vellaichamy Ganesan

2D Materials-Based Sensors: Technology and Applications
Vinod Kumar Khanna

Metal Organic Framework Derived Materials: Design Strategies and Applications
Gomathi Nageswaran, Varsha M V, Arun Kumar Rajasekaran and M Shashank Rao

Hydrogen Production, Storage, and Utilization: Technologies and Applications
Abbas Tcharkhtchi, Hamidreza Vanaei, Albert Lucas and Sedigheh Farzaneh

MXenes for Energy Storage Applications: Emerging Characteristics, Compositions, and Synthesis Methods
Muhammad Rafique, M. Bilal Tahir, and Saira Anwar

Multi-scale and Multifunctional Coatings and Interfaces for Tribological Contacts
Ajit Behera, Kuldeep K Saxena, Dipen Kumar Rajak and Shankar Sehgal

Nanotechnology in Green Energy Generation
Ahmed Thabet Mohamed

Thermo-Acoustics of Nanofluids and Transfer Processes
Shriram S. Sonawane and Manjakuppam Malika

Hybrid Nanofluids: Heat and Mass Transfer Processes
Shriram S. Sonawane and Manjakuppam Malika

Applications of Hybrid Nanofluids in Science and Engineering
Edited by A. K. Pandey, H. Upreti, O. D. Makinde, and A. J. Chamkha

Solar Cells Development and Fabrication
Edited by Shivani Dhall, Kapil Sood and Vinay Gupta

Nanomaterials for Electrochemical Sensing
Edited by Syed Shahabuddin, Rama Gaur and Nandini Mukherjee

For more information about this series, please visit: www.routledge.com/Emerging-Materials-and-Technologies/book-series/CRCEMT

Nanomaterials for Electrochemical Sensing

Edited by
Syed Shahabuddin, Rama Gaur,
and Nandini Mukherjee

CRC Press
Taylor & Francis Group
Boca Raton London New York

CRC Press is an imprint of the
Taylor & Francis Group, an **informa** business

Designed cover image: Shutterstock

First edition published 2025
by CRC Press
2385 NW Executive Center Drive, Suite 320, Boca Raton FL 33431

and by CRC Press
4 Park Square, Milton Park, Abingdon, Oxon, OX14 4RN

CRC Press is an imprint of Taylor & Francis Group, LLC

ISBN: 9781032458885 (hbk)
ISBN: 9781032654898 (pbk)
ISBN: 9781032654874 (ebk)

DOI: 10.1201/9781032654874

Typeset in Times
by codeMantra

Contents

Preface

This book provides a thorough overview of the most recent developments and advancements in the field of nanomaterials while also providing an exciting look at what is still to come. This text is intended for postgraduate and undergraduate researchers, students, and professionals in the research industries who are interested in learning about the fascinating role of nanomaterials for electrochemical sensing, an area that has seen incredible progress in the last few years.

In the realm of electrochemical sensing, nanomaterials have changed the game by enabling unparalleled sensitivity and selectivity in the detection and analysis of chemical species. Their special qualities, like their large surface area, increased catalytic activity, and adjustable electrical characteristics, have created new possibilities for the development of incredibly dependable and effective sensing platforms.

We have collected a compilation of state-of-the-art research papers and knowledgeable inputs from top scientists and engineers in the field in this book. With an emphasis on a particular facet of nanomaterials for electrochemical sensing, each chapter offers a comprehensive look at the synthesis, characterization, and uses of these extraordinary materials.

This book covers a wide range of issues that highlight the multidisciplinary nature of nanomaterials research, from the development of novel nanomaterials like graphene, carbon nanotubes, metal nanoparticles, and metal-organic frameworks to their integration into diverse sensing systems. It investigates the synergistic marriage of materials science, electrochemistry, and nanotechnology, among other disciplines, to push the limits of sensitivity, selectivity, and real-time monitoring.

We hope this book will be an invaluable resource for professionals, researchers, and students alike, encouraging them to investigate the enormous possibilities of nanomaterials for electrochemical sensing and help drive its further development. Through exchanging the most recent findings and encouraging cooperation, we can all push innovation and take on urgent global issues.

It's important to recognize that perfection is often an unattainable goal, especially when it comes to creating written works. Authors and editors alike may find themselves constantly striving for improvement even after multiple rounds of review and editing. While the chapters and the final draft of a book may have undergone comprehensive scrutiny, there is always room for refinement and modification to enhance the quality and quantity of the content.

Therefore, as readers of this book, your feedback is not only welcomed but also encouraged. We value the input of students, researchers, scientists, and professors who can offer valuable insights and suggestions regarding any errors, omissions, or important information that may have been overlooked. Your recommendations can help us improve and enhance the overall quality of this book.

About the Editors

Syed Shahabuddin earned his M.Sc. degree in materials chemistry from Jamia Millia Islamia, India, in 2011 and his Ph.D. degree in polymer chemistry from the University of Malaya, Malaysia, in September 2016. He served as an Assistant Manager in Samtel Avionics Limited for almost 3 years in the research and development of avionics-grade displays. Formerly, he worked as a Senior Research Fellow (senior lecturer) at the Research Centre for Nano-Materials and Energy Technology (RCNMET), Sunway University, Malaysia, for more than 2 years. Currently, he is working as an Assistant Professor at Pandit Deendayal Petroleum University, Gandhinagar, Gujarat. He has published more than 150 research articles in international journals of repute and presented 30 papers at international/national conferences. He has successfully completed three research projects as Principal Investigator (PI) and twelve research projects as Co-Investigator (Co-I) with different institutes in India and Malaysia. He is a member of the Royal Society of Chemistry (MRSC), life member of Society for Materials Chemistry (SMC) and a reviewer of many high-impact journals. Eight students have completed PhD under his guidance. His current research focuses on the synthesis of nanomaterials, 2D-MXene, graphene, conducting polymer nanocomposites for water treatment, photocatalysis, supercapacitors, DSSCs, nanofluids for solar thermal applications, waste lubricant oil refining, and phase change materials.

Rama Gaur is an Assistant Professor at the School of Energy Technology, PDEU Gandhinagar, Gujarat. She received her Ph.D. degree in chemistry from IIT Roorkee in 2017. She is a talented and versatile researcher who has made significant contributions in various fields. With a background in technology and a passion for innovation, she has established herself as a creative problem solver and a forward-thinking researcher. She has a deep understanding of cutting-edge technologies and their potential to drive positive change in society.

Her expertise extends to fields such as the synthesis of nanoscale materials with interesting and unique morphologies by simple and economical chemical approaches; shape- and size-dependent optical, magnetic, and electrochemical properties; and applications in photocatalysis, optoelectronics, electrochemical sensing, electrocatalytic reduction/oxidation, solar energy conversion, energy storage devices, supercapacitors, water splitting, and environmental remediation. She has a knack for identifying patterns and trends, enabling them to develop innovative solutions to complex problems. Her ability to think outside the box and her natural curiosity have led her to explore diverse domains, ranging from healthcare to environment and beyond. She has received the DST-SERB-TARE Grant in collaboration with IIT Gandhinagar and several institute research grants. She has authored more than 50 research articles and 10 book chapters and has edited a book. She has served as an Assistant Professor at the National Institute of Technology Hamirpur and the Indian Institute of Information Technology, Una (Himachal Pradesh). She has vast experience in teaching and research. She has been awarded with Senior Science Laureate Award for 2022 and 2023 by Vikram Sarabhai Community Science Centre, Gujarat, India.

Apart from her technical expertise, she also has outstanding leadership and communication abilities. She has a talent for energizing and encouraging groups, creating a cooperative atmosphere that allows ideas to flourish. She is a dependable mentor and advisor to others around them because of her warm and personable demeanour.

Nandini Mukherjee earned her bachelor's (B.Sc. Hons. in chemistry) degree from Scottish Church College, University of Calcutta, in 2012 and her master's degree in chemistry from Banaras Hindu University in 2014. She earned her Ph.D. degree in bioinorganic and medicinal chemistry from the Indian Institute of Science, Bangalore, in 2020. Currently, she is working as an Assistant Professor in the Department of Chemistry, Dinhata College, Cooch Behar Panchanan Barma University, West Bengal, India. Prior to joining Barma University, she also worked as an Assistant Professor in the Department of Chemistry, Pandit Deendayal Energy University (PDEU), Gandhinagar, Gujarat, and as a Research Associate in IISc Bangalore. She has more than 5 years of research experience in organic synthesis and coordination chemistry and published her research in international peer-reviewed journals. Her current research interest focuses on the design and synthesis of molecules that have applications in the field of chemosensing of toxins and monitoring biological phenomena; she is also working to develop cost-effective nanomaterials for application in medicinal chemistry and renewable energy sector.

Contributors

Muhammad Shakeel Ahmad
Higher Institution Centre of Excellence (HICoE)
UM Power Energy Dedicated Advanced Centre (UMPEDAC)
University of Malaya
Kuala Lumpur, Malaysia

Waqar Ahmed
Centre of Micro-Nano System
School of Information Science and Technology
Fudan University
Shanghai, China

Md. Aurangzeb
Department of Chemical Engineering
School of Energy Technology
Pandit Deendayal Energy University
Gandhinagar, India

Vijay Bahadur
Department of Science
Alliance University
Bengaluru, India
and Department of Pharmaceutical and Pharmacology
College of Pharmacy
University of Houston
Houston, Texas

Nor Kartini Abu Bakar
Department of Chemistry
Universiti Malaya
Kuala Lumpur, Malaysia

Sharvari Bhat
Amity Institute of Pharmacy
Amity University
Noida, India

Ashish Kumar Bhui
Department of Chemistry
Government Shaheed Kaushal Yadav College
Chhattisgarh, India

Meetkumar Y. Chaudhary
Department of Chemistry
Pandit Deendayal Energy University
Gandhinagar, India

Urvashi Chauhan
Department of Electronics & Communication Engineering
Parul University
Gujarat, India

Himanshu Chhabra
Department of Mechatronics Engineering
Parul University
Gujarat, India

Manoj S Dikkatwar
Amity Institute of Pharmacy
Amity University
Noida, India

Prince Jain
Department of Mechatronics Engineering
Parul University
Gujarat, India

Dashrathbhai B. Kanzariya
Department of Chemistry
Pandit Deendayal Energy University
Gandhinagar, India

Pritesh Keshari
Department of Chemistry
Indian Institute of Technology (BHU)
Varanasi, India

Aemun Khalid
Higher Institution Centre of Excellence (HICoE)
UM Power Energy Dedicated Advanced Centre (UMPEDAC)
University of Malaya
Kuala Lumpur, Malaysia

Ruchi Khare
Department of Biological Science & Engineering
Maulana Azad National Institute of Technology
Bhopal
Madhya Pradesh, India

Vinod Khatri
Department of Chemistry
TDL Govt. College for Women
Haryana, India

Manish Kumar
Department of Information and Communications Technology
Pandit Deendayal Energy University
Gandhinagar, India

Habibun Nabi Muhammad Ekramul Mahmud
Department of Chemistry
Universiti Malaya
Kuala Lumpur, Malaysia

Rahul Kumar Maurya
Amity Institute of Pharmacy
Amity University
Noida, India

Shirsendu Mitra
Department of Chemical Engineering
School of Energy Technology
Pandit Deendayal Energy University
Gandhinagar, India

Noman Nawaz
College of Electronic Information and Optical Engineering
Nankai University
Tianjin, P. R. China

Ankita Ojha
Department of Chemistry
Maharaja College (VKSU)
Arrah, India

Tapan K. Pal
Department of Chemistry
Pandit Deendayal Energy University
Gandhinagar, India

Parmanand
Department of Chemistry
Kirori Mal College
University of Delhi
Delhi, India

Badri Parshad
Wellman Center for Photomedicine
Massachusetts General Hospital
Harvard Medical School
Boston, Massachusetts

Shatruhan Singh Rajput
Division of Pulmonary and Critical Care Medicine
Massachusetts General Hospital
Harvard Medical School
Boston, Massachusetts

Bushra Rashid
Faculty of Defence Science and Technology
National Defence University of Malaysia
Kuala Lumpur, Malaysia

Rosy
Department of Chemistry
Indian Institute of Technology (BHU)
Varanasi, India

Subhankar Roy
Department of Chemical Engineering
School of Energy Technology
Pandit Deendayal Energy University
Gandhinagar, India

Bharti Saini
Department of Chemical Engineering
School of Energy Technology
Pandit Deendayal Energy University
Gandhinagar, India

Aarti Sharma
SVKM's Narsee Monjee Institute of Management Studies (deemed to be University)
School of Technology Management and Engineering
Kharghar, India

Paawan Sharma
Department of Information and Communications Technology
School of Technology
Pandit Deendayal Energy University
Gandhinagar, India

Rachna Sharma
Department of Applied Sciences and Humanities
TULAS Institute
Dehradun, India

Rahul Shrivastava
Department of Biological Science & Engineering
Maulana Azad National Institute of Technology
Bhopal
Madhya Pradesh, India

Vivek Shrivastava
Amity Institute of Pharmacy
Amity University
Noida, India

Sanvedna Shukla
Department of Chemistry
Government Digvijay PG Autonomous College
Rajnandgaon, India

Vipin Shukla
Department of Electrical Engineering
School of Energy Technology
Pandit Deendayal Energy University
Gandhinagar, India

Ankush Kumar Singh
Department of Chemistry
Indian Institute of Technology (BHU)
Varanasi, India

Avinash Singh
Department of Chemistry
SRM University Delhi-NCR
Sonepat, India

Avinash Kumar Singh
Department of Electrical and Computer Engineering
University of Houston
Houston, Texas

Manvendra Pratap Singh
Cutaneous Biology Research Center
Massachusetts General Hospital
Harvard Medical School
Boston, Massachusetts

Nanthini Sridewi
Department of Maritime Science and Technology
National Defence University of Malaysia
Kuala Lumpur, Malaysia

Ravi Tejasvi
Department of Chemical Engineering
School of Energy Technology
Pandit Deendayal Energy University
Gandhinagar, India

Alok Shiomurti Tripathi
Department of Pharmacology
Era College of Pharmacy
Era University
Lucknow, India

Preeti Yadav
Department of Chemistry
Maitreyi College
University of Delhi
New Delhi, India

Mohammad Yasir
Amity Institute of Pharmacy
Amity University
Noida, India

1 Introduction to Electrochemical Sensors

Syed Shahabuddin, Rama Gaur, and Nandini Mukherjee

1.1 INTRODUCTION

1.1.1 Overview of Electrochemical Sensors

A sensor is a device that reacts to a physical stimulus, such as temperature, illumination, noise, force, magnetic fields, or motion. It then conveys an electric signal resulting from this reaction, serving the purpose of gauging a change in an inherent property of the material it's made of [1]. The term "sensor" is rooted in the Latin word "sentire," which means to perceive. In essence, sensors possess the quality of perceiving their surroundings to establish a connection. Notably, electrochemical sensors belong to a category of chemical sensors where an electrode is employed as a transducer component while interacting with an analyte.

Contemporary electrochemical sensors utilize a range of characteristics to identify different factors in our daily experiences, encompassing physical, chemical, and biological aspects. Illustrative instances comprise overseeing the environment, health-related measurements, and instrumental sensors associated with machines such as cars, aircraft, mobile devices, and technological equipment. Over the past few decades, advancements in microelectronics and microengineering have significantly contributed to modern sensing systems [2]. This progress has chiefly involved the creation of even tinier sensors endowed with heightened sensitivity and specificity, all while reducing expenses for both manufacturing and upkeep.

Within this framework, the integration of nanoscience and nanotechnology (N&N) concepts and techniques plays a significant role in the rapid advancement of nanosensors. This is primarily due to the profound impact of the surface-to-volume ratio, a key characteristic of nanoscale dimensions. However, this context falls short of meeting the increasing expectations of societies marked by social inequalities. This situation emphasizes the challenge of democratizing knowledge and embracing emerging technologies, including N&N, which has become a prominent issue for 21st-century science. Consequently, social and political considerations have become more prominent within the scientific domain [3].

Presently, the paradigms governing research and development in electrochemical sensors facilitate the exploration of novel materials, diverse sample applications, innovative manufacturing techniques, and strategies for enhancing selectivity and detection limits. In tandem with N&N, these paradigms foster interdisciplinary collaborations, underscoring the role of scientists in the transformative scientific advancements of recent decades.

The historical roots of electrochemical sensor usage trace back to the 1950s when industrial oxygen monitoring was initiated, driven by labour regulations aimed at safeguarding worker health in confined spaces, necessitating the detection of toxic gases and fuels [4]. This spurred intensive research into electrochemical sensors capable of selectively detecting various gases. Notably, Leland C. Clark introduced the oxygen sensor concept, employing two electrodes within a cell separated by an oxygen-permeable membrane and electrolyte solution. Oxygen diffused through the membrane, underwent reduction at the indicator electrode, and generated a current proportional to the oxygen concentration in the sample. The Clark oxygen sensor found wide applications in fields such as medicine, environment, and industrial monitoring. In 1963, during the Cold War era, a distinct oxygen

DOI: 10.1201/9781032654874-1

sensor was created for monitoring water quality in the former Soviet Union. Despite commercial popularity, Clark oxygen sensors yielded unstable electric current signals, necessitating frequent precalibration, which posed usability challenges [5].

The turning point for the contemporary evolution of electrochemical sensors arrived with the invention of the scanning tunnelling microscopy (STM) in 1981 and the atomic force microscopy (AFM) in 1986 by the Binnig and Rohrer teams [6,7]. These groundbreaking advancements in scientific instrumentation enabled the development of innovative products and processes founded on the capacity of these instruments to visualize material structures at dimensions as small as 1.0×10^{-9}m (1 nm), ushering in the realm of N&N. Nanoscience swiftly became an expansive interdisciplinary research domain, built on diverse materials (polymers, ceramics, metals, semiconductors, composites, and biomaterials) structured at the nanoscale to yield structures such as nanoparticles, nanotubes, nanofibres, and nanospheres. Leveraging material behaviour at the nanoscale, fresh avenues of research and applications emerged for electrochemical sensors and electroanalytical techniques.

1.1.2 Scope of the Chapter

The scope of this chapter is to provide a brief understanding of the various electroanalytical techniques and a complete understanding of the electrochemical sensors. A detailed insight into the types of sensors, transduction mechanisms, designing and construction of sensors, their calibration methods, and performance parameters has been provided to get a comprehensive understanding of electrochemical sensors. This information may help researchers, scientists, and industrialists to use electrochemical sensors for a wide range of applications, such as environmental monitoring, food quality and safety, industrial process monitoring, and biomedical and clinical diagnostics.

1.2 FUNDAMENTALS OF ELECTROCHEMISTRY

1.2.1 Electrochemical Reactions and Redox Processes

The chemical reaction consisting of a charge transfer process across the electrode-electrolyte interface is called an electrochemical reaction. The transfer of electrons is the most common charge transfer process, which involves the oxidation and reduction processes. When the electrons are transferred from the electrode to the electrolyte, the process is called reduction, and if the electrons are transferred from the electrolyte to the electrode, the process is called oxidation. This phenomenon is termed as oxidation-reduction or redox reaction and can be expressed by equation 1.1 [8]:

$$O + \text{ne}^- \leftrightarrow R\ (i) \tag{1.1}$$

where O, R, and n are oxidant, reductant, and no. of electrons on the electrode surface, respectively.

The above-mentioned equations indicate that the charge transfer is associated with chemical transformations such as the transformation of O to R and vice versa. The charge transfer suggests the transport of electrons or charge, which indicates the electrical current. Thus, information about the chemical transformation can be gathered with the help of electrical current. The electrical current depends on the direction of the electron either from or towards the electrode transfer, which can be negative or positive. The two electrochemical reactions can also happen at the same time in a way that their electrical current is equal but opposite, which indicates that the net charge is zero. The excess charge which is not compensated by the opposite charge transfer reactions can be termed as net charge. The formula for the net charge can be given by equation 1.2:

$$I_{\text{net}} = I_{\text{ox}} + I_{\text{red}} \tag{1.2}$$

where I_{net} is net electrical current, I_{ox} is oxidation electrical current, and I_{red} is reduction electrical current. The electrons carry the electrical current which flows through the electrical system [8].

1.2.2 Electrochemical Cell Components

1.2.2.1 Working Electrode

The working electrode is the most important component of the electrochemical cell. The reaction occurs on its surface. When the potential difference is applied across the working electrode and reference electrode, reactions occur on the surface of the working electrode. Gold, mercury, carbon, and platinum are the common materials that are used in the making of the working electrodes. Choosing a material for the working electrode is a crucial step. The material should exhibit redox behaviour, be fast enough, and favour electron transfer easily. The material should limit the fouling of electrodes easily. Among the various materials, platinum is the most opted material, because it exhibits electrochemical inertness and it can easily be moulded and fabricated in different forms. There are four major types of working electrodes: ultramicroelectrode (UME), rotating ring-disc electrode (RDE), rotating disc electrode, hanging mercury drop electrode (HMDE) [9], and dropping mercury electrode (DME). An UME is a type of electrode that has a length of less than 20 µm. Miniaturizing may result in the enhancement of some properties, the current density increases, and variation in the mass transport of the electroactive species [10]. As the name suggests, the rotating disc electrode rotates, resulting in the convection of reactants towards the surface of the electrode and driving the product away from the surface [11]. DME is useful in the determination of both organic and inorganic ions. The analyte of interest should show electroactivity with mercury at its potential range. In DME, during the scan, every time a drop falls, it gets replaced by a new drop acting as a fresh electrode. Figure 1.1 depicts the three-electrode system. Working electrodes are useful in measuring conductivity, electroplating, electrolysis, etc.

1.2.2.2 Reference Electrode

For most electrochemical experiments, it is required to maintain one of the electrodes in an electrochemical cell at a constant potential [13]. In other words, an electrode that has a stable and known electrode potential is known as the reference electrode [14]. Reference electrodes are useful to control the potential (voltage) of the working electrode [15]. The potential of the reference electrode is known or fixed. Reference electrodes have a potential of zero volts. Reference electrodes are mainly included in reversible types of electrodes [16]. A reference electrode has a known electrode potential and is stable. Its high stability is achieved by employing the redox system, which must contain

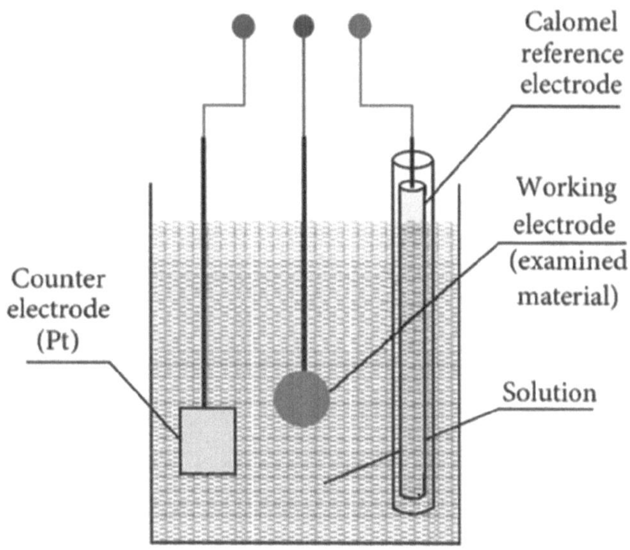

FIGURE 1.1 Schematic of three-electrode systems [12].

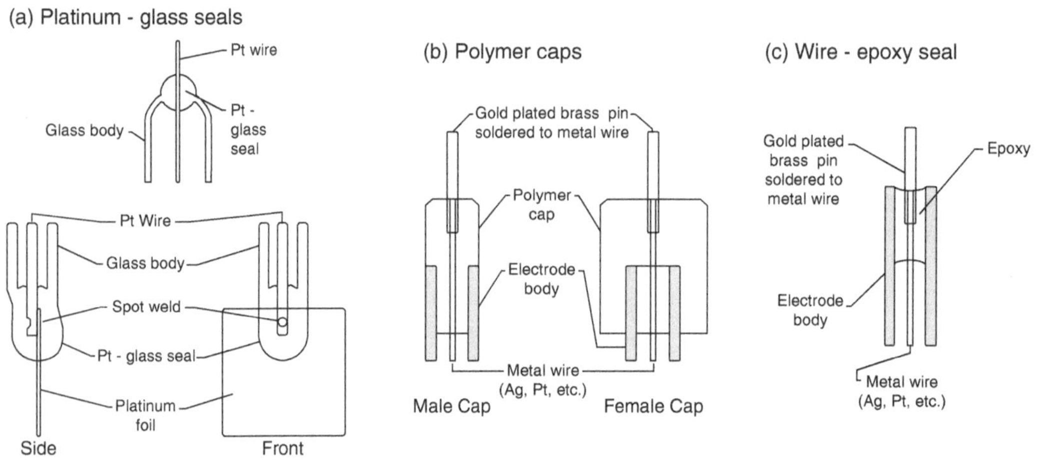

FIGURE 1.2 Schematic representation of the active component of the reference electrode.

Source: Reprinted from Smith, T. J., and K. J. Stevenson, "Reference electrodes." ***Handbook of Electrochemistry*** **2007: 73–110, Copyright (2023), with permission from Elsevier.**

saturated concentrations in each of the participating solutions of the reaction. There are several applications for reference electrodes, but the electrochemical cell is by far their most crucial. Here, it serves as a half-cell in the electrochemical cell to make it possible to ascertain the cell potential of the other half. The basic component of reference electrode is shown in Figure 1.2 [17].

Reference electrodes can be classified as aqueous, calomel, non-aqueous, and own-constructing. The reference point of an electrochemical cell can be established by measuring the reference electrode and the working electrode. The most important use of reference electrodes is in the electrochemical cell. The most common reference electrodes are the calomel electrode, the saturated calomel electrode (SCE), the silver-silver chloride electrode, and the standard hydrogen electrode (SHE) [18].

1.2.2.3 Counter Electrode

The counter electrode, also known as the auxiliary electrode, offers a process to apply input voltage to the working electrode. Generally, the voltage of the counter electrode is not monitored but is changed to balance the ongoing reaction on the working electrode [14]. These electrodes are utilized to complete a circuit and allow charge to flow [19,20]. As a result, they must be made of inert materials such as carbon or platinum, and their size must be significantly bigger than the working electrode to avoid current limitations [21]. It can also be used in other reactions where a current is expected to flow. The auxiliary electrode is used to make a connection to the electrolyte for the purpose of applying a current to the working electrode. Sometimes in microelectrode studies, a counter electrode is also used as a reference electrode, when the current is very low [22] (Figure 1.3).

The material used to make an auxiliary electrode must be an inert material such as graphite or a noble metal such as gold or platinum. This is necessary to prevent it from dissolving in the electrolyte. The most suitable material for a counter electrode is a substance that can cause electrochemical processes involving reactants in an electrolyte but does not degrade and contaminate the electrolyte solution.

1.2.3 Electrochemical Techniques

1.2.3.1 Potentiometry

Potentiometry, a widely used electroanalytical method, works by measuring the system's potential against time while there is no current flowing through it [23]. This method has gained tremendous

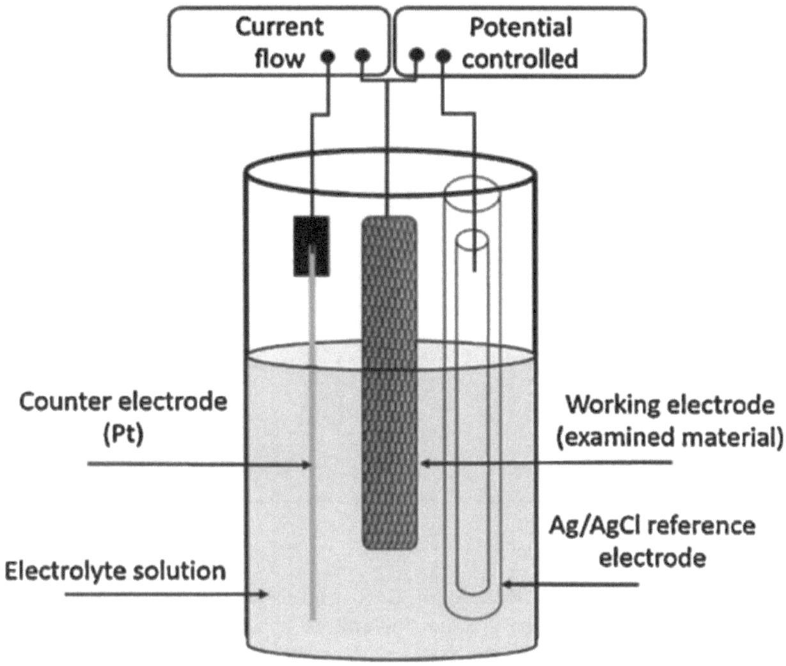

FIGURE 1.3 Counter electrode setup during the measurement of cyclic voltammetry.

Source: Reprinted from Sahito, I. A., et al., "Graphene coated cotton fabric as textile structured counter electrode for DSSC." ***Electrochimica Acta*** **173 (2015): 164–171, Copyright (2023), with permission from Elsevier.**

attention among researchers owing to its benefits such as good stability and reproducibility, high selectivity, cost-effectiveness, short response time, and simple preparation [24–26]. These benefits have made potentiometry useful in different fields such as agriculture, environmental science, medicine, and industries [27,28].

In potentiometric analysis, two types of electrodes are used: indicator and reference (or standard) electrodes, which complete the electrochemical cell (Figure 1.4). The indicator electrode (E_{ind}) is dipped in the electrolyte solution whose potential is to be determined in comparison with the reference electrode (E_{ref}) of a constant potential [23]. The reference electrode is separately placed and connected to the electrolyte solution via a salt bridge. At the end of the salt bridge, the potential is generated. If the amounts of anions and cations in the bridge solution are roughly equal, these two potentials cancel out each other. The resulting net potential difference across the salt bridge is measured in E_j. Both electrodes are connected to the voltameter, which determines the potential difference between the electrodes that is generated by varying concentrations of ions (E_{cell}) [29]. The potential difference is calculated by the equation 1.3:

$$E_{\text{cell}} = E_{\text{ind}} - E_{\text{ref}} + E_j \tag{1.3}$$

1.2.3.2 Amperometry

Amperometry is an electroanalytical method that determines changes in current that occur in an electrochemical cell during oxidation and reduction reactions when the potential is held constant [30]. A schematic representation of the amperometric setup is presented in Figure 1.5 [31]. For amperometric titration, at constant potential, a change in current is determined as soon as the titrant is added to the analyte solution. At the end point, clear changes are observed. Different cases are depicted in Figure 1.6 [32]. Figure 1.6a shows an amperometric curve when the product

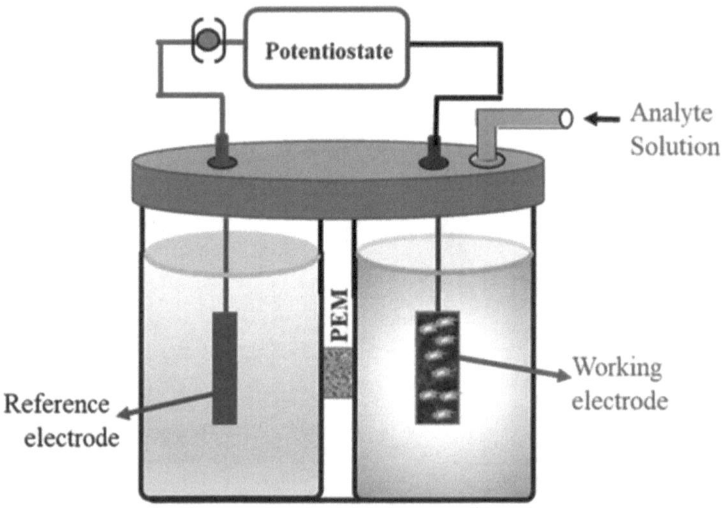

FIGURE 1.4 Diagram depicting potentiometer.

Source: Reprinted by permission from Kushwaha, C. S., and S. K. Shukla. "Non-enzymatic potentiometric malathion sensing over chitosan-grafted polyaniline hybrid electrode." *Journal of Materials Science* 54.15 (2019): 10846–10855, Copyright (2023), with permission from Springer.

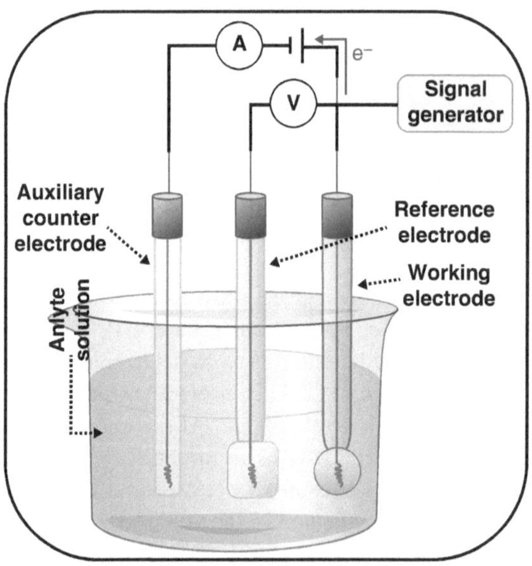

FIGURE 1.5 Schematic representation of amperometer [31].

and titrant are not reducible but the titrand is reducible. For instance, the titration of Pb^{2+} and SO_4^{2-} results in the formation of the insoluble product $PbSO_4$ [32]. Figure 1.6b depicts an amperometric curve when the product and titrand are not reducible but the titrant is reducible, for instance, titration of reducible 8-hydroxyquinolone with Mg^{2+} [32]. Figure 1.6c shows an amperometric curve when the product does not reduce but the titrant and titrand reduce, for instance, the titration of $K_2Cr_2O_7^{2-}$ with Pb^{2+} [32].

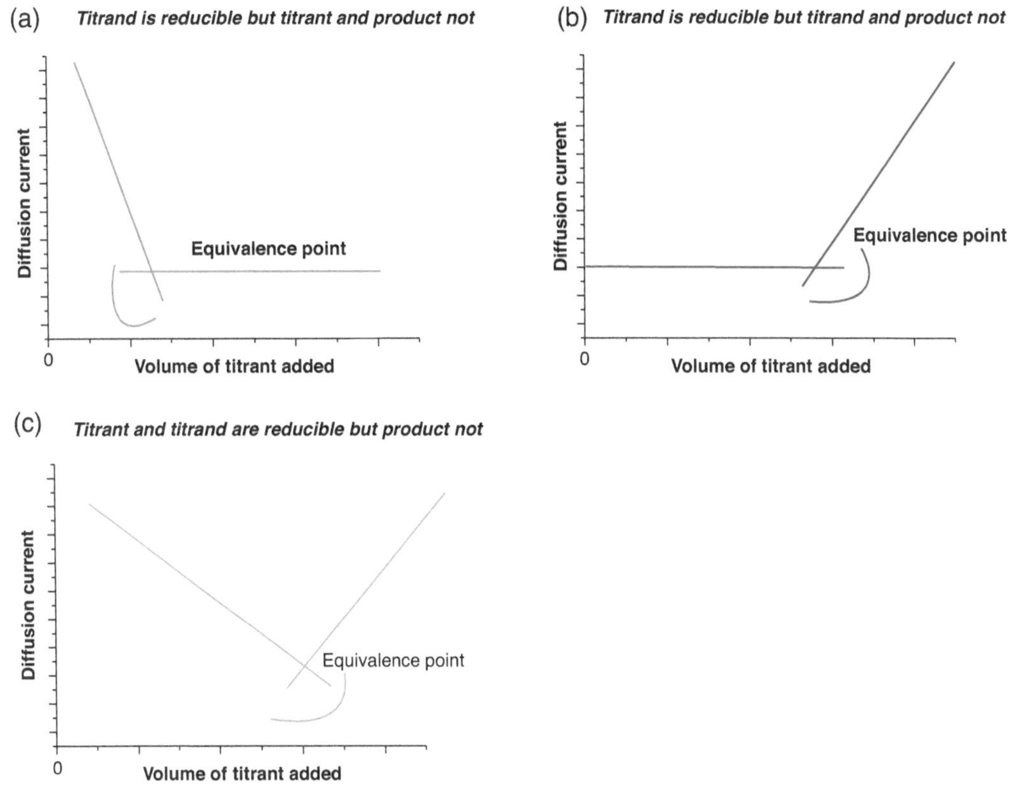

FIGURE 1.6 (a) Product and titrand are not reducible, and titrant is reducible; (b) product and titrand are not reducible, and titrant is reducible; and (c) product does not reduce, but titrant and titrand reduce [32].

The amperometric method possesses a number of advantages, including the ability to carry out the titration quickly and the use of this method in situations when the potentiometric technique is inadequate due to dilution issues [33,34]. These advantages have made amperometry useful in various applications such as sensors, chlorine concentration in biological fluids, and wastewater analysis [32,35].

1.2.3.3 Voltammetry

Voltammetry belongs to the class of electroanalytical techniques employed in analytical chemistry and in a variety of industrial processes. Voltammetry uses current measurements as the potential varies to gain information about an analyte. The output of the voltammetric experiment is a voltammogram, a plot of the current against the potential of the working electrode. The half-cell reactivity of an analyte is examined in a voltammetry investigation. Current as a function of applied voltage is studied by voltammetry, where $I=F(E)$ curve is called a voltammogram [30].

There are several voltammetric electrodes or cells introduced that are employed in both business and academic settings. These electrodes are known as sensors, and they are used to analyse different kinds of analytes in diverse mediums. Examples include oxygen electrodes and glucose sensors. Figure 1.7 depicts a regular electrochemical cell of voltammetry [36]. Voltammetry has evolved quite quickly, and various varieties with excellent selectivity, sensitivity, and efficiency have been established. Examples of a few categories include linear sweep, polarography, cyclic, staircase, square wave, normal pulse, stripping, and direct current (DC) [30,37,38]. Some of the advantages of voltammetry over potentiometric titrations include easy handling, quick measurements,

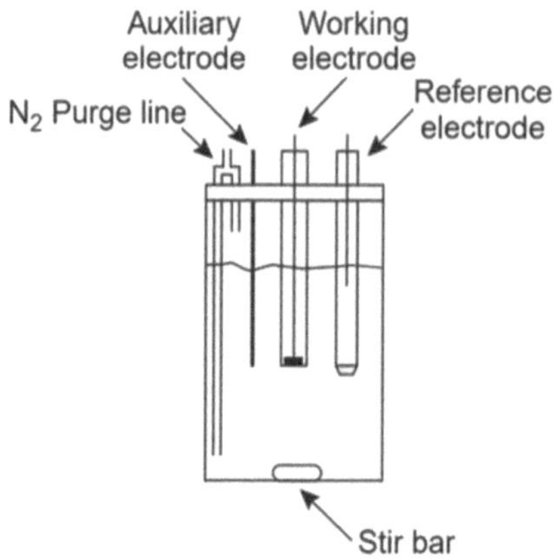

FIGURE 1.7 Schematic representation of electrochemical cell used in voltammetry.

Source: Reprinted with permission from Ref. [36].

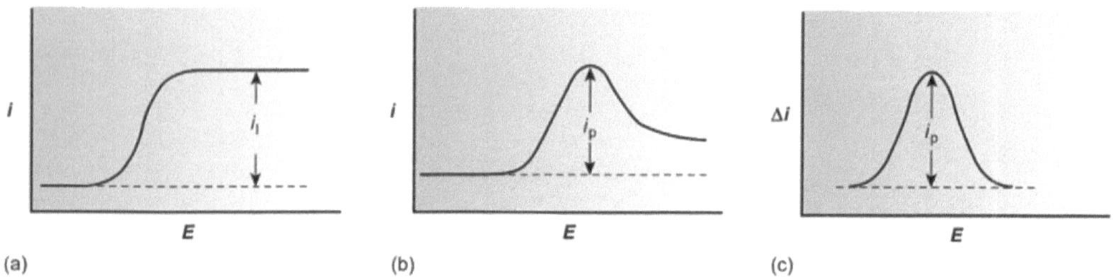

FIGURE 1.8 Three possible curves of voltammogram.

Source: Reprinted with permission from Ref. [36].

instantaneous dialysis, simple economy, easier analysis and modelling of data, and in situ measurements (redox potential measurements as a function of pressure or temperature) [39,40]. Figure 1.8 shows the three possible voltammogram curves [36].

1.2.3.4 Impedance Spectroscopy

Electrochemical impedance spectroscopy (EIS), also known as the AC impedance method, is a method that studies the response of a system with respect to a small amplitude periodic AC signal. It was first used to calculate the in-AC polarography [41,42] and double-layer capacitance [43], and nowadays, it is employed to describe electrode processes and interfaces of the complex. The name impedance spectroscopy was later chosen since the measurements are performed at various AC frequencies. The system response analysis reveals details about the interface, its structure, and the reactions occurring there [44].

The interaction of matter-(redox species)-electrode in a usual electrochemical cell involves the charge and mass transfer from the bulk solution to the surface of the electrode, electrolyte resistance, and the concentration of electroactive species. All of these characteristics are represented by an electrical circuit composed of capacitors, resistances, or constant phase elements coupled in parallel or series to produce an equivalent circuit, as seen in Figure 1.9. The EIS might thus be used to investigate charge transfer, mass transfer, and diffusion processes. As a result, the EIS can investigate intrinsic material qualities or specific processes that may influence an electrochemical system's conductance, resistance, or capacitance. As the resistance measured in DCs directly obeys Ohm's law, the impedance varies from resistance. To measure the impedance response, a slight signal excitation is used [45].

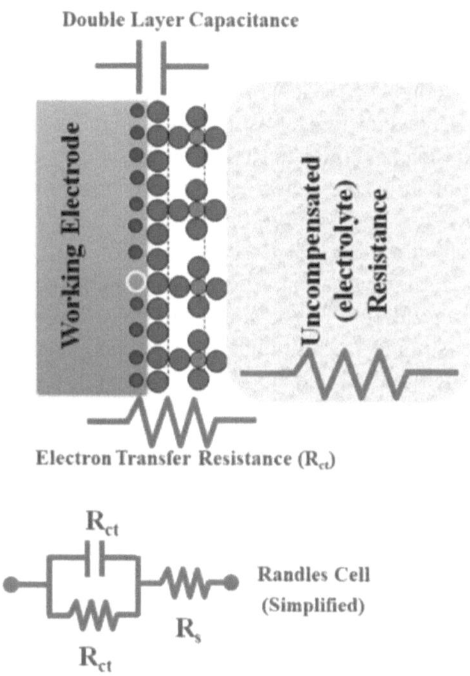

FIGURE 1.9 Schematic representation of EIS circuits along with the redox reaction occurring on the working electrode surface in a usual electrochemical cell. R_s, electrolyte resistance; R_{ct}, charge transfer resistance; and C_{dl}, capacitance double layer [45].

1.3 WORKING PRINCIPLES OF ELECTROCHEMICAL SENSORS

1.3.1 Types of Electrochemical Sensors

1.3.1.1 Potentiometric Sensors

Potentiometric sensors function primarily by assessing analyte concentration through the measurement of potential difference variations between the working and reference electrodes at different analyte concentrations [46]. Ion-selective electrodes (ISEs), such as the commonly known pH meter, are an example of such sensors. These sensors have also been adapted for pathogen detection [47].

Potentiometric sensors have garnered significant interest as a versatile platform for tracking various analytes in diverse matrices through both in situ and ex situ methodologies.

While pH monitoring remains the most common application of potentiometric sensors, their utility extends to a wide range of fields, such as monitoring healthcare, ensuring food quality, and conducting water and environmental analysis.

Despite the abundance of promising potentiometric sensors that have been published or made commercially available, there is still room for enhancing their performance in terms of reproducibility, stability, and sensitivity, particularly when used to monitor specific analytes in complex and real-world samples. By harnessing the potential of nanomaterials and emerging technologies, these enhancements can be achieved.

Also, there is a necessity for potentiometric sensors that prioritize testing without the need for sample treatment, making them ideal for applications in health, wastewater management, and food quality assurance. The incorporation of miniaturized potentiostats can reduce testing costs, facilitate in situ testing, and simplify the operation, thereby potentially increasing the adoption of potentiometric sensors when these features are integrated. This is especially relevant for situations where simultaneous monitoring of multiple analytes is crucial [48].

1.3.1.2 Amperometric Sensors

Amperometric sensors are highly sensitive analytical systems that function by measuring the current resulting from the gain (reduction) or loss (oxidation) of electrons during an electrochemical reaction involving an electroactive substance [49]. An example of a simple amperometric sensor is the Clark oxygen electrode, which is employed to determine oxygen concentration [50].

In amperometry, the cell potential of the electrochemical sensor remains constant while the current is measured. This method has given rise to various terminologies used by different scientists due to the diverse ways in which the current is generated. To estimate the quantity of an analyte or an electroactive species under investigation, scientists use the correlation between the generated current and the analyte's concentrations. This is accomplished through a current plot, often referred to as a voltammogram, with its highest peak indicating the maximum amount of current generated [51]. The detection limit for such sensors can easily reach pM or lower.

1.3.1.3 Voltammetric Sensors

Voltammetry is an important electroanalytical technique that involves measuring the current at a working electrode while applying a potential within a limited range. This range is determined by factors such as the electrode material, supporting electrolyte, and reference electrode. Essentially, it entails measuring the current in response to a potential waveform applied to the electrode. The current is directly proportional to the concentration of the target compound present in the solution. Voltammetric sensors have been effectively utilized for the sensitive analysis of various pollutants, including heavy metals, herbicides, pesticides, dyes, and pharmaceutical compounds, in environmental samples [52].

1.3.1.4 Conductometric Sensors

Electrochemical sensors of this kind operate by measuring the current conducted between two electrodes and are designed to detect changes in resistance, which reflects an analyte's ability to impede the flow of electrical charge. These conductometric sensors have garnered increasing demand due to their versatile applications. They are frequently employed to assess the quality of various analytes, such as oil.

In an electrochemical setup, an analyte's conductivity determines its capacity to permit electrical current to flow between the electrodes. Analytes with high conductivity (low resistance) enable a more substantial current to pass through, while analytes with low conductivity (high resistance) impede the flow. Consequently, conductometric sensors can differentiate between two analytes based on their conductance, making it possible to distinguish, for instance, between clean and contaminated oil, as these substances hinder charge flow differently.

The success of conductometric sensors in a wide array of applications can be attributed to ongoing efforts to enhance their sensitivity. This progress allows for the direct observation of changes in conductance occurring on the electrode's surface. Furthermore, advancements in technology enable these sensors to measure the conductance of multiple analytes by modifying the sensitive surface of the electrode with materials that enhance sensitivity. Nanostructures have also played a role in enhancing sensor performance, making conductometric sensors appealing for various potential applications [53].

1.3.1.5 Impedance-based Sensors

Impedance-based sensors are particularly useful as biosensors. Impedance-based electrochemical biosensors assess alterations in both resistive and reactive characteristics at the solid/liquid interface of electrodes by measuring the current response when a small AC voltage is applied [54]. Consequently, these biosensors represent an advancement over conductometric biosensors. They have gained recognition as exceptionally sensitive, low-power devices capable of real-time sensing. Impedance-based sensors are well-suited for point-of-care (POC) applications due to their cost-effectiveness, ease of miniaturization, and the availability of straightforward instrumentation.

In the context of impedance biosensors, a three-electrode system offers prolonged stability when compared to a two-electrode system. This extended stability facilitates the reproducible real-time detection necessary for POC applications. The surface of the working electrode can be modified by applying coatings of functionalization molecules and biorecognition molecules. The functional molecules form self-assembled monolayers (SAMs) on the electrode's surface. This functionalization step serves to prevent direct interaction between analytes and the sensor substrate, providing a stable interface for the attachment of biorecognition molecules, facilitating charge transfer, and supporting analyte-binding events.

1.3.2 Transduction Mechanisms

1.3.2.1 Ion-Selective Electrodes (ISEs)

ISEs are broadly of two types: solid-contact (SC) and liquid-contact. SC-ISEs typically feature a straightforward sandwich structure, consisting of a conductive substrate at the base, an intermediate SC layer, and the sensing membrane on top [55]. In some representations, SC-ISEs can be symbolically denoted as GC/GR/ISM, with "GC" and "GR" representing glassy carbon and graphene, respectively.

In SC-ISEs, the sensing membrane (ISM) serves as the recognition element responsible for the selective chemisorption of primary ions, allowing for the quantification of their activity. The SC in SC-ISEs acts as the ion-to-electron transducer, converting the input ionic signal into a measurable electrical potential signal. Figure 1.10 shows the ion-to-electron transduction process observed in redox capacitance-based SC-ISEs. The primary ion I^+ reversibly binds to an ionophore (L) at the ISM solution interface involving the reaction: $I^+(aq)+L(m) \rightleftarrows I^+L(m)$, where (aq) denotes aqueous phase and (m) denotes membrane phase. I^+ ion activity leads to charge separation at this interface, creating interfacial potential (φ_{a1}). Also, I^+ and the lipophilic anionic sites R^- can reversibly transfer across the SC|ISM interface, driving the local faradaic process in the SC layer. As a result, electrons flow to the conductive substrate, and the electrical potential difference is measured by the equipment.

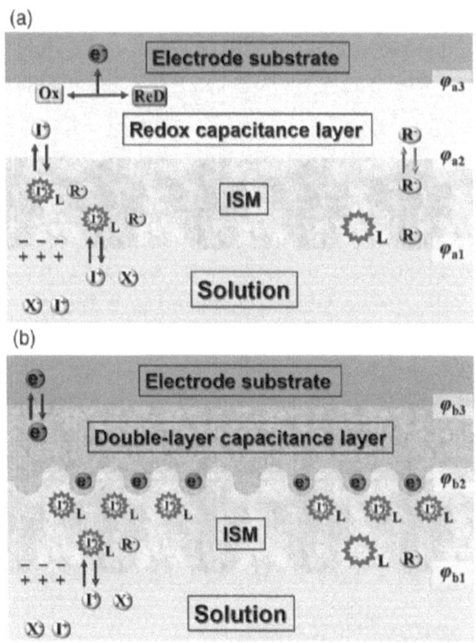

FIGURE 1.10 Schematic representation of (a) an SC-ISE with a redox capacitance layer and (b) an SC-ISE with a double-layer capacitance layer. The cation-selective membrane contains an electrically neutral ionophore (L) and a lipophilic anionic site (R). I^+ is the primary ion. X^- is the anion in solution.

Source: Reproduced with permission from Ref. [55].

Maintaining a well-defined pathway for ion-to-electron transduction is crucial for the stability and reproducibility of SC-ISEs, owing to the distinct conductive properties of the membranes and substrates. In traditional liquid-contact ISEs, an internal Ag/AgCl reference electrode and a filling solution with constant activities of primary ions and chloride enable a reversible redox reaction (AgCl(s)+e $\rightleftharpoons$ Ag(s)+Cl(aq)) that achieves equilibrium in charge transition from ionic to electronic conductivity. However, in the original form of SC-ISEs, such as carbon-contacting ISEs, this transduction process is "blocked" due to the poorly defined interfacial potential between the sensing membrane and the underlying metal conductor. In this setup, ions and electrons are confined to their respective phases.

In contrast, the well-defined ion-to-electron transduction processes in SC-ISEs are linked to the presence of redox/double-layer capacitance at the interface between the sensing membrane and the SC [56]. This capacitance is determined by the inherent properties of SC functional materials that serve as ion-to-electron transducers [57]. These materials can be described as asymmetric capacitors based on redox/double-layer capacitance, allowing for charging and discharging over a small but finite measuring current during potentiometric measurements.

1.3.2.2 Enzyme-based Sensors (Biochemical Sensors)

A biosensor is an analytical device designed to analyse a sample in the presence of a specific target analyte [58]. Typically, a biosensor consists of two main components: a biological component, often referred to as a molecular recognition element, and a physicochemical detector component or transducer. The recognition elements in a biosensor are affixed or immobilized onto the surface of transducers, enabling them to interact with target molecules in the sample without requiring the addition of reagents.

During operation, the specific interactions between the target analyte and the recognition elements lead to physicochemical changes on the surface of the transducer. These changes are then detected and converted into measurable signals. These signals can be used to determine the quantity of the analyte present in the sample.

Biosensors are typically classified based on the biological component employed, which may include enzymes, antibodies, nucleic acids, or cells. Alternatively, they can be classified based on the type of transducer utilized, which may include electrochemical, optical, mass-based, or piezoelectric transducers.

1.3.2.3 Gas Sensors

Electrochemical gas sensors operate by employing oxidation-reduction (redox) reactions to measure the concentration of gases. In these sensors, the gas molecules targeted for detection undergo an oxidative reaction at a sensing electrode. This reaction generates both ions and electrons.

The ions produced during this process are transferred to a counter electrode through an electrolyte, while the electrons are transferred to the same counter electrode via an external electrical circuit. At the counter electrode, a reduction reaction occurs as a result of the electron transfer, thus completing the redox reaction cycle. The measurement of the resulting electrical current or potential difference is used to determine the concentration of the gas being detected.

The reader is directed to Chapter 7 of this book for a deeper understanding of the transduction mechanism.

1.4 DESIGN AND CONSTRUCTION OF ELECTROCHEMICAL SENSORS

1.4.1 Materials Used in Sensor Fabrication

1.4.1.1 Electrode Materials

Sensors can be categorized into two primary types: non-flexible (often referred to as rigid sensors) and flexible sensors [59]. Rigid sensors offer advantages such as cost-effectiveness and low power consumption, but they are limited in terms of flexibility. This lack of flexibility restricts their use in applications such as health monitoring where sensors need to endure continuous bending, stretching, or pressure. In contrast, flexible sensors excel in deformable environments. They maintain their electrical properties even when bent or stretched and can utilize these deformations to detect motion. This is particularly valuable in robotics and human motion detection. Polymers are the most commonly used flexible materials in sensor fabrication due to their high flexibility and resilience under bending stresses, as well as their ease of manufacturing. Furthermore, this flexibility facilitates the development of intricate shapes, leading to reductions in mass, volume, and cost. For example, highly sensitive and recyclable touch sensors made from paper composed of cellulose nanofibres have been produced using an economical method known as inkjet printing [60]. Polymers such as polydimethylsiloxane, poly(N-isopropylacrylamide), and polyvinyl acetate find extensive application in flexible sensor technology. Carbon and silver are the most commonly employed nanomaterials in flexible sensors due to their exceptional electrical properties. Flexible sensors have numerous advantages over their rigid counterparts. They exhibit excellent sensing capabilities even under harsh bending stresses of up to 1500 $\mu\varepsilon$ [61]. Moreover, some of these sensors can withstand 8000 bending cycles without compromising their sensing capabilities.

Flexible sensors also possess improved thermal and mechanical properties, and they are typically lighter in weight compared to rigid sensors. However, their use can be hampered by the relatively low electrical conductivity of the flexible materials. To address this limitation, nanomaterials such as carbon, silver, and copper are being integrated into flexible components to enhance the electrical characteristics of these devices [62].

Advancements in additive manufacturing (AM) have played a pivotal role in propelling the flexible sensor industry forward. AM technology enables the easy formation of complex shapes by controlling computer-aided design (CAD) files, providing a versatile platform for creating innovative sensor designs. Researchers are also focusing on self-healing sensors since self-healing properties ensure sensor stability in harsh conditions, enhancing accuracy and reliability and unlocking new possibilities for innovative sensor applications.

Sensors play a vital role in connecting humans to the Internet, facilitating the monitoring of phenomena like the real-time heart status of an elderly individual, thereby enabling early detection of emergencies. Figure 1.11 provides an illustration of the application of flexible sensors in today's digital age [63].

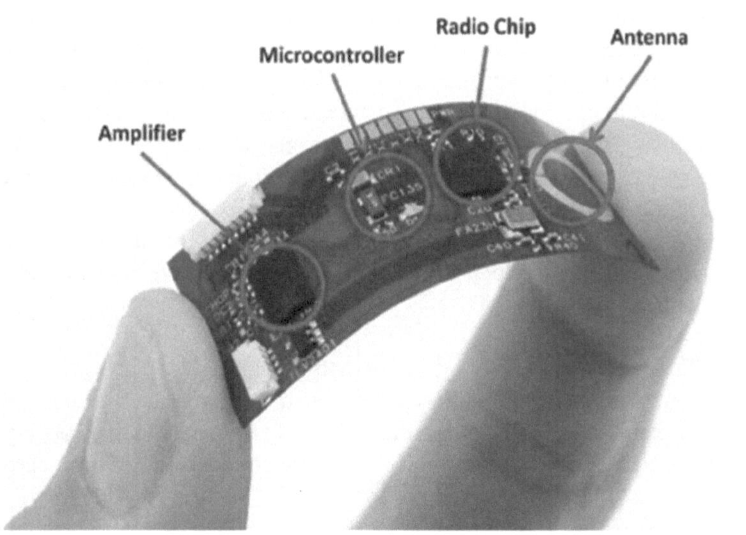

FIGURE 1.11 Flexible wireless ECG sensor with a fully functional microcontroller by IMEC. Developments in the field of flexible electronics are expected to lead to the advent of smaller, lighter, and more comfortable wearable systems. (Courtesy of IMEC, The Netherlands).

Source: Reproduced with permission from Ref. [63] under creative common license copyright @2012.

Some researchers are concentrating on self-healing flexible sensors. These sensors possess the unique ability to restore their mechanical and electrical properties following damage, such as cracks, scratches, or cuts. Self-healing capabilities extend the sensor's lifespan, resulting in material and cost savings and reducing environmental harm linked to the disposal of waste generated during manufacturing processes. Moreover, self-healing properties ensure sensor stability in harsh conditions, enhancing accuracy and reliability and unlocking new possibilities for innovative sensor applications.

1.4.1.2 Membranes and Coatings

Membrane, a semi-permeable barrier, is an important part of the electrochemical sensor development process due to various reasons. Membranes serve as a physical barrier between the working electrode and the sample solution, safeguarding the electrode from direct contact with the sample. This protection is essential to ensure the electrode's longevity and stability. The following section highlights the major advantages of introducing a membrane in a sensor construct.

Selective Permeability: Membranes can be tailored to be selectively permeable, allowing specific ions or molecules to pass through while blocking others. This selectivity is vital for electrochemical sensors to accurately detect a particular analyte amidst potential interfering substances.

Compatibility with Samples: In certain instances, the sample may not be compatible with the electrode material. Membranes prevent the sample from directly interacting with the electrode, preventing chemical reactions that could disrupt the electrochemical measurement.

Analyte Concentration: Membranes can concentrate the target analyte at the electrode's surface. By controlling the analyte's transport through the membrane, it can accumulate at the electrode, heightening the sensor's sensitivity and enabling the detection of low concentrations.

Gas Separation: In gas sensors, membranes can selectively transport specific gases while excluding others. For instance, in oxygen sensors, a membrane can be designed to exclusively transport oxygen.

Biological Compatibility: In biosensors, membranes are frequently used to immobilize biorecognition elements (like enzymes or antibodies) that interact with the analyte. The membrane provides a stable environment for these biological components.

Differential Measurements: Some electrochemical sensors employ membranes to separate the sample solution from reference electrodes. This separation ensures the reference electrode's potential remains stable, which is critical for accurate measurements.

Miniaturization and Microfabrication: In miniaturized and microfabricated electrochemical sensors, membranes are used to control the movement of ions or molecules precisely, enabling accurate measurements in small-scale devices.

Environmental and Safety Applications: Membranes act as protective shields in situations where the sensor must be shielded from harsh or hazardous environments. This allows the sensor to operate safely and effectively.

In the context of electrochemical sensors, a membrane's main role is to provide physical separation of the sensor electrode/surface from the analyte solution and ensure selective permeability/diffusion of analytes. On the other hand, coating of a suitable material is done onto the working electrode (as a separate step) surface as a layer mainly to modify the surface properties of the electrode to improve its performance.

Coating is important for electrochemical sensors for several reasons as discussed below [64].

Enhancing Selectivity: Coatings are instrumental in boosting the selectivity of an electrochemical sensor. By modifying the electrode's surface with a particular coating, it becomes more attuned to the desired analyte, reducing the impact of interfering substances present in the sample.

Amplifying Sensitivity: Coatings can increase the sensitivity of the sensor by promoting superior adsorption of the target analyte on the electrode surface. This results in more substantial variations in the electrochemical signal in response to changes in analyte concentration.

Ensuring Stability: Coatings act as a shield, safeguarding the electrode and prolonging its operational life. Electrodes are prone to degradation over time due to chemical reactions, fouling, and other factors. Coatings serve as a protective layer, preventing direct contact between the electrode material and the sample, ultimately enhancing the sensor's stability and durability.

Improved Electron Transfer: Certain coatings facilitate electron transfer between the analyte and the electrode, leading to quicker response times and more accurate measurements.

Minimizing Fouling: Coatings help in reducing the fouling of the electrode surface, which can occur when contaminants in the sample adhere to the electrode, diminishing its responsiveness to the target analyte.

Biocompatibility: In the context of biosensors, coatings are vital for ensuring the compatibility of the electrode with biological samples. Coatings create a biocompatible environment for enzymes or biomolecules employed in the sensor.

Recognition Element Immobilization: Coatings are used in some sensors, such as biosensors, to immobilize recognition elements such as enzymes or antibodies on the electrode surface. This immobilization enables specific interactions with the target analyte.

Control of Diffusion: Coatings help in controlling the diffusion of analytes to the electrode surface, ensuring that the analyte reaches the electrode in a regulated and predictable manner, a critical factor for precise measurements.

Tailoring Properties: Depending on the application and specific needs, coatings can be designed with distinct characteristics such as hydrophobicity, hydrophilicity, charge, or other properties that are advantageous for the sensor's performance.

To optimize the performance of electrochemical sensors, coating with suitable material/molecules is crucial for various applications including environmental monitoring, medical diagnostics, and industrial process control. Coatings can be of several types including silica, ionomers, polymers, small molecules, and biomolecules.

When materials are scaled down to the nanoscale, molecular, and surface forces, such as Van der Waals, steric, hydrophobic, and capillary forces gain greater significance compared to bulk phenomena [65]. This phenomenon is particularly evident in nanofibrous membranes (NFMs), where the porous structure enhances accessibility and reduces diffusion resistance for reactive reagents. Consequently, these nanostructured fibres are ideal for applications such as catalysis, nanoreactors, and adsorption [66]. Electrospun NFMs primarily consist of empty spaces such as pores and channels, and nylon-6 NFMs, for example, have a significantly lower density (90%–95% less) than thicker films [67]. The potential to fill these voids with gases, solvents, or reactive chemical and biochemical substances opens up numerous applications, especially with respect to sensors. Within this 3-D "empty" nanostructure, the interaction between surface functionalities of the fibres and diffusing chemical or biochemical species is expected to be maximized. Furthermore, NFMs exhibit excellent solvent permeability, with increased porosity reducing mass transport resistance as fluids flow through the empty channels [68]. Therefore, electrodes coated with NFMs offer virtually no barrier to the diffusion of analytes towards the electrode surface, making them highly attractive for sensing and biosensing applications.

1.5 ELECTROCHEMICAL SENSOR CALIBRATION AND PERFORMANCE EVALUATION

1.5.1 Calibration Methods

Calibrating a device can be a complex process, and there are various methods for accomplishing it. Calibration laboratories rely on their expertise, specialized equipment, an understanding of the inaccuracy of the device under test (DUT), and the time necessary to make accurate determinations.

The process involves comparing the DUT's output with a known standard or reference and then making necessary adjustments to minimize any discrepancies.

The equipment used in calibration labs is essential for accurately measuring and adjusting the DUT. These instruments are typically highly accurate and regularly calibrated themselves to ensure reliability.

Understanding the inaccuracy of the DUT is crucial in calibration. It helps identify how far the device's measurements deviate from the true values, enabling calibration technicians to make appropriate adjustments.

Additionally, the amount of time required for calibration depends on the complexity of the device, the number of measurements needed, and the precision required. Some calibrations may be relatively quick, while others may take more time to ensure high accuracy.

In summary, calibration labs combine their knowledge, specialized equipment, an understanding of DUT inaccuracies, and a time-efficient approach to provide precise and reliable calibrations for a wide range of devices.

There are three different calibration processes commonly employed.

1.5.1.1 Single-Point Calibration

The quickest method among the three calibration techniques is the zero-point adjustment. This calibration approach is typically performed within the lower 20% of the transducer's measurement range. It involves using a single point to calculate the difference between the reference value and the reading from the DUT, creating an offset correction.

For gauge transducers, zeroing is relatively straightforward as it involves venting the device to the atmosphere on both the pressure and reference ports. In contrast, zeroing an absolute transducer may require the instrument to be evacuated to a vacuum, with the specific procedure depending on the full-span pressure range of the transducer.

One-point calibration is particularly suitable for transducers that exhibit a consistent offset throughout their measurement range. This adjustment applies universally to all points within the range, making it a practical choice for such transducers.

Single-point calibration is typically employed in sensors where the primary consideration is minimizing costs rather than achieving the utmost accuracy.

1.5.1.2 Two-Point Calibration

Another common calibration procedure is the zero and span adjustment, a two-point calibration method that incorporates the same zero-point calibration process. However, it includes an additional step where the instrument is pressurized to the top 20% of its measurement range to obtain the span or second point reading.

The span adjustment acquired in this process is used to create a multiplier that is applied at every point within the measured pressure range. This allows for a correction that accounts for deviations from the ideal linear response of the transducer.

Zero and span adjustment is typically performed on transducers that exhibit both a zero error and a linear drift throughout their measurement range. This calibration technique is effective for correcting the zero-point error as well as maintaining linearity over the entire range of the transducer.

1.5.1.3 Multipoint Calibration

While it's common for transducers to exhibit zero shifts or span drift, there are instances when they display inconsistent linearity across their entire measurement range. Sometimes, there may be no detected offset at the zero or span point, yet errors can still be present at various points within the range. In such cases, a multipoint adjustment is required to correct for these non-linear behaviours.

Performing a multipoint adjustment involves using anywhere from 3 to 11 reference points, with each set of points treated as individual two-point calibrations. This method aims to provide the most accurate calibration results by addressing the non-linearities at various points within the range. However, it's important to note that a multipoint adjustment is the most time-consuming calibration process due to the need to evaluate and correct for deviations at multiple points.

1.5.2 Analytical Performance Parameters

1.5.2.1 Sensitivity

Sensitivity is a crucial parameter used to describe the efficiency of electrochemical transducers in converting energy and is considered a key indicator of transducer performance [69]. In simple terms, sensitivity is defined as the ratio of an output quantity to an input quantity. It provides insight into how effectively a transducer can convert input signals or energy into an observable output response.

1.5.2.2 Selectivity

Selectivity refers to the capability of a particular analytical method or technology to identify and quantify a specific compound even when other potentially interfering compounds are present. The term "interference" is typically reserved for substances or analytes that are not the target of the analysis and can lead to false increases, decreases, or obscuration of the response from the target analyte(s). Interference can affect the reliability of results in all methods, including traditional fixed laboratory methods, unless appropriate corrective actions are taken.

These corrective actions might involve using cleanup methods before analysis to eliminate interfering substances, particularly in some organic analyses. Alternatively, they could involve employing wavelength scans to detect and correct for spectral interferences, as is the case in certain metal analyses.

The term "specificity" conveys a similar concept to "selectivity," but to an even greater degree. A test with a high degree of specificity not only resists interference from non-target compounds but also provides unique responses that enable the differentiation of closely related compounds from the specific target analyte. In contrast, non-specific methods are designed to yield a response that combines the individual responses of all target analytes present in the sample. This is common in immunoassay test kits, where the term "cross-reactivity" is used to describe the summation of all responses from the various target analytes into a single, aggregated response.

1.5.2.3 Limit of Detection (LOD)

The concept of detection capability pertains to the lowest amount of an analyte that can be reasonably expected to be detected with a certain level of confidence. Typically, when quantitative results are reported near the detection limit, there is a considerable degree of uncertainty associated with those results. In such cases, numerical results should be considered unreliable, reflecting the inherent nature of analytical chemistry rather than a problem with the laboratory conducting the analysis.

In contrast, the quantification capability represents the lowest amount of analyte that can be reliably measured and reported as a quantitative result, backed by a specific level of statistical confidence. Different organizations may have varying rules for how laboratories determine these limits and the level of statistical confidence assigned to each. Different combinations of these limits often have different names, which can lead to confusion.

It's important for data users to understand that the concepts of detection and quantification, though seemingly straightforward, are complex in real-world analytical systems. This complexity is especially pronounced in the analysis of trace pollutant levels in environmental contexts, where concentrations may be in parts per million (ppm) or lower. Detection and quantification limits estimated by laboratories are often derived from idealized scenarios and matrices, and they can be misleading when extrapolated to real-world settings without accounting for deviations from the ideal.

These discrepancies can result in both false-positive and false-negative detections, leading to decision errors. Data users should be cautious about placing undue confidence in a single numerical result, as it represents a point estimate, and the true value could statistically fall within a range, rather than being precisely as reported. For instance, a result reported as "5.5" should not be interpreted as an exact value; it is merely an estimate, and the actual value may fall anywhere between 3 and 8 with a certain degree of statistical uncertainty.

1.5.2.4 Response Time

Response time, in the context of sensors, is the duration it takes for the sensor to transition its output signal from its initial state in the surrounding air (known as the baseline) to a specific percentage of the final or saturated response value [70]. The most widely adopted definition of response time employs the 90% mark of the final response.

1.5.2.5 Stability and Longevity

Ensuring the stability of sensors, especially in remote locations with fluctuating temperatures, has been a significant challenge [71]. Sensor durability is often defined by its shelf life, making the development of sensors that can operate reliably for extended periods crucial. Long-term stability can be particularly challenging when dealing with nanomaterials, as issues such as aggregation and flaking of nanomaterial-modified layers can arise.

Recent advancements have demonstrated that incorporating sol-gel materials, ceramics, and nanomaterials can significantly enhance sensor stability. This is especially important in scenarios where maintaining stability over extended periods is a key requirement.

One common challenge in sensor stability is maintaining the stability of metallic nanoparticles and quantum dots. These materials are prone to aggregation and can be sensitive to various environmental factors, including thermal heating, UV exposure, and the ionic strength of aqueous media. To address this issue, researchers often employ strategies such as creating core-shell structures and applying stabilizing agents to prevent colloid aggregation or degradation. Additionally, efforts are made to control the configuration of ligand shells to protect quantum dots from degradation. These strategies aim to ensure the long-term stability and reliability of sensors employing nanomaterials. Chapter 2 elaborates on the practical aspects of electrochemical sensing.

1.6 APPLICATIONS OF ELECTROCHEMICAL SENSORS

1.6.1 Environmental Monitoring

Environmental monitoring electrochemical sensors are designed to measure and assess various aspects of the environment, including air quality, water purity, and soil composition. These monitoring systems have the purpose of both determining the present state of an environment and identifying trends in environmental parameters over time.

Toxic contaminants and pathogens have the potential to be released into the air, soil, and water, posing significant environmental and health concerns. For instance, air pollutants such as sulphur dioxide (SO_2), nitrogen dioxide (NO_2), carbon monoxide (CO), and volatile organic compounds (VOCs) can originate from vehicle emissions, industrial facilities such as refineries and power plants, and laboratory processes.

Contaminants in soil and water can be of various types, including microbiological contaminants such as pathogenic bacteria, viruses, and other microorganisms that pose health risks; radioactive contaminants such as radionuclides released into the environment from nuclear facilities or improper disposal; inorganic contaminants such as heavy metals (e.g. lead, mercury, arsenic) and other inorganic compounds that can have toxic effects; and synthetic organic compounds including pesticides, herbicides, and industrial chemicals.

Some contaminants can persist for extended periods and migrate through large areas of soil, eventually reaching water resources. This can present ecological and human health risks, as these contaminants can enter the food chain, affecting both aquatic ecosystems and communities that rely on these water resources. Environmental sensors play a crucial role in monitoring three primary environmental conditions: air, soil, and water. Air quality monitoring involves air sampling, where parameters such as gas concentrations, ambient temperature, and humidity are measured. Soil monitoring is performed to evaluate the soil quality and assess factors such as erosion, salinity, and contamination. Environmental sensors are used to monitor parameters such as the presence of oxygen, nutrients, oils, pesticides, and metals in water. The electrochemical sensors used for environmental monitoring are broadly of two types: enzymatic sensors and non-enzymatic sensors [72]. The reader is especially directed to Chapters 4–9 and 13 of this book for more details and examples of electrochemical sensors for environmental monitoring.

1.6.2 Biomedical and Clinical Diagnostics

Biomedical diagnostics involve the use of biosensor devices to provide information related to specific health conditions or diseases. These medical biosensors have the capability to diagnose a wide range of diseases and health conditions, including diabetes, cardiovascular issues, infectious diseases, and cancer. Traditionally, biomedical diagnosis involved sending collected samples (such as blood, urine, or genetic samples) to off-site analytical laboratories. However, the advent of POC biosensor devices has revolutionized this approach, enabling biomedical diagnosis at the time and place of patient care. Electrochemical test strips, for example, are used to monitor not only blood glucose but also non-blood diabetic management, lactic acid, and cholesterol. The reader is especially directed to Chapters 7, 8, 10, 14, 16, and 18 of this book for more details and examples of electrochemical sensors for biological monitoring.

1.6.3 Food Quality and Safety

Current food sensors predominantly rely on colorimetric or electrochemical detection mechanisms [73]. Many commercial food sensors indirectly assess food quality by monitoring environmental factors such as time and temperature. However, recent advancements in sensing mechanisms and materials offer the potential to create more selective sensors that can directly detect chemicals, toxins, or pathogens within food packages.

The majority of electrochemical sensors used in food monitoring employ voltammetric and impedimetric measurements, which require an external energy source. There are also promising opportunities to develop precise and selective chemiresistive food sensors that do not rely on batteries for power. Chapter 10 explores the role of electrochemical sensors in food quality monitoring.

1.6.4 Industrial Process Monitoring

Process instrumentation and sensors are employed to monitor and control variables such as temperature, pressure, flow, and level in industrial and manufacturing settings. These systems play a crucial role in ensuring safety, efficiency, and productivity. Process instrumentation systems typically consist of sensors that measure the process variables, along with instrumentation components such as transmitters, controllers, and displays. These components work together to gather data from the sensors, process it, and display it for monitoring and control. Process sensors have evolved significantly since their inception in 1883. While the first sensors were designed to measure temperature, today's sensors can measure an array of variables, including pressure, level, flow, and viscosity. Modern process sensors are increasingly enhanced with artificial intelligence (AI) and machine learning (ML) capabilities. These advanced technologies enable sensors to provide more sophisticated insights and make real-time adjustments in response to changing conditions. More details about the integration of AI/ML in electrochemical sensor performance can be found in Chapter 11 of this book.

1.6.5 Energy and Fuel Cell Applications

Electrochemical sensors can be used to monitor energy storage and harvesting. Chapter 7 discusses various examples of such sensors.

1.6.6 Wearable and Point-of-Care Devices

Wearable sensors have emerged as powerful tools for real-time monitoring of human physiological metrics outside the clinical setting [74]. These sensors, often integrated with various technologies, including micro-electromechanical, biological, and chemical sensors, provide valuable data for monitoring health and diagnosing diseases. Some key points related to wearable sensors for POC diagnostics include the following: Wearable sensors can be integrated with technologies such as ECG, EEG, and EMG for comprehensive health monitoring. Wrist-mounted electrochemical fitness trackers, which monitor parameters such as body temperature, heart rate, blood pressure, step count, and energy expenditure, have become increasingly common [75].

1.7 CONCLUSION

Electrochemical sensors are devices that are used to quantify and detect the concentration of particular chemical species in a sample. In recent times, researchers have directed their attention towards this realm of investigation due to the numerous advantages, such as high selectivity and sensitivity, detection of very low concentrations, cost-effectiveness, fast response time, ease of use, long life span, and a plethora of other benefits. The current chapter provides an overview of the different electroanalytical techniques, types of sensors, transduction mechanisms, designing and construction of sensors, their calibration methods, and performance parameters. Understanding these aspects is crucial for the effective development and application of electrochemical sensors in various analytical and sensing fields.

REFERENCES

1. Simões, F.R. and M.G. Xavier, Electrochemical sensors. *Nanoscience and Its Applications*, 2017. **1**: pp. 155–178.
2. Brugger, J.J.N., Nanotechnology impact on sensors. *Nanotechnology*, 2009. **20**(43): pp. 430206.
3. Latour, B., *Pandora's Hope: Essays on the Reality of Science Studies*. 1999: Harvard University Press, Cambridge, MA.
4. Islam, S.K. and M.R. Haider, *Sensors and Low Power Signal Processing*. 2009: Springer Science & Business Media, Berlin, Germany.
5. Nei, L., Some milestones in the 50-year history of electrochemical oxygen sensor development. *ECS Transactions*, 2007. **2**(25): pp. 33.
6. Binnig, G., C.F. Quate, and C. Gerber, Atomic force microscope. *Physical Review Letters*, 1986. **56**(9): pp. 930.
7. Binnig, G., et al., Surface studies by scanning tunneling microscopy. *Physical Review Letters*, 1982. **49**(1): pp. 57.
8. Westbroek, P., Fundamentals of electrochemistry. *Analytical Electrochemistry in Textiles*, 2005. **2**: pp. 3–36.
9. Ferreira, M.A. and A.A. Barros, Determination of As (III) and arsenic (V) in natural waters by cathodic stripping voltammetry at a hanging mercury drop electrode. *Analytica Chimica Acta*, 2002. **459**(1): pp. 151–159.

10. Forster, R.J. and T.E. Keyes, *Ultramicroelectrodes*, in C.G. Zoski (Ed.) *Handbook of Electrochemistry*. 2007, Elsevier, Amsterdam, The Netherlands. pp. 155–171.
11. Jarju, J.J., M.C. Figueiredo, and Y.V. Kolen'ko, Electrocatalysis using nanomaterials. *Frontiers of Nanoscience*, 2021. **18**: pp. 343–420.
12. Raja, A.S., S. Rajendran, and P. Satyabama, Inhibition of corrosion of carbon steel in well water by DL-Phenylalanine-Zn^{2+} System. *Journal of Chemistry*, 2013. **3**(2): pp. 91–104.
13. Roscher, J. and R.J.E. Holze, Reference electrodes. *Encyclopedia*, 2023. **3**(2): pp. 478–489.
14. Bard, A.J., L.R. Faulkner, and H.S. White, *Electrochemical Methods: Fundamentals and Applications*. 2022: John Wiley & Sons, New York.
15. Bard, A.J. and M.V. Mirkin, *Scanning Electrochemical Microscopy*. 2001: CRC Press, Boca Raton, FL.
16. Papavinasam, S., Electrochemical polarization techniques for corrosion monitoring, in L. Yang (Ed.) *Techniques for Corrosion Monitoring*. 2021, Elsevier, Amsterdam, The Netherlands. pp. 45–77.
17. Smith, T.J. and K.J. Stevenson, Reference electrodes, in C.G. Zoski (Ed.) *Handbook of Electrochemistry*. 2007, Elsevier, Amsterdam, The Netherlands. pp. 73–110.
18. Inzelt, G., A. Lewenstam, and F. Scholz, (Eds.) *Handbook of Reference Electrodes*. Vol. 541. 2013, Springer, Berlin, Germany.
19. Honeychurch, K.C., Printed thick-film biosensors. *Electronic and Optical Materials*, 2012. **2**: pp. 366–409.
20. Sahito, I.A., et al., Graphene coated cotton fabric as textile structured counter electrode for DSSC. *Electrochimica Acta*, 2015. **173**: pp. 164–171.
21. Kissinger, P. and W.R. Heineman, *Laboratory Techniques in Electroanalytical Chemistry, Revised and Expanded*. 2018: CRC Press, Boca Raton, FL.
22. Cottis, R., Electrochemical methods. *Shreir's Corrosion*, 2010. **2**: pp. 1341–1373.
23. Isildak, Ö. and O. Özbek, Application of potentiometric sensors in real samples. *Critical Reviews in Analytical Chemistry*, 2021. **51**(3): pp. 218–231.
24. Özbek, O., et al., The use of crown ethers as sensor material in potentiometry technique. *Organic Communications*, 2021. **14**(3): pp. 809–814.
25. Shawish, H.M.A., S.M. Saadeh, and S.T. Al-kahlout, PVC membrane, coated-wire, and carbon-paste electrodes for potentiometric determination of vardenafil hydrochloride in tablet formulations and urine samples. *Sensors International*, 2022. **3**: pp. 100175.
26. Özbek, O., Potentiometric PVC membrane ion–selective electrode for the determination of Sr (II) ions. *Sensors International*, 2022. **3**: pp. 100185.
27. Özbek, O., et al., Synthesis and sensor properties of a phenol derivative molecule: Potentiometric determination of silver (I) ions. *Electrocatalysis*, 2022. **13**(4): pp. 486–493.
28. Özbek, O. and C. Berkel, Recent advances in potentiometric analysis: Paper–based devices. *Sensors International*, 2022. **3**: pp. 100189.
29. Fiedler, D.A. and F. Scholz, Electrochemical studies of solid compounds and materials, in A.M. Fritz Scholz, et al., (Eds.) *Electroanalytical Methods: Guide to Experiments and Applications*. 2005, Springer, Berlin, Germany. pp. 211–222.
30. Shah, N., M.B. Arain, and M. Soylak, Historical background: Milestones in the field of development of analytical instrumentation, in M. Soylak (Ed.) *New Generation Green Solvents for Separation and Preconcentration of Organic and Inorganic Species*. 2020, Elsevier, Amsterdam, The Netherlands. pp. 45–73.
31. Curulli, A., Electrochemical biosensors in food safety: Challenges and perspectives. *Molecules*, 2021. **26**(10): pp. 2940.
32. Amine, A. and H. Mohammadi, Amperometry. *Reference Module in Chemistry, Molecular Sciences and Chemical Engineering*, 2018. **10**: pp. 51.
33. Amer, M., Mathematical model of chloride concentration in human blood. *Journal of Medical Engineering & Technology*, 2006. **30**(1): pp. 25–30.
34. Golovko, D.A., et al., Determination of submillimolar concentration of ferrate (VI) in alkaline solutions by amperometric titration. *Central European Journal of Chemistry*, 2011. **9**: pp. 808–812.
35. Xu, Q., et al., *Amperometric sensors*, in A. Barhoum and Z. Altintas (Eds.) *Fundamentals of Sensor Technology*. 2023, Elsevier, Amsterdam, The Netherlands. pp. 123–145.
36. Harvey, D., *Modern Analytical Chemistry*. 2000: McGraw Hill, New York.
37. Skoog, D.A., F.J. Holler, and S.R. Crouch, *Principles of Instrumental Analysis*. 2017: Cengage Learning, New York.
38. Lubert, K.H. and K. Kalcher, History of electroanalytical methods. *Electroanalysis*, 2010. **22**(17–18): pp. 1937–1946.

39. Léger, C., An introduction to electrochemical methods for the functional analysis of metalloproteins, in R.R. Crichton and R.O. Louro (Eds.) *Practical Approaches to Biological Inorganic Chemistry*. 2013, Elsevier, Amsterdam, The Netherlands. pp. 179–216.
40. Gillès de Pélichy, L.D. and E.T. Smith, Redox properties of mesophilic and hyperthermophilic rubredoxins as a function of pressure and temperature. *Biochemistry*, 1999. **38**(24): pp. 7874–7880.
41. Breyer, B. and H.H. Bauer, Alternating current polarography and tensammetry. *Chemical Analysis*, 1963. **13**: pp. 288.
42. Reeves, R. and A.M. Bond, *Modern Polarographic Methods in Analytical Chemistry*. 1980: Elsevier, Amsterdam, The Netherlands.
43. Grahame, D.C., The electrical double layer and the theory of electrocapillarity. *Chemical Reviews*, 1947. **41**(3): pp. 441–501.
44. Lasia, A., *Electrochemical impedance spectroscopy and its applications*, in B.E. Conway, J.O. Bockris, and R.E. White (Eds.) *Modern Aspects of Electrochemistry*. 2002, Springer, New York. pp. 143–248.
45. Magar, H.S., R.Y. Hassan, and A. Mulchandani, Electrochemical impedance spectroscopy (EIS): Principles, construction, and biosensing applications. *Sensors*, 2021. **21**(19): pp. 6578.
46. Zdrachek, E. and E. Bakker, Potentiometric sensing. *Analytical Chemistry*, 2020. **93**(1): pp. 72–102.
47. Sadak, O., Potentiometric sensors, in A. Barhoum and Z. Altintas (Eds.) *Fundamentals of Sensor Technology*. 2023, Elsevier, Amsterdam, The Netherlands. pp. 99–121.
48. Bi, H. and X. Han, Chemical sensors for environmental pollutant determination, in K. Mitsubayashi, O. Niwa, and Y. Ueno (Eds.) *Chemical, Gas, and Biosensors for Internet of Things and Related Applications*. 2019, Elsevier, Amsterdam, The Netherlands. pp. 147–160.
49. Thévenot, D.R., et al., Electrochemical biosensors: Recommended definitions and classification. *Analytical Letters*, 2001. **34**(5): pp. 635–659.
50. Luppa, pp.B., L.J. Sokoll, and D.W. Chan, Immunosensors—principles and applications to clinical chemistry. *Clinica Chimica Acta*, 2001. **314**(1–2): pp. 1–26.
51. Chaubey, A. and B. Malhotra, Mediated biosensors. *Biosensors and Bioelectronics*, 2002. **17**(6–7): pp. 441–456.
52. Hussain, C.M. and R. Kecili, Electrochemical techniques for environmental analysis, in C.M. Hussain and R. Keçili (Eds.) *Modern Environmental Analysis Techniques for Pollutants*. 2020, Elsevier, Amsterdam, The Netherlands. pp. 199–222.
53. Patolsky, F., G. Zheng, and C.M. Lieber, *Nanowire-Based Biosensors*. 2006: ACS Publications, Washington, DC.
54. Chakraborty, A., D.N. Tibarewala, and A. Barui, Impedance-based biosensors, in K. Pal et al., (Eds.) *Bioelectronics and Medical Devices*. 2019, Elsevier, Amsterdam, The Netherlands. pp. 97–122.
55. Shao, Y., Y. Ying, and J. Ping, Recent advances in solid-contact ion-selective electrodes: Functional materials, transduction mechanisms, and development trends. *Chemical Society Reviews*, 2020. **49**(13): pp. 4405–4465.
56. Hu, J., A. Stein, and pp. Bühlmann, Rational design of all-solid-state ion-selective electrodes and reference electrodes. *TrAC Trends in Analytical Chemistry*, 2016. **76**: pp. 102–114.
57. Crespo, G.A., et al., Transduction mechanism of carbon nanotubes in solid-contact ion-selective electrodes. *Analytical Chemistry*, 2009. **81**(2): pp. 676–681.
58. Nguyen, H.H., et al., Immobilized enzymes in biosensor applications. *Materials*, 2019. **12**(1): pp. 121.
59. Nyabadza, A., et al., Review of materials and fabrication methods for flexible nano and micro-scale physical and chemical property sensors. *Applied Sciences*, 2021. **11**(18): pp. 8563.
60. Ling, H., et al., Transparent, flexible and recyclable nanopaper-based touch sensors fabricated via inkjet-printing. *Green Chemistry*, 2020. **22**(10): pp. 3208–3215.
61. Xuan, W., et al., High sensitivity flexible Lamb-wave humidity sensors with a graphene oxide sensing layer. *Nanoscale*, 2015. **7**(16): pp. 7430–7436.
62. Wen, N., et al., Emerging flexible sensors based on nanomaterials: Recent status and applications. *Journal of Materials Chemistry A*, 2020. **8**(48): pp. 25499–25527.
63. Patel, S., et al., A review of wearable sensors and systems with application in rehabilitation. *Journal of Neuroengineering and Rehabilitation*, 2012. **9**(1): pp. 1–17.
64. Scampicchio, M., et al., Electrospun nonwoven nanofibrous membranes for sensors and biosensors. *Electroanalysis*, 2012. **24**(4): pp. 719–725.
65. Min, Y., et al., The role of interparticle and external forces in nanoparticle assembly. *Nature Materials*, 2008. **7**(7): pp. 527–538.
66. Kriegel, C., et al., Fabrication, functionalization, and application of electrospun biopolymer nanofibers. *Critical Reviews in Food Science and Nutrition*, 2008. **48**(8): pp. 775–797.

67. Scampicchio, M., et al., Electrospun nanofibers as selective barrier to the electrochemical polyphenol oxidation. *Electrochemistry Communications*, 2008. **10**(7): pp. 991–994.
68. Gibson, pp., H. Schreuder-Gibson, and D. Rivin, Transport properties of porous membranes based on electrospun nanofibers. *Colloids and Surfaces A: Physicochemical and Engineering Aspects*, 2001. **187**: pp. 469–481.
69. Indrayanto, G., Validation of chromatographic methods of analysis: Application for drugs that derived from herbs. *Profiles of Drug Substances, Excipients and Related Methodology*, 2018. **43**: pp. 359–392.
70. Nguyet, T.T., et al., Novel synthesis of a PANI/ZnO nanohybrid for enhanced NO_2 gas sensing performance at low temperatures. *Journal of Electronic Materials*, 2023. **52**(1): pp. 304–319.
71. Pandey, R., et al., Challenges and future prospects in bioelectrochemical sensors. *Multifaceted Bio-Sensing Technology*, 2023. **4**: pp. 99–110.
72. Heo, J.H., et al., Sensor design strategy for environmental and biological monitoring. *EcoMat*, 2023. **5**(5): pp. e12332.
73. Oveissi, F., et al., *Sensors for food quality and safety*, in pp. Juliano, et al., (Eds.) *Food Engineering Innovations across the Food Supply Chain*. 2022, Elsevier, Amsterdam, The Netherlands. pp. 389–410.
74. Mondal, S., et al., Wearable sensing devices for point of care diagnostics. *ACS Applied Bio Materials*, 2020. **4**(1): pp. 47–70.
75. Parrilla, M., et al., Wearable wristband-based electrochemical sensor for the detection of phenylalanine in biofluids. *Biosensors and Bioelectronics*, 2022. **197**: pp. 113764.

2 Contemporary Techniques of Fabrication of Electrochemical Sensors

Ravi Tejasvi

2.1 INTRODUCTION

The basic principle of electrochemical sensing is using an electrochemical reaction to assess the analyte in a sample. The sensing materials used in electrochemical sensors are typically electrodes made of a conductive material and are polarisable by the applied voltage. The analyte electrochemically reacts with the material of the electrode and causes electrochemical disturbances. These *disturbances* are sensed as the perturbations in the current or voltage at the electrode. Several types of electrochemical reactions can be used for sensing. The most common are oxidation-reduction reactions, in which the analyte is oxidised or reduced at the electrode surface. Another type of reaction is the proton-transfer reaction, in which protons are transferred between the electrode and the analyte. In both cases, the reaction rate depends on the analyte's concentration, which allows for the detection and quantification of the analyte. The miniaturised reaction locale for such sensing is called an electrochemical sensor.

Electrochemical sensors are widely used in sensing the target analytes in various applications, ranging from biomedical diagnostics to controlling industrial processes. The fabrication of electrochemical sensors involves several key steps, including the selection of the sensing material, the design and construction of the sensor platform, and the integration of the sensor with electronic components.

Atypically, the first step in fabricating an electrochemical sensor is selecting the sensing material. The sensing material should selectively interact with the specific analyte and produce a measurable response. This response can be an electrical signal such as a perturbation in the electrode voltage or the current. Commonly used sensing materials include metals, metal oxides, and conducting polymers. The specific application and the sensor's desired sensitivity and selectivity often guide the sensing material's selection. The second step is when the sensor platform is designed and constructed. The sensor platform serves as the skeletal support for the sensing material and is responsible for the proper functioning of the sensor. The specific application guides the design of the sensor platform, and it can vary from simple to complex. Commonly used sensor platforms include screen-printed, microfabricated, and carbon-based electrodes. The final step of electrochemical sensor fabrication occurs when the electrodes and the skeletal sensor platform are integrated with the electrical circuitry for data processing and a power source. The electrical circuitry is used for measuring electrochemical perturbation data generated by the analyte's reaction within the electrochemical sensor. The circuitry can vary depending on the specific application but typically includes a power source, a data acquisition system, and a microcontroller.

The techniques used to fabricate electrochemical sensors are varied and often tailored to the specific application. With the advancement of technology, the field of electrochemical sensor fabrication continues to evolve, and new techniques are being developed to improve sensor performance. All fabrication techniques put a pre-specified sensing material on the electrode platform, which is done through physical deposition, chemical deposition, or printing. This book chapter is an attempt to discuss these fabrication techniques of the electrochemical sensors comprehensively.

DOI: 10.1201/9781032654874-2

2.2 PHYSICAL VAPOUR DEPOSITION (PVD)

PVD techniques have been widely used to fabricate electrochemical sensors because they can produce thin films with precise control over film thickness and composition. PVD techniques involve converting material from a solid or liquid state into a vapour, which is then transferred towards a substrate to form a thin film. The target material is either bombarded with ions or heated to evaporate. It results in the sputtering or evaporation of material from the target material, which then deposits onto a substrate to form a thin film. The thickness and composition of the deposited film are dependent upon the evaporation energy, the temperatures of the target material, and the substrate. The film thickness and the deposition rate also depend upon the distance the vapours must diffuse through, i.e., the distance between the target and the substrate. There are several PVD techniques, each with unique advantages and disadvantages. The most widely used PVD techniques for fabricating electrochemical sensors are discussed below.

2.2.1 Thermal Evaporation

Thermal evaporation is a PVD technique that involves high-temperature evaporation of target material, diffusion-based transport of the evaporated material towards the substrate, and condensation at the substrate to form a thin film. The substrate is formatted so that only the design template for the electrode material is exposed towards the incoming deposition. Later, the mask is peeled off to reveal the deposited pattern. Thermal evaporation has been used to fabricate various electrochemical sensors, including pH, glucose, and oxygen sensors. This technique is well-suited for manufacturing sensors based on metal oxides and other inorganic materials. Figure 2.1 shows a typical schematic of the mechanism of the thermal evaporation-based deposition technique. There are various thermal evaporation-based techniques for fabricating electrochemical sensors, as given below.

2.2.1.1 Resistance Heating

Resistance heating is the most widely used thermal evaporation technique for fabricating electrochemical sensors. It involves passing an electric current through a metal filament or wire, which heats up and evaporates the material. This technique is relatively simple and inexpensive, but it can be challenging to control the film thickness and composition precisely.

2.2.1.2 Electron Beam Heating

Electron beam heating is another thermal evaporation technique for fabricating the electrochemical sensors. It involves heating the material by bombarding it with a high-energy beam of electrons.

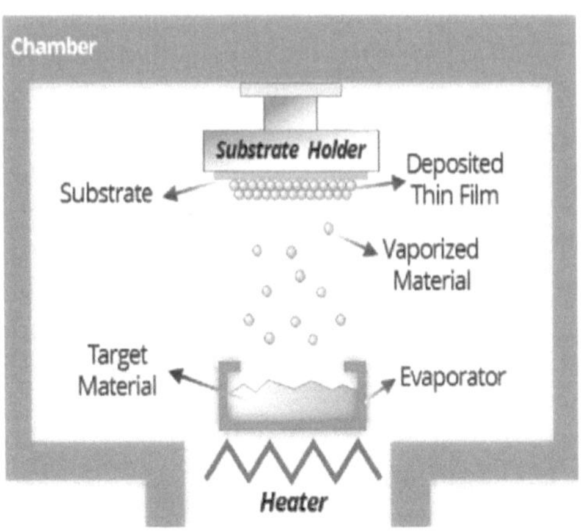

FIGURE 2.1 A general mechanism of thermal evaporation-based deposition technique [1].

This technique can achieve higher evaporation rates than resistance heating and is particularly useful for evaporating high-melting-point materials. However, it is more expensive and complex than resistance heating.

2.2.1.3 Flash Evaporation

Flash evaporation is a relatively new thermal evaporation technique for fabricating the electrochemical sensors. It involves heating the material by passing a high-energy electrical current through it, resulting in rapid and intense heating. This technique can achieve high evaporation rates and precise film thickness and composition control. However, it is a nascent technique and, hence, not yet widely used to fabricate electrochemical sensors.

Thermal-evaporation-based techniques have several advantages, including their relative simplicity, low cost, and ability to fabricate sensors based on a wide range of materials. However, they also have some limitations, including the potential for structural and chemical changes in the deposited material and the challenge of scaling up to large-scale production processes. Nevertheless, research in thermal evaporation-based techniques for fabricating electrochemical sensors is ongoing. New advances in this area can potentially improve the performance and accessibility of electrochemical sensors.

2.2.2 Sputtering

Sputtering is widely used to fabricate electrochemical sensors due to its high sensitivity, selectivity, and low cost. Sputtering involves the bombardment of a target material with ions, resulting in the sputtering of atoms from the target material, which then deposit onto a substrate to form a thin film. Some of the sputtering techniques of fabricating electrochemical sensors are discussed here.

2.2.2.1 Direct Current (DC) Sputtering

DC sputtering is a PVD technique. In this technique, a DC-powered electron gun is used to produce an ion beam from the chamber gas. This ion beam then hits the target material, resulting in the sputtering of atoms from the surface of the target material. The sputtered particles then deposit onto a grounded substrate forming a thin film. DC sputtering is flexible to be performed under various conditions, including different pressures, temperatures, and gas compositions. The DC sputtering process is performed in a vacuum chamber having a partial vacuum between 10^{-5} and 10^{-3} Pa. The target material is usually a metal or semiconductor placed in the chamber facing the substrate. A high-voltage potential difference is applied between the target and the substrate, creating an electrical field that accelerates ions from the target towards the substrate. These ions hit the surface of the target material and eject the charged surface particles. These charged particles move towards the grounded substrate surface and get deposited onto it.

Out of several variations of DC sputtering, DC magnetron sputtering is the most widely used mode. It uses the magnetic field as a lens to focus the plasma onto the target. It increases the ionisation of the sputtered atoms, resulting in higher deposition rates and improved film quality. DC diode sputtering is a more straightforward mode that does not require a magnetic field and is typically used for depositing thin films of simple materials. DC unbalanced magnetron sputtering is a mode that uses a magnetic field and an unstable DC voltage, resulting in improved control over the ionisation of the sputtered atoms and improved film quality. DC sputtering is a simple and reliable technique but has the disadvantage of being less efficient than other sputtering techniques. Figure 2.2a shows a typical schematic of the mechanism of the DC/radio frequency (RF) sputtering-based deposition technique.

2.2.2.2 Radio Frequency (RF) Sputtering

RF sputtering is a variation of DC sputtering in which an RF power source is used instead of a high-voltage DC power source. RF sputtering involves bombarding a target material with ions accelerated by an RF electric field, typically oscillating at a frequency of 13.56 MHz. The ions are

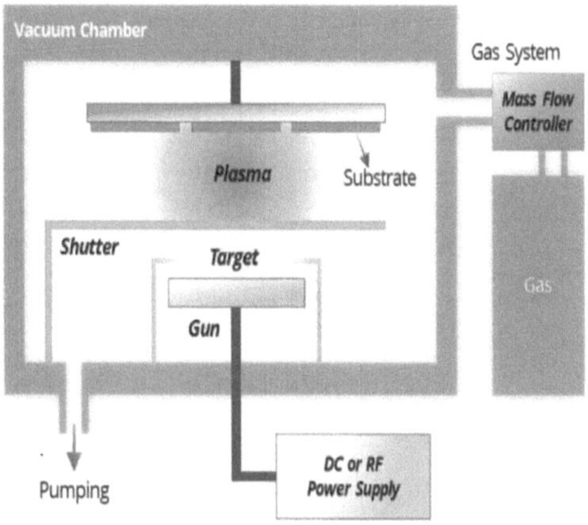

FIGURE 2.2a Schematic of the mechanism of DC/RF sputtering [1].

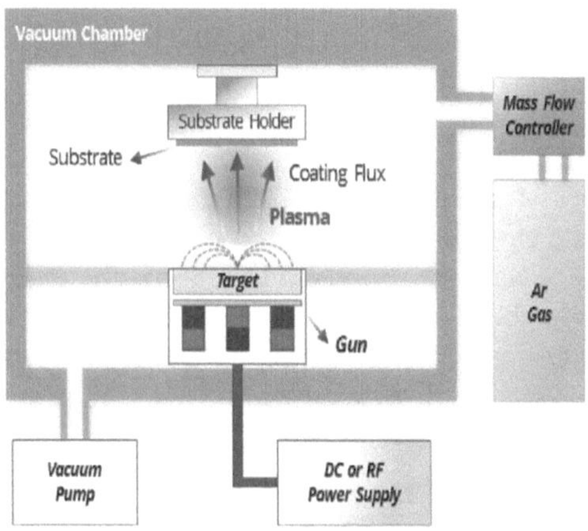

FIGURE 2.2b Schematic of the mechanism of RF magnetron sputtering [1].

the chamber gas atoms ionised by an RF electric field and accelerated towards the target material. These ions transfer their kinetic energy to the target surface by means of collision, which results in the particles getting *sputtered* out of the target material's surface. These *sputtered* atoms travel to the substrate and undergo condensation-polymerisation to form a thin film there. The RF sputtering process has three main stages: ionisation, acceleration, and sputtering. In the ionisation stage, the RF electric field ionises a small fraction of the chamber gas atoms. The ionisation is typically achieved by a process known as Penning ionisation.

Sometimes a magnetic field is set to assist in achieving the ionisation, and then, the sputtering process is known as RF magnetron sputtering. Figure 2.2b shows a typical schematic of the mechanism of the RF magnetron-sputtering-based deposition technique. In the acceleration stage, the RF electric field accelerates electron-gun-ejected ions towards the target material. The acceleration is typically achieved by a process known as diode sputtering. Here, the RF electric field accelerated the ions towards the target material. The acceleration is critical for controlling the film properties and optimising the sensor performance, as the kinetic energy of the ions determines the sputtering yield and the film properties. In the sputtering stage, the ions hit the surface of the target material and transfer their kinetic energy to the surface atoms, resulting in the sputtering of particles from the surface of the target material. Sputtering yield, the number of atoms sputtered per incident ion, is used to compare various sputtering processes for their efficacies.

The choice of the target material is critical in determining the film properties and optimising the sensor performance. Also, the choice of target material depends on the desired film properties and the application of the electrochemical sensor. Typically, DC sputtering is used with pure metal targets, whereas RF sputtering is the sputtering process of choice for ceramic and dielectric target materials. Gold and silver are commonly used to fabricate amperometric sensors, while platinum is frequently used to fabricate potentiometric sensors. Ceramic target materials are also widely used to fabricate electrochemical sensors due to their high chemical stability and good mechanical properties. Alumina and zirconia are commonly used to fabricate amperometric and potentiometric sensors, respectively.

2.2.2.3 Pulsed DC Sputtering

Pulsed DC sputtering is a variation of DC sputtering in which the DC voltage is applied in pulses rather than continuously. Pulsed DC sputtering is more efficient than DC sputtering and can be used to deposit thin films with improved film uniformity. However, pulsed DC sputtering is more complex than DC sputtering and requires specialised equipment. Pulsed DC sputtering involves the application of a pulsed DC power source to the target. The technique improves the target material's utilisation and film uniformity compared to DC sputtering. Additionally, it allows for the deposition of non-conductive materials by interrupting the power supply to the target during the sputtering process.

The sputtering-based fabrication technologies offer a high deposition rate, the ability to deposit a variety of materials, and the ability to control the composition and properties of the deposited films. Despite the advantages of sputtering-based fabrication technologies for the fabrication of electrochemical sensors, some challenges still need to be addressed. For example, film thickness and composition control can be an issue. Another challenge is the integration of sputtering-based fabrication technologies into large-scale production processes. Additionally, developing new materials and techniques for sputtering-based fabrication technologies to improve the performance of electrochemical sensors is an area of active research.

2.2.3 Atomic Layer Deposition (ALD)

ALD is a self-limiting growth process that allows for precise film thickness and composition control, making it an attractive option for fabricating electrochemical sensors. The ALD process involves sequentially exposing a substrate to two or more precursors, usually in a chamber under ultra-high vacuum conditions. Figure 2.3 shows a typical schematic of the mechanism of the steps of the ALD technique.

The first precursor is a metal-containing compound, while the second is a reactant gas. When the first precursor is introduced into the chamber, it adsorbs onto the substrate surface, forming a monolayer. The excess of the first precursor is purged, and the substrate is further exposed to the second precursor, which forms a thin film by reaction with the adsorbed first precursor monolayer. The process is then repeated to build up the desired thickness of the film. One of the critical advantages of ALD is its ability to control the thickness and composition of the deposited film precisely. It is achieved by controlling the cycles of precursor vapour exposure and purging and precursor vapour concentrations. It allows the ALD to have precise control of film thickness, which is critical for the high performance of sensors.

Additionally, the ability to use different precursors and reaction conditions allows for the precise control of film composition, which can be used to optimise the performance of the electrochemical sensor. Another advantage of ALD-based techniques is depositing films on a variety of substrate materials, which makes ALD-based techniques suitable for fabricating a wide range of electrochemical sensors. ALD-based techniques can also deposit films at low temperatures, which can be beneficial for preserving sensitive substrates. However, ALD-based techniques

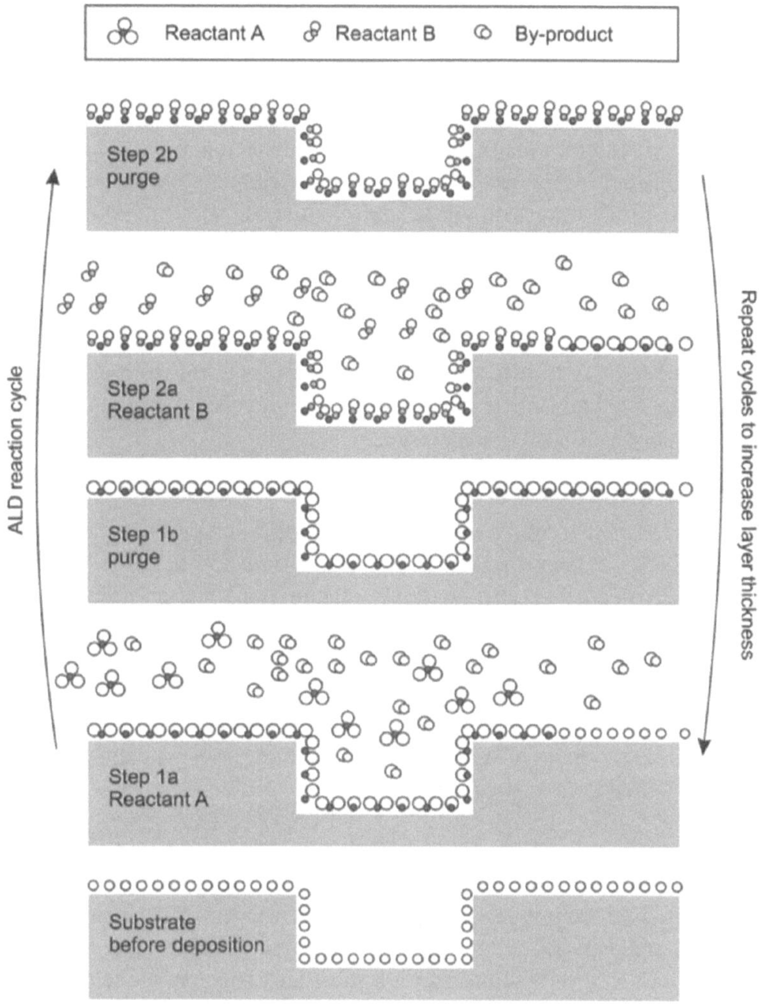

FIGURE 2.3 Schematic of the mechanism of atomic layer deposition [2].

suffer from the high equipment cost required. Additionally, the process can be slow, making it less suitable for large-scale production of electrochemical sensors. Moreover, in ALD-based techniques, the choice of precursors is limited, limiting the film composition and properties that can be obtained.

2.2.4 Pulsed Laser Deposition (PLD)

PLD is a PVD technique that involves irradiating a target material with a pulsed laser beam, which causes material ablation from the target. It forms a plasma plume containing atoms, ions, and clusters of the target material. The plume is then transported towards the substrate and deposited as a thin film. The process parameters, such as laser fluence, pulse duration, and repetition rate, can be precisely controlled to produce thin films with specific properties. Several PLD-based techniques have been developed for fabricating electrochemical sensors including single-pulse PLD, multi-pulse PLD, raster-scan PLD, laser-induced forward transfer (LIFT), and laser-assisted chemical vapour deposition (CVD) (LACVD).

2.2.4.1 Single-Pulse PLD

Single-pulse PLD involves irradiating the target with a single laser pulse, followed by the immediate deposition of the ablated material onto the substrate. This technique helps fabricate thick films, but the quality of the deposited film is typically poor. Multi-pulse PLD involves irradiating the target with multiple laser pulses, followed by the immediate deposition of the ablated material onto the substrate. This technique helps fabricate high-quality thin films but is relatively slow. Raster-scan PLD involves irradiating the target with a laser pulse while the substrate is raster-scanned underneath the target. This technique forms large-area thin films with high uniformity.

2.2.4.2 Laser-Induced Forward Transfer (LIFT)

LIFT is a recently developed PLD-based technique for fabricating electrochemical sensors. LIFT involves using a pulsed laser to transfer a small amount of material from a target to a substrate, forming a thin film. The main advantage of LIFT is its high precision, making it well-suited for the small-scale fabrication of electrochemical sensors.

2.2.4.3 Laser-Assisted CVD (LACVD)

LACVD is a variant of PLD that uses a laser to generate a plasma plume containing precursors of the target material, which are then transported to the substrate and deposited to form a thin film. LACVD has been used to deposit various materials, including metals, ceramics, and polymers. The main advantage of LACVD is its ability to deposit a wide range of materials, making it well-suited for fabricating electrochemical sensors with specific properties.

PLD-based techniques can deposit high-quality thin films with precise film thickness, composition, and microstructure control. However, PLD is a relatively slow process, which makes it less well-suited for large-scale production. Additionally, the cost of laser equipment can be a barrier to widespread adoption.

2.3 CHEMICAL VAPOUR DEPOSITION (CVD)

CVD is another widely used technique for fabricating electrochemical sensors because it deposits high-quality, uniform, and conformal thin films on various substrates. CVD is a chemical process involving gases' reactions to form solid films on a substrate. This technique offers several advantages over other thin-film fabrication techniques, including (1) depositing a wide range of materials, (2) high deposition rates, and (3) precise control of film features. Figure 2.4 shows a typical schematic of the steps of the CVD technique. CVD can be classified into two subcategories based on the pressure of the reaction precursor gas, as described below.

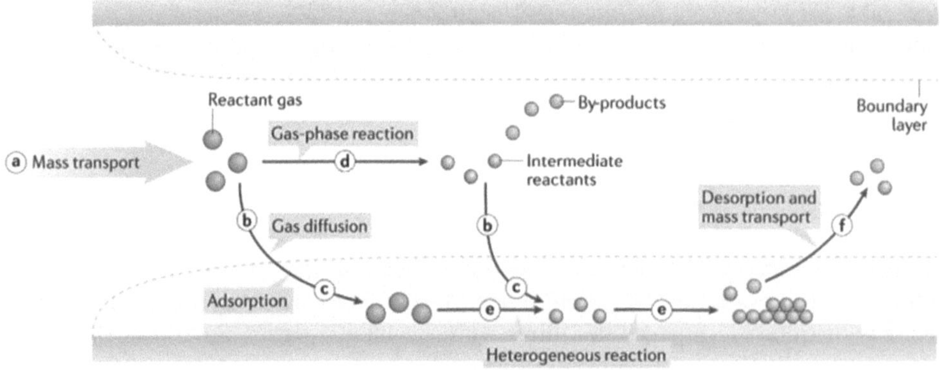

FIGURE 2.4 Schematic of the mechanism of steps involved in typical CVD [3].

2.3.1 Low-Pressure CVD (LPCVD)

LPCVD is a CVD technique performed at low pressures, typically in the 0.1–10 Torr range. The technique introduces a gaseous precursor into a reactor chamber, typically a volatile liquid or gas. The precursor is then heated to a temperature at which it decomposes, producing reactive species that deposits a thin film on the substrate. The pressure in the reactor chamber is maintained at a low level to promote the formation of high-quality, uniform, and conformal films. Several variants of LPCVD techniques are commonly used for fabricating electrochemical sensors, including thermal LPCVD, plasma-assisted LPCVD (PALPCVD), and remote-plasma-assisted LPCVD (RPALPCVD).

2.3.1.1 Thermal LPCVD

Thermal LPCVD is the most straightforward and widely used LPCVD technique. In thermal LPCVD, the precursor is heated to a high temperature, typically between 600°C and 1000°C, to promote the decomposition of the precursor and the formation of the thin film. Thermal LPCVD is generally used for the fabrication of metal and dielectric films.

2.3.1.2 Plasma-Assisted LPCVD (PALPCVD)

PALPCVD is an LPCVD technique in which plasma is used to enhance the decomposition of the precursor and the formation of the thin film. In PALPCVD, plasma is generated in the reactor chamber by applying a high-frequency voltage to the precursor. The plasma enhances the precursor's decomposition and the thin film's formation, resulting in improved film quality and better control of film thickness and composition. PALPCVD is typically used for the fabrication of semiconductors and metal films.

2.3.1.3 Remote-Plasma-Assisted LPCVD (RPALPCVD)

RPALPCVD is an LPCVD technique in which a remote plasma source is used to enhance the decomposition of the precursor and the formation of the thin film. In RPALPCVD, plasma is generated *ex situ* and introduced into the chamber through a showerhead or a porous membrane. The remote plasma source enhances the precursor's decomposition and the thin film's formation, resulting in improved film quality and better control of film thickness and composition. RPALPCVD is typically used for the fabrication of semiconductors and metal films.

LPCVD-based techniques produce thin films which are uniformly thick, composition-wise homogeneous, and relatively low cost. LPCVD-based techniques are compatible with many substrates and can fabricate various materials, including metals, dielectrics, and semiconductors. However, there are also some limitations to LPCVD-based techniques, including the need for high-temperature processing, which can be a barrier to using specific substrates, and the need for expensive equipment. Additionally, LPCVD-based techniques can be time-consuming and may require hazardous chemicals.

2.3.2 Atmospheric Pressure CVD (APCVD)

APCVD is a CVD technique carried out at ambient pressure, as opposed to the extreme pressure conditions used in traditional CVD techniques. APCVD introduces a gaseous precursor into a reactor at atmospheric pressure. It is then decomposed by heat or plasma to form a thin film of the desired material on a substrate. The main advantage of APCVD over LPCVD is its ability to restrict the flash evaporation of liquid reactants. It controls the vapour formation and produces uniform thin films at moderate temperatures and low costs. Several types of APCVD techniques have been developed to fabricate electrochemical sensors. These include thermal CVD, plasma-enhanced CVD (PECVD), and microwave-assisted CVD (MWCVD).

2.3.2.1 Thermal CVD

Thermal CVD is the most widely used APCVD technique for fabricating electrochemical sensors. In thermal CVD, the precursor is heated to a high temperature, typically above 600°C, to decompose and form a thin film. The main advantage of thermal CVD is its ability to produce high-quality thin films at relatively low costs. However, the high temperatures required for thermal CVD can cause thermal damage to the substrate and limit the materials that can be used.

2.3.2.2 Plasma-Enhanced CVD (PECVD)

PECVD is an APCVD technique that uses plasma to decompose the precursor and form a thin film. PECVD is advantageous mainly in producing uniform and homogeneous thin films at relatively low temperatures. However, PECVD can be more capital-expensive than thermal CVD due to the high cost of the plasma equipment.

2.3.2.3 Microwave-Assisted CVD (MWCVD)

MWCVD is an APCVD technique, like PECVD, but uses microwaves to decompose the precursor and form a thin film. MWCVD is also advantageous in producing uniform and homogeneous thin films at relatively low temperatures. However, MWCVD can be more complex and challenging to control than thermal CVD or PECVD.

APCVD-based electrochemical sensors have several advantages over other types of electrochemical sensors. For example, they have high sensitivity and selectivity, can be produced at relatively low temperatures and low costs, and are compatible with a wide range of materials. However, there are also some limitations to APCVD-based electrochemical sensors. For example, controlling film thickness and composition can be difficult, and integrating APCVD-based fabrication technologies into large-scale production processes can be challenging.

2.4 INK-BASED PRINTING

Several methods use the simplistic approach of printing a template onto the substrate using ink made of the coating nanomaterial. Although these techniques have successfully transcended into fully automated versions, some are still practiced as manual jobs. Various ink-based printing techniques are being used worldwide due to the technique's broad adaptability. The most common of those techniques are described below.

2.4.1 Continuous Inkjet (CIJ) Printing

CIJ is a cost-effective method of electrochemical sensors' fabrication. It is a non-contact printing technique. It uses a high-pressure pump to generate a continuous stream of ink microdroplets. The microdroplets are then deflected by an electric field and directed towards the substrate, where they are deposited to form patterns. The microdroplet formation and deflections are controlled by the properties of the ink and the electric field, which can be adjusted to achieve high resolution and fine patterns.

CIJ-based fabrication techniques can be divided into two main categories: piezoelectric and thermal inkjet printing. In piezoelectric CIJ, the mechanical deformation of a piezoelectric crystal generates the microdroplets. In contrast, in thermal CIJ, microdroplets are generated by the thermal expansion of a resistor within the ink chamber. Both techniques have their advantages and disadvantages and are suitable for different applications. Various CIJ techniques, which are currently being used for fabricating electrochemical sensors, are discussed below.

2.4.1.1 Direct Inkjet Printing (DIP)

DIP is the most common CIJ technique for fabricating electrochemical sensors. In DIP, the ink microdroplets are directed onto the substrate, forming a thin film. The ink microdroplets can comprise various materials, including metals, polymers, and ceramics, making DIP a versatile technique for fabricating electrochemical sensors.

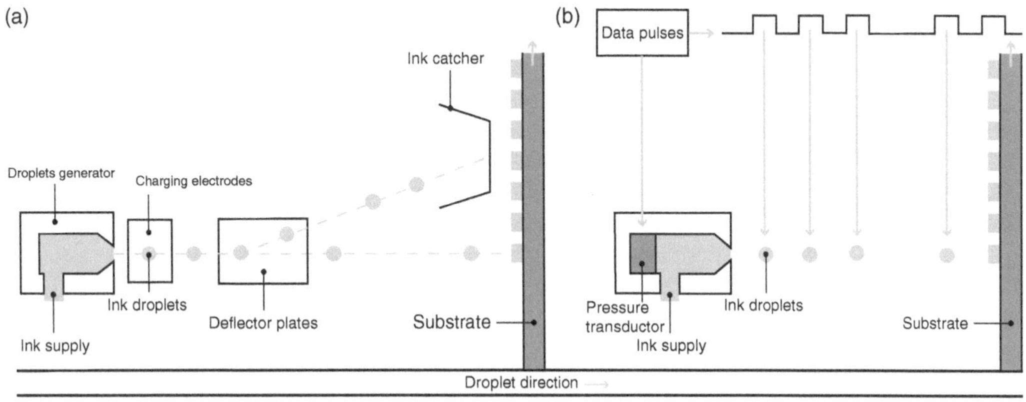

FIGURE 2.5 Schematic of the mechanisms of (a) CIJ and (b) DOD printing mechanisms [4].

2.4.1.2 Drop-on-Demand Inkjet Printing (DOD)

DOD is a variant of the CIJ printing technique. It first generates microscopic ink microdroplets by a piezoelectric or thermal actuator, which is triggered by an electrical signal. Then, it selectively deposits ink microdroplets onto the substrate. It allows for precise microdroplet size and flow rate control, making DOD an attractive option for fabricating electrochemical sensors with high spatial resolution. Atypical schematic of the mechanisms of CIJ and DOD printing mechanisms is shown in Figures 2.5a and b, respectively.

2.4.1.3 Spray Deposition Inkjet Printing (SDI)

SDI is a CIJ technique involving spraying the microdroplets of ink onto the substrate. The ink microdroplets are generated by an *atomiser* (high-pressure nozzle) and directed onto the substrate where they form a thin film. SDI is a fast and efficient technique for fabricating electrochemical sensors with high coverage and uniformity.

CIJ-based techniques allow for the precise control of the microdroplet size, flow rate, and ink composition with high throughput. It makes the technique suitable for the large-scale fabrication of electrochemical sensors. CIJ-based techniques are also cost-effective and more accessible to researchers and manufacturers. However, CIJ requires stable microdroplet generation and accurate placement to ensure consistent film formation. Hence, the instability of any component leads to variability in the fabricated sensor's performance. Moreover, as the viscosity of the ink determines the microdroplet size and ink flow rate, a mismatch between the ink viscosity and the dispersal mechanisms can result in impaired sensor performance.

2.4.2 Aerosol Jet Printing (AJP)

AJP is a novel method for fabricating electrochemical sensors, which involves using a focused stream of liquid microdroplets to deposit patterns of material on a surface. This technique is advantageous in terms of high resolution, high throughput, and the ability to deposit materials in complex geometries. The AJP process is based on the principle of electrohydrodynamic atomisation, which uses a high-frequency, high-voltage electric field to generate microdroplets from a liquid precursor. Figure 2.6 shows a typical schematic of the mechanism of the AJP technique. The steps can be termed as (1) atomisation, (2) gravity-enforced sorting, (3) focusing of jet, and (4) deposition.

The AJP process begins with the generation of microdroplets from a liquid precursor by applying a high-frequency, high-voltage electric field to the liquid, which causes the fluid to break into microdroplets. These microdroplets are then focused into a stream by an air or gas stream, which is directed through a nozzle. The nozzle shapes the microdroplets into a particular pattern, which is then directed onto a substrate. The liquid in the microdroplets then evaporates, leaving a thin

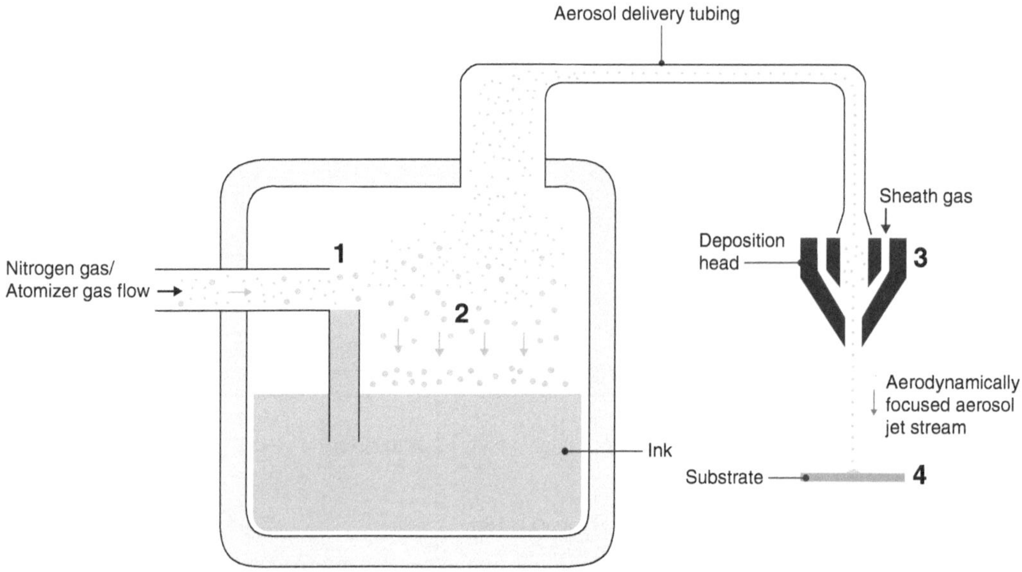

FIGURE 2.6 Working principle of a compressor nebuliser-based AJP [4].

material film on the substrate. The key parameters that control the AJP process include the microdroplet size, the microdroplet velocity, the nozzle geometry, and the substrate temperature. The liquid precursor viscosity, the nozzle geometry, and the electric field strength control the microdroplet size. The microdroplet velocity is governed by the air or gas (carrier) stream. The nozzle geometry controls the pattern of the microdroplets, while the substrate temperature controls the evaporation rate of the microdroplets.

AJP can deposit a wide range of materials, including nanoparticles of metals, ceramics, polymers, and even biomaterials. For example, AJP has been used to deposit metals such as gold, silver, and copper, commonly used to fabricate electrochemical sensors. AJP has also been used to deposit ceramic materials such as titanium dioxide, which is used to fabricate gas sensors. In addition, AJP has been used to deposit polymers such as polyaniline, which is used to fabricate biosensors. Despite these advantages, AJP lacks control of film thickness and composition. The thickness and composition of the deposited film are critical factors in determining the performance of the electrochemical sensor. However, controlling these factors can be difficult due to the complexity of the AJP process.

2.4.3 Screen Printing

Screen printing is a versatile and cost-effective technique for fabricating electrochemical sensors. It is a form of stencil printing in which a stencil or mesh transfers ink or other materials onto a substrate. Screen printing is gaining wide acceptance as the method of choice for fabricating electrochemical sensors due to its ability to deposit various materials, including conductive inks, insulating inks, and enzymes, with high precision and uniformity. The basic principle of screen printing is to force a fluid material through a stencil or mesh onto a substrate. The stencil or mesh is used to define the shape and dimensions of the deposited film, which can then be used to fabricate electrochemical sensors with high precision and repeatability. The screen-printing process can be divided into three main steps: (1) preparation of the stencil or mesh, (2) preparation of the ink or paste, and (3) printing of the ink or paste onto the substrate. Figures 2.7a and b show the typical schematic of the mechanisms of flatbed and rotary screen printing, respectively.

The stencil or mesh is a critical component of the screen-printing process, as it determines the shape and dimensions of the deposited film. The stencil or mesh can be prepared using photolithography,

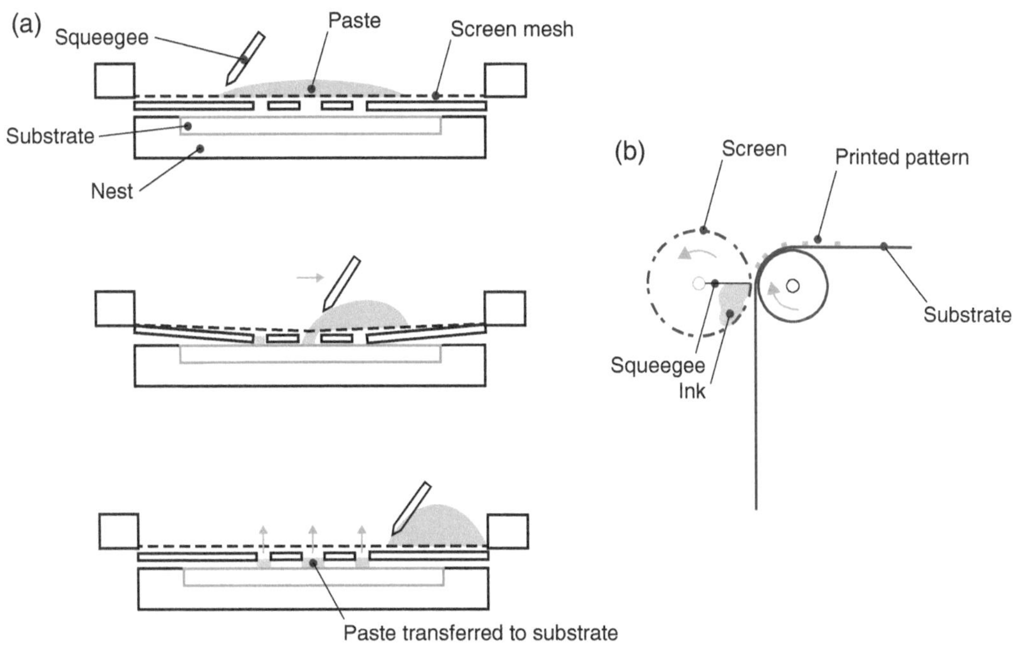

FIGURE 2.7 Working principles of screen-printing variants [4].

laser cutting, and chemical etching techniques. The ink or paste used in the screen-printing process must have the appropriate viscosity and rheology to force through the stencil or mesh quickly. The ink or paste can be prepared using various physical dispersion techniques. The choice of ink or paste will depend on the material being deposited and the desired properties of the electrochemical sensor. For example, conductive inks, such as carbon-based and metal-based inks, are typically used to fabricate electrodes, while insulating inks, such as polymeric and ceramic inks, are used to manufacture insulating layers. Enzymes, such as glucose oxidase and lactate oxidase, can also be screen-printed onto the substrate to create biosensors. The final step in the screen-printing process is printing the ink or paste onto the substrate. It can be done manually or using an automated screen-printing machine. Several screen-printing techniques have been developed for the fabrication of electrochemical sensors. These techniques include direct screen printing, inkjet screen printing, and microcontact screen printing.

2.4.3.1 Direct Screen Printing

Direct screen printing is the most basic technique for fabricating electrochemical sensors using screen printing. In this technique, a stencil is prepared with the desired pattern of the sensor, and then, ink or other materials are deposited onto the stencil using a squeegee. The ink or materials are then forced through the openings in the stencil and onto the substrate, forming the desired pattern. This technique is simple and inexpensive, but it can be challenging to precisely control the sensor pattern and thickness.

2.4.3.2 Inkjet Screen Printing

Inkjet screen printing is a more advanced technique that uses an inkjet printer to deposit ink or other materials onto the stencil. In this technique, inkjet printers deposit ink or other materials onto the stencil in a precise pattern. The stencil is then placed on top of the substrate, and the ink or materials are forced through the openings in the stencil and onto the substrate, forming the desired pattern. This technique allows for more precise sensor pattern and thickness control, but it is more expensive than direct screen printing.

2.4.3.3 Microcontact Screen Printing

Microcontact screen printing is a technique that uses a modified screen-printing process to fabricate electrochemical sensors with microscopic dimensions. This technique prepares a stencil with openings much smaller than those used in conventional screen-printing techniques. The stencil is then placed on top of the substrate, and ink or other materials are deposited onto the stencil using a squeegee. The ink or materials are then forced through the openings in the stencil and onto the substrate, forming the desired pattern. This technique helps fabricate electrochemical sensors with microscopic dimensions, but precisely controlling the sensor pattern and thickness can be challenging.

2.4.3.4 Screen Printing, Followed by Sintering

Screen printing followed by electroplating deposits a thin material film onto a substrate through a stencil or mesh. Then, the substrate is heated to a high temperature to sinter the material. The sintered material can then be used as the active material for the electrochemical sensor. **Screen printing followed by electroplating** is a similar technique that uses the screen-printed layer as the active layer to invite and adhere to the electroplated second layer.

One of the main advantages of screen-printing-based techniques for fabricating electrochemical sensors is their ability to deposit various materials with high precision and uniformity. It makes fabricating sensors with multiple layers and complex structures, such as electrodes, insulating layers, and enzymes, in a single process possible. Additionally, screen printing is a relatively low-cost technique, making it suitable for large-scale production of sensors. However, screen-printing-based techniques for fabricating electrochemical sensors also have some limitations. One limitation is the difficulty in controlling the thickness and composition of the deposited material or ink. It can lead to variations in the sensor's performance, particularly for sensors that require precise thickness and composition control. Additionally, screen printing can be limited by the size and shape of the substrate, as it is not suitable for fabricating sensors on irregular or small surfaces.

2.5 LITHOGRAPHY

Lithography is a technique that uses a patterned mask to selectively expose a photosensitive material, which is then developed and etched to form a desired pattern. The patterned mask can be a physical mask, such as a stencil, or a digital mask, such as a computer-generated image. Lithography-based techniques are widely used to fabricate electrochemical sensors due to their high resolution, aspect ratio, and accuracy. Currently, a variety of lithography-based techniques are available and are discussed below.

2.5.1 Photolithography

Photolithography-based techniques have been widely used to fabricate electrochemical sensors due to their ability to provide high resolution and precise control over the sensor's structure and composition. Photolithography is a process in which a photosensitive material, called a resist, is coated onto a substrate and then exposed to a patterned light source. The exposed areas of the resist are then selectively removed to create the desired pattern, which can be used to define the sensor's structure and composition. The basic principle of photolithography is using light to selectively expose a photosensitive material, such as photoresist, which is then developed to reveal the underlying substrate. The substrate is then etched to form the desired pattern. The process is typically performed in several steps, including substrate preparation, photoresist coating, mask alignment, exposure, development, and etching.

Substrate preparation is the first step in the photolithography process. The substrate is cleaned and treated to remove any contaminants, such as dust or oils, that may affect the quality of the sensor. Photoresist coating is then performed to apply a thin layer of photoresist to the substrate. The photoresist is then spun to ensure a uniform thickness. The next step is mask alignment, which involves the alignment of the photomask with the substrate. The photomask contains the pattern that will be transferred to the substrate. Exposure is then performed to expose the photoresist to light selectively. The light is typically provided by a light source, such as an ultraviolet (UV) lamp, and is passed through the photomask to reach the substrate. Development is the next step in the process, which involves the removal of the unexposed areas of the photoresist to reveal the underlying substrate. Etching is then performed to remove the exposed areas of the substrate to form the desired pattern. Several photolithography techniques have been developed to fabricate electrochemical sensors, including contact, proximity, and projection photolithography.

2.5.1.1 Contact Photolithography

Contact photolithography or contact printing is a simple and low-cost photolithography technique in which a photomask, patterned transparency, is placed in contact with the resist-coated substrate. The patterned light source is then used to expose the resist, creating the desired pattern. Contact photolithography is often used to fabricate simple sensors, such as glucose sensors, but it is limited by its low resolution and poor alignment accuracy.

2.5.1.2 Proximity Photolithography

Proximity photolithography or near-field photolithography is a more advanced photolithography technique in which the patterned light source is positioned a small distance from the resist-coated substrate, typically on a few micrometres. It can provide higher resolution and better alignment accuracy than contact photolithography but is more complex and costly. Proximity photolithography is often used to fabricate more complex sensors, such as biosensors and gas sensors.

2.5.1.3 Projection Photolithography

Projection photolithography also known as stepper or scanner photolithography is the most advanced photolithography technique, and it can provide the highest resolution and alignment accuracy. In this technique, the patterned light source is projected onto the resist-coated substrate through a lens system, creating the desired pattern. Projection photolithography is often used to fabricate high-precision sensors, such as micro-electromechanical systems (MEMS) and integrated circuits (ICs).

Photolithography-based techniques provide high resolution and precise control over the sensor's structure and composition, which is critical for achieving high sensitivity and selectivity. Additionally, they are compatible with a wide range of materials, including metals, ceramics, and polymers, which allows for fabricating a wide range of sensors. However, photolithography-based techniques can be complex and costly, hindering widespread adoption. Additionally, they may require hazardous materials, such as solvents and acids, which can be a safety concern.

2.5.2 Electron Beam Lithography (EBL)

EBL is a powerful technique for fabricating high-precision and resolution electrochemical sensors. EBL uses a focused electron beam, and based on the design, selectively exposes the photosensitive material to form patterns on the carrier surface. These patterns can then be transferred from the carrier to the substrate to create structures with dimensions on the nanometre scale. EBL has been widely used for fabricating electrochemical sensors due to its ability to create high-resolution patterns, which can be used to improve the sensitivity and selectivity of the sensors. Additionally, EBL is being used to fabricate electrochemical sensors based on the composites of conductive polymers, metals, and ceramics.

EBL systems consist of an electron source, a beam-shaping and focusing system, a pattern generator, and a stage for holding the substrate. The electron source is typically a thermionic cathode or a field emission cathode, which generates a beam of electrons. The beam-shaping and focusing system is used to control the electron beam's size, shape, and intensity. The pattern generator is used to create the desired patterns on the substrate, and the stage is used to position the substrate in the correct location for exposure.

In EBL, electrons are generated by the electron gun and are then focused by an electron lens onto a target material. The electrons in the beam have high kinetic energy, which allows them to penetrate the target material and cause chemical reactions that result in the formation of patterns. The exposure process causes a chemical change in the resist, which can then be developed to create a patterned resist mask. The patterned resist mask is then used to transfer the pattern to the underlying substrate through a process such as etching. Moreover, EBL can be performed using various EBL systems, including scanning EBL (SEBL) and projection EBL (PEBL). In SEBL, the electron beam is scanned over the target material, whereas in PEBL, the electron beam is projected onto the target material.

EBL has several advantages over other lithography techniques, such as high resolution, significant depth of focus, and the ability to pattern a wide range of materials. However, EBL has limitations, such as high cost and complexity, low throughput, and a limited field of view.

1. Top of form
2. Bottom of form

2.5.3 Nanoimprint Lithography (NIL)

NIL is a direct-write patterning technique that uses a mould, or stamp, to transfer a pattern onto a substrate. NIL is based on the principle of pattern transfer by mechanical deformation. The stamp is typically silicon or glass and coated with a thin-release material layer, such as polydimethylsiloxane (PDMS). The substrate, typically made of a polymer or glass, is then coated with a thin layer of resist, and the stamp is pressed onto the substrate with a controlled force, transferring the pattern onto the substrate. The resist is then developed, and the substrate is etched to create the desired pattern. The two variants of NIL, thermal NIL and UV-NIL, are being widely used at present. Thermal NIL requires heating the material to a temperature above its glass transition temperature, while UV-NIL requires exposing the material to UV light. The choice of NIL technique depends on the material's properties and the desired pattern resolution. Atypical steps of thermal NIL, UV-NIL, and soft lithography are shown in Figures 2.8a–c, respectively.

One of the main advantages of NIL is the process's high resolution and high throughput. NIL can create structures with dimensions in the 10 nm size range. It makes the technique vital for developing susceptible and selective electrochemical sensors. Additionally, NIL is a cost-effective technique as it does not require expensive equipment or complex processes. NIL also allows for fabricating large-area patterns, essential for developing large-scale electrochemical sensors. Still, despite the advantages of NIL-based techniques, some challenges need to be addressed. One of the main challenges is the alignment of the stamp with the substrate. Misalignment can result in the formation of defects, which can affect the performance of the electrochemical sensor. Additionally, the stamp quality can also affect the quality of the pattern transferred onto the substrate. The stamp must be flat and smooth to ensure the formation of high-quality patterns. Another challenge is the development of new materials and techniques for NIL-based fabrication of electrochemical sensors. For example, using new resists and etchants is a promising approach for developing more sensitive and selective electrochemical sensors. However, these materials can be challenging to process using NIL and may require developing new NIL techniques.

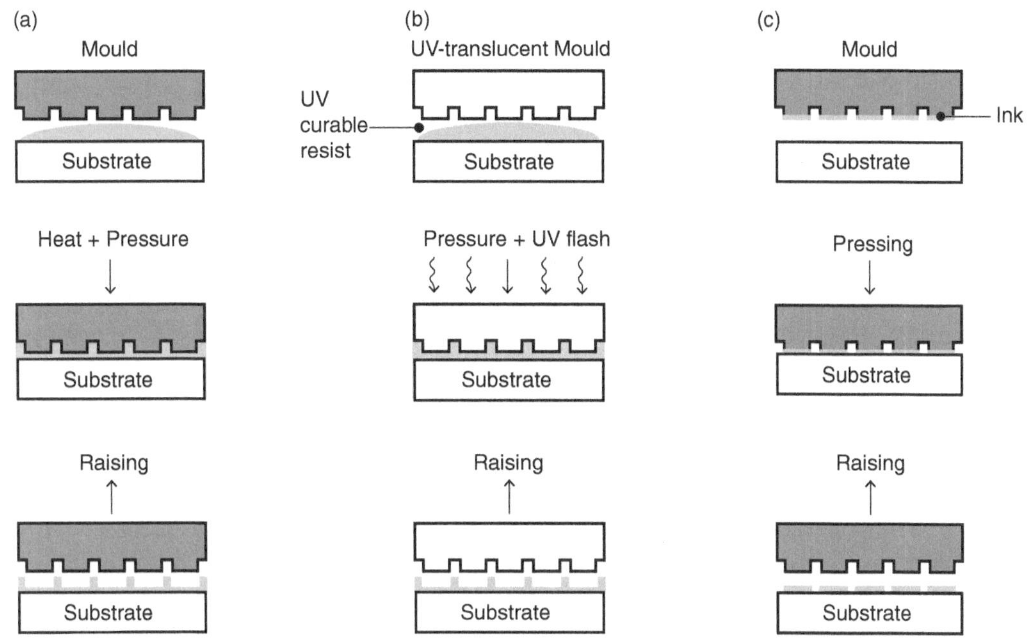

FIGURE 2.8 Working principles of some variants of nanoimprint lithography [4].

2.6 ELECTROSPINNING

Electrospinning is an inexpensive technique for fabricating electrochemical sensors, which has gained considerable attention in recent years. The technique involves the creation of a high electric field that electrostatically draws a polymer solution or a melt through a small nozzle, forming fine fibres. These fibres can then be deposited onto a substrate to form a thin film. The unique properties of electrospun fibres, such as high surface area-to-volume ratio, high porosity, and tunable pore size, make them ideal for electrochemical sensors.

Electrospinning is based on the principle of electrostatic repulsion. A polymer solution or melt is fed into a high-voltage electric field, which causes the polymer to be drawn out of the nozzle and into the electric field. The electric field then stretches the polymer, causing it to become thinner and thinner until it breaks into fibres. These fibres are then deposited onto a substrate to form a thin film. Electrospinning can be divided into two main stages: the jet formation stage and the fibre formation stage. In the jet formation stage, the electric field draws the polymer solution or the melt-out of the nozzle. The electric field stretches and breaks the polymer jet into fibres in the fibre formation stage. The diameter of the fibres can be controlled by adjusting the electric field strength, the polymer concentration, the nozzle diameter, and the distance between the nozzle and the substrate.

A wide range of polymers can be used for electrospinning, including synthetic polymers, such as polyvinyl alcohol, polyethylene oxide, polyvinylpyrrolidone, and natural polymers, such as collagen and gelatine. The choice of polymer will depend on the desired properties of the electrochemical sensor, such as conductivity and biocompatibility. In addition to polymers, various other materials, such as metals, ceramics, and biomaterials, can be incorporated into electrospun fibres to improve the performance of the electrochemical sensor. For example, incorporating metal nanoparticles into electrospun fibres can improve the sensor's conductivity.

Electrospinning-based techniques have been widely used to fabricate various electrochemical sensors, such as glucose, pH, and DNA because electrospun fibres' high surface area and porosity provide high sensitivity and selectivity in the fabricated sensor. However, controlling fibre diameter and alignment is still a crucial issue to be solved.

2.7 CONCLUSIONS

The development of electrochemical sensors requires advanced fabrication techniques. These techniques are critical to having precise control over the properties of the sensing materials and electrodes. The choice of fabrication technique depends on the required performance and the intended application of the sensor. Various techniques, such as screen printing, lithography, and thin-film deposition, have been extensively used to fabricate electrochemical sensors. The screen-printing technique is a widely used low-cost technique that allows for the simultaneous deposition of multiple layers of ink, which can be used to print various sensing materials and electrodes. Lithography is a technique that uses light to transfer a pattern onto a substrate, which can be used to create high-resolution patterns of electrodes and sensing materials. Thin-film deposition techniques allow for the deposition of thin layers of materials with precise control over the thickness, composition, and structure of the films. These techniques, with their distinct pros and cons, allow to produce miniaturised electrochemical sensors that offer high sensitivity, selectivity, and fast response time. Figures 2.9–2.11 may assist in understanding the application area and selecting the most suitable deposition technique. Figure 2.9 shows the various fields of application for different printing techniques. Figure 2.10 compares printing technologies based on their speed vs. obtainable resolutions. Figure 2.11 shows a comparison of film thicknesses obtained through different printing methods. With the advent of novel fabrication techniques, there is great potential for developing advanced electrochemical sensors that offer high sensitivity, selectivity, and flexibility.

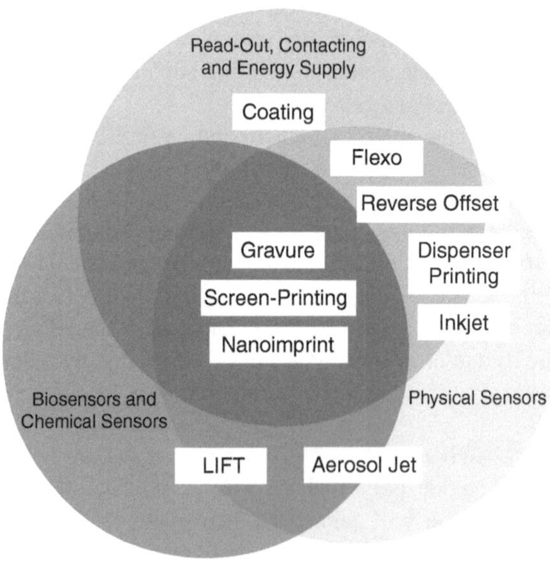

FIGURE 2.9 Various printing-based sensor fabrication techniques, differentiated for their fields of applications [4].

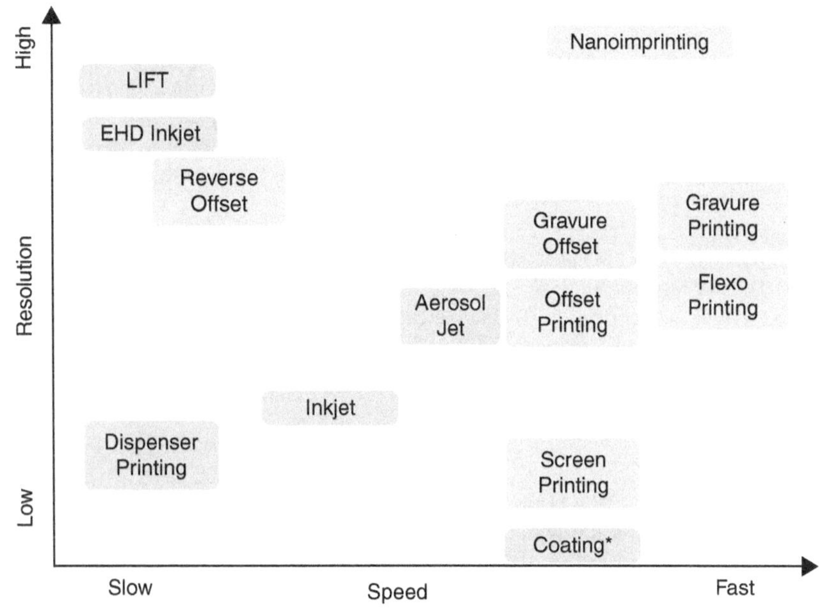

FIGURE 2.10 Various printing-based sensor fabrication techniques, differentiated for their printing speed vs. resolution [4].

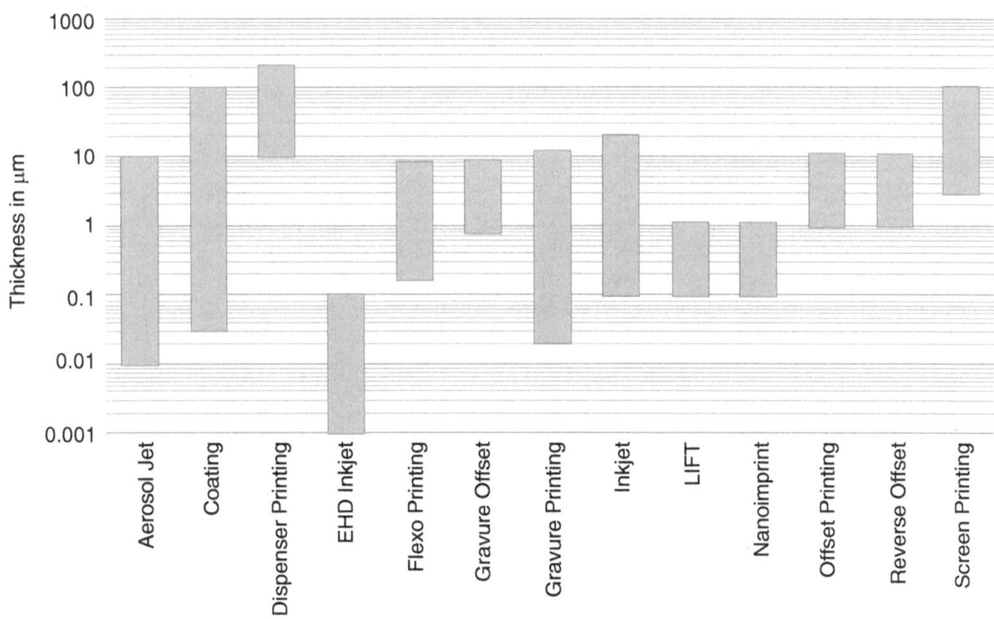

FIGURE 2.11 Various printing-based sensor fabrication techniques, differentiated for their film thicknesses of different printing methods [4].

REFERENCES

1. Park, S.-I.; Quan, Y.-J.; Kim, S.-H.; Kim, H.; Kim, S.; Chun, D.-M.; Lee, C. S.; Taya, M.; Chu, W.-S.; Ahn, S.-H. A Review on Fabrication Processes for Electrochromic Devices. *Int. J. Precis. Eng. Manuf. Technol.*, 2016, 3 (4), 397–421. https://doi.org/10.1007/s40684-016-0049-8.
2. Miikkulainen, V.; Leskelä, M.; Ritala, M.; Puurunen, R. L. Crystallinity of Inorganic Films Grown by Atomic Layer Deposition: Overview and General Trends. *J. Appl. Phys.*, 2013, 113 (2), 021301. https://doi.org/10.1063/1.4757907.
3. Sun, L.; Yuan, G.; Gao, L.; Yang, J.; Chhowalla, M.; Gharahcheshmeh, M. H.; Gleason, K. K.; Choi, Y. S.; Hong, B. H.; Liu, Z. Chemical Vapour Deposition. *Nat. Rev. Methods Prim.*, 2021, 1 (1), 5. https://doi.org/10.1038/s43586-020-00005-y.
4. Zikulnig, J.; Kosel, J. Fabrication Technologies for Flexible Printed Sensors. In Roger Narayan; & Kazunori Ikebukuro (eds) *Encyclopedia of Sensors and Biosensors*; Elsevier, Amsterdam, The Netherlands, 2023; Vol. 1–4, pp. 33–50. https://doi.org/10.1016/B978-0-12-822548-6.00010-8.

FURTHER READING

Laser Induced Forward Transfer | Thin Films and Interfaces | Paul Scherrer Institut (PSI) https://www.psi.ch/en/lmx-interfaces/lift (accessed Aug 6, 2023).

Ahn, S. H.; Guo, L. J. High-Speed Roll-to-Roll Nanoimprint Lithography on Flexible Plastic Substrates. *Adv. Mater.*, 2008, 20 (11), 2044–2049. https://doi.org/10.1002/adma.200702650.

Ahn, S.; Cha, J.; Myung, H.; Kim, S.; Kang, S. Continuous Ultraviolet Roll Nanoimprinting Process for Replicating Large-Scale Nano- and Micropatterns. *Appl. Phys. Lett.*, 2006, 89 (21), 213101. https://doi.org/10.1063/1.2392960.

Allen, M.; Lee, C.; Ahn, B.; Kololuoma, T.; Shin, K.; Ko, S. R2R Gravure and Inkjet Printed RF Resonant Tag. *Microelectron. Eng.*, 2011, 88 (11), 3293–3299. https://doi.org/10.1016/j.mee.2011.08.010.

Arias, A. C.; MacKenzie, J. D.; McCulloch, I.; Rivnay, J.; Salleo, A. Materials and Applications for Large Area Electronics: Solution-Based Approaches. *Chem. Rev.*, 2010, 110 (1), 3–24. https://doi.org/10.1021/cr900150b.

Arshak, K. I.; Ansari, F.; Collins, D. Analysis of Thick Film Strain Resistors on Stainless Steel and Ceramic Substrates. *Int. J. Electron.*, 2012, 179, 50–55.

Arshak, K. I.; Ray, A. K.; Hogarth, C. A.; Collins, D. G.; Ansari, F. An Analysis of Polymeric Thick-Film Resistors as Pressure Sensors. *Sensors Actuators A Phys.*, 1995, 49 (1–2), 41–45. https://doi.org/10.1016/0924-4247(95)01012-P.

Backes, C.; Abdelkader, A. M.; Alonso, C.; Andrieux-Ledier, A.; Arenal, R.; Azpeitia, J.; Balakrishnan, N.; Banszerus, L.; Barjon, J.; Bartali, R.; et al. Production and Processing of Graphene and Related Materials. *2D Mater.*, 2020, 7 (2), 022001. https://doi.org/10.1088/2053-1583/ab1e0a.

Banks, D. P.; Grivas, C.; Mills, J. D.; Eason, R. W.; Zergioti, I. Nanodroplets Deposited in Microarrays by Femtosecond Ti: Sapphire Laser-Induced Forward Transfer. *Appl. Phys. Lett.*, 2006, 89 (19), 2386921. https://doi.org/10.1063/1.2386921.

Beck, M. Nanoelectrochemical Transducers for (Bio-) Chemical Sensor Applications Fabricated by Nanoimprint Lithography. *Microelectron. Eng.*, 2004, 73–74, 837–842. https://doi.org/10.1016/S0167-9317(04)00230-8.

Belsey, K. E.; Parry, A. V. S.; Rumens, C. V.; Ziai, M. A.; Yeates, S. G.; Batchelor, J. C.; Holder, S. J. Switchable Disposable Passive RFID Vapour Sensors from Inkjet Printed Electronic Components Integrated with PDMS as a Stimulus Responsive Material. *J. Mater. Chem. C*, 2017, 5 (12), 3167–3175. https://doi.org/10.1039/C6TC05509E.

Bernacka-Wojcik, I.; Wojcik, P. J.; Aguas, H.; Fortunato, E.; Martins, R. Inkjet Printed Highly Porous TiO_2 Films for Improved Electrical Properties of Photoanode. *J. Colloid Interface Sci.*, 2016, 465, 208–214. https://doi.org/10.1016/j.jcis.2015.11.070.

Bosco, R.; Van Den Beucken, J.; Leeuwenburgh, S.; Jansen, J. Surface Engineering for Bone Implants: A Trend from Passive to Active Surfaces. *Coatings*, 2012, 2 (3), 95–119. https://doi.org/10.3390/coatings2030095.

Bould, D.C.; Claypole, T.C.; Bohan, M.F.J. An Investigation into Plate Deformation in Flexographic Printing. *Proc. Inst. Mech. Eng. Part B J. Eng. Manuf.*, 2004, 218 (11), 1499–1511. https://doi.org/10.1243/0954405042418428.

Amruth, C.; Colella, M.; Griffin, J.; Kingsley, J.; Scarratt, N.; Luszczynska, B.; Ulanski, J. Slot-Die Coating of Double Polymer Layers for the Fabrication of Organic Light Emitting Diodes. *Micromachines*, 2019, 10 (1), 53. https://doi.org/10.3390/mi10010053.

Cai, F.; Chang, Y.-H.; Wang, K.; Zhang, C.; Wang, B.; Papapolymerou, J. Low-Loss 3-D Multilayer Transmission Lines and Interconnects Fabricated by Additive Manufacturing Technologies. *IEEE Trans. Microw. Theory Tech.*, 2016, 64 (10), 3208–3216. https://doi.org/10.1109/TMTT.2016.2601907.

Chang, J.; Lee, S.; Lee, K. B.; Lee, S.; Cho, Y. T.; Seo, J.; Lee, S.; Jo, G.; Lee, K.; Kong, H.-S.; et al. Overlay Accuracy on a Flexible Web with a Roll Printing Process Based on a Roll-to-Roll System. *Rev. Sci. Instrum.*, 2015, 86 (5), 055108. https://doi.org/10.1063/1.4921495.

Chen, H. L.; Chuang, S. Y.; Cheng, H. C.; Lin, C. H.; Chu, T. C. Directly Patterning Metal Films by Nanoimprint Lithography with Low-Temperature and Low-Pressure. *Microelectron. Eng.*, 2006, 83 (4–9), 893–896. https://doi.org/10.1016/j.mee.2006.01.095.

Choi, Y.-M.; Kim, K.-Y.; Jo, J.; Lee, T.-M. Effect of PDMS Blanket Deformation on Printability in Reverse-Offset Printing. *Trans. Korean Soc. Mech. Eng. B*, 2014, 38 (8), 709–714. https://doi.org/10.3795/KSME-B.2014.38.8.709.

Choi, Y.-M.; Lee, E.; Lee, T.-M. Mechanism of Reverse-Offset Printing. *J. Micromech. Microeng.*, 2015, 25 (7), 075019. https://doi.org/10.1088/0960-1317/25/7/075019.

Choi, Y.-M.; Lee, E.-S.; Lee, T.-M.; Kim, K.-Y. Optimisation of a Reverse-Offset Printing Process and Its Application to a Metal Mesh Touch Screen Sensor. *Microelectron. Eng.*, 2015, 134, 1–6. https://doi.org/10.1016/j.mee.2014.12.007.

Chou, S. Y.; Keimel, C.; Gu, J. Ultrafast and Direct Imprint of Nanostructures in Silicon. *Nature*, 2002, 417 (6891), 835–837. https://doi.org/10.1038/nature00792.

Chou, S. Y.; Krauss, P. R.; Renstrom, P. J. Nanoimprint Lithography. *J. Vac. Sci. Technol. B Microelectron. Nanom. Struct. Process. Meas. Phenom.*, 1996, 14 (6), 4129–4133. https://doi.org/10.1116/1.588605.

Clemens, W.; Krumm, J.; Blache, R. Printed RFID and Smart Objects for New High Volume Applications. In Cantatore, E. (eds) *Applications of Organic and Printed Electronics*; Springer, Boston, MA, 2013; pp. 115–132. https://doi.org/10.1007/978-1-4614-3160-2_6.

de Vos, M.; Torah, R.; Tudor, J. Dispenser Printed Electroluminescent Lamps on Textiles for Smart Fabric Applications. *Smart Mater. Struct.*, 2016, 25 (4), 045016. https://doi.org/10.1088/0964-1726/25/4/045016.

Dou, B.; Whitaker, J. B.; Bruening, K.; Moore, D. T.; Wheeler, L. M.; Ryter, J.; Breslin, N. J.; Berry, J. J.; Garner, S. M.; Barnes, F. S.; et al. Roll-to-Roll Printing of Perovskite Solar Cells. *ACS Energy Lett.*, 2018, 3 (10), 2558–2565. https://doi.org/10.1021/acsenergylett.8b01556.

Dumond, J. J.; Mahabadi, K. A.; Yee, Y. S.; Tan, C.; Fuh, J. Y. H.; Lee, H. P.; Low, H. Y. High Resolution UV Roll-to-Roll Nanoimprinting of Resin Moulds and Subsequent Replication via Thermal Nanoimprint Lithography. *Nanotechnology*, 2012, 23 (48), 485310. https://doi.org/10.1088/0957-4484/23/48/485310.

Farraj, Y.; Bielmann, M.; Magdassi, S. Inkjet Printing and Rapid Ebeam Sintering Enable Formation of Highly Conductive Patterns in Roll to Roll Process. *RSC Adv.*, 2017, 7 (25), 15463–15467. https://doi.org/10.1039/C7RA00967D.

Fukuda, K.; Takeda, Y.; Yoshimura, Y.; Shiwaku, R.; Tran, L. T.; Sekine, T.; Mizukami, M.; Kumaki, D.; Tokito, S. Fully-Printed High-Performance Organic Thin-Film Transistors and Circuitry on One-Micron-Thick Polymer Films. *Nat. Commun.*, 2014, 5 (1), 4147. https://doi.org/10.1038/ncomms5147.

Gelinck, G. H.; Huitema, H. E. A.; van Veenendaal, E.; Cantatore, E.; Schrijnemakers, L.; van der Putten, J. B. P. H.; Geuns, T. C. T.; Beenhakkers, M.; Giesbers, J. B.; Huisman, B.-H.; et al. Flexible Active-Matrix Displays and Shift Registers Based on Solution-Processed Organic Transistors. *Nat. Mater.*, 2004, 3 (2), 106–110. https://doi.org/10.1038/nmat1061.

Grigaliūnas, V.; Tamulevičius, S.; Muehlberger, M.; Jucius, D.; Guobienė, A.; Kopustinskas, V.; Gudonytė, A. Nanoimprint Lithography Using IR Laser Irradiation. *Appl. Surf. Sci.*, 2006, 253 (2), 646–650. https://doi.org/10.1016/j.apsusc.2005.12.166.

Guo, L. J. Nanoimprint Lithography: Methods and Material Requirements. *Adv. Mater.*, 2007, 19 (4), 495–513. https://doi.org/10.1002/adma.200600882.

Guo, L. J.; Ahn, S. H. High-Speed Roll to Roll Nanoimprint Lithography on Flexible Plastic substrates. *Adv Mater.*, 2008, 9 (11), 2044–2049.

Haatainen, T.; Majander, P.; Riekkinen, T.; Ahopelto, J. Nickel Stamp Fabrication Using Step and Stamp Imprint Lithography. *Microelectron. Eng.*, 2006, 83 (4–9), 948–950. https://doi.org/10.1016/j.mee.2006.01.038.

Häffner, M.; Heeren, A.; Fleischer, M.; Kern, D. P.; Schmidt, G.; Molenkamp, L. W. Simple High Resolution Nanoimprint-Lithography. *Microelectron. Eng.*, 2007, 84 (5–8), 937–939. https://doi.org/10.1016/j.mee.2007.01.020.

Hales, S.; Tokita, E.; Neupane, R.; Ghosh, U.; Elder, B.; Wirthlin, D.; Kong, Y. L. 3D Printed Nanomaterial-Based Electronic, Biomedical, and Bioelectronic Devices. *Nanotechnology*, 2020, 31 (17), 172001. https://doi.org/10.1088/1361-6528/ab5f29.

Harrey, P. M.; Ramsey, B. J.; Evans, P. S. A.; Harrison, D. J. Capacitive-Type Humidity Sensors Fabricated Using the Offset Lithographic Printing Process. *Sensors Actuators B Chem.*, 2002, 87 (2), 226–232. https://doi.org/10.1016/S0925-4005(02)00240-X.

Harsa'nyi, G. Polymer Thick-Film Technology: A Possibility to Obtain Very Low Cost Pressure Sensors? *Sensors Actuators A Phys.*, 1991, 27 (1–3), 853–857. https://doi.org/10.1016/0924-4247(91)87098-N.

Hauser, H.; Michl, B.; Kübler, V.; Schwarzkopf, S.; Müller, C.; Hermle, M.; Bläsi, B. Nanoimprint Lithography for Honeycomb Texturing of Multicrystalline Silicon. *Energy Procedia*, 2011, 8, 648–653. https://doi.org/10.1016/j.egypro.2011.06.196.

Hay, G. I.; Southee, D. J.; Evans, P. S. A.; Harrison, D. J.; Simpson, G. B.; Wood, J. A Time–Strain Monitoring System Fabricated via Offset Lithographic Printing. *Smart Mater. Struct.*, 2009, 18 (1), 015015. https://doi.org/10.1088/0964-1726/18/1/015015.

Hay, G. I.; Evans, P. S. A.; Harrison, D. J.; Southee, D.; Simpson, G.; Harrey, P. M. Characterization of Lithographically Printed Resistive Strain Gauges. *IEEE Sens. J.*, 2005, 5 (5), 864–871. https://doi.org/10.1109/JSEN.2005.845209.

Hay, G. I.; Southee, D. J.; Evans, P. S. A.; Harrison, D. J.; Simpson, G.; Ramsey, B. J. Examination of Silver–Graphite Lithographically Printed Resistive Strain Sensors. *Sensors Actuators A Phys.*, 2007, 135 (2), 534–546. https://doi.org/10.1016/j.sna.2006.10.003.

Hernandez-Sosa, G.; Bornemann, N.; Ringle, I.; Agari, M.; Dörsam, E.; Mechau, N.; Lemmer, U. Rheological and Drying Considerations for Uniformly Gravure-Printed Layers: Towards Large-Area Flexible Organic Light-Emitting Diodes. *Adv. Funct. Mater.*, 2013, 23 (25), 3164–3171. https://doi.org/10.1002/adfm.201202862.

Hiroshima, H.; Komuro, M. Control of Bubble Defects in UV Nanoimprint. *Jpn. J. Appl. Phys.*, 2007, 46 (9B), 6391–6394. https://doi.org/10.1143/JJAP.46.6391.

Hong, H.; Jiang, L.; Tu, H.; Hu, J.; Yan, X. Formulation of UV Curable Nano-Silver Conductive Ink for Direct Screen-Printing on Common Fabric Substrates for Wearable Electronic Applications. *Smart Mater. Struct.*, 2021, 30 (4), 045001. https://doi.org/10.1088/1361-665X/abe4b3.

Hösel, M.; Søndergaard, R. R.; Angmo, D.; Krebs, F. C. Comparison of Fast Roll-to-Roll Flexographic, Inkjet, Flatbed, and Rotary Screen Printing of Metal Back Electrodes for Polymer Solar Cells. *Adv. Eng. Mater.*, 2013, 15 (10), 995–1001. https://doi.org/10.1002/adem.201300011.

Hoth, C. N.; Schilinsky, P.; Choulis, S. A; Balasubramanian, S.; Brabec, C. J. Solution-Processed Organic Photovoltaics. In Cantatore, E. (eds) *Applications of Organic and Printed Electronics: A Technology-Enabled Revolution*; Springer, Boston, MA, 2013; pp. 27–56. https://doi.org/10.1007/978-1-4614-3160-2_2.

Hughes, R. A.; Menumerov, E.; Neretina, S. When Lithography Meets Self-Assembly: A Review of Recent Advances in the Directed Assembly of Complex Metal Nanostructures on Planar and Textured Surfaces. *Nanotechnology*, 2017, 28 (28), 282002. https://doi.org/10.1088/1361-6528/aa77ce.

Hwang, J.-H.; Mendoza, K. L.; Jang, K.; Hwang, S.; Kim, N. Optimisation of Reverse Offset Roll to Plate (RO-R2P) Through Contact Angle. *J. Control Syst. Eng.*, 2013, 1 (3), 67–72. https://doi.org/10.18005/JCSE0103001.

Hwang, S.-Y.; Hong, S.-H.; Jung, H.-Y.; Lee, H. Fabrication of Roll Imprint Stamp for Continuous UV Roll Imprinting Process. *Microelectron. Eng.*, 2009, 86 (4–6), 642–645. https://doi.org/10.1016/j.mee.2008.11.055.

Hyun, W. J.; Secor, E. B.; Hersam, M. C.; Frisbie, C. D.; Francis, L. F. High-Resolution Patterning of Graphene by Screen Printing with a Silicon Stencil for Highly Flexible Printed Electronics. *Adv. Mater.*, 2015, 27 (1), 109–115. https://doi.org/10.1002/adma.201404133.

Jiang, W.; Liu, H.; Ding, Y.; Shi, Y.; Yin, L.; Lu, B. Investigation of Pattern Coating on Mould Roller in Roller-Reversal Imprint Process. *Microelectron. Eng.*, 2009, 86 (12), 2412–2416. https://doi.org/10.1016/j.mee.2009.05.003.

Jung, M.; Kim, J.; Noh, J.; Lim, N.; Lim, C.; Lee, G.; Kim, J.; Kang, H.; Jung, K.; Leonard, A. D.; et al. All-Printed and Roll-to-Roll-Printable 13.56-MHz-Operated 1-Bit RF Tag on Plastic Foils. *IEEE Trans. Electron Devices*, 2010, 57 (3), 571–580. https://doi.org/10.1109/TED.2009.2039541.

Kaestner, M.; Rangelow, I. W. Scanning Probe Lithography on Calixarene towards Single-Digit Nanometer Fabrication. *Int. J. Extrem. Manuf.*, 2020, 2 (3), 032005. https://doi.org/10.1088/2631-7990/aba2d8.

Kamyshny, A.; Magdassi, S. Inkjet Printing. In Buchholz, F.L. (eds) *Kirk-Othmer Encyclopedia of Chemical Technology*; John Wiley & Sons, Inc., Hoboken, NJ, 2013; pp. 1–21. https://doi.org/10.1002/0471238961.inkjmagd.a01.

Kettle, J.; Lamminmäki, T.; Gane, P. A Review of Modified Surfaces for High Speed Inkjet Coating. *Surf. Coatings Technol.*, 2010, 204 (12–13), 2103–2109. https://doi.org/10.1016/j.surfcoat.2009.10.035.

Khan, Y.; Thielens, A.; Muin, S.; Ting, J.; Baumbauer, C.; Arias, A. C. A New Frontier of Printed Electronics: Flexible Hybrid Electronics. *Adv. Mater.*, 2020, 32 (15), 1905279. https://doi.org/10.1002/adma.201905279.

Kim, D.; Moon, J. Highly Conductive Ink Jet Printed Films of Nanosilver Particles for Printable Electronics. *Electrochem. Solid-State Lett.*, 2005, 8 (11), J30. https://doi.org/10.1149/1.2073670.

Kim, H.; Lee, E.; Choi, Y.-M.; Kwon, S.; Lee, S.; Jo, J.; Lee, T.-M.; Kang, D. Development of a Precision Reverse Offset Printing System. *Rev. Sci. Instrum.*, 2016, 87 (1), 015102. https://doi.org/10.1063/1.4939237.

Kim, M.; Bon Koo, J.; Baeg, K.-J.; Jung, S.-W.; Ju, B.-K.; You, I.-K. Top-Gate Staggered Poly(3,3‴-Dialkyl-Quarterthiophene) Organic Thin-Film Transistors with Reverse-Offset-Printed Silver Source/Drain Electrodes. *Appl. Phys. Lett.*, 2012, 101 (13), 133306. https://doi.org/10.1063/1.4755878.

Kim, M.; You, I.-K.; Han, H.; Jung, S.-W.; Kim, T.-Y.; Ju, B.-K.; Koo, J. B. Organic Thin-Film Transistors with Short Channel Length Fabricated by Reverse Offset Printing. *Electrochem. Solid-State Lett.*, 2011, 14 (8), H333. https://doi.org/10.1149/1.3591435.

Kim, Y.; Jang, S.; Oh, J. H. High-Resolution Electrohydrodynamic Printing of Silver Nanoparticle Ink via Commercial Hypodermic Needles. *Appl. Phys. Lett.*, 2015, 106 (1), 014103. https://doi.org/10.1063/1.4905387.

Kim, Y. Y.; Yang, T.; Suhonen, R.; Välimäki, M.; Maaninen, T.; Kemppainen, A.; Jeon, N. J.; Seo, J. Gravure-Printed Flexible Perovskite Solar Cells: Toward Roll-to-Roll Manufacturing. *Adv. Sci.*, 2019, 6 (7), 1802094. https://doi.org/10.1002/advs.201802094.

Kooy, N.; Mohamed, K.; Pin, L. T.; Guan, O. S. A Review of Roll-to-Roll Nanoimprint Lithography. *Nanoscale Res. Lett.*, 2014, 9 (1), 320. https://doi.org/10.1186/1556-276X-9-320.

Korvink, J. G.; Smith, P. J.; Shin, D. Y. *Inkjet-Based Micromanufacturing*. Wiley-VCH Verlag GmbH & Co. KGaA: Weinheim, Germany, 2012. https://doi.org/10.1002/9783527647101.

Koutake, M.; Katayama, Y. Reverse Offset Printing and Specialised Inks for Organic TFTs. In *2014 International Conference on Electronics Packaging (ICEP)*. IEEE, Toyama, Japan, 2014; pp 279–282. https://doi.org/10.1109/ICEP.2014.6826701.

Krainer, S.; Smit, C.; Hirn, U. The Effect of Viscosity and Surface Tension on Inkjet Printed Picoliter Dots. *RSC Adv.*, 2019, 9 (54), 31708–31719. https://doi.org/10.1039/C9RA04993B.

Krebs, F. C.; Fyenbo, J.; Jørgensen, M. Product Integration of Compact Roll-to-Roll Processed Polymer Solar Cell Modules: Methods and Manufacture Using Flexographic Printing, Slot-Die Coating and Rotary Screen Printing. *J. Mater. Chem.*, 2010, 20 (41), 8994. https://doi.org/10.1039/c0jm01178a.

Kusaka, Y.; Kanazawa, S.; Ushijima, H. Design Rules for Vertical Interconnections by Reverse Offset Printing. *J. Micromech. Microengin.*, 2018, 28 (3), 035003. https://doi.org/10.1088/1361-6439/aaa1d6.

Kusaka, Y.; Kanazawa, S.; Yamamoto, N.; Ushijima, H. Direct Observation of Microcontact Behaviours in Pattern-Generation Step of Reverse Offset Printing. *J. Micromechan. Microengin.*, 2017, 27 (1), 015003. https://doi.org/10.1088/0960-1317/27/1/015003.

Kusaka, Y.; Sugihara, K.; Koutake, M.; Ushijima, H. Rewetting of Semi-Dried Ink Patterns by Vapour Annealing for Developing a Reflow Process in Reverse Offset Printing. *J. Micromechan. Microengin*, 2017, 27 (1), 017002. https://doi.org/10.1088/0960-1317/27/1/017002.

Kusaka, Y.; Sugihara, K.; Ushijima, H. Push-Pull Process for Contact Defect-Free Patterning in Reverse Offset Printing. *Flex. Print. Electron.*, 2016, 1 (4), 045004. https://doi.org/10.1088/2058-8585/aa5164.

Lan, S.; Song, J.-H.; Lee, M. G.; Ni, J.; Lee, N. K.; Lee, H.-J. Continuous Roll-to-Flat Thermal Imprinting Process for Large-Area Micro-Pattern Replication on Polymer Substrate. *Microelectron. Eng.*, 2010, 87 (12), 2596–2601. https://doi.org/10.1016/j.mee.2010.07.021.

Lebib, A.; Chen, Y.; Bourneix, J.; Carcenac, F.; Cambril, E.; Couraud, L.; Launois, H. Nanoimprint Lithography for a Large Area Pattern Replication. *Microelectron. Eng.*, 1999, 46 (1–4), 319–322. https://doi.org/10.1016/S0167-9317(99)00094-5.

Lebib, A.; Chen, Y.; Cambril, E.; Youinou, P.; Studer, V.; Natali, M.; Pépin, A.; Janssen, H. M.; Sijbesma, R. P. Room-Temperature and Low-Pressure Nanoimprint Lithography. *Microelectron. Eng.*, 2002, 61–62, 371–377. https://doi.org/10.1016/S0167-9317(02)00485-9.

Lee, H.; Hong, S.; Yang, K.; Choi, K. Fabrication of 100 nm Metal Lines on Flexible Plastic Substrate Using Ultraviolet Curing Nanoimprint Lithography. *Appl. Phys. Lett.*, 2006, 88 (14), 143112. https://doi.org/10.1063/1.2193653.

Lee, J.; Park, S.; Choi, K.; Kim, G. Nano-Scale Patterning Using the Roll Typed UV-Nanoimprint Lithography Tool. *Microelectron. Eng.*, 2008, 85 (5–6), 861–865. https://doi.org/10.1016/j.mee.2007.12.059.

Lee, T.-M.; Han, H.-S.; Kim, B.; Kwak, S.-W.; Noh, J.-H.; Kim, I. Roll Offset Printing Process Based on Interface Separation for Fine and Smooth Patterning. *Thin Solid Films*, 2013, 548, 566–571. https://doi.org/10.1016/j.tsf.2013.09.023.

Lee, T.-M.; Noh, J.-H.; Kim, C. H.; Jo, J.; Kim, D.-S. Development of a Gravure Offset Printing System for the Printing Electrodes of Flat Panel Display. *Thin Solid Films*, 2010, 518 (12), 3355–3359. https://doi.org/10.1016/j.tsf.2009.10.017.

Leyland, N. S.; Evans, J. R. G.; Harrison, D. J. Lithographic Printing of Force-Sensitive Resistors. *J. Mater. Sci. Mater. Electron.*, 2002, 13 (7), 387–390. https://doi.org/10.1023/A:1016032325024.

Li, D.; Lai, W.-Y.; Zhang, Y.-Z.; Huang, W. Printable Transparent Conductive Films for Flexible Electronics. *Adv. Mater.*, 2018, 30 (10), 1704738. https://doi.org/10.1002/adma.201704738.

Li, X.; Tian, J.; Garnier, G.; Shen, W. Fabrication of Paper-Based Microfluidic Sensors by Printing. *Colloids Surfaces B Biointerfaces,* 2010, 76 (2), 564–570. https://doi.org/10.1016/j.colsurfb.2009.12.023.

Liang, J.; Tong, K.; Pei, Q. A Water-Based Silver-Nanowire Screen-Print Ink for the Fabrication of Stretchable Conductors and Wearable Thin-Film Transistors. *Adv. Mater.*, 2016, 28 (28), 5986–5996. https://doi.org/10.1002/adma.201600772.

Liang, R. C.; Hou, J.; Zang, H.; Chung, J.; Tseng, S. Microcup® Displays: Electronic Paper by Roll-to-Roll Manufacturing Processes. *J. Soc. Inf. Disp.*, 2003, 11 (4), 621. https://doi.org/10.1889/1.1825690.

Lim, H.; Choi, K.-B.; Kim, G.; Park, S.; Ryu, J.; Lee, J. Roller Nanoimprint Lithography for Flexible Electronic Devices of a Sub-Micron Scale. *Microelectron. Eng.*, 2011, 88 (8), 2017–2020. https://doi.org/10.1016/j.mee.2011.02.018.

Liu, L.; Zhang, Y.; Wang, W.; Gu, C.; Bai, X.; Wang, E. Nanosphere Lithography for the Fabrication of Ultranarrow Graphene Nanoribbons and On-Chip Bandgap Tuning of Graphene. *Adv. Mater.*, 2011, 23 (10), 1246–1251. https://doi.org/10.1002/adma.201003847.

Lorenz, A.; Gredy, C.; Senne, A.; Beyer, S.; Yao, Y.; Papet, P.; Ufheil, J.; Reinecke, H.; Clement, F. Flexo-Printed Busbarless Solar Cells for Multi-Wire Interconnection. *Energy Procedia*, 2016, 98, 46–60. https://doi.org/10.1016/j.egypro.2016.10.080.

Lund, P. D.; Halme, J.; Hashmi, G.; Asghar, I.; Miettunen, K. Application of Dye-Sensitized and Perovskite Solar Cells on Flexible Substrates. *Flex. Print. Electron.*, 2018, 3 (1), 013002. https://doi.org/10.1088/2058-8585/aaabc9.

Lupo, D.; Clemens, W.; Breitung, S.; Hecker, K. OE-A Roadmap for Organic and Printed Electronics. In *AIMCAL R2R Conference USA 2018- Roll 2 Roll Web Coating and Finishing with SPE's FlexPackCon.* Association of International Metallizers, Coaters, and Laminators, Phoenix, AZ, 2013; pp 1–26. https://doi.org/10.1007/978-1-4614-3160-2_1.

Mäkelä, T.; Jussila, S.; Kosonen, H.; Bäcklund, T. G.; Sandberg, H. G. O.; Stubb, H. Utilising Roll-to-Roll Techniques for Manufacturing Source-Drain Electrodes for All-Polymer Transistors. *Synth. Met.*, 2005, 153 (1–3), 285–288. https://doi.org/10.1016/j.synthmet.2005.07.140.

Mäkelä, T.; Haatainen, T.; Ahopelto, J. Roll-to-Roll Printed Gratings in Cellulose Acetate Web Using Novel Nanoimprinting Device. *Microelectron. Eng.*, 2011, 88 (8), 2045–2047. https://doi.org/10.1016/j.mee.2011.02.016.

Mäkelä, T.; Haatainen, T.; Majander, P.; Ahopelto, J. Continuous Roll to Roll Nanoimprinting of Inherently Conducting Polyaniline. *Microelectron. Eng.*, 2007, 84 (5–8), 877–879. https://doi.org/10.1016/j.mee.2007.01.131.

Margalit, N.; Xiang, C.; Bowers, S. M.; Bjorlin, A.; Blum, R.; Bowers, J. E. Perspective on the Future of Silicon Photonics and Electronics. *Appl. Phys. Lett.*, 2021, 118 (22). https://doi.org/10.1063/5.0050117.

Maury, P.; Turkenburg, D.; Stroeks, N.; Giesen, P.; Barbu, I.; Meinders, E.; van Bremen, A.; Iosad, N.; van der Werf, R.; Onvlee, H. Roll-to-Roll UV Imprint Lithography for Flexible Electronics. *Microelectron. Eng.*, 2011, 88 (8), 2052–2055. https://doi.org/10.1016/j.mee.2011.02.022.

Merino, S.; Retolaza, A.; Juarros, A.; Landis, S. A New Way of Manufacturing High Resolution Optical Encoders by Nanoimprint Lithography. *Microelectron. Eng.*, 2007, 84 (5–8), 848–852. https://doi.org/10.1016/j.mee.2007.01.024.

Nguyen, H. A. D.; Lee, C.; Shin, K.-H. Approach to Optimising Printed Conductive Lines in High-Resolution Roll-to-Roll Gravure Printing. *Robot. Comput. Integr. Manuf.*, 2017, 46, 122–129. https://doi.org/10.1016/j.rcim.2017.01.003.

Ning, H.; Tao, R.; Fang, Z.; Cai, W.; Chen, J.; Zhou, Y.; Zhu, Z.; Zheng, Z.; Yao, R.; Xu, M.; et al. Direct Patterning of Silver Electrodes with 2.4 Mm Channel Length by Piezoelectric Inkjet Printing. *J. Colloid Interface Sci.*, 2017, 487, 68–72. https://doi.org/10.1016/j.jcis.2016.10.016.

Noh, J.; Yeom, D.; Lim, C.; Cha, H.; Han, J.; Kim, J.; Park, Y.; Subramanian, V.; Cho, G. Scalability of Roll-to-Roll Gravure-Printed Electrodes on Plastic Foils. *IEEE Trans. Electron. Packag. Manuf.*, 2010, 33 (4), 275–283. https://doi.org/10.1109/TEPM.2010.2057512.

Nomura, K.; Kusaka, Y.; Ushijima, H.; Nagase, K.; Ikedo, H.; Mitsui, R.; Takahashi, S.; Nakajima, S.; Iwata, S. Continuous Fine Pattern Formation by Screen-Offset Printing Using a Silicone Blanket. *J. Micromech. Microengin.*, 2014, 24 (9), 095021. https://doi.org/10.1088/0960-1317/24/9/095021.

Palavesam, N.; Marin, S.; Hemmetzberger, D.; Landesberger, C.; Bock, K.; Kutter, C. Roll-to-Roll Processing of Film Substrates for Hybrid Integrated Flexible Electronics. *Flex. Print. Electron.*, 2018, 3 (1), 014002. https://doi.org/10.1088/2058-8585/aaaa04.

Pardo, D. A.; Jabbour, G. E.; Peyghambarian, N. Application of Screen Printing in the Fabrication of Organic Light-Emitting Devices. *Adv. Mater.*, 2000, 12 (17), 1249–1252. https://doi.org/10.1002/1521-4095(200009)12:17<1249::AID-ADMA1249>3.0.CO;2–Y.

Park, J.; Shin, K.; Lee, C. Mechanical Aspects of the Chemical Mechanical Polishing Process: A Review. *Int. J. Precis. Eng. Manuf.*, 2016, 17, 525–536.

Park, J.-U.; Hardy, M.; Kang, S. J.; Barton, K.; Adair, K.; Mukhopadhyay, D. K.; Lee, C. Y.; Strano, M. S.; Alleyne, A. G.; Georgiadis, J. G.; et al. High-Resolution Electrohydrodynamic Jet Printing. *Nat. Mater.*, 2007, 6 (10), 782–789. https://doi.org/10.1038/nmat1974.

Park, S.; Choi, K.; Kim, G.; Lee, J. Nanoscale Patterning with the Double-Layered Soft Cylindrical Stamps by Means of UV-Nanoimprint Lithography. *Microelectron. Eng.*, 2009, 86 (4–6), 604–607. https://doi.org/10.1016/j.mee.2008.12.074.

Perret, C. Characterization of 8-in. Wafers Printed by Nanoimprint Lithography. *Microelectron. Eng.*, 2004, 73–74, 172–177. https://doi.org/10.1016/S0167-9317(04)00094-2.

Prudenziati, M.; Morten, B.; Taroni, A. Characterization of Thick-Film Resistor Strain Gauges on Enamel Steel. *Sensors and Actuators*, 1981, 2 (C), 17–27. https://doi.org/10.1016/0250-6874(81)80025-X.

Pudas, M.; Hagberg, J.; Leppävuori, S. Printing Parameters and Ink Components Affecting Ultra-Fine-Line Gravure-Offset Printing for Electronics Applications. *J. Eur. Ceram. Soc.*, 2004, 24 (10–11), 2943–2950. https://doi.org/10.1016/j.jeurceramsoc.2003.11.011.

Ramsey, B. J.; Evans, P. S. A.; Harrison, D. A Novel Circuit Fabrication Technique Using Offset Lithography. *J. Electron. Manuf.*, 1997, 07 (01), 63–67. https://doi.org/10.1142/S0960313197000075.

Rivadeneyra, A.; Fernández-Salmerón, J.; Agudo-Acemel, M.; López-Villanueva, J. A.; Capitan-Vallvey, L. F.; Palma, A. J. Printed Electrodes Structures as Capacitive Humidity Sensors: A Comparison. *Sensors Actuators A Phys.*, 2016, 244, 56–65. https://doi.org/10.1016/j.sna.2016.03.023.

Schift, H.; Kristensen, A. Nanoimprint Lithography. In Wanf, M. (eds) *Springer Handbook of Nanotechnology*; InTech, New York, 2017; pp. 113–142. https://doi.org/10.1007/978-3-662-54357-3_5.

Shin, K.-H.; Nguyen, H. A. D.; Park, J.; Shin, D.; Lee, D. Roll-to-Roll Gravure Printing of Thick-Film Silver Electrode Micropatterns for Flexible Printed Circuit Board. *J. Coatings Technol. Res.*, 2017, 14 (1), 95–106. https://doi.org/10.1007/s11998-016-9844-y.

Singh, M.; Haverinen, H. M.; Dhagat, P.; Jabbour, G. E. Inkjet Printing-Process and Its Applications. *Adv. Mater.*, 2010, 22 (6), 673–685. https://doi.org/10.1002/adma.200901141.

Singh, R.; Singh, E.; Nalwa, H. S. Inkjet Printed Nanomaterial Based Flexible Radio Frequency Identification (RFID) Tag Sensors for the Internet of Nano Things. *RSC Adv.*, 2017, 7 (77), 48597–48630. https://doi.org/10.1039/C7RA07191D.

Sneck, A.; Mäkelä, T.; Alastalo, A. Reverse-Offset for Roll-to-Roll High-Resolution Printing. *Flex. Print. Electron.*, 2018, 3 (1), 014001. https://doi.org/10.1088/2058-8585/aa9f00.

Sohn, K.-J.; Park, J. H.; Lee, D.-E.; Jang, H.-I.; Lee, W. Il. Effects of the Process Temperature and Rolling Speed on the Thermal Roll-to-Roll Imprint Lithography of Flexible Polycarbonate Film. *J. Micromech. Microengin.*, 2013, 23 (3), 035024. https://doi.org/10.1088/0960-1317/23/3/035024.

Southee, D. J.; Hay, G. I.; Evans, P. S.; Harrison, D. J. Flexible Dot-Matrix Display Manufacture by Offset Lithography. *Proc. Inst. Mech. Eng. Part B J. Eng. Manuf.*, 2008, 222 (8), 943–948. https://doi.org/10.1243/09544054JEM939.

Southee, D.; Hay, G. I.; Evans, P. S. A.; Harrison, D. J. Lithographically Printed Voltaic Cells – a Feasibility Study. *Circuit World*, 2007, 33 (1), 31–35. https://doi.org/10.1108/03056120710723698.

Sun, Y.; Zhang, Y.; Liang, Q.; Zhang, Y.; Chi, H.; Shi, Y.; Fang, D. Solvent Inkjet Printing Process for the Fabrication of Polymer Solar Cells. *RSC Adv.*, 2013, 3 (30), 11925. https://doi.org/10.1039/c3ra22659j.

Taki, K.; Ito, H. Roll-to-Roll UV Nanoimprinting Process. *Seikei-Kakou*, 2018, 30 (12), 616–619. https://doi.org/10.4325/seikeikakou.30.616.

Tan, H. Roller Nanoimprint Lithography. *J. Vac. Sci. Technol. B Microelectron. Nanom. Struct.*, 1998, 16 (6), 3926. https://doi.org/10.1116/1.590438.

Taniguchi, J.; Koga, K.; Kogo, Y.; Miyamoto, I. Rapid and Three-Dimensional Nanoimprint Template Fabrication Technology Using Focused Ion Beam Lithography. *Microelectron. Eng.*, 2006, 83 (4–9), 940–943. https://doi.org/10.1016/j.mee.2006.01.101.

Teichler, A.; Perelaer, J.; Schubert, U. S. Inkjet Printing of Organic Electronics – Comparison of Deposition Techniques and State-of-the-Art Developments. *J. Mater. Chem. C*, 2013, 1 (10), 1910. https://doi.org/10.1039/c2tc00255h.

Unno, N.; Taniguchi, J. Fabrication of the Metal Nano Pattern on Plastic Substrate Using Roll Nanoimprint. *Microelectron. Eng.*, 2011, 88 (8), 2149–2153. https://doi.org/10.1016/j.mee.2011.02.006.

Wei, Y.; Torah, R.; Li, Y.; Tudor, J. Dispenser Printed Capacitive Proximity Sensor on Fabric for Applications in the Creative Industries. *Sensors Actuators A Phys.*, 2016, 247, 239–246. https://doi.org/10.1016/j.sna.2016.06.005.

Wilkinson, N. J.; Smith, M. A. A. A.; Kay, R. W.; Harris, R. A. A Review of Aerosol Jet Printing—a Non-Traditional Hybrid Process for Micro-Manufacturing. *Int. J. Adv. Manuf. Technol.*, 2019, 105 (11), 4599–4619. https://doi.org/10.1007/s00170-019-03438-2.

Xia, Y.; Whitesides, G. M. Soft Lithography. *Angew. Chemie Int. Ed.*, 1998, 37 (5), 550–575. https://doi.org/10.1002/(SICI)1521-3773(19980316)37:5<550::AID-ANIE550>3.3.CO;2–7.

Yamamoto, T.; Numakura, D. Advanced Fine-Line Thick-Film Conductors with High Conductivity and Soldering Capability Built by Screen-Printing. *Trans. Japan Inst. Electron. Packag.*, 2009, 2 (1), 40–45. https://doi.org/10.5104/jiepeng.2.40.

Yin, G.; Wang, W.; Niu, H.; Ding, Y.; Zhang, D.; Zhang, J.; Liu, G.; Wang, S.; Zhang, H. Jasmonate-Sensitivity-Assisted Screening and Characterization of Nicotine Synthetic Mutants from Activation-Tagged Population of Tobacco (Nicotiana Tabacum L.). *Front. Plant Sci.*, 2017, 8, 237452. https://doi.org/10.3389/fpls.2017.00157.

Zapka, W. *Handbook of Industrial Inkjet Printing*. Wiley-VCH Verlag GmbH & Co. KGaA: Weinheim, Germany, 2017; Vol. 1–2. https://doi.org/10.1002/9783527687169.

Zeng, P.; Tian, B.; Tian, Q.; Yao, W.; Li, M.; Wang, H.; Feng, Y.; Liu, L.; Wu, W. Screen-Printed, Low-Cost, and Patterned Flexible Heater Based on Ag Fractal Dendrites for Human Wearable Application. *Adv. Mater. Technol.*, 2019, 4 (3), 1800453. https://doi.org/10.1002/admt.201800453.

3 Analytical Methods for Electrochemical Sensors

Ankush Kumar Singh, Pritesh Keshari, and Rosy

3.1 INTRODUCTION

The design and development of electrochemical sensors have been the most rapidly growing front of analytical chemistry. Out of all the sensing techniques ranging from spectroscopic, chromatographic, or chemical tools, electrochemical analysis has timely proven its utility in both qualitative and quantitative analyses of food, water, or medical diagnostic and forensic domains. Modern electrochemical analyses, with their compendium of selective and sensitive techniques, have revolutionized the field of analytical chemistry. Due to diversity in the combinations of parameters, excitation signals, and the working electrodes, a vast array of techniques, including polarography [1–3], voltammetry [4–6], chronoamperometry [7], potentiometry, are known.

Based on the electroanalytical method used and the electrochemical property measured, the electrochemical sensors have been classified into three main categories: (1) conductometric, (2) potentiometric, and (3) amperometric/voltammetric sensors. This chapter precisely aims to provide an overview of each category of an electrochemical sensor, the basic fundamental principle involved, their applications, and shortcomings. The last section provides a practical guide to determine the details of the electrochemical reaction of interest from the experimental studies and data obtained using voltammetry.

This chapter will be instrumental in giving an in-depth understanding of the analytical methods involved in electrochemical sensing. Furthermore, the content of this chapter will help the reader to design the electrochemical experiments as per the needs of the problem statement.

3.2 ELECTRO-ANALYTICAL TECHNIQUES: BRIEF INTRODUCTION

Broadly, the electroanalytical techniques have been classified based on the electrochemical changes that arise as a response to the external stimuli related to the nature or quantity of the analyte. Depending on whether the observed change is resistance, potential, or current as a function of changing concentration of the analyte, the method is known as conductometric, potentiometric, or amperometric/voltammetric, respectively.

These techniques can further be classified into interfacial or non-interfacial depending on whether the response signal is directly or indirectly related to the presence of the analyte on the electrode surface (sensory unit) or away from the electrode surface.

The methods which lead to a response signal without involving any electrode polarization and are independent of electron exchange fall under the non-interfacial category, e.g. conductometry. On the other hand, in potentiometry or voltammetry, the potential or current arises because of the changes in the analyte concentration at the surface of the electrode. Thus, both techniques fall in the interfacial category. Figure 3.1 illustrates the classification of electroanalytical methods. The next few sections will summarize the major electroanalytical methods and their applications in sensing. However, majorly, the sensing techniques use interfacial techniques. Thus, special emphasis has been given to voltammetric techniques.

DOI: 10.1201/9781032654874-3

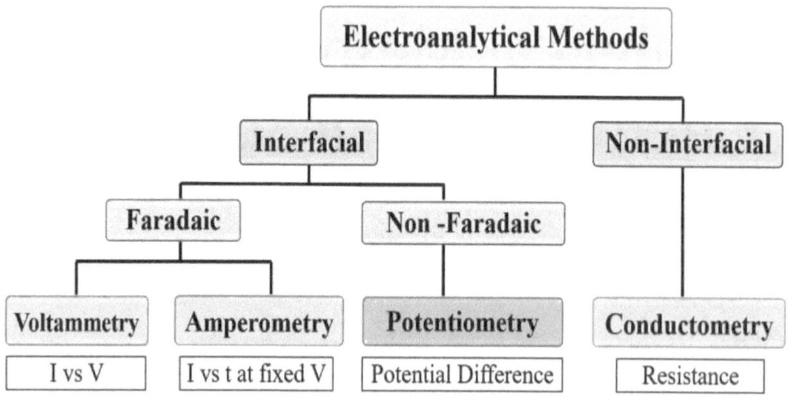

FIGURE 3.1 Classification of electroanalytical methods.

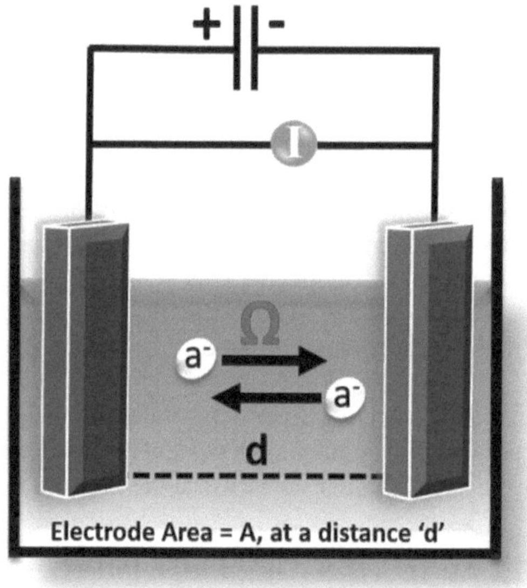

FIGURE 3.2 A typical diagram of two electrode system used in conductometry.

3.2.1 Conductometry

Conductometry is an electroanalytical technique in which changes in the electrical conductivity, or the resistivity, of the sample lying between two electrodes are monitored (Figure 3.2). Since the analysis relies on the changes in the bulk medium, this technique falls in the category of non-interfacial electroanalytical methods. Based on the measurement of altered conductivity/resistivity concerning the presence or absence of an analyte, the changes in the experimental environment are analysed without undergoing any faradaic process. Using the measured conductivity, in a two-electrode system, the molar concentration of the ion can be calculated using Kohlrausch and Ohm's law.

Primarily, conductometric titrations were used for determining the ionic species and monitoring the progress of the reaction targeting the mobility of a specific ion, which is expected to be replaced by less mobile ions during the course of the reaction, leading to a decrease in conductivity. Theoretically, conductometry is a non-selective, non-interfacial electroanalytical technique and thus is severely challenged by the presence of interfering species. Nevertheless, attributed to the dependence of conductivity on a number of factors such as the type of ions, the concentration

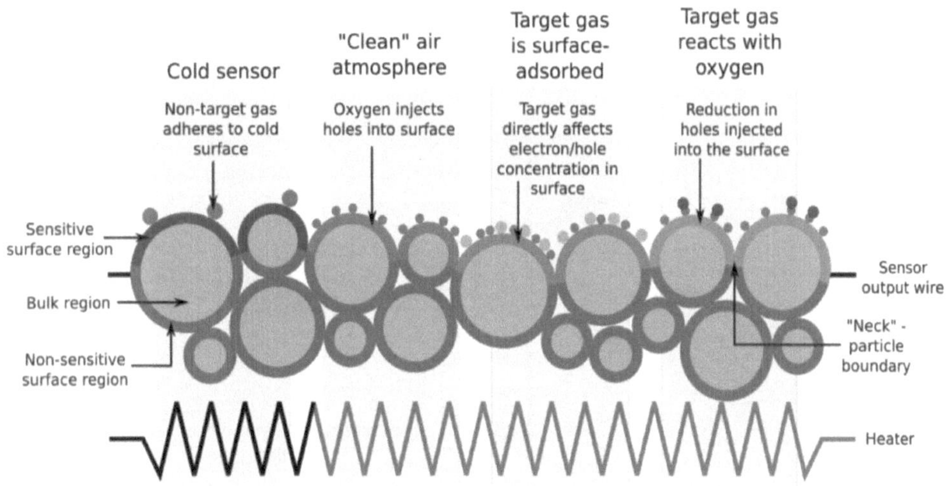

FIGURE 3.3 Diagram of the various types of interaction between atmospheric gases and a MOS sensor surface. In the leftmost region, the sensor is unpowered (and exhibits the base resistance). The three other regions of the diagram describe different processes that actually occur simultaneously to varying degrees. The sensor's output is the resistance across the whole of the sensor material, which forms a resistor network with contributions from both the bulk and surface regions (although the non-sensitive surface will have similar properties to the bulk). This model of the sensor material also explains the wide variation in base resistance between individual sensors of the same type, as the random nature of the surface geometry means an equally random network of resistances.

Source: Reproduced with permission from Ref. [8].

of ions, temperature, and mobility of ions, the experiments can be carefully planned to impart the specificity for the desirable outcome.

Lately, with the advancement of instrumentation, the changes in the conductivity of the solid substrates could be precisely measured, which led to the development of resistive sensors. Most of the gas and humidity sensors lie in this category, which relied on the changes in the conductivity of the surface-modified solid substrate when exposed to the analyte. Due to the chemisorption of specific gas, changes in the band structure are induced, leading to altered conductivity. Majorly, a substrate is coated with a thin layer of n-type metal oxide. On exposure to the air, gas gets chemisorbed on the metal oxide surface, leading to surface acceptor or donor bands and thus ultimately changing the conductivity of the substrate. "The presence of reducing gases increases the conductivity of the material, while the opposite is observed for oxidizing gases." Metal oxide gas sensors (Figure 3.3) and alcohol sensors are one of the most widely used examples of conductometric or resistive sensors. However, these sensors have found their way to commercialization ascribed to easy fabrication and low cost. Nevertheless, their durability is a matter of concern that usually is overlooked due to cheap replacement possibilities. Additionally, due to a lack of selectivity, these chemo-resistive sensors failed to penetrate the point-of-care sensor regime. Consequently, other electroanalytical techniques with appreciable selectivity were looked for such applications.

3.2.2 Potentiometry and Nernst Equation

Potentiometry is another electroanalytical technique that monitors the potential difference between two electrodes arising because of the selective arrangement of analyte ions near the working electrode under the conditions of no current flow [9,10]. Thus, in contrast to conductometry, potentiometry is an interfacial method. Potentiometry is non-faradaic in nature and responds to the existence

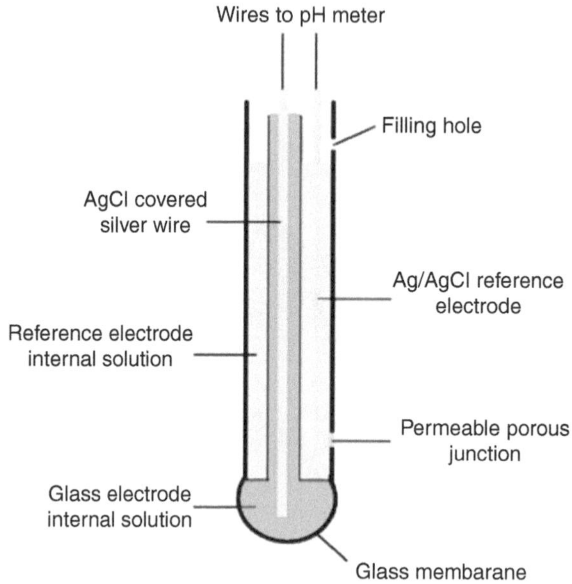

FIGURE 3.4 Typical glass electrode for pH measurements.

Source: Reproduced with permission from Ref. [15].

of the desired species on the electrode surface without relying on the redox reaction at the interface between two materials. The heart of potentiometry is the ion-selective electrodes (ISEs) [11–14]. The selectivity, accuracy, and sensitivity of the potentiometric method rely primarily on the nature of ISEs. Therefore, a lot of progress in designing various types of ISE is going on to bring potentiometric sensors closer to robust, calibration-free, and miniature devices. In addition to conventional glass membranes, solid-state, surface-engineered polymer membranes, liquid membranes, and compounds or enzyme-based ISEs have been developed in the recent past. One of the most primitive yet most commercialized examples of a potentiometric sensor is a pH meter that analyses the activity of the H^+ ions in the solution using glass membrane ISE as shown in Figure 3.4. By comparing the H^+ ion activity on the two sides of the glass surface, where the concentration of the inner solution is fixed, a pH meter normally evaluates the activity of the H^+ ion in the sample solution. The responses are calculated using the potential-concentration relationship given by the Nernst equation to translate the potential difference (V) of the solution on each side of the glass surface to a pH scale:

$$E = E_0 - \frac{RT}{nF}\ln\left(\frac{[\mathrm{OXD}]}{[\mathrm{RED}]}\right)$$

Here, E represents the cell potential (V); E_0 represents the standard cell potential (V) of a half-reaction; R is the universal gas constant; T is the temperature (K); n stands for the number of electrons involved in the half-reaction; F is the Faraday constant (C mol^{-1}); [RED] = activity of the reduced species; and [OXD] = activity of the oxidized species.

Based on their application, potentiometric sensor systems can be divided into two categories: traditional (medical, biology, and ecology) and contemporary (electronic tongue). Potentiometry is one of the most sensitive and selective electrochemical techniques currently available when used at low levels in constrained samples. Potentiometry has been recognized as a potent transducer for biosensing. With the aid of an enzyme electrode, one can potentiometrically detect the tiny molecules involved in an enzymatic reaction. Potentiometric sensors are desirable tools for the creation of a variety of electronic tongues that have a wide range of uses, from the identification of food to environmental monitoring and medical diagnostics.

Conductive polymers have been the most auspicious transducer for chemical sensors over the past 10 years. When coupled with particular ion-recognition sites such as complexing ligands and ionophores, conducting polymers offer a wide range of choices for the creation of all-solid-state potentiometric ion sensors. Although more than two dozen ISE-based potentiometric sensors are commercialized, the major restriction is the selection of dopant and ionophore matrix of the membrane. Interference due to surface adsorption and liquid junction potential are additional drawbacks that retard the performance of such devices due to frequent calibration requirements.

3.2.3 Amperometry and Cottrell Equation

Amperometry is one of the dynamic analysis methods for the identification and quantification of electroactive species present in the liquid and gaseous state. In this technique, the changes in the electrical current in response to a specific constant potential are measured as a function of time. Selectivity of the amperometry lies in the characteristic property of redox potentials, which is a signature of a specific analyte species [7,10,16,17]. Therefore, the application of an appropriate potential on a suitable electrode results in steady-state anodic/cathodic current due to oxidation/reduction of the electroactive species. The current so produced is directly proportional to the concentration of the analyte. Under the conditions of diffusion-controlled mass transfer, the observed current is correlated with the concentration of the analyte using the Cottrell equation [18–20] as given below:

$$I(t) = \frac{nFAC^0\sqrt{D}}{\sqrt{\pi}\sqrt{t}}$$

where n, F, A, C^0, D, and t stand for the number of electrons involved, Faraday constant, concentration of analyte at the electrode interface, diffusion coefficient, and time, respectively.

The chronoamperometry technique is quite selective in general and allows for screening the oxidation, reduction, or complete redox behaviour of one or multiple analytes with careful selection of applied potential pulses as shown in Figure 3.5. Therefore, it contributes significantly to the electrochemical biosensing of complex matrices. For example, in Figure 3.5, the application of 0.6 V leads

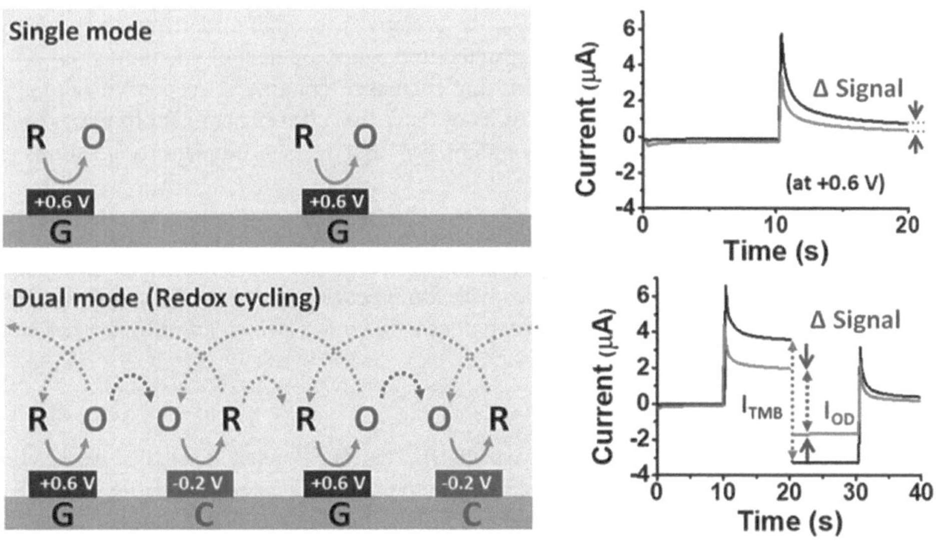

FIGURE 3.5 Chronoamperometry-based redox cycling according to the potential of generator and collector at two different potentials. The potential conditions could result in the iterative redox reaction (redox cycling) for a single molecule.

Source: Taken with permission from Ref. [21].

to a current signal due to the oxidation of R to O, whereas in the dual mode, an application of both positive (+0.6 V) and negative (−0.2 V) potentials resulted in current responses corresponding to both oxidation and reduction of analyte species. The advantages of controlled-potential techniques facilitate high sensitivity and selectivity, a wider linear concentration range, portable and cost-effective instrumentation, etc. The great sensitivity and broad linear range of an amperometric sensor make it more desirable. However, this technique can be applied effectively only if the qualitative information of the analyte is available beforehand. So, in order to choose the operating potentials, preliminary analysis details of the sample solution are necessary, and thus, this technique cannot be directly applied to assay an unknown solution. To overcome these problems, instead of measuring the current in response to the constant potential, the current is recorded while scanning the potential from one pre-set value to another, and the technique is termed voltammetry. The redox activity thus observed hints towards the presence of possible analytes in the solution and thus serves the purpose of both preliminary and advanced analysis. The invention of voltammetry has revolutionized electrochemical sensor applications, and thus, it is discussed in detail in the next section.

3.3 VOLTAMMETRY AND RANDLES–SEVCIK EQUATION

Historically, voltammetry is a more sophisticated version of another similar technique called polarography, which was invented in 1922 by Czech scientist Jaroslav Heyrovsky [2,5]. On the basis of several experiments, "Heyrovský observed that the current-voltage curves obtained with the dropping mercury electrode represented qualitative and quantitative relationships characteristic of the solution undergoing electrolysis. As the applied voltage becomes greater, the current increases not continuously but in steps, reaching limiting values corresponding to the different cations or other reducible groups in the solution. On plotting the values for voltage and current, he obtained usually S-shaped curves in which the position of the polarographic curve or wave (voltage) indicated the qualitative composition of the solution and the height of the curve (current) determined its contents quantitatively" [22]. The idea of recording current while scanning a continuous range of potentials has revolutionized the world of electrochemical sensing to such an extent that Heyrovsky's invention was recognized by a Nobel Prize in Chemistry in 1959. Since then, polarography has become a mature analytical method capable of adjusting to ever-increasing demands of sensitivity and selectivity. To simplify the electrodics and increase the sensitivity, dropping mercury electrodes were replaced by solid working electrodes, which resulted in the development of voltammetry.

Voltammetry involves measuring and analysing current originating in response to the applied range of potential. The relation between the units of these two electrochemical properties, namely volt (V) and ampere (A), inspired the term "voltammetry," and the measurement connotations of the suffix "metry" completed the name:

$$\text{Voltammetry} = \text{Volt} + \text{Am} + \text{Metry}$$

Depending on how the applied potential varies with time, i.e., depending on the nature of the excitation signal, several different voltammetric techniques are known, which are summarized below.

3.3.1 Linear Sweep Voltammetry (LSV)

The linear sweep voltammetry involves measuring the current flowing through a working electrode when the potential is swiped linearly [23–25] in the defined range as shown in Figure 3.6. In response, the current increases gradually corresponding to the potential at which the species begins to undergo a redox process. Initially, a small, nearly constant current known as capacitive current arises because of the surface adsorption of ions near the electrode surface. As the potential is swiped further, a slight increase in the capacitive current is observed till the point when the applied potential reaches close to the standard reduction potential of the analyte species. At such potentials,

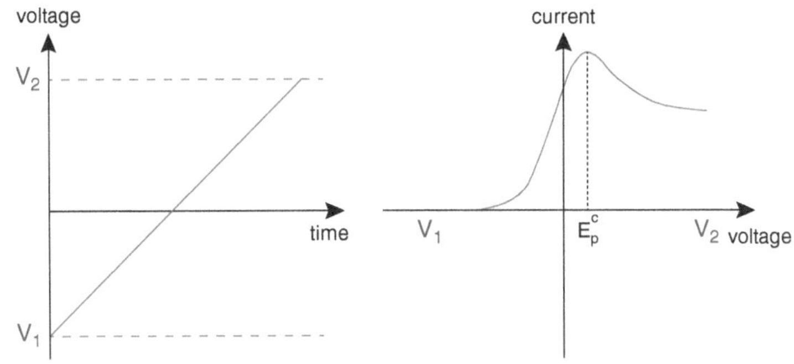

FIGURE 3.6 Typical excitation signal and observed response of linear sweep voltammetry.

a gradual increase in current is witnessed, ascribed to the charge transfer in a faradaic process reaching maxima. The peak so observed thus hints towards the presence of a specific analyte as per the value of redox potential observed. A more advanced version of LSV is cyclic voltammetry (CV) where the potential wave of both positive and negative magnitude is applied in one cycle.

3.3.2 Cyclic Voltammetry (CV)

CV [17,25–31] has emerged as an important, versatile, and widely used electroanalytical technique. Even though it does not hold much significance in quantitative analysis, its capability to provide illustrious information about the nature, thermodynamics, and kinetics of the redox behaviour of a species over a wide potential window marks its effective and extensive use not only in physical, organic and inorganic chemistry but also in many diverse fields like most areas of analytical chemistry, biochemistry, etc. On closely surveying the literature, one can observe a few reports showing the applications of CV in the field of sensing in the late 1980s. However, several studies demonstrating the use of CV for investigating the mechanistic aspects of CO_2 reduction [32], dopamine measurements [33,34], and other such applications [35–37] have been reported in the early 1990s. CV has benchmarked its utility in qualitative analysis, and thus, even to date every electrochemical investigation begins with the preliminary CV studies either for interrogating the redox chemistry or for calculating any physical quantity like diffusion coefficient, number of electrons involved. Thus, the fundamentals of the technique used for the present investigation have been listed below.

For a CV experiment, the typical excitation signal is a linear potential ramp from an initial potential "E_i" to a final potential "E_f" and then switches back to the initial potential, thereby making a triangular wave [4,26,28,38] as shown in Figure 3.7. The current response so obtained when plotted as a function of applied potential gives the cyclic voltammogram from which a number of information can be acquired. The important characteristics of the voltammograms are the peak potentials and

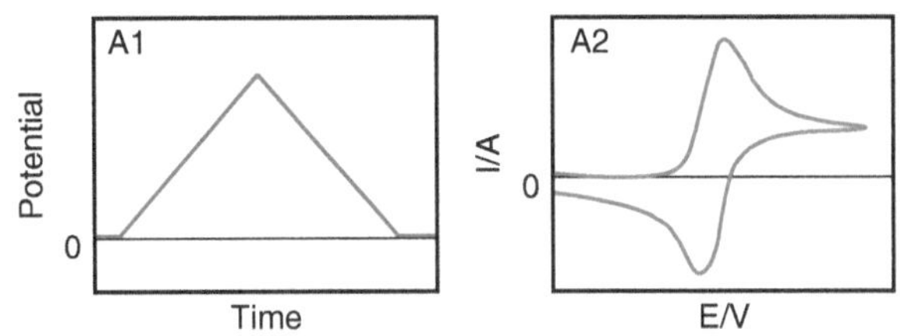

FIGURE 3.7 Typical excitation signal in triangular waveform and the resulting cyclic voltammogram for a reversible redox process.

Source: Taken from Ref. [39].

peak currents. A redox couple that exchanges electrons rapidly with the working electrode is said to exhibit a reversible redox process. For such reversible systems, the number of electrons involved in a chemical reaction can be calculated using the difference in the anodic (E_{pa}) and cathodic potentials (E_{pc}) as per the following expression:

$$\Delta E_p / V = [E_{pa} - E_{pc}] = 0.059/n$$

As a result, for a reversible process involving one electron exchange, the difference between the two peaks lies near to 0.059 V, whereas for an irreversible system, the peak separation is usually larger, which may be attributed to the sluggish electron transfer.

Similarly, the peaks obtained in CV are attributed to the reduction and oxidation of the analyte species, and for a reversible reaction, the current varies as a function of its concentration that can be predicted by the Randles–Sevcik equation [26,38]:

$$i_p = 0.4463 n^{3/2} F^{3/2} A C^0 (v D_0 / RT)^{1/2}$$

where n is the number of electrons transferred in the redox event, F is the Faraday's constant, A is the electrode surface area (cm^2), υ is the rate at which potential is being varied (V s^{-1}), D_0 is the analyte diffusion coefficient (cm^{2}s^{-1}), and C^0 is the analyte bulk concentration (mol cm^{-3}).

Analytical chemistry has witnessed extensive use of CV for the analysis of redox systems, but its merits are of a realm only for the qualitative aspects; moreover, it also suffers from certain drawbacks like a much pronounced charging current, abrupt results at higher scan rates, the effect of slow electron transfer on the redox process. Thus, in order to mark the niche of electrochemical analysis, more sensitive pulse techniques are used, which give the best quantitative analysis with a high level of sensitivity.

3.3.3 Pulse Voltammetry

Over the course of time, several different kinds of potential modulation, including more than just a plain staircase ramp, have been attempted in an effort to increase both speed and sensitivity of the process. The ability of the pulse technique to discriminate between the charging and faradaic current is the fundamental reason for its enhanced sensitivity and thus is given an advantage over the other voltammetric techniques. Majorly, two pulse methods, namely differential pulse and square-wave voltammetry, have the most widespread application and thus are discussed one by one in the subsequent sections.

3.3.3.1 Differential Pulse Voltammetry

Differential pulse voltammetry (DPV) [4–6,17,25,40] is an advanced technique that allows selectively to screen the current arising solely from the faradaic process. The excitation signal of DPV consists of a pulse of relatively low amplitude (between 10 and 100 mV), superimposed over a base potential that increases steadily in small increments. Two readings of the current are taken for each pulse: the first reading is taken just before the pulse is applied, which provides the contribution from non-faradaic background current, and the second reading is taken near the end of the pulse before the potential starts decreasing, which provides the maximum faradaic current. The resulting voltammogram is obtained by recording the difference between the currents at two points vs. the base potential [41,42] (Figure 3.8).

3.3.3.2 Square Wave Voltammetry

Due to improvisation in digital and analogue electronics, square wave voltammetry (SWV) [4–6,40,43–46] has found its practical utility in routine quantitative analyses. The past literature has revealed several reports where SWV resulted in highly sensitive [47] and specific quantification

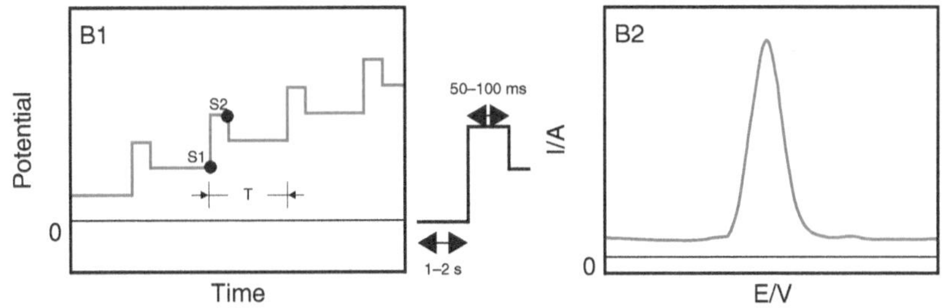

FIGURE 3.8 Typical excitation signal and observed response of differential pulse voltammetry.

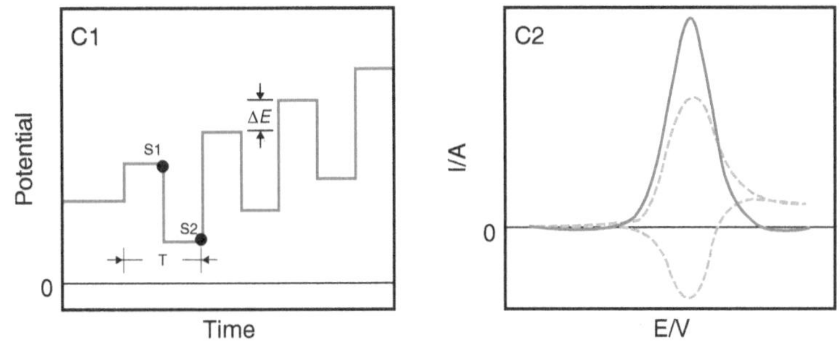

FIGURE 3.9 Typical excitation signal and observed response of square wave voltammetry.

Source: Taken from Ref. [39].

of molecules up to the femtomolar range. For an SWV experiment, a symmetrical square wave pulse of amplitude "E_{SW}" is superimposed on the staircase wave in such a way that the forward pulse of the square wave coincides with the step of the staircase signal as shown in Figure 3.9.

In SWV, the current is sampled at two stages of the square wave pulse, and the difference between the forward and reverse current is taken as the net current, which is directly proportional to the analyte concentration.

> The faradaic currents that flow during the positive and negative potential steps are opposite in polarity while the polarity of the charging current, which is mostly due to the ramp voltage, does not change. Thus, the component of the output current due to the faradaic signal is amplified while the component due to the charging current is reduced [48].

Due to the aforementioned sampling procedure, the response current of the excitation signal has a negligible contribution from the charging current, and thus, the peak current is mainly because of faradaic processes. As a result, much enhanced sensitivity and a lower limit of detection (LOD) can be realized using SWV. Consequently, SWV is the most widely used voltammetric technique at present and has been deployed in almost every field where electrochemistry has some practical aspect.

3.4 ELECTROANALYTICAL METHODS FOR SENSING

This section aims to provide a practical guide for planning the experiments and analysing the observed data to extract fundamental details about the targeted electrochemical process. For a proper analysis, it is important to first record the data correctly, which can be ensured by keeping the following points in mind.

1. **The working electrode is clean and doesn't interfere in the chosen potential range.** Individual working electrodes have their own cleaning procedures; for example, a glassy carbon electrode is polished using alumina slurry on a microcleaning cloth pad [26], a graphite electrode is scratched on an emery paper, and indium tin oxide (ITO)-coated glass is washed with soap followed by sonication in isopropyl alcohol. As the working electrode hosts the redox process and is responsible for the charge transfer, therefore, a strict cleaning procedure should be followed to ensure a clean electrode/electrolyte interface free from any dust, impurity, or pre-deposited contaminant.
2. **The mass transfer is solely controlled by the diffusion of the analyte.** The Randles–Sevcik equation, which relates the peak current with the concentration of the analyte, is applicable only under conditions where the mass transfer is diffusion controlled. As per theory, the mass transfer can take place in three ways: migration (flow of ions under electrostatic pull), convection (movement of ions ascribed to mechanical disturbances), and diffusion (movement of ions under the influence of concentration gradient). Since, out of the three possible ways, only the diffusion process is controlled by the concentration of analyte, thus, it is the desired process. Under the conditions where migration or convection interferes with the mass transfer process, a portion of current arises because of the electrostatic pull or mechanical disturbances, and thus, inaccurate current-to-concentration relations will be devised, leading to faulty analysis. So, to ensure only diffusion-controlled mass transfer, an excess of supporting electrolyte (KCl, phosphate buffer, etc.) is used, which balances the electrostatic interactions by preferably migrating towards the electrode. As a result, the main analyte ion doesn't experience any electrostatic pull and thus moves to the electrode under the influence of the concentration gradient.
3. **The solution matrix is conducting and doesn't offer any resistance to flow of ions.** For a proper analysis, the free movement of ions should be certified by taking conducting solutions as supporting electrolyte. If a poorly conducting matrix is taken, the mass transfer will be impeded, leading to IR drop and inaccurate analysis.
4. **The matrix is stable in the applied working potential window** Before performing the main electrochemical studies, the blank measurement, i.e. the I-V curve in the solution having a composition closest to the sample matrix without analyte, should be recorded. The blank measurement is crucial to confirm that the experimental setup including the working electrode, supporting electrolytes, sample matrix doesn't undergo any side reaction in the working potential window that can lead to possible interference in the primary analysis. If some variable current responses are observed, the potential window, scan rate, or sensitivity can be tuned.
5. **Reproducible blank measurements are recorded before the primary studies.** In addition to recording the blank voltammogram as stated in the previous point, it is important to have reproducible and nearly overlapping blank voltammograms to ensure that the subsequent application of potential does not lead to any adsorption complications and surface fouling of electrodes. In addition, the minimum standard deviation of blank measurements is important to have a sensitive analysis, so the experimental conditions should be carefully chosen to have overlapping blank measurements without involving any background side processes.

Once the experimental conditions and parameters are decided, the measurements can be recorded to find out desirable details. A few of these are discussed step by step in the following paragraphs.

3.4.1 How to Calculate Electrochemical Surface Area?

Electrochemical surface area is very different from the geometrical surface area and has a huge role to play during sensing. Since for a diffusion-controlled process, the peak current is directly proportional to the electrode surface area, several methods have targeted tuning it by using surface modifications for obtaining larger currents and thus higher sensitivity. So, calculating electrochemical surface area is an effective way to compare the modified and unmodified sensors and also to know if an increased surface area is the primary reason for increased sensitivity or if some specific chemical affinities are also important.

As per the Randles–Sevcik equation,

$$i_p = 0.4463n^{3/2}F^{3/2}C\left(\frac{vD}{RT}\right)^{1/2} A$$

where i_p is the current (Amperes), n is the number of electrons transferred in the redox event, F is the Faraday's constant (C mol^{-1}), A is the electrode surface area (cm^2), υ is the rate at which potential is being varied (V s^{-1}), D_0 is the analyte diffusion coefficient ($cm^2 s^{-1}$), and C^0 is the analyte bulk concentration (mol cm^{-3}).

By putting the values of R, F, and T (generally at 25°C), the equation reduces to

$$i_p = 2.29\times10^5 n^{3/2} ACD^{1/2} v^{1/2}$$

The constant with a value of 2.69×10^5 has units of C $mol^{-1} V^{-1/2}$.

In order to calculate the electrochemical surface area of the working electrode, current at different scan rates is recorded in a solution of a fixed concentration of any standard redox couple like $K_3[Fe(CN)_6]$ whose diffusion coefficient is well documented [49–51]. By plotting the peak current (i_p) vs. $\upsilon^{1/2}$, the surface area can be obtained from the slope as related below:

$$\text{Slope} = 2.29\times10^5 n^{3/2} ACD^{1/2}$$

The only unknown in the slope is "A" as the D for $K_3[Fe(CN)_6]$ is known to be $7.4\times10^{-6} cm^2 s^{-1}$, the number of electrons involved (n) is 1, and the measurements are carried out at constant concentration.

As an example, the calculation of the electrochemical surface area of ITO is shown below. To calculate the electroactive surface area of the ITO, cyclic voltammograms were recorded at different scan rates ranging from 10 to 150 mV s^{-1} in an equimolar combination of 1 mM $K_3[Fe(CN)_6]$ solution and 1 M KCl using ITO as the working electrode. The voltammograms observed are shown in Figure 3.10. The peak current was found to increase with the increasing scan rate. The variation

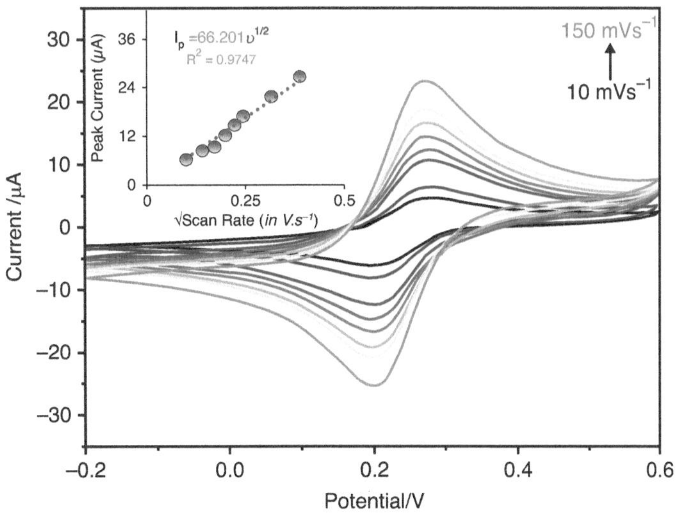

FIGURE 3.10 Cyclic voltammograms observed for 0.5 mM solution of $K_3[Fe(CN_6)]$ with varying scan rates from 10 to 150 mV s^{-1} recorded using ITO. Insets represent the corresponding I_p vs. $\upsilon^{1/2}$ linear plots.

of peak current corresponding to the oxidation peak (or reduction peak) as a function of $\upsilon^{1/2}$ was then plotted and is shown in the inset.

Using the Randles–Sevcik equation and the slope of the observed linear regression equation, the following relation can be used for calculating the electrode surface area "*A*":

$$\text{Slope} = 2.29\times10^5(\text{A.s.mol}^{-1}\text{V}^{-1/2})A(0.5\text{ mM})(7.4\times10^{-6}\text{cm}^2\text{s}^{-1})^{1/2}$$

As seen from the linear plot, the slope is 66.201 ($\mu A^{-1} V^{1/2} s^{-1/2}$). By substituting the value of the slope and taking care of the units, we get

$$66.201\times10^{-6}(A/V^{1/2}\,\text{s}^{-1/2}) = 2.69\times10^5(\text{A s mol}^{-1}\text{V}^{-1/2})A(0.5\times10^{-6}\text{mol cm}^{-3})$$

$$(7.4\times10^{-6}\text{cm}^2\,\text{s}^{-1})^{1/2}$$

Now, on rearranging and solving the equation, we get

$$A = 66.201\times10^{-6}/0.000366$$

$$= 0.181\text{ cm}^2$$

So, the electroactive surface area of the ITO electrode was found to be 0.181 cm^2.

3.4.2 How to Determine the Reversibility and Irreversibility of a Redox Process?

CV offers an easy approach to calculate the $E°$ for a reversible electrochemical reaction, as indicated in Figure 3.11, and corresponds to the potential halfway between the two observed peaks (*C* and *F*).

The diffusion of the analyte to and from the electrode separates the two peaks. Peak-to-peak separation (E_p), which measures the difference between the anodic and cathodic peak potentials, is

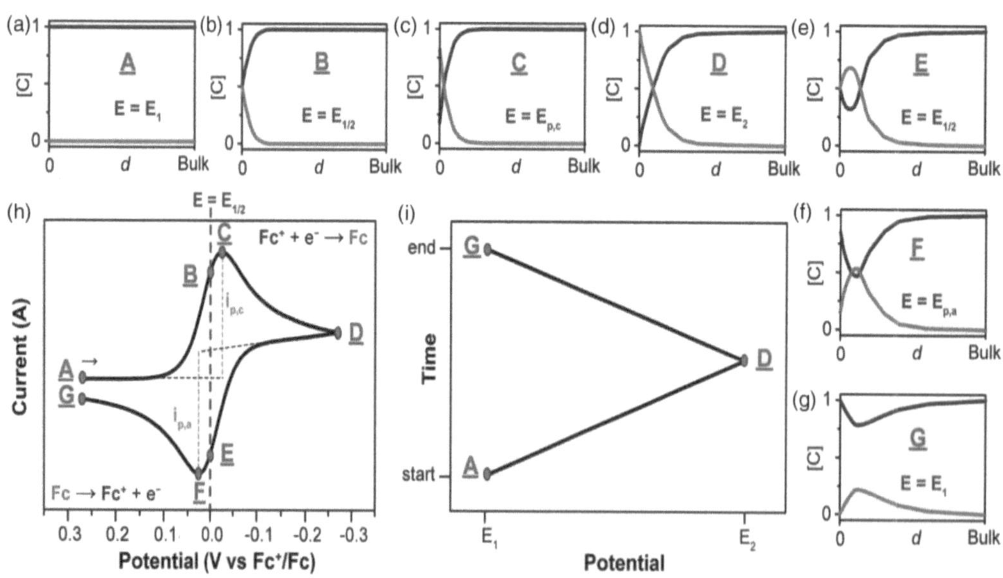

FIGURE 3.11 (a–g) Concentration profiles (mM) for Fc^+ (blue) and Fc (green) at various points during the voltammogram (where Fc^+ represents the ferrocenium ion, while Fc represents the ferrocene).

Source: Taken with permission from Ref. [26].

57 mV at 25°C (2.22 *RT/F*) and the width at half-maximum on the forward scan of the peak is 59 mV if the reduction process is chemically and electrochemically reversible [26]. Whether the analyte is stable upon reduction and can afterwards be reoxidized is indicated by chemical reversibility. The kinetics of electron transport between the electrode and the analyte are referred to as electrochemical reversibility. When the barrier to electron transport is minimal (there is electrochemical reversibility), the Nernstian equilibrium is reached right away when the applied voltage changes. Therefore, the term "Nernstian" is frequently used to describe electrochemically reversible processes with rapid electron transfers that obey the Nernst equation.

On the contrary, the process is referred to as "electrochemically irreversible" when electron transfer at the electrode surface is slow in comparison with mass transport. Due to impeded charge transfer, additional work is needed to carry out the subsequent reduction (oxidation) reactions. In electrochemistry, work is related to the applied potentials, and thus, to do the extra work, more negative (positive) potentials are applied to carry out such processes, resulting in larger ΔE_p. Therefore, peak-to-peak separation greater than 57 mV predicts a quasi-reversible process. The absence of a cathodic/anodic peak, on the other hand, indicates an electrochemically irreversible reaction.

3.4.3 How to Evaluate the Nature of the Electrochemical Process?

Is the analyte diffusing freely or adsorbed on the electrode? Is the charge transfer dominated by diffusion-controlled pathways or do adsorption complications co-exist? These questions could be understood by observing the nature of the electrochemical process. From the basic understanding of interfacial processes, it can be understood that the scan rate, i.e. the rate at which potential is being applied to the electrode or electrons are being supplied (or taken) from the electrode, will have a significant impact on the thickness of the diffusion layer formed at the electrode/electrolyte interface. Therefore, the rate at which the potential is applied and the resulting current can be used to understand the nature of the interfacial charge transfer process.

Under the condition where the analyte is solely and reversibly diffusing to the electrode surface from the bulk or vice versa, the peak current (i_p) is supposed to vary linearly with $\upsilon^{1/2}$ obeying the Randles–Sevcik equation [26].

However, any deviation from the linearity signifies that either the process is quasi-reversible or charge transfer is having some adsorption complications on the solid electrode surface. Under such a situation, mass transfer is usually faster than charge transfer. It was observed that an electrode-adsorbed species follows the linearity with scan rate (υ) in accordance with the following expression:

$$i_p = \left[\frac{(nF)^2}{4RT}\right] vA\Gamma^*$$

where r^* represents the surface coverage of the adsorbed species (mol cm^{-2}) and υ denotes the scan rate.

Further analysis of the peak-to-peak separation in the voltammogram helps us to distinguish between quasi-reversible process and adsorbed electrode. If the peak-to-peak separation shifts with changes in the scan rate, the process is considered as quasi-reversible, whereas for a reversible diffusion-controlled process, E_p remains constant with changing scan rates.

For example, the nature of the electrochemical oxidation of dopamine on the ITO surface was evaluated by recording the square wave voltammograms for 150 μM DA in pH-7.4 buffer at different frequencies ranging from 5 to 50 Hz. The observed variation of peak current in response to the frequency was found to be related to the following regression equations (Figure 3.12):

$$I_p(\mu A) = 0.1010f\left[2-50\text{ Hz}\right] + 4.695\ R^2 = 0.9656$$

$$I_p(\mu A) = 0.880f^{1/2}\left[2-50\text{ Hz}\right] + 3.102\ R^2 = 0.9828$$

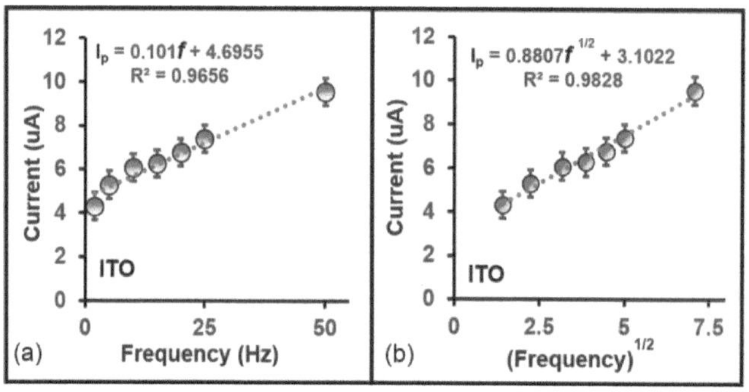

FIGURE 3.12 The linear relation between (a) I_p Vs (f) and (b) I_p Vs $(f)^{½}$ recorded in 150 μM DA in pH-7.4 buffer with increasing frequencies from 5 to 50 Hz using bare ITO.

A more linear variation of I_p vs. $(f)^{1/2}$ compared to I_p vs. f was observed from the R^2 values. The linearity of I_p vs. $(f)^{1/2}$ indicates the involvement of diffusion-controlled mass transfer obeying the Randles–Sevcik equation.

3.4.4 How to Analyse the Number of Electron Transfer Involved in a Redox Process?

The separation between the peak potentials can be used to calculate the number of electrons transferred (n) in the electrode reaction for a reversible pair:

$$\Delta E_p = |E_{\text{pa}} - E_{\text{pc}}| = 2.303\ RT/nF$$

A more accurate way to calculate the number of electrons involved in the diffusion-controlled, the reversible electrochemical process is to record the variation of peak current with respect to changing pH. As per the Nernst equation, the slope of E_p vs. pH will be equal to $0.059/n$, and from there, n, i.e. the number of electrons involved, can be calculated.

However, for a process that has adsorption complications, the above equations fail to predict the electrochemical behaviour. Under such conditions, Laviron's equation is used to calculate the transfer coefficient (α) and the number of electrons involved. According to Laviron's equation [52], the slope of the plot E_p vs. log v gives the value of αn:

$$E_p = E + (RT/\alpha nf)[(RTk_s/\alpha nf) - \ln v]$$

where α is the electron transfer coefficient, k_s is the standard rate constant of the surface reaction, v is the scan rate, n is the electron transfer number, and E is the formal potential. k_s and αn values can be concluded from the intercept and slope of the linear plot of E_p with respect to ln v, if the value of E is known.

3.4.5 How to Quantify an Analyte?

In order to quantify an analyte, the current is first recorded for different concentrations of the analyte using any voltammetric method. An increase in peak current should be observed with the increasing concentration of the analyte. From the recorded data, a calibration curve can be obtained by plotting the peak current against concentration, leading to a linear correlation (linear regression line, $Y(i_p) = mX\,(\text{Conc}) + C$) between peak current vs concentration of the analyte. Once the relation is devised, the voltammogram of an unknown solution is recorded to have a corresponding current

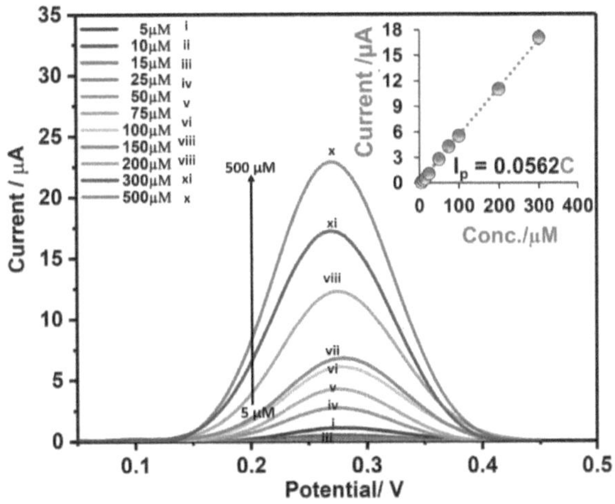

FIGURE 3.13 Concentration-dependent study: Square wave voltammogram recorded for bare ITO for blank, 5, 10, 15, 25, 50, 75, 100, 150, 200, 300, 500 µM dopamine concentration recorded using bare ITO. Insets include the corresponding calibration plots for dopamine concentration from 5 to 300 µM.

value (Y) from which the concentration (X) of the unknown solution can be calculated using the established relationship.

The slope of the calibration plot is also considered a measure of sensitivity for the measurement protocol. Consequently, the slope value is used to estimate the LOD which is calculated using the standard formula [53]:

$$\text{LOD} = 3S_D/b$$

where b is the slope of the calibration curve and S_D is the standard deviation of three or five blank current outputs.

For a better understanding, an example of the quantitative estimation of dopamine is cited below. Figure 3.13 shows the concentration study and the respective calibration plot of dopamine observed using ITO as the working electrode. In the concentration range of 5–500 uM, the linear regression equation was found to be $I_p = 0.0562\,\text{C}$ showing the slope to be 0.0562 µA µM^{-1}. The standard deviation was next calculated by recording five square-wave voltammograms in the blank solution. To do so, the current obtained (mostly charging current) in the blank voltammogram at the peak potential of dopamine was noted, and the standard deviation formula was applied. Standard deviation (S_D) was found to be 0.002055. By substituting both the values, i.e. slope and standard deviation, in the given formula, the LOD was calculated to be 109.7 nM as follows:

$$\text{LOD} = 3\left(0.002055\,\mu\text{A}\right)/0.0562\left(\mu\text{A}\,\mu\text{M}^{-1}\right)$$

$$\text{LOD} = 0.1097\,\mu\text{M} = 109.7\text{ nM}$$

3.5 CONCLUSIONS

This chapter gives a concise overview of various interfacial and non-interfacial electroanalytical techniques and their application in sensing. We've briefly discussed the application of conductometric and potentiometric techniques using metal oxide-based gas sensors and pH electrodes as real-world examples. The advantages of techniques relying on faradaic processes over those that investigate non-interfacial processes are discussed, emphasizing the superiority of amperometric and voltammetric methods for selective sensing of analytes. Most importantly, a comprehensive guide has been provided for organizing the voltammetric sensing experiments and analysing the observed data. Overall, this chapter provides an overview of the capabilities of electroanalytical methods for in-depth fundamental analysis of electrochemical processes as well as detailed qualitative and quantitative sensing applications. With all of the listed dos and don'ts, we believe that this chapter will not only give readers a better understanding of electroanalytical sensing techniques but also give them a chance to put what they have learnt into practice.

REFERENCES

1. M. Heyrovský, Polarography - Past, present, and future, *Journal of Solid State Electrochemistry* 15 (2011) 1799–1803. https://doi.org/10.1007/s10008-010-1244-0.
2. J. Barek, A.G. Fogg, A. Muck, J. Zima, Polarography and voltammetry at mercury electrodes, *Critical Reviews in Analytical Chemistry* 31 (2001) 291–309. https://doi.org/10.1080/20014091076776.
3. Y. Kumar, R. Pandey, A review of polarographic techniques, *International Journal of Creative Research Thoughts* 6 (2018) 948–964.
4. C. Batchelor-Mcauley, E. Kätelhön, E.O. Barnes, R.G. Compton, E. Laborda, A. Molina, Recent advances in voltammetry, *ChemistryOpen* 4 (2015) 224–260. https://doi.org/10.1002/open.201500042.
5. S. Mendoza, E. Bustos, J. Manríquez, L.A. Godínez, Voltammetric techniques, *Agricultural and Food Electroanalysis* 3 (2015) 21–48. https://doi.org/10.1002/9781118684030.ch2.
6. R. Gulaboski, The future of voltammetry, *Macedonian Journal of Chemistry and Chemical Engineering* 41 (2022) 224–260. https://doi.org/10.20450/mjcce.2022.2555.
7. J.R. Stetter, J. Li, Amperometric gas sensors - A review, *Chemical Reviews* 108 (2008) 352–366. https://doi.org/10.1021/cr0681039.
8. P.J.D. Peterson, A. Aujla, K.H. Grant, A.G. Brundle, M.R. Thompson, J. Vande Hey, R.J. Leigh, Practical use of metal oxide semiconductor gas sensors for measuring nitrogen dioxide and ozone in urban environments, *Sensors (Switzerland)* 17 (2017) 1653. https://doi.org/10.3390/s17071653.
9. G. Chihara, Clinical applications of infrared spectroscopy. II. Identification of pathologic concretions and other substances, *Clinical Chemistry* 6 (1961) 453–465. https://doi.org/10.5111/bunkou.9.181.
10. N.R. Stradiotto, H. Yamanaka, M.V.B. Zanoni, Electrochemical sensors: A powerful tool in analytical chemistry, *Journal of the Brazilian Chemical Society* 14 (2003) 159–173. https://doi.org/10.1590/S0103-50532003000200003.
11. G. Dimeski, T. Badrick, A.S. John, Ion selective electrodes (ISEs) and interferences: A review, *Clinica Chimica Acta* 411 (2010) 309–317. https://doi.org/10.1016/j.cca.2009.12.005.
12. M.A. Arnold, M.E. Meyerhoff, Ion-selective electrodes, *Analytical Chemistry* 56 (1984) 20–48. https://doi.org/10.1021/ac00269a003.
13. M.E. Meyerhoff, W.N. Opdycke, Ion-selective electrodes, *Advances in Clinical Chemistry* 25 (1986) 1–47. https://doi.org/10.1016/S0065-2423(08)60123-7.
14. E. Bakker, E. Pretsch, The new wave of ion-selective electrodes, *Analytical Chemistry* 74 (2002) 420A–426A.
15. G. Orellana, C. Cano-Raya, J. López-Gejo, A.R. Santos, Online monitoring sensors, in: Wilderer P. (ed.) *Treatise on Water Science*, Elsevier, Amsterdam, The Netherlands, 2011: pp. 221–261. https://doi.org/10.1016/B978-0-444-53199-5.00059-2.
16. C. Del Gratta, G.L. Romani, MEG: Principles, methods, and applications, *Biomedical Technology* 44 (1999) 11–23. https://doi.org/10.1515/bmte.1999.44.s2.11.
17. K. Kawagoe, J. Zimmerman, R.M. Wightman, Principles of voltammetry and microelectrode surface states, *Journal of Neuroscience Methods* 48 (1993) 225–240.
18. J. Baranwal, B. Barse, G. Gatto, G. Broncova, A. Kumar, Electrochemical sensors and their applications: A review, *Chemosensors* 10 (2022). https://doi.org/10.3390/chemosensors10090363.
19. T. Pajkossy, Electrochemistry at fractal surfaces, *Journal of Electroanalytical Chemistry and Interfaci al Electrochemistry* 300 (1991) 1–11. https://doi.org/10.1016/0022-0728(91)85379-4.
20. E. Kätelhön, R.G. Compton, Testing and validating electroanalytical simulations, *Analyst* 140 (2015) 2592–2598. https://doi.org/10.1039/C4AN02276A.
21. G.Y. Lee, J.H. Park, Y.W. Chang, S. Cho, M.J. Kang, J.C. Pyun, Chronoamperometry-based redox cycling for application to immunoassays, *ACS Sensors* 3 (2018) 106–112. https://doi.org/10.1021/acssensors.7b00681.

22. J. Heyrovsky, Biographical, complet. Dictionary of Scientific Biography Encyclpedia. 21 Dec. 2022 <https//Www.Encyclopedia.Com>. (n.d.). https://www.encyclopedia.com/people/history/historians-miscellaneous-biographies/jaroslav-heyrovsky (accessed December 30, 2022).
23. G. Bontempelli, N. Dossi, R. Toniolo, *Linear Sweep and Cyclic☆*, Elsevier Inc., Amsterdam, The Netherlands, 2016. https://doi.org/10.1016/b978-0-12-409547-2.12200-0.
24. J. González-Velasco, The linear sweep voltametric method: An application to the study of reversible and irreversible processes, *Electroanalysis* 6 (1994) 711–724. https://doi.org/10.1002/ELAN.1140060902.
25. C.A. Marsden, M.H. Joseph, Z.L. Kruk, N.T. Maidment, R.D. O'Neill, J.O. Schenk, J.A. Stamford, In vivo voltammetry—Present electrodes and methods, *Neuroscience* 25 (1988) 389–400. https://doi.org/10.1016/0306-4522(88)90247-3.
26. N. Elgrishi, K.J. Rountree, B.D. McCarthy, E.S. Rountree, T.T. Eisenhart, J.L. Dempsey, A practical beginner's guide to cyclic voltammetry, *Journal of Chemical Education* 95 (2018) 197–206. https://doi.org/10.1021/acs.jchemed.7b00361.
27. J.F. Rusling, S.L. Suib, Characterizing materials with cyclic voltammetry, *Advanced Materials* 6 (1994) 922–930. https://doi.org/10.1002/adma.19940061204.
28. R.G. Compton, C.E. Banks, Cyclic voltammetry at macroelectrodes, *Understanding Voltammetry* 4 (2007) 922–930. https://doi.org/10.1142/9789812779809_0004.
29. P.T. Kissinger, W.R. Heineman, Cyclic voltammetry, *Journal of Chemical Education* 60 (1983) 702–706. https://doi.org/10.1021/ED060P702.
30. G.A. Mabbott, An introduction to cyclic voltammetry, *Journal of Chemical Education* 60 (1983) 697–702. https://doi.org/10.1021/ED060P697.
31. A.B. Bocarsly, Cyclic voltammetry, *Materials Characterization* 2 (2012) 1–14. https://doi.org/10.1002/0471266965.COM050.PUB2.
32. M. Fujihira, Y. Hirata, K. Suga, Electrocatalytic reduction of CO_2 by nickel(II) cyclam: Study of the reduction mechanism on mercury by cyclic voltammetry, polarography and electrocapillarity, *Journal of Electroanalytical Chemistry and Interfacial Electrochemistry* 292 (1990) 199–215. https://doi.org/10.1016/0022-0728(90)87336-I.
33. J.B. Zimmerman, R. Mark Wightman, Simultaneous electrochemical measurements of oxygen and dopamine in vivo, *Analytical Chemistry* 63 (1991) 24–28. https://doi.org/10.1021/AC00001A005/ASSET/AC00001A005.FP.PNG_V03.
34. S. Hafizi, Z.L. Kruk, J.A. Stamford, Fast cyclic voltammetry: Measurement of dopamine in the presence of its biological precursors and metabolites, *Journal of Electroanalytical Chemistry and Interfacial El ectrochemistry* 283 (1990) 125–133. https://doi.org/10.1016/0022-0728(90)87384-V.
35. H. Yang, D.O. Wipf, A.J. Bard, Application of rapid scan cyclic voltammetry to a study of the oxidation and dimerization of N,N-dimethylaniline in acetonitrile, *Journal of Electroanalytical Chemistry* 331 (1992) 913–924. https://doi.org/10.1016/0022-0728(92)85014-T.
36. D. V. Heyd, D.A. Harrington, Platinum oxide growth kinetics for cyclic voltammetry, *Journal of Electroanalytical Chemistry* 335 (1992) 19–31. https://doi.org/10.1016/0022-0728(92)80229-W.
37. H. Yang, A.J. Bard, The application of fast scan cyclic voltammetry. Mechanistic study of the initial stage of electropolymerization of aniline in aqueous solutions, *Journal of Electroanalytical Chemistry* 339 (1992) 423–449. https://doi.org/10.1016/0022-0728(92)80466-H.
38. V. Gau, S.C. Ma, H. Wang, J. Tsukuda, J. Kibler, D.A. Haake, Electrochemical molecular analysis without nucleic acid amplification, *Methods* 37 (2005) 73–83. https://doi.org/10.1016/J.YMETH.2005.05.008.
39. J. Liu, Y. Xu, S. Liu, S. Yu, Z. Yu, S.S. Low, Application and progress of chemometrics in voltammetric biosensing, *Biosensors* 12 (2022) 494. https://doi.org/10.3390/BIOS12070494.
40. N. Voltammetry, D. Voltammetry, and S. Voltammetry, EC-Lab–Application Note# 67 Introduction to pulsed voltammetric techniques: DPV. NPV and SWV I–Introduction II–Theoretical Description EC-Lab–Application Note, 67 (2019) 1–5.
41. F. Crespi, Differential pulse voltammetry: Evolution of an in vivo methodology and new chemical entries, a short review, *Journal of New Developments in Chemistry* 2 (2020) 20. https://doi.org/10.14302/ISSN.2377-2549.JNDC-20-3298.
42. H. Göhr, Modern electrochemistry. An introduction to an interdisciplinary area, *Angewandte Chemie* 84 (1972) 72–73. https://doi.org/10.1002/ANGE.19720840217.
43. V. Mirceski, R. Gulaboski, Recent achievements in square-wave voltammetry a review, *Macedonian Journal of Chemistry and Chemical Engineering* 33 (2014) 1–12.
44. V. Mirceski, R. Gulaboski, M. Lovric, I. Bogeski, R. Kappl, M. Hoth, Square-wave voltammetry: A review on the recent progress, *Electroanalysis* 25 (2013) 2411–2422. https://doi.org/10.1002/elan.201300369.

45. J.G. Osteryoung, M.M. Schreiner, Recent advances in pulse voltammetry, *Critical Reviews in Analytical Chemistry* 19 (1988) S1–S27. https://doi.org/10.1080/15476510.1988.10401465.
46. V. Mirceski, S. Skrzypek, L. Stojanov, Square-wave voltammetry, *ChemTexts* 4 (2018) 10. https://doi.org/10.1007/s40828-018-0073-0.
47. D. De Souza, S.A.S. Machado, L.A. Avaca, Square wave voltammetry. Part I: Theoretical aspects, *Química Nova* 26 (2003) 81–89.
48. R.K. Franklin, S.M. Martin, T.D. Strong, R.B. Brown, Chemical and biological systems: Chemical sensing systems for liquids, in: Franklin R.K. (ed.) *Reference Module in Materials Science and Materials Engineering*, Elsevier, Amsterdam, The Netherlands, 2016. https://doi.org/10.1016/B978-0-12-803581-8.00549-X.
49. S.S. Beshahwored, Cyclic voltammetric (Cv) evaluation of electro chemical behaviors of K3 Fe (CN) 6. and uric acid, *International Research Journal of Natural Science* 8 (2020) 52–59.
50. D. Coelho, G.M. Luiz, S.A.S. Machado, Estimating the electrochemically active area: Revisiting a basic concept in electrochemistry, *Journal of the Brazilian Chemical Society* 32 (2021) 1912–1917. https://doi.org/10.21577/0103-5053.20210080.
51. Rosy, S.K. Yadav, B. Agrawal, M. Oyama, R.N. Goyal, Graphene modified Palladium sensor for electrochemical analysis of norepinephrine in pharmaceuticals and biological fluids, *Electrochimica Acta* 125 (2014) 622–629. https://doi.org/10.1016/j.electacta.2014.01.160.
52. E. Laviron, General expression of the linear potential sweep voltammogram in the case of diffusionless electrochemical systems, *Journal of Electroanalytical Chemistry* 101 (1979) 19–28. https://doi.org/10.1016/S0022-0728(79)80075-3.
53. B. Uslu, S.A. Ozkan, Electroanalytical methods for the determination of pharmaceuticals: A review of recent trends and developments, *Electrochemistry* 44 (2011) 2644–2702. https://doi.org/10.1080/00032719.2011.553010.

4 Carbon-Based Materials for Electrochemical Sensing

Bharti Saini, Shirsendu Mitra, and Ravi Tejasvi

4.1 INTRODUCTION

Carbon-based materials have become important participants in the field of electrochemical sensing because of their remarkable qualities and wide range of applications. These materials, which are mostly made of carbon atoms, are excellent choices for a variety of sensing applications due to their exceptional electrical conductivity, chemical stability, and large surface area. Carbon-based materials can be categorized based on their structures and dimensionalities, which can affect their qualities. Because of their remarkable electrochemical qualities, carbon nanomaterials including graphene, graphene oxide (GO), carbon nanotubes (CNTs), crystalline diamond, nanofibers, and diamond-like carbon (DLC) have found extensive use (Figure 4.1). Since the unique carbon-derived nanostructures have features incomprehensible in bulk materials, their potential in sensing applications is undeniable. As a result, they are able to function over a wider temperature and dynamic range and with increased sensitivity and selectivity in challenging conditions [1]. Graphene is made up of a single layer of hexagonally organized carbon atoms. Because of its superior mechanical, electrical, and thermal characteristics, graphene is a great material for electrochemical sensors. Analytics ranging from gases and ions to proteins can be detected, thanks to its exceptional surface area and electrical conductivity [2]. Graphene sheets are rolled into cylindrical nanostructures called CNTs, which have attracted a lot of attention because of their exceptional mechanical strength and electrical conductivity. By utilizing their distinct structural and electrical characteristics, CNT-based sensors are used to detect gases, biomolecules, and heavy metals. Similar to CNTs but with a more disorganized structure, carbon nanofibers (CNFs) have a high surface area and good electrical conductivity. Because of these characteristics, CNFs can be used for a wide range of sensing applications, including biological diagnostics and environmental monitoring [3]. Because of its excellent electrical conductivity, graphite, a layered form of carbon, is frequently used in electrode designs that have been modified to detect a variety of analytics, including drugs, neurotransmitters, and environmental pollutants. With a thin powder of amorphous carbon particles, carbon black has a high electrical conductivity and surface area. Analytics including glucose, dopamine, and hydrogen peroxide have been detected via electrochemical sensors that use modified carbon black electrodes. Because of its chemical stability, glassy carbon, an amorphous carbon material, is widely used as an electrode material in electrochemical sensors and finds use in a variety of fields [4]. Because of its wide porosity and huge surface area, activated carbon is especially well-suited for applications using adsorption-based sensing. It has applications in the detection of organic substances, poisons, and contaminants [5]. In order to create novel carbon-based composites, researchers have also investigated the combination of carbon-based materials with other functional entities, such as metal nanoparticles, polymers, and metal oxides. By fusing improved functions with the inherent qualities of carbon materials, these composites increase the sensitivity and selectivity of sensors [6]. Carbon-based materials can be functionalized and modified to improve their sensing capabilities and enable certain interactions between the analytic and sensor surfaces. These changes increase sensitivity while reducing interference, which makes the materials much more flexible and suited to complex sensing requirements. The needs of the sensing application determine which particular carbon-based material should be used. The best option is

DOI: 10.1201/9781032654874-4

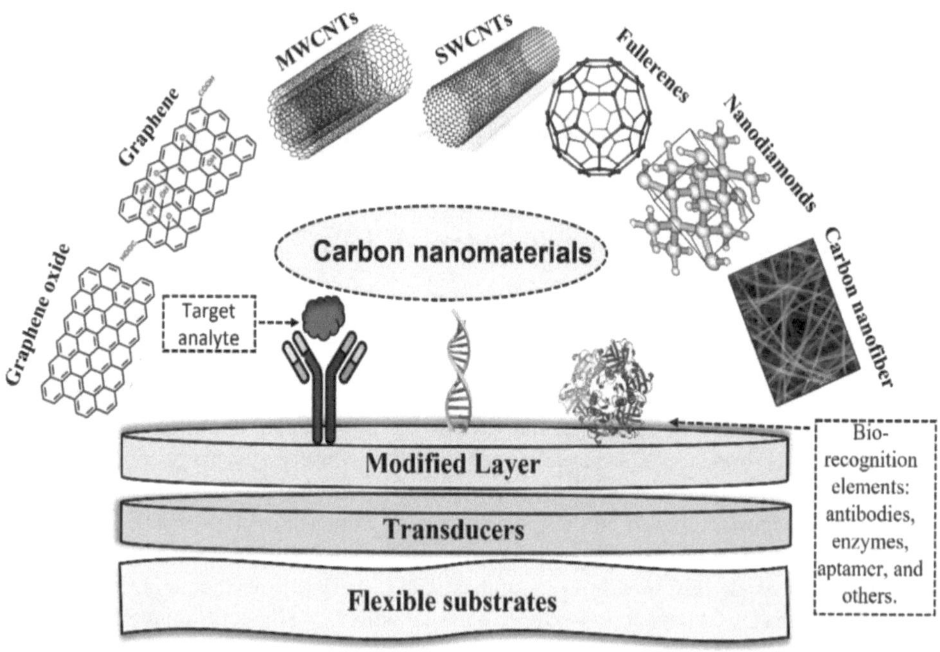

FIGURE 4.1 Different types of carbon materials used for electrochemical sensing applications.

Source: The figure is reprinted with permission from Ref. [4], copyright, Royal Society of Chemistry.

determined by elements including the analyte's type, the required sensitivity levels, and the current environmental circumstances. Carbon-based materials are essential components that drive the development of electrochemical sensing technologies in a larger sense. Their function encompasses a wide range of industries, including industrial process control, environmental monitoring, and healthcare. Carbon-based materials are well-positioned to continue influencing electrochemical sensing through further research and innovation, enabling the creation of state-of-the-art sensors that push the limits of detection and analysis [7].

Because of this, the recent introduction and utilization of carbon-derived nanomaterials for electrochemical sensing applications, as well as any possible ramifications, are the main topics of this chapter. Furthermore, a number of advancements in electrochemical sensors have been made possible by the exceptional qualities of carbon nanomaterials, including their huge surface-to-volume ratio, high conductivity, and electron mobility at ambient temperature. This chapter tries to highlight the many synthesis techniques and applications of carbon-based sensors in various disciplines, such as gas sensing, environmental monitoring, electrochemical biosensors, and biomolecule detection.

4.2 TYPES OF CARBON-BASED MATERIALS

4.2.1 Carbon Nanotubes (CNTs)

CNTs have sparked a lot of interest in potential future uses due to their superior mechanical strength, chemical stability, and unique field emission and electronic transport capabilities. CNTs have a growing range of applications, including field emission devices, nanoscale transistors, scanning microscopy tips, and composite material components [3]. In analytical and industrial electrochemistry, there is widespread recognition of several unique characteristics of CNTs, including their large edge or basal plane ratio, high surface area, and fast electrode transfer kinetics. The composition heterogeneity of CNTs allows for structural stiffness combined with ballistic conductivity. For the creation of microelectrodes for cell and tissue sensors, they are therefore perfect. Additionally, the development of biodegradable CNTs has a bright future for studies pertaining to CNTs [5]. Nevertheless, the inherent insolubility of the substance significantly impedes the application process. The chemical functionalization of CNTs has attracted significant attention in recent decades as a means to mitigate this drawback and expand their potential applications. This approach has led to the development of diverse hybrid materials with distinct applications. It is worth mentioning that there has been a significant advancement in the utilization of functionalized CNTs as electrochemical sensors [6]. CNTs can be functionalized with organic groups such as oxygen functional groups, small organic molecules, polymers, DNA, protein; inorganic groups such as metal nanoparticles metal oxide; and organic–inorganic hybrids.

4.2.2 Graphene and Graphene Derivatives

Graphene represents the latest addition to the family of multi-dimensional carbon nanomaterials, which includes fullerenes as zero-dimensional (0D) materials, single-walled CNTs (SWCNTs) as one-dimensional (1D) nanomaterials, and graphite as a three-dimensional (3D) material. A two-dimensional (2D) sheet of carbon atoms arranged hexagonally and joined by sp2 bonds is known as graphene. Graphene serves as a viable solution for addressing the void in 2D carbon nanomaterials. The remarkable characteristics of graphene, which include a tunable band gap, room-temperature Hall effect, a very large surface area ($2630\,m^2 g^{-1}$) that is double that of SWCNTs, high mechanical strength (200 times greater than steel), high elasticity, and high thermal conductivity, are due to these bonds and this electron configuration [7]. Because of its vast surface area, low cost, and extremely high 2D electrical conductivity, graphene is a perfect material for electrochemistry. Graphene has some benefits compared to CNTs. Graphene is free of metallic contaminants such as CNTs, and it is produced from graphite, which is readily available and inexpensive. The electrical conductivity of graphene is subject to variation based on the specific procedures employed for its synthesis or treatment, as well as the morphology of the resultant graphene particles. The electrical conductivity of the graphene particles was determined to be $64\,mS\ cm^{-1}$, exhibiting a significant improvement of nearly sixtyfold compared to SWCNTs [8].

4.2.3 Carbon Nanofibers

CNFs exhibit cylindrical nanostructures that possess various stacking configurations of graphene sheets, including stacked platelet, ribbon, or herringbone arrangements. The lengths of these entities are typically on the scale of micrometers, while their diameter ranges from tens of nanometers to many hundreds of nanometers. The mechanical strength and electrical properties of CNFs exhibit similarities to those of CNTs, while offering the advantage of controllable size and graphite ordering. The main differentiating feature of CNFs and CNTs is in the arrangement of graphene sheets with different forms, resulting in a greater number of edge sites on the outer wall of CNFs compared to CNTs [9]. The electrospinning method is utilized to fabricate cellulose nanofibers (CNFs) that possess notable characteristics such as a large surface area, porosity, homogeneity, and mechanical strength [10]. In recent years, a number of research have provided evidence supporting the notion that CNFs possess highly desirable qualities for the advancement of electrochemical-based sensors.

4.2.4 Carbon Quantum Dots (CQDs)

The accidental discovery of carbon nanodots (CDs) was made by Xu et al. in 2004 during the process of purifying SWCNTs synthesized from soot using the arc-discharge method [11]. The results of their research prompted a significant number of investigations into the synthesis of these nanomaterials, subsequently resulting in notable progress in their characteristics and potential uses. Colloidal quantum dots (CQDs) are commonly observed as quasi-spherical nanoparticles that consist of amorphous to nanocrystalline cores. These cores are mostly composed of graphitic or turbostratic carbon, which is characterized by sp2 hybridization. Additionally, the CQDs exhibit the presence of graphene and GO sheets that are cemented together through diamond-like sp3 hybridized carbon insertions. Oxidized carbon quantum dots (CQDs) have a significant presence of carboxyl functional groups on their surface. The oxygen level in the oxidized carbon quantum dots (CQDs) varies between 5 and 50 weight percent, depending on the specific synthesis pathway employed [12]. CQDs have garnered considerable attention from the academic research community owing to their intriguing characteristics, such as their small size and cost-effectiveness. Furthermore, significant advancements have been made due to their notable solubility, robust luminosity, and frequent designation as carbon nanolights. CQDs exhibit notable biological characteristics, such as little toxicity and compatibility with living systems, which endow them with considerable potential for utilization in several fields, including biosensing, bioimaging, optronics, catalysis, and drug administration [13].

4.3 SYNTHESIS METHODS

4.3.1 Chemical Vapor Deposition (CVD)

CVD is a commonly employed technique in material processing [14]. The primary utilization of this technology revolves around the deposition of solid thin-film coatings onto various surfaces. However, it is also employed for the production of bulk materials and powders with high levels of purity. Additionally, it facilitates the fabrication of composite materials through the implementation of infiltration techniques. The technique has been employed for the deposition of a diverse array of materials. In its most basic form, CVD entails the introduction of one or more precursor gases into a chamber where one or more heated objects are present, with the aim of applying a coating. Chemical reactions take place in close proximity to and onto the heated surfaces, leading to the formation of a thin film that is deposited onto the surface. This phenomenon is concomitant with the generation of chemical by-products, which are subsequently expelled from the chamber together with unreacted precursor gases. The process is conducted in both hot-wall and cold-wall reactors, operating at pressures ranging from sub-torr levels to over-atmospheric pressures. This can be done with or without the presence of carrier gases, and the temperatures normally fall within the range of 200°C–1600°C (Figure 4.2a). Additionally, there exists a range of accelerated CVD techniques that employ plasmas, ions, photons, lasers, hot filaments, or combustion reactions to augment deposition rates and/or reduce deposition temperatures.

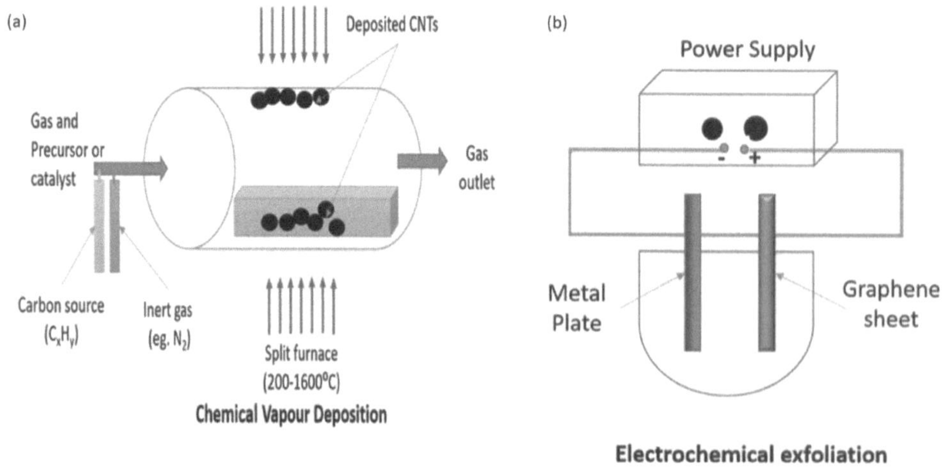

FIGURE 4.2 (a) Schematic diagram of chemical vapor deposition setup. (b) Synthesis process in electrochemical exfoliation method.

CVD possesses several notable benefits as a technique for the deposition of thin films. One of the main benefits is that CVD films exhibit a high degree of conformity, meaning that the film thickness on the sidewalls of features is similar to the thickness on the top surface. This implies that films own the skill to conform to intricately contoured surfaces, surrounding both internal and lower regions of structures, hence permitting the entire filling of high-aspect-ratio voids and other geometric characteristics. Alternatively, it is worth noting that physical vapour deposition (PVD) methods, such as sputtering or evaporation, typically necessitate an unhindered path between the target surface and the deposition source. One further benefit of CVD is its competency to deposit materials with exceptional clarity, along with its ability to deposit a wide range of materials. Additional benefits include comparatively rapid deposition rates and the notable advantage that CVD techniques frequently necessitate lower vacuum levels in comparison with PVD methods. CVD is associated with several drawbacks. Ideally, it is desirable for the precursors to exhibit volatility at temperatures close to ambient conditions. However, the utilization of metal-organic precursors has alleviated this predicament. Certain precursors for CVD can exhibit significant levels of toxicity, such as $Ni(CO)_4$, explosiveness, as seen in B_2H_6, or corrosiveness, as observed in $SiCl_4$. The hazardous byproducts resulting from CVD reactions include carbon monoxide (CO), hydrogen gas (H_2), and hydrogen fluoride (HF). Certain precursors, particularly those of the metal-organic nature, can potentially incur significant expenses. Another significant drawback pertains to the deposition process of the films, which typically involves extreme temperatures. These limitations impose constraints on the kind of substrates that are suitable for coating. Furthermore, this phenomenon gives rise to tensions in films that are applied to substances with distinct coefficients of thermal expansion. Consequently, these tensions have the potential to induce mechanical instabilities within the deposited films. CVD processes can be classified based on the application type, method, and reactor employed, as well as the precursor and chemical reaction utilized [15]. The technique of CVD is utilized in the large-scale manufacture of CNTs. This process involves the utilization of hydrocarbon or other carbon-containing precursors, which are combined with a catalyst. Subsequently, the CNTs are deposited onto a substrate.

4.3.2 Electrochemical Exfoliation

Electrochemical exfoliation constitutes an integral component within the expanding repertoire of graphene synthesis techniques. The electrochemical approach is associated with wet chemical exfoliation methods, such as the modified Hummers process employed for the synthesis of GO. In contrast to the aforementioned techniques that frequently depend on aggressive oxidizing agents, electrochemical approaches leverage the conductive properties of graphite to facilitate the intercalation of molecules within the interlayer spaces of graphene (Figure 4.2b). The utilization of graphite as an electrode enables the introduction of either a positive charge or a negative charge to the material, promoting the intercalation of ions with opposite charges and facilitating the process of exfoliation [16].

One of the primary objectives in the field of graphene synthesis is to achieve the production of graphene that possesses a high degree of purity, characterized by low oxygen content, a reduced number of layers, and a substantial lateral dimension. Nevertheless, the process of graphite exfoliation in liquid electrolytes under ambient conditions often results in the incorporation of oxygen (O) into the carbon lattice of graphene, leading to the fragmentation of the exfoliated graphene into smaller fragments. Electrochemical exfoliation techniques have exhibited a certain degree of control over the structural characteristics of graphene materials [17]. Electrochemically exfoliated graphene has the capability to be synthesized using several types of electrolytes, including ionic liquids, acidic media, and inorganic salts. Among these options, it has been discovered that inorganic salts are capable of generating graphene with significant lateral dimensions and a reduced quantity of oxygen functional groups when compared to alternative sorts of electrolytes [18].

4.3.3 Hydrothermal and Solvothermal Methods

Hydrothermal and solvothermal procedures are extensively employed methodologies for the synthesis of carbon-based materials, enabling precise control over their structures, sizes, and properties. The aforementioned techniques entail the utilization of elevated temperature and pressure conditions to facilitate chemical interactions between precursor substances within a liquid media. The primary distinction between hydrothermal and solvothermal procedures pertains to the selection of solvent.

Hydrothermal methods employ water as the solvent, while solvothermal methods utilize organic solvents. Hydrothermal synthesis is a chemical process that uses water as the solvent and entails performing reactions under conditions of elevated temperatures and pressures within a sealed vessel, sometimes referred to as an autoclave. The chemical purity of powder materials in hydrothermal synthesis is enhanced as a result of commencing the process using precursors of high purity. This approach demonstrates notable efficacy in the synthesis of carbon-based materials, enabling precise control over their morphologies, crystallinities, and surface chemistries. The reduction of GO and the subsequent synthesis of diverse carbon-based nanomaterials, such as CNTs and carbon nanospheres, can be accomplished through the use of hydrothermal conditions.

Solvothermal synthesis is a process that bears resemblance to hydrothermal synthesis, however, utilizing nonaqueous solutions as the primary reaction media in place of water at relatively high temperatures. The utilization of this approach provides enhanced manipulation of the reaction parameters, as distinct solvents have the potential to influence the solubility, kinetics, and shape of the carbon-based compounds being synthesized. The solvothermal synthesis method is frequently employed in the production of carbon nanomaterials that exhibit distinct structures and characteristics. One illustrative instance involves the utilization of solvothermal techniques to generate carbon quantum dots, which are carbon nanoparticles at the nanoscale exhibiting distinctive optical characteristics.

4.3.4 Pyrolysis Techniques

Pyrolysis is a thermochemical process characterized by the degradation of matter into smaller components through high-temperature treatment in an oxygen-free environment. Examples of carbon materials with technological significance that are prepared through the process of pyrolysis include graphene, CNTs, carbon fibers (CFs), DLC coatings, and other industrial carbons such as glass-like carbon (GC) and graphite [19]. There are various methods to produce carbon material by pyrolysis such as fixed bed pyrolysis, spray pyrolysis, fluidized bed pyrolysis.

Zhu et al. conducted a study involving the implementation of single-stage pyrolysis and the growth of multi-walled CNTs (MWCNTs). In this experiment, a fixed bed horizontal tube furnace was utilized to house fresh bamboo cubes. The reactor underwent an argon purge, followed by sealing and subsequent heating at a heating rate of 1°C per minute until reaching a temperature range of 1000°C–1500°C. Argon and ethanol were utilized as purge agents and promoters during the purging process, operating at a specified temperature range of 1000°C–1500°C. The transmission electron microscopy (TEM) images provided evidence that the silicates found in the sample had a significant impact on the process of nucleation and subsequent growth [20]. Spray pyrolysis exhibits certain advantages compared to the fixed bed pyrolysis technique. It enables the potential for continuous operation and occasionally eliminates the need for a meticulously prepared substrate. During this procedure, vaporous precursors and catalyst particles are atomized within a high-temperature reaction zone in the presence of an inert carrier gas. Hydrocarbons undergo catalytic decomposition, resulting in the formation of carbon nanomaterial and the liberation of H_2. The deposition process can take place either on the walls of a tube reactor or on a substrate that is positioned inside the tube [19].

A comprehensive study was conducted by Son et al. to investigate the synthesis of MWCNTs using fluidized bed pyrolysis. The impact of various carbon sources, including methane (CH_4), C_2H_2, C_2H_4, and C_2H_6, was assessed. The experimental results demonstrated that CNTs synthesized from CH_4 exhibited superior crystallinity and a smaller diameter compared to CNTs synthesized from other feedstocks. However, the carbon conversion efficiency was found to be lower in the case of CH_4. The procedure involved utilizing a stainless steel reactor equipped with a gas distributor located at the lower section. The volumetric flow rates of hydrogen (H_2), nitrogen (N_2), and various carbon precursors were regulated using mass flow controllers. These flow rates were directed over a gas distributor, where catalyst particles composed of iron (Fe), molybdenum (Mo), and magnesium oxide (MgO) were loaded. The catalyst particles underwent fluidization using a combination of H_2, N_2, and a carbon source gas. The fluidization process occurred within a temperature range of 873–1173 K [21].

4.4 APPLICATIONS IN ELECTROCHEMICAL SENSORS

4.4.1 Environmental Monitoring

Due to the inherent limitations of field measurements compared to laboratory measurements, as well as the various factors such as complexities and other connections that can affect the response and potentially vary over time, it is often more advantageous to utilize diagnostic sensors, specifically alarm sensors, for field applications. Instead of providing measurements for specific chemical components, these measurements indicate the overall response to multiple components or demonstrate a characteristic pattern that represents the entire mixture. This concept is closely related to pattern recognition and electronic tongues.

4.4.1.1 Water Contamination

Kukla et al., in their studies, utilized a combination of multi-electrode sensor arrays made from various materials alongside pattern recognition techniques to create a unique identification for a specific water sample. Modifications in the water composition result in variations in the fingerprint, which serves as the fundamental characteristic of an alarm-type sensor. Enzyme inhibition can be utilized to evaluate total metal ion concentrations [23]. This approach exhibits potential for future applicability within the respective domain. In a traditional manner, electrochemical sensors have been employed in aquatic systems alongside depth profiling and 2D surface mapping of the oceans [24]. This tool is primarily utilized for investigating biogeochemical processes, but it can also effectively identify the impacts of pollution near discharge points and monitor temporal variations. The assessment of wastewater toxicity has been conducted utilizing a microbial sensor that involves the attachment of oxygen-consuming bacteria to an electrode. The overall toxicity of the wastewater hinders the normal functioning of the bacteria [25]. Alternatively, there have been proposals to employ lipid-based sensors as a universal toxicity detector. This phenomenon is grounded in the principle that cations elevate the membrane potential of negatively charged polarized membranes, whereas anions diminish the membrane potential of positively polarized membranes. Therefore, the quantification of the combined concentration of cations and anions can be achieved using multichannel potentiometric sensors [26].

4.4.1.2 Detection of Heavy Metals

Solvent oxidation or reduction, supporting electrolyte breakdown, electrode surface oxidation or reduction, and potential electrode dissolution determine voltammetry potential ranges on electrode materials. Chemically unreactive electrode material is usually required for electroanalytical measurements. Since the 1920s, liquid mercury has been used in metal cation reduction processes. It was first used as a dropping mercury electrode and thereafter as a hanging/static one. Mercury's oxidation limits its practical potential, confining research to reductions. Stripping voltammetry for heavy metals reduces and accumulates the metal at the electrode over time. After accumulating, metal is reoxidized [27]. Mercury electrodes, usually thin films or mercury salts, have been used to reduce hydrogen evolution [28]. Numerous environmental monitoring applications have used stripping voltammetry [28], along with several mercury reduction methods. Preconcentration helps achieve detection limits far lower than environmental laws. These techniques can also quantify labile or weakly complexed species rather than measure the overall amount. A recent study showed the environmental importance of filtering the analyte and switching the medium to accurately evaluate heavy metal speciation in sewage treatment plant effluent [29]. This strategy may not work long term. Mercury thin films have replaced liquid mercury, yet their formation is still necessary. Mercuric ions are used in this technique in small amounts. However, this strategy may grow unpopular with the community in the future. Therefore, mercury-free sensor materials must be used. Carbon materials have been widely used in numerous applications for decades, thanks to significant advances. These materials are effective in studying positive potential oxidation processes and can be used to deposit mercury thin films. Thick carbon films, boron-doped diamonds [30], and carbon resistor-derived carbon film electrodes [31] have shown promise at negative potentials. Certain carbon compounds may detect ecologically harmful ions including lead, cadmium, and copper. Quantifying zinc ions at submicromolar quantities is difficult. Because zinc ions are reduced at negative potentials, most electrode materials generate hydrogen.

4.4.2 Electrochemical Biosensors

Electrochemical biosensor is a versatile unit that functions on fundamental electrochemistry to detect analytes utilizing suitable bioreceptors. The inception of electrochemical biosensor took place in the 1960s when a group of researchers for the first time developed biosensor by functionalizing enzyme onto a working electrode substrate and recorded variable responses based on concentrations of analyte. Over time, the development of electrochemical biosensor evolved significantly. In the early days, around the 1970s, the development of only enzyme-based electrochemical sensors took place. Further development around the 1980s happened in terms of improvements and miniaturization of the electrochemical biosensor device. Owing to the advent of nanotechnology in the 1990s, the concept of nanotechnology incorporation took place in electrochemical biosensors. Following the 2000s, the development of electrochemical biosensors took place in terms of microelectronic integration, smartphone integration, compact portable device development, and the latest is the integration of artificial intelligence and machine learning. Among different varieties of biosensors, electrochemical biosensors are well known for their accuracy, specificity, and applications in different parts of biomedical engineering. Electrochemical sensors are based on specific redox reactions that take place on a specially designed electrode commonly known as the working electrode (Figure 4.3). Designing an electrochemical biosensor involves the synthesis/fabrication of an appropriate working electrode that can carry out the chosen redox reaction with the target analyte. A typical example is electrochemical glucose sensors, where glucose oxidase is attached to a suitable working electrode, which allows lock-and-key type glucose and glucose oxidase redox reaction generating quantifiable electronic change on the working electrode, which is used for the preparation of the calibration curve. A host of materials is preferred for the fabrication of working electrodes that include indium tin oxide (ITO)-coated glass/polymer substrate, gold nano/composite, carbonaceous materials like graphene/CNTs/graphite, silicon substrate [32], alumina [33], magnetic micro/nanoparticles [34], molecularly imprinted polymer [35], nanoporous gold supported porous cobalt oxide [36], nanowire [37], zinc oxide microspheres [38], sol-gel materials [39], lipid bilayer [40], transition metal dichalcogenides [41]. Carbonaceous materials have gained additional importance owing to their excellent conductivity, scope of chemical modifications, environmentally friendly nature, and more. Figure 4.3 shows that different kinds of redox reactions happen on electrochemical working electrodes on diverse carbon materials.

It is important to note that electrochemical sensors are categorized into different subgroups based on different kinds of bioreceptors such as enzymes, antibodies, cells, DNA, and aptamers, which are immobilized on the working electrode to quantify the target analyte (Figure 4.4). A few instances are modified working electrodes used for the detection of the targeted antigen via specific

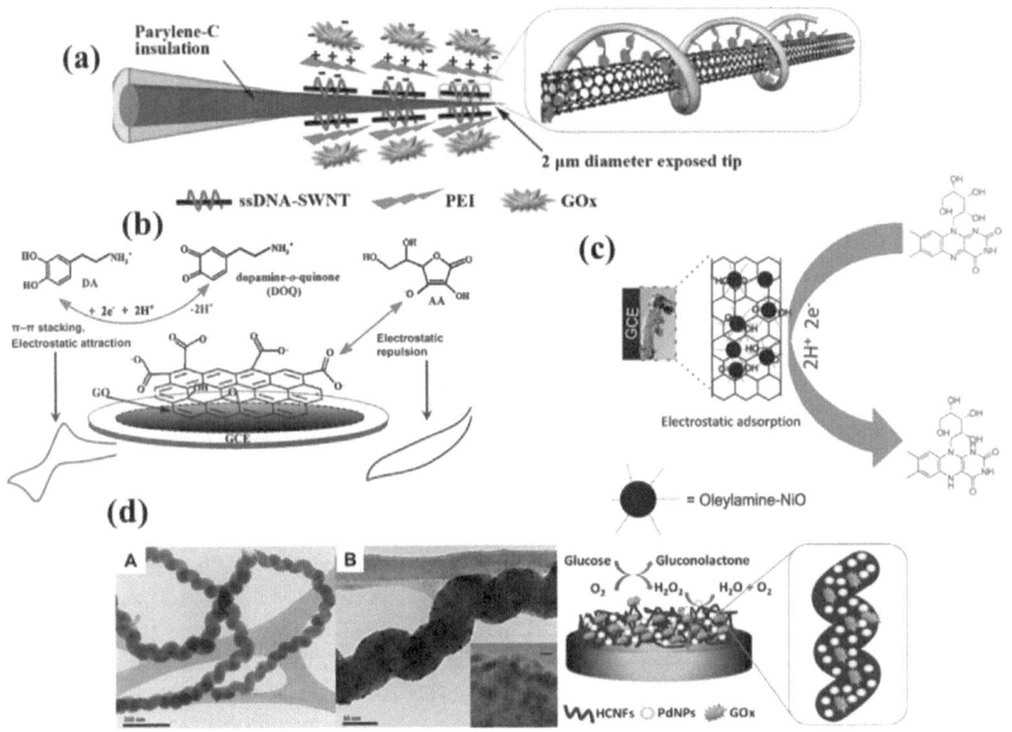

FIGURE 4.3 Different kinds of redox reactions that happen on electrochemical working electrodes on diverse carbon materials.

Source: The image is reprinted with permission from Ref. [22] copyright American Chemical Society.

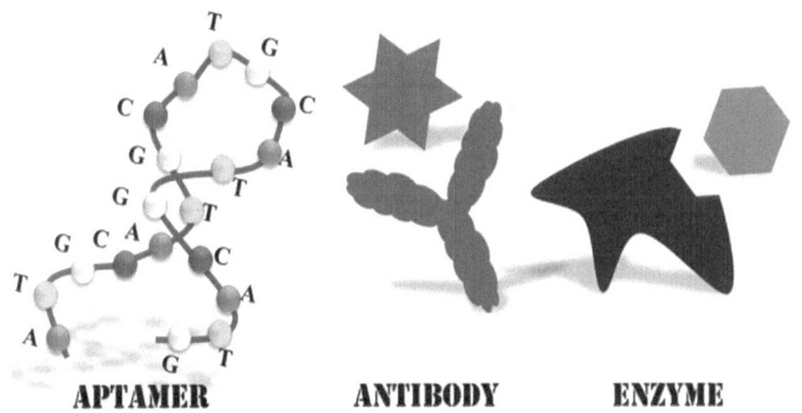

FIGURE 4.4 Different kinds of bioreceptors used for specific interactions with targeted analyte to design sensitive electrochemical biosensors.

antibody–antigen interaction, nucleic acids for the detection of HIV [42], detection of various types of cancers via protein–protein interaction [43,44], detection of dengue using specific receptors like aptamers to capture the virus [45], etc. We must recall again that for the design of an electrochemical sensor, the fabrication of a superlative, specific, and sensitive working electrode is a must. Recently, advancements have also been made in the fields of development of flexible [46] and disposable [47] electrochemical sensors or clinical applications to boost the healthcare sectors.

4.4.2.1 Enzymatic Biosensors

Enzymatic biosensors are the ones that work on the enzyme and substrate interaction chemistry. Enzyme–substrate reactions are very specific, which enables designing robust enzymatic biosensors. There are multiple reports that show enzyme-functionalized working electrodes for the detection of a wide spectrum of biomarkers that include cholesterol [48], glucose [49], ketones [50], urea [51], hemoglobin [52], HIV [42], various kinds of cancer [43,44], leukemia [53], detection of dengue [45]. Like other bioreceptors, the fabrication of enzymatic electrochemical sensors needs the designing of a working electrode that contains functionalized enzymes. The enzymes are proteins by chemical nature, and they are sensitive to temperature and contamination and hence special. The fabrication technique of the working electrodes can be broadly classified into two categories: (1) non-covalent bonding-based attachment via electrostatic interaction between substrate material and the enzyme [54] and (2) attachment of enzyme via amide coupling chemistry [55]. It can further be mentioned that the technique of covalent attachment is more acceptable compared to the non-specific electrostatic attachment.

The importance of enzymatic electrochemical biosensors [56,57] lies in their sensitivity, high selectivity, ease of real-time monitoring, excellent response time, portability, low cost of fabrication, biocompatibility, and minimal environmental waste generation. Detection of glucose [49] using glucose oxidase, cholesterol [48] by cholesterol oxidase, alcohols [58] by alcohol dehydrogenase, tri-glyceraldehyde [59] by lipase, lactose [60] by lactase, urea [61] by urease, uric acid [62] by uricase, and bacterial pathogen [63] or drug detection by suitable enzymes are a few of the widely used electrochemical biosensors. Despite the fact that enzymatic biosensors are one of the well-established methods, research and development in this field are still ongoing with attempts to create smart enzymatic biosensors with the usage of modern-day nanocomposites, nanoplasmonics [64], and nanobiotechnology [65]. With further progression, the incorporation of artificial intelligence and data analytics into the electrochemical enzymatic sensors is creating a new pathway of advanced detection. The latest development also shows real-time monitoring of analytes in living cells via the functionalization of enzymes onto microneedles [66] or micro-patches [67], which act as a suitable working electrode for valuable electrochemical signal recording. Besides the use of carbonaceous enzymes, carbonaceous materials are also being used as suitable conductive electrode material [68–70]. For example, for the immobilization of enzymes, there are several carbonaceous materials which are in use. In the early days, active charcoal and graphite were used extensively for making working electrodes. As time progressed, several allotropes of carbon became popular and they became popular materials for electrodes. CNTs [71], graphite [72], graphene dots [73], carbon dots [74], and carbon soot [75] were exploited as efficient electrode materials.

Of late, plasmon-enhanced carbon nanocomposites and surface-enhanced Raman spectroscopy (SERS) active carbon composites have proved themselves as the most reliable electrode materials for enzyme-linked electrochemical biosensors.

4.4.2.2 Immunosensors

Immunosensors [76–78] are the type of sensors that relies on specific interaction between an antigen and an antibody. Antigens are substances that the immune system recognizes as harmful or foreign, which prompt immune systems. Our human body has the ability to synthesize appropriate proteins, known as antibodies, that can combat foreign proteins. The synthesized antibodies by our immune system help in destruction of the foreign biomaterials and thereby cure diseases naturally. However, there are human malconditions where natural immunity fails to combat the disease or it makes the effect prolonged. Medical science always recommends early diagnosis of an ailment so that appropriate actions can be taken before the condition becomes severe.

For the fabrication of electrochemical biosensors, an appropriate electrode is necessary wherein sensitive antibodies can be immobilized. Antibodies are sensitive toward heat and foreign contamination, and hence, the fabrication of antibody-attached working electrode needs to be done at a controlled temperature and in a clean area. There is a host of electrode fabrication techniques, and among them, attachment of antibodies on ITO-coated surface [79,80] exploiting sialylation chemistry is one of the popular ones and there are multiple previous reports. Of late, the fabrication and functionalization of carbon-based nanocomposites [81], carbonaceous plasmonic nanoparticles [82], graphene nanocomposites [83,84], CNTs [85], carbonaceous metal nanowires [86] are being used as host materials for antibody attachment.

In this regard, we must mention that carbon-based materials are extensively utilized for the design and fabrication of such electrochemical immunosensors owing to their notable material properties that include high surface area, excellent electrical conductivity, biocompatibility, and trivial functionalization. As carbon-based materials possess excellent material properties, they enhance the performance and sensitivity of immunosensors. For example, few carbonaceous materials, such as graphene, CNTs, carbon quantum dots, graphite, are recognized for their suitable electrical conductivity. Further to this, carbon-based materials possess the ease of functionalizing antibodies onto their surfaces, for having several modification chemistries. Another important factor for choosing carbon-based materials as immunosensors is their large surface area. Large surface areas allow excellent reaction and diffusion during the specific interactions between antigens and antibodies, which is the cause behind immunosensing. Other notable properties of carbon-based materials are biocompatibility, rapid response, regeneration, excellent stability, etc. Conclusively, carbon-based materials are the most versatile platform for designing electrochemical immunosensors with enhanced sensitivity, specificity, and performance.

4.4.2.3 Aptamer-Based Sensors

Aptameric biosensors [87–89] are another area of research that has been gaining pace since the last two decades. Briefly, aptamers are DNA fragments composed of carbon materials, with a specific sequence of nucleotides that can show specific interactions with the target analyte. Scientists have developed several techniques to synthesize aptamer for a specific target analyte. There are multiple processes to synthesize aptamers. Among them, systematic evolution of ligands by exponential enrichment (SELEX) [90] and CRISPR-modified SELEX [91] are the most specific ones.

Like antibodies, aptamers are also very sensitive materials, and they need to be kept at controlled temperatures as directed by the manufacturer. For the fabrication of aptameric electrochemical sensors, aptamers are needed to be attached onto a suitable working electrode substrate. There are multiple well-developed techniques for the attachment of aptamer onto the electrode surface. In general, the manufacturer of aptamers creates a sulfur-ended aptamer to make a provision for attachment onto electrode surface via sulfur chemistry. For example, gold has a nice covalent interaction with sulfur atom, and this chemistry is widely used for the attachment of aptamers with gold-coated or gold nanoparticle-coated electrodes. In one of the techniques, gold is sputter-coated onto a suitable substrate and then the thin layer of gold is utilized for making a bonding with sulfur-ended aptamer. Another method involves attaching plasmonic gold nanoparticles onto a glass substrate using APTES ((3-Aminopropyl)triethoxysilane)-based silanization chemistry, and further attachment of sulfur-ended aptamer is done by exploiting gold nanoparticle–sulfur chemistry.

The aptameric electrochemical sensors have advantages in terms of excellent specificity, faster interaction, ease of aptamer design, and ease of attachment chemistry. To date, scientists have developed aptamers for biomolecules [92], viruses [93], bacteria [94,95], heavy metal ions [96], and more. These developed aptamers created the pathway to develop excellent biosensors for the detection of several antigens, proteins, biomolecules, metal ions such as sodium/potassium/calcium, heavy metals such as arsenic/mercury, and more.

4.5 FUTURE PERSPECTIVES AND CHALLENGES

As carbon-based materials continue to evolve, there remain several challenges and opportunities in the field of electrochemical sensing. Further research is needed to enhance the sensitivity, selectivity, and stability of carbon-based sensors. Integration with nanotechnology, molecular engineering, and signal amplification strategies could lead to innovative sensor designs for real-time and point-of-care applications.

4.6 CONCLUSION

The distinctive characteristics of carbon nanomaterials have significantly contributed to the advancement and progression of electrochemical sensors and biosensors. Both the novel and modified carbon-based probes frequently exhibit improved analytical performance compared to conventional non-nanostructured electrochemical systems. The electrocatalytic potential of carbon nanomaterial-modified electrodes and their excellent analytical performance support their use in modern electrochemical systems. These qualities include CNTs' increased active surface area and diamond and diamond-like surfaces' fouling resistance. Carbon-based materials like graphene and CNTs are sensitive and selective to heavy metals and organic contaminants. These materials' large surface areas and excellent electrical conductivity allow the fabrication of sensitive sensors that can detect even trace amounts of contaminants in complex environmental samples. However, there is still ample opportunity for further investigation into the utilization of carbon-based materials in electrochemical sensors. Despite the progress made, there are still obstacles to overcome in comprehending and advancing the application of carbon within the field of electrochemistry.

REFERENCES

1. A.C. Power, B. Gorey, S. Chandra, J. Chapman, *Nanotechnology Reviews*, 7 (2018) 19–41.
2. D. Chen, L. Tang, J. Li, *Chemical Society Reviews*, 39 (2010) 3157–3180.
3. A. Merkoçi, M. Pumera, X. Llopis, B. Pérez, M. Del Valle, S. Alegret, *TrAC Trends in Analytical Chemistry*, 24 (2005) 826–838.
4. A. Khan, E. DeVoe, S. Andreescu, *Sensors & Diagnostics*, 2 (2023) 529–558.
5. K. Bala, D. Sharma, N. Gupta, *ChemElectroChem*, 6 (2019) 274–288.
6. C. Gao, Z. Guo, J.-H. Liu, X.-J. Huang, *Nanoscale*, 4 (2012) 1948–1963.
7. M. Pumera, A. Ambrosi, A. Bonanni, E.L.K. Chng, H.L. Poh, *TrAC Trends in Analytical Chemistry*, 29 (2010) 954–965.
8. E.B. Bahadır, M.K. Sezgintürk, *TrAC Trends in Analytical Chemistry*, 76 (2016) 1–14.
9. J. Huang, Y. Liu, T. You, *Analytical Methods*, 2 (2010) 202–211.
10. M. Adabi, R. Saber, R. Faridi-Majidi, F. Faridbod, *Materials Science and Engineering: C*, 48 (2015) 673–678.
11. X. Xu, R. Ray, Y. Gu, H.J. Ploehn, L. Gearheart, K. Raker, W.A. Scrivens, *Journal of the American Chemical Society*, 126 (2004) 12736–12737.
12. S.Y. Lim, W. Shen, Z. Gao, *Chemical Society Reviews*, 44 (2015) 362–381.
13. H.V. Nguyen, L. Richtera, A. Moulick, K. Xhaxhiu, J. Kudr, N. Cernei, H. Polanska, Z. Heger, M. Masarik, P. Kopel, *Analyst*, 141 (2016) 2665–2675.
14. Y.M. Manawi, K. Ihsanullah, A. Samara, T. Al-Ansari, M.A. Atieh, *Materials*, 11 (2018) 822.
15. H.O. Pierson, *Handbook of Chemical Vapor Deposition: Principles, Technology and Applications*, William Andrew, Norwich, 1999.
16. P. Yu, S.E. Lowe, G.P. Simon, Y.L. Zhong, *Current Opinion in Colloid & Interface Science*, 20 (2015) 329–338.
17. F. Liu, C. Wang, X. Sui, M.A. Riaz, M. Xu, L. Wei, Y. Chen, *Carbon Energy*, 1 (2019) 173–199.
18. F. Sharif, A.S. Zeraati, P. Ganjeh-Anzabi, N. Yasri, M. Perez-Page, S.M. Holmes, U. Sundararaj, M. Trifkovic, E.P. Roberts, *Carbon*, 157 (2020) 681–692.
19. M.U. Zahid, E. Pervaiz, A. Hussain, M.I. Shahzad, M.B.K. Niazi, *Materials Research Express*, 5 (2018) 052002.
20. J. Zhu, J. Jia, F.L. Kwong, D.H.L. Ng, S.C. Tjong, *Biomass and Bioenergy*, 36 (2012) 12–19.
21. S.Y. Son, Y. Lee, S. Won, D.H. Lee, S.D. Kim, S.W. Sung, *Industrial & Engineering Chemistry Research*, 47 (2008) 2166–2175.
22. J.N. Tiwari, V. Vij, K.C. Kemp, K.S. Kim, *ACS Nano*, 10 (2016) 46–80.
23. A. Kukla, N. Kanjuk, N. Starodub, Y.M. Shirshov, *Sensors and Actuators B: Chemical*, 57 (1999) 213–218.
24. M. Taillefert, G.W. Luther III, D.B. Nuzzio, *Electroanalysis: An International Journal Devoted to Fundamental and Practical Aspects of Electroanalysis*, 12 (2000) 401–412.
25. J.-D. Liao, S.-H. Wang, D.-J. Hsu, *Sensors and Actuators B: Chemical*, 72 (2001) 167–173.
26. H. Sakai, S. Iiyama, K. Toko, *Sensors and Actuators B: Chemical*, 66 (2000) 251–255.

27. K. Brainina, E. Neyman, *Electroanalytical Stripping Methods*, John Wiley & Sons, Hoboken, NJ, 1994.
28. K.Z. Brainina, N. Malakhova, N.Y. Stojko, *Fresenius' Journal of Analytical Chemistry*, 368 (2000) 307–325.
29. E.A. Alvaréz, M.C. Mochón, J.J. Sánchez, M.T. Rodriguéz, *Electroanalysis: An International Journal Devoted to Fundamental and Practical Aspects of Electroanalysis*, 10 (1998) 917–920.
30. G.S. Reeder, W.R. Heineman, *Sensors and Actuators B: Chemical*, 52 (1998) 58–64.
31. C.M. Brett, L. Angnes, H.D. Liess, *Electroanalysis: An International Journal Devoted to Fundamental and Practical Aspects of Electroanalysis*, 13 (2001) 765–769.
32. E. Souteyrand, J. Martin, C. Martelet, *Sensors and Actuators B: Chemical*, 20 (1994) 63–69.
33. B.T.T. Nguyen, A.E.K. Peh, C.Y.L. Chee, K. Fink, V.T. Chow, M.M. Ng, C.-S. Toh, *Bioelectrochemistry*, 88 (2012) 15–21.
34. Y. Xu, E. Wang, *Electrochimica Acta*, 84 (2012) 62–73.
35. R. Gui, H. Jin, H. Guo, Z. Wang, *Biosensors and Bioelectronics*, 100 (2018) 56–70.
36. X.-Y. Lang, H.-Y. Fu, C. Hou, G.-F. Han, P. Yang, Y.-B. Liu, Q. Jiang, *Nature Communications*, 4 (2013) 1–8.
37. A.K. Wanekaya, W. Chen, N.V. Myung, A. Mulchandani, *Electroanalysis: An International Journal Devoted to Fundamental and Practical Aspects of Electroanalysis*, 18 (2006) 533–550.
38. X. Lu, H. Zhang, Y. Ni, Q. Zhang, J. Chen, *Biosensors and Bioelectronics*, 24 (2008) 93–98.
39. J. Wang, *Analytica Chimica Acta*, 399 (1999) 21–27.
40. H.T. Tien, A. Ottova, *Electrochimica Acta*, 43 (1998) 3587–3610.
41. Y.-H. Wang, K.-J. Huang, X. Wu, *Biosensors and Bioelectronics*, 97 (2017) 305–316.
42. D. Zhang, Y. Peng, H. Qi, Q. Gao, C. Zhang, *Biosensors and Bioelectronics*, 25 (2010) 1088–1094.
43. S.N. Topkaya, M. Azimzadeh, M. Ozsoz, *Electroanalysis*, 28 (2016) 1402–1419.
44. J. Wang, *Biosensors and Bioelectronics*, 21 (2006) 1887–1892.
45. J. Cecchetto, F.C. Carvalho, A. Santos, F.C. Fernandes, P.R. Bueno, *Sensors and Actuators B: Chemical*, 213 (2015) 150–154.
46. A. Reddy, B.B. Narakathu, M. Atashbar, M. Rebros, E. Hrehorova, M. Joyce, *Sensors*, 14 (2010) 1596–1600.
47. B.C. Janegitz, J. Cancino, V. Zucolotto, *Journal of Nanoscience and Nanotechnology*, 14 (2014) 378–389.
48. A. Wisitsoraat, P. Sritongkham, C. Karuwan, D. Phokharatkul, T. Maturos, A. Tuantranont, *Biosensors and Bioelectronics*, 26 (2010) 1514–1520.
49. S. Wang, Q. Zhang, R. Wang, S. Yoon, *Biochemical and Biophysical Research Communications*, 311 (2003) 572–576.
50. M.R. Cavallari, G.S. Braga, M.F. da Silva, J.E. Izquierdo, L.G. Paterno, E.A. Dirani, I. Kymissis, F.J. Fonseca, *IEEE Sensors Journal*, 17 (2017) 1971–1980.
51. G. Das, H.H. Yoon, *International Journal of Nanomedicine*, 10 (2015) 55.
52. T.S. Anirudhan, S. Alexander, *European Polymer Journal*, 97 (2017) 84–93.
53. A.A. Ensafi, M. Taei, H. Rahmani, T. Khayamian, *Electrochimica Acta*, 56 (2011) 8176–8183.
54. J.E. Contreras-Naranjo, O. Aguilar, *Biosensors*, 9 (2019) 15.
55. A. John, L. Benny, A.R. Cherian, S.Y. Narahari, A. Varghese, G. Hegde, *Journal of Nanostructure in Chemistry*, 11 (2021) 1–31.
56. N.J. Ronkainen, H.B. Halsall, W.R. Heineman, *Chemical Society Reviews*, 39 (2010) 1747–1763.
57. R.S. Freire, C.A. Pessoa, L.D. Mello, L.T. Kubota, *Journal of the Brazilian Chemical Society*, 14 (2003) 230–243.
58. N. Roy, S. Mitra, N.M. Das, N. Mandal, D. Bandyopadhyay, H.B. Nemade, T.K. Mandal, *IEEE Sensors Journal*, 20 (2019) 2278–2286.
59. M. Labib, E.H. Sargent, S.O. Kelley, *Chemical Reviews*, 116 (2016) 9001–9090.
60. A.R. de Brito, R.S. de Jesus, I.M. de Carvalho Tavares, F.N. Silva, N.B. Santana, S.P.B. Ferrao, M. Bilal, A. de Santana Santos, L.C. Salay, J.R. de Oliveira, *Surfaces and Interfaces*, 22 (2021) 100839.
61. S.M.U. Ali, Z.H. Ibupoto, S. Salman, O. Nur, M. Willander, B. Danielsson, *Sensors and Actuators B: Chemical*, 160 (2011) 637–643.
62. F. Muhammad, G. Dik, S. Kolak, K.K. Gedik, B. Bakar, A. Ulu, B. Ateş, *Electrochimica Acta*, 439 (2023) 141675.
63. M. Xu, R. Wang, Y. Li, *Talanta*, 162 (2017) 511–522.
64. S.K. Maji, *ACS Applied Nano Materials*, 2 (2019) 7162–7169.
65. T. Tite, E.A. Chiticaru, J.S. Burns, M. Ioniţă, *Journal of Nanobiotechnology*, 17 (2019) 1–22.
66. L. Zhao, Z. Wen, F. Jiang, Z. Zheng, S. Lu, *RSC Advances*, 10 (2020) 6163–6171.

67. X. Xuan, H.S. Yoon, J.Y. Park, *Biosensors and Bioelectronics*, 109 (2018) 75–82.
68. D. Kato, O. Niwa, *Analytical Sciences*, 29 (2013) 385–392.
69. A. Smart, A. Crew, R. Pemberton, G. Hughes, O. Doran, J. Hart, *TrAC Trends in Analytical Chemistry*, 127 (2020) 115898.
70. Z. Wen, J. Li, *Journal of Materials Chemistry*, 19 (2009) 8707–8713.
71. J.H. Luong, S. Hrapovic, D. Wang, *Electroanalysis: An International Journal Devoted to Fundamental and Practical Aspects of Electroanalysis*, 17 (2005) 47–53.
72. Á. Torrinha, C.G. Amorim, M.C. Montenegro, A.N. Araújo, *Talanta*, 190 (2018) 235–247.
73. P. Zheng, N. Wu, *Chemistry–An Asian Journal*, 12 (2017) 2343–2353.
74. J. Wang, J. Qiu, *Journal of Materials Science*, 51 (2016) 4728–4738.
75. J. Xu, J. Tao, L. Su, J. Wang, T. Jiao, *Materials*, 14 (2021) 3987.
76. A.A. Ansari, A. Kaushik, P.R. Solanki, B.e.D. Malhotra, *Electroanalysis: An International Journal Devoted to Fundamental and Practical Aspects of Electroanalysis*, 21 (2009) 965–972.
77. Y. Lin, Y. Peng, J. Di, *Sensors and Actuators B: Chemical*, 220 (2015) 1086–1090.
78. S.-E. Kim, Y.J. Kim, S. Song, K.-N. Lee, W.K. Seong, *Sensors and Actuators B: Chemical*, 278 (2019) 103–109.
79. A. Kaushik, P.R. Solanki, A.A. Ansari, S. Ahmad, B.D. Malhotra, *Electrochemistry Communications*, 10 (2008) 1364–1368.
80. M.B. Dos Santos, S. Azevedo, J. Agusil, B. Prieto-Simón, C. Sporer, E. Torrents, A. Juárez, V. Teixeira, J. Samitier, *Bioelectrochemistry*, 101 (2015) 146–152.
81. K.-J. Huang, D.-J. Niu, W.-Z. Xie, W. Wang, *Analytica Chimica Acta*, 659 (2010) 102–108.
82. S. Alim, J. Vejayan, M.M. Yusoff, A. Kafi, *Biosensors and Bioelectronics*, 121 (2018) 125–136.
83. G. Manasa, R.J. Mascarenhas, S.J. Malode, N.P. Shetti, *Biosensors and Bioelectronics: X*, 11 (2022) 100189.
84. P. Sharma, S.K. Tuteja, V. Bhalla, G. Shekhawat, V.P. Dravid, C.R. Suri, *Biosensors and Bioelectronics*, 39 (2013) 99–105.
85. J. Wang, *Electroanalysis: An International Journal Devoted to Fundamental and Practical Aspects of Electroanalysis*, 17 (2005) 7–14.
86. N. Zanato, L. Talamini, E. Zapp, D. Brondani, I.C. Vieira, *Electroanalysis*, 29 (2017) 1820–1827.
87. L. He, R. Huang, P. Xiao, Y. Liu, L. Jin, H. Liu, S. Li, Y. Deng, Z. Chen, Z. Li, *Chinese Chemical Letters*, 32 (2021) 1593–1602.
88. X. Ma, Y. Jiang, F. Jia, Y. Yu, J. Chen, Z. Wang, *Journal of Microbiological Methods*, 98 (2014) 94–98.
89. A.K. Cheng, D. Sen, H.-Z. Yu, *Bioelectrochemistry*, 77 (2009) 1–12.
90. J. Wu, Y. Zhu, F. Xue, Z. Mei, L. Yao, X. Wang, L. Zheng, J. Liu, G. Liu, C. Peng, *Microchimica Acta*, 181 (2014) 479–491.
91. J.C. Rosch, E.H. Neal, D.A. Balikov, M. Rahim, E.S. Lippmann, *Cellular and Molecular Bioengineering*, 13 (2020) 559–574.
92. M. Hasanzadeh, N. Shadjou, M. de la Guardia, *TrAC Trends in Analytical Chemistry*, 89 (2017) 119–132.
93. A.K. Sari, Y.W. Hartati, S. Gaffar, *Analytical and Bioanalytical Electrochemistry*, 14 (2022) 127–143.
94. R.B. Jamal, S. Shipovskov, E.E. Ferapontova, *Sensors*, 20 (2020) 5561.
95. M. Basak, S. Mitra, M. Gogoi, S. Sinha, H.B. Nemade, D. Bandyopadhyay, *ACS Applied Bio Materials*, 5 (2022) 5321–5332.
96. Z. Chen, M. Xie, F. Zhao, S. Han, *Foods*, 11 (2022) 1404.

5 Metal Oxide Nanomaterials Based Electrochemical Sensors

Aarti Sharma

5.1 INTRODUCTION

Electrochemical sensors based on nanomaterials have gained enormous attention as powerful analytical tools compared to conventional analytical methods available such as high-performance liquid chromatography (HPLC), atomic absorption spectroscopy (AAS), gas chromatography–mass spectroscopy (GC–MS), inductively coupled plasma atomic emission spectroscopy (ICP-AES), inductively coupled plasma mass spectroscopy (ICP-MS), and quantitative polymerase chain reaction. The advantages of using nanomaterial-based electrochemical sensors are their rapid response time, low limit of detection (LOD), selectivity, sensitivity, portability, cost-effectiveness, and stability. The electrochemical cells mostly consist of a working electrode (WE), also known as a sensing electrode, a reference electrode (RE), and a counter electrode (CE). The WEs are modified with nanomaterials for the recognition of analyte molecules. In an electrochemical sensor, when the potential is applied between WE and CE, a redox reaction occurs at the solid/liquid interface of the WE. The corresponding current response is measured with respect to the applied voltage. As shown in the schematic diagram (Figure 5.1),

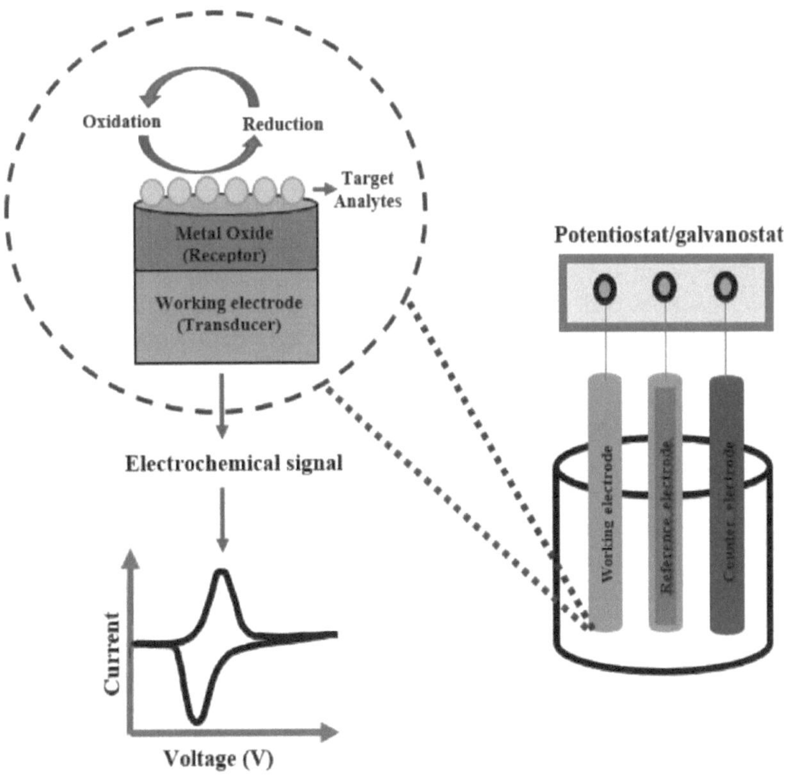

FIGURE 5.1 Illustrative schematic diagram of metal oxide electrochemical sensors.

DOI: 10.1201/9781032654874-5

there are two main components of the electrochemical sensor: firstly, the chemical recognition system, which is responsible for identifying the analyte molecules; secondly, the physiochemical transducer, which converts the chemical interactions at the solid/liquid interface of the WE into detectable electrical signals monitored by modern instruments [1]. The electrochemical sensors are classified on the basis of the electroanalytical technique used to measure the chemical interactions at the WE. The three main categories of electrochemical sensors are (1) potentiometric sensors, (2) amperometric/voltammetric sensors, and (3) impedance/conductometric sensors [2]. A range of analytical techniques used as target recognition and for signal transduction paved ways for developing an array of electrochemical assays for fast, specific, and cost-effective determination of relevant analytes [3]. The most commonly employed electroanalytical techniques in developing electrochemical sensors are voltammetry and amperometry. Voltammetry monitors the change in current (I) versus applied potential (V), whereas amperometry measures the change in current with time at constant potential [4,5]. The conventional electrodes have slow surface kinetics, which affects the sensitivity and selectivity of the electrodes for lower-concentration analytes. The slow kinetic reaction of analytes at the solid/liquid interface of the electrode requires a high potential to proceed at a high rate. With suitable electrocatalysts, the slow electrode reaction can be increased. Therefore, modification of the WE surface with nanomaterials plays a vital role in the development of new-generation smart electrochemical sensors [6,7].

The immense development in designing and synthesizing advanced nanomaterials and subsequently modifying the WEs with these has led to the significant enhancement of electrochemical sensor performance, sensitivity, selectivity, rate of detection, low detection limit, miniaturization, low cost, and quick response time [4,5,8]. A large number of nanomaterials, including quantum dots, carbonaceous nanomaterials (graphene, reduced graphene oxide (GO), carbon nanotubes (CNTs), etc.), novel metal nanomaterials, metal oxide nanomaterials, zeolites, nanocomposites have been extensively studied as electrochemical sensors [9]. The main advantage of using metal oxide nanomaterials as electrocatalysts is their large surface-to-volume ratio due to their small size as compared to bulk counterparts. Also, structural, electronic, and surface properties at the nanoscale level improve significantly, facilitating fast electron transfer at the solid/liquid reaction interface for better catalytic properties [10].

Metal oxide nanomaterials such as ZnO, TiO_2, Fe_3O_4, CuO, MgO, NiO are garnering a lot of attention in the field of electrochemical sensors. Metal oxide nanomaterials are the most versatile group of semiconductor nanomaterials with unique and tunable optical, electrical, magnetic, mechanical, structural, thermal, and chemical properties. These properties are exploited for a wide range of applications, such as catalysts [11], photocatalysts [12], sensors [13], antibacterial agents [14], fuel cells, ion batteries [15], solar cells [16], pyroelectric, piezoelectric, ferroelectric, and magnetic devices [17], supercapacitors [18], fuel cells [19], photoelectrochemical cells [20], and many more [21].

Metal oxide nanomaterials have a large surface area and surface energy, high binding sites, and high porosity, which helps in the adsorption of target analytes and facilitates the redox reaction [22]. Metal oxide nanomaterials of different shapes and morphology, viz. spherical, nanorods, nanowires, nanocubes, nanobelts, nanotubes, nanoflowers, prisms, have been developed [23]. Notably, the properties and applications of metal oxide nanomaterials were found to be dependent on their size, shapes, and textural morphologies. Several synthetic protocols have been developed for synthesizing a wide range of shapes and textural morphologies of different types of metal oxide nanomaterials. Some of the significant methods are the hydrothermal process, solvothermal process, sol-gel process, electrochemical deposition, pulsed laser deposition process, chemical precipitation, sonochemical process, and thermal decomposition [24]. In fact, the properties of metal oxide can be further modified by doping or by forming heterostructures [25], hybrid structures [26], or composite structures [27] with other metal oxides and carbonaceous or polymer materials for enhanced electrochemical properties. Another thing to consider is that the performance of metal oxide nanomaterials as electrocatalysts is dependent on surface properties such as density of surface states, electronic structures, adsorption–desorption parameters, catalytic activity, and the electro-physical properties such as electroconductivity, oxygen gas diffusion, band gap, and type of conductivity [28]. Therefore, the systematic understanding of the factor that controls metal oxide nanomaterial reactivity, selectivity, and sensitivity is imperative. Therefore, significant research has been made in designing the metal oxide-based electrode material with improved catalytic activity and conductivity for signal amplification. Also, to enhance selectivity and sensitivity of metal oxide electro catalysts, the facile interaction of chemical and biological reagents with the electro catalysts can be improved through the immobilization of functional moieties or recognition elements. Moreover, the method used in modifying the electrode by metal oxide nanomaterials also holds the key aspect in the performance of electrochemical sensors [29]. Methods such as electrodeposition, drop casting, chemical modification are used to modify the electrode [30].

The preceding discussion includes a comprehensive overview of the current state of the research in using metal oxide nanomaterial in constructing electrochemical sensors. Firstly, a brief introduction to some of the most explored metal oxide nanomaterials, their properties, and applications is discussed. Then, we focus on reviewing electrochemical sensors based on metal oxide nanomaterials for the analysis of diverse target analytes (heavy metal ions, organic pollutants (OPs), biomolecules, toxic gases).

5.2 METAL OXIDE NANOMATERIALS FOR ELECTROCHEMICAL SENSOR APPLICATION

5.2.1 Iron Oxide

Iron oxide nanomaterials (magnetite (Fe_3O_4), maghemite (γ-Fe_2O_3), and hematite (α-Fe_2O_3)) are extensively explored in electrochemical sensing. Applications of iron oxides depend on their ability to redox (reduction and oxidation) cycle between +2 and +3 oxidation states [31]. The oxidation state of iron in Fe_2O_3 is Fe^{3+}, whereas in Fe_3O_4 it has two oxidation states such as Fe^{2+} and Fe^{3+}. The surface charge transfer in Fe_3O_4 is due to rapid electron transfer between two oxidation states of iron, forming unbound electrons for electrocatalytic reaction at the electrode solid/liquid interface.

Iron oxide nanomaterials are rigorously studied as electrochemical sensors for the detection of many organic and inorganic analytes such as heavy metals (Pb [32,33], Cd [33], Zn [34]), NO_2^- and NO_3^- [35]), glucose, H_2O_2, many organic molecules (urea NO_2^- [36], dopamine (DA) [37], bisphenol A [38]). The doping from transition metal and rare earth ions (Mn [39], Co [40], Ni [41], Pt-Pd [42]) into iron oxide has also enhanced its electrochemical sensing.

Coupling iron oxide with carbonaceous materials (CNTs, GO) also showed high electrocatalytic efficiency toward many analytes [43,44]. Using carbonaceous materials along with metal oxides favors electrocatalytic activity by preventing agglomeration of nanomaterials, resulting in increased surface-to-volume ratio, better electron transfer, excellent mechanical and thermal strength, high electrical conductivity, and stability.

5.2.2 Titanium Dioxide

TiO_2 nanomaterials hold great potential as electrode material in photochemical and electrochemical processes. They possess some remarkable properties such as a large band gap (3.2 eV), strong oxidizing ability, high dielectric constant, low production cost, unique morphology, high surface area, nontoxicity, chemical stability [45]. Different morphologies of TiO_2 have been explored in designing electrochemical sensors such as nanoparticles, nanofibers, nanoneedles, and nanotubes. It has been well-studied that TiO_2 nanotube arrays facilitate the incorporation of biomolecules due to their rigid structure, large surface area, and negative surface charge [46]. The major drawback of pure TiO_2 as an electrochemical sensor is its low solubility and instability of the TiO_2 nanofilms on the electrode surface, resulting in low sensitivity. Novel metal-modified TiO_2 or doped TiO_2 nanoparticles have shown improved electrochemical sensing efficiency [47]. Titanium dioxide–silicon carbide nanocomposite decorated with palladium nanoparticles (Pd@TiO_2-SiC/glassy carbon electrode (GCE)) was used as an electrode for efficiently detecting hydroquinone and bisphenol A. TiO_2 integrated with reduced GO (rGO) has been widely testified in various applications with improved properties and better electrochemical activity efficiency. Titanium dioxide/rGO (TiO_2/rGO) nanocomposites displayed excellent sensitivity and selectivity for the determination of Pb^{2+} and Cu^{2+} [48].

5.2.3 Copper Oxide

Among various metal oxides, CuO is the best naturally occurring intrinsically *p*-type oxide and has properties such as large surface area, biocompatibility, nontoxicity, simple synthesis [49]. Many studies have extensively investigated CuO for the electrochemical detection of organic gases [50], hydrogen sulfide (H_2S) [51], carbon monoxide (CO) [52], nitrogen oxide (NO_2) [53], ethanol gas [54], ammonia (NH_3) [55], HCN [56], and humidity [57]. Coating of noble metal catalysts on the surface of CuO nanostructures or doping CuO has significantly enhanced its sensing properties. Li et al. [58] constructed an ultrasensitive electrochemical sensor by coating Ag nanoparticles onto CuO porous nanobelts for the determination of DA. Coupling CuO with carbonaceous materials such as graphene [59], mesoporous carbons [60], CNTs [61], and carbon nanofibers [62] can enhance the electrochemical performance by improving the electron charge transfer between support matrix and analytes [63].

5.2.4 Zinc Oxide

ZnO is an extensively studied metal oxide with a wide band gap (3.37 eV) and binding energy (60 meV) at room temperature. It has unique physiochemical properties due to the quantum confinement effect at the nanorange, biocompatibility, low synthesis cost, nontoxicity, good electrochemical activity, and thermal/mechanical stability [64]. The wide band gap nanomaterial has the ability to operate at high temperatures, environmental and radiation stability, mechanical robustness, and thermal/mechanical stability [65,66]. The structure and morphology of electrode material play a vital role in electrochemical sensing. Different structures and morphologies of ZnO material can be fabricated by manipulating the operating conditions such as temperature, time, reducing agents, presence of surfactant or capping agents, and pH [67]. The non-centrosymmetric wurtzite configuration of ZnO with polar crystalline faces essential for crystal growth, facilitates the synthesis of different morphologies of ZnO nanomaterials such as nanoplates, nanospheres, nanowires, nanorods, and nanocubes [68]. Ahmed et al. studied a series of ZnO structures with different morphologies such as hexagonal bowl-like microparticles, cones, prisms, nanotablets, synthesized by the solvo/hydrothermal method for electrochemical sensing of amine solution [69]. The electrochemical sensitivity of ZnO can further be improved by doping, coating, or coupling it with carbonaceous materials. Porous GO with ZnO microspheres was synthesized and used for detecting DA [70].

5.2.5 Vanadium Oxide

Vanadium pentoxide (V_2O_5) nanoparticles are widely being studied in Li-ion batteries, supercapacitors, and gas sensors. V_2O_5 possesses redox behavior ($V^{5+/4+}$) due to which it can facilitate electrocatalytic reactions. Vanadium V^{5+} ions have a d^0 electronic configuration which acts as an active site for the adsorption of analyte molecules and catalytic reaction [71]. Furthermore, V_2O_5 is an *n*-type oxide whose electrical conductivity increases on interacting with reducing gases. The V^{5+} ions are reduced to V^{4+}, with electronic conductivity of the order of 0.5 S cm^{-1} at room temperature, making it more sensitive to gas detection [72].

Different V_2O_5 forms including nanorods, nanotubes, nanowires, nanofibers, nanobelts, nanotubes have shown great sensitivity for the determination of oxidizing gases, such as NO_x [73,74], NH_3 [75], and ethanol [76]. Vanadium oxide polypropylene carbonate/indium tin oxide (ITO) electrodes are used for the detection of heavy metals such as Cu^{2+}, Pb^{2+}, Cd^{2+}, and Hg^{2+} [77].

5.3 APPLICATION OF METAL OXIDE NANOMATERIALS IN ELECTROCHEMICAL SENSORS

5.3.1 Heavy Metal Detection

Heavy metals are non-degradable, toxic, carcinogenic, mutagenic, and teratogenic, posing a great threat to human health. Some heavy metals pose a great threat to living organisms even in trace amounts. The toxicity of heavy metals depends on their oxidation state. In the human body, metal ions can be converted into more toxic forms interfering with metabolic processes and causing bio-toxic compounds [78,79]. Mercury (Hg), cadmium (Cd), arsenic (As), chromium (Cr), zinc (Zn), nickel (Ni), copper (Cu), and lead (Pb) are some heavy metals that are very toxic even in trace amount. Heavy metal ions are released into the environment through industrial discharge, mining activities, and household wastewater [79]. Therefore, the determination of heavy metal ions in environmental samples has attracted a great deal of attention. Heavy metal ions are electroactive and undergo redox reactions; therefore, they can be determined by electrochemical sensing methods by measuring changes in the current, potential, capacitance, or impedance. Metal oxide nanomaterials with unique and tunable physiochemical properties are highly suitable for the detection of heavy metal ions. Fe_3O_4 nanoparticles functionalized with terephthalic acid (TA) modified GCE, which was used to selectively detect metal ions such as Hg, Pb, and Cd ions [80]. Dimercaptosuccinic acid-functionalized Fe_3O_4 nanomaterials have been researched for the determination of many metal ions such as Cd, Cu, Ag, and Pb in natural water [81].

Narouei et al. [82] demonstrated the rapid easy-to-use electrochemical method for the selective detection of Pb^{2+} in the presence of interfering ions such as Cu^{2+}, As^{3+}, Zn^{2+}, Cd^{2+}, and Cr^{3+} by TiO_2 nanoparticles when measured in a potential range of −0.4 to −0.6 V. Lui et al. [83] fabricated high surface area ZnO nanofibers for the determination of Cd^{2+} ions. The ZnO nanofibers have a hydroxyl group on their surface that can strongly bind through electrostatic interaction with Cd^{2+} ions, resulting in an enhanced electrochemical response. Belcovici et al. [84] prepared ZnO nanoparticles for the detection of heavy metal ions such as Cu^{2+}, Pb^{2+}, Cd^{2+}, and Fe^{2+}. Metal oxide and carbonaceous entities (graphene, carbon dots, etc.) nanocomposites show enhanced electrochemical sensing performance toward heavy metal ion detection. Mao et al. [85] reported a GCE modified with nano-TiO_2 and multi-walled CNTs composites dispersed in a novel cationic surfactant (3-dehydroabietylamine-2-hydroxypropyl trimethylammonium chloride [DHAHPTMA]) for the determination of trace mercury(II) ions by linear anodic stripping voltammetry. The designed electrode showed a high selectivity for Hg detection in the presence of interfering metal ions (Pb^{2+}, Cu^{2+}, Cd^{2+}, Zn^{2+}, Ni^{2+}, Mn^{2+}). The modified electrode showed good reproducibility of results and stability for the determination of Hg(II) in river water and industrial wastewater samples. Doping or addition of impurities in metal oxide nanomaterials matrix results in increased sensitivity and selectivity toward heavy metal ion detection. Recently, Fe-doped TiO_2 nanomaterials were synthesized for efficient electrochemical detection of Hg^{2+} ions. Fe doping of TiO_2 tuned the concentration of oxygen vacancies in TiO_2 nanoparticles, which led to the formation of surface hydroxyl groups (–OH) in water, which accelerated the capture of Hg(II) in aqueous medium. Furthermore, it was observed that doping also resulted in the Ti^{3+} formation in TiO_2. Fe^{2+}/Fe^{3+} and Ti^{3+}/Ti^{4+} cycles speeded the redox of Hg(II) ions, favoring the electrochemical detection [86]. Karthik et al. [87] prepared cobalt-doped ZnO/rGO (Co:ZnO/rGO) nanorods, which showed excellent sensitivity for Cd^{2+} and Pb^{2+} ions, with a detection limit of 0.94 μg L^{-1} (Cd (II)) and 0.83 μg L^{-1} (Pb^{2+})

in the concentration range of 10–90 μg L^{-1}. Dong et al. [88] synthesized 3D porous core-shell vanadium oxide microspheres by assembling many nanoparticles by hydrothermal process. They observed that active sites on core-shell vanadium oxide microspheres with abundant ion transport channels effectively reduced the ion diffusion distance and favored the detection of thallium (Tl^{3+}) and indium (In^{3+}) ions.

5.3.2 Organic Pollutant Detection

OPs are highly toxic hydrophobic organic compounds with highly stable chemical structures that can persist in the environment for a really long time. The OPs are derived from agricultural and industrial activities, such as pharmaceutical wastes, the textile industry, pesticides, disinfectants, fungicides, and insecticides. These compounds can bioaccumulate in organisms to a level that can be harmful to human health and the environment [88]. OPs cause immunotoxicity, damage to the liver, kidney, blood, nervous system, cardiovascular system, and immune system, retard reproduction, and also induce congenital disabilities. Therefore, the quick detection of these OPs in the environment is an important issue [89]. Many efforts have been put into developing a reliable method that is fast, analytically sensitive, selective, reproducible, and easy to use for the detection of OPs. For the detection of OPs, electrochemical sensors have emerged as an attractive method [90]. However, electrochemical sensors for the determination of OPs are mostly biosensors, which are enzyme-based and therefore suffer from many limitations such as high cost, denaturation of enzyme, and low reproducibility [91]. Under adverse environmental conditions such as pH, temperature, and chemical change, enzymes are unstable, and enzyme denaturation happens, which limits biosensors' practical application. For example, in the case of acetylcholinesterase (AChE)-based biosensors for the detection of organophosphates, the presence of various interfering metal ions can affect the enzyme selectivity, resulting in unreliable results [92]. Therefore, non-enzymatic electrochemical sensors as a reliable method are strongly advocated.

Many metal oxide nanomaterials have been successfully used for designing electrochemical sensors for OPs. For example, CuO nanoballs have been used for electrochemical determination of 2,4,6-trichlorophenol (TCP). CuO nanoballs modified GCE (CuO/Nafion/GCE) showed potential as an efficient electrochemical sensor for 2,4,6-TCP. The concentration range for 2,4,6-TCP was from 1 to 120 μM with LOD value of 0.046 μM. The proposed electrochemical sensor was successfully used for real water sample analysis as well [93]. Nurdin et al. [94] synthesized graphene-modified TiO_2 for the electrochemical-based phenol sensing. Furthermore, it has been observed that metal oxides with hybrid or heterogeneous structures or composites showed better electrochemical properties for OP detection with better sensitivity and selectivity. An electrochemical detection of the OP lindane was carried out by CuO-MnO_2. The voltammogram showed a single irreversible peak confirming reduction by the six-electron process to form benzene. The concentration range for lindane was set from 1 to 700 mM with a LOD value of 4.8 nM and sensitivity value of 0.12 mA mM cm^2 [95]. Tian et al. [96] studied CuO-TiO_2 nanocomposites for the detection of organophosphate pollutant (methyl parathion) electrochemically. The modified electrode showed good sensitivity in the detection range of 0–2000 ppb with LOD of 1.21 ppb [96]. Zhao et al. [97] synthesized CuO/rGO nanocomposites which exhibited a higher electrocatalytic activity toward catechol oxidation. Huo et al. [98] developed a hybrid structure of copper oxide nanowire and single-walled CNT (CuO NW/SWCNT)-based non-enzymatic electrochemical sensor. A differential pulse voltammetry (DPV) technique was used for the detection of insecticide malathion. The set concentration range was 0–40 ppb with LOD value of 0.1 ppb (0.3 nM). The developed sensor showed good selectivity of malathion detection even in the presence of interfering analytes, good stability, and reproducibility. Sangamithirai et al. [99] recently reported g-C_3N_4/V_2O_5 (VOCN) nanocomposites modified GCE as an effective electrochemical sensor with high sensitivity and selectivity for detecting nitroaromatics (4-nitrophenol (4-NP), 2,4-dinitrophenol (DNP), dinitrotoluene (DNT), and 2,4,6-trinitrophenol (TNP)) in the presence of interfering metal ions. The designed electrode showed LOD values as

0.85, 3.3, 3.4, and 11.0 nM in a linear range between 1.0 nM and 100.0 μM, 10.0 nM and 100.0 μM, 10.0 nM and 100.0 μM, and 30.0 nM and 100.0 μM for 4-NP, DNP, DNT, and TNP, respectively. Rajesh et al. [100] synthesized rGO intercalated–vanadium pentoxide (Gr–V_2O_5) nanosheets. The Gr-V_2O_5 nanosheet-modified GCE (Gr-V_2O_5/GCE) was used for the detection of 3,5-dichlorophenol (DCP). The modified electrode showed high selectivity with LOD value of 5 nM and sensitivity of 0.3577 μA μM^{-1}.

Another important OP of concern is per- and polyfluoroalkyl substances (PFAS). PFAS are short- and long-chain fluorinated alkyl acids. The two most explored and studied members of the PFAS family are perfluorooctanoic acid (PFOA) and perfluorooctane sulfonic acid (PFOS) [101]. Fenti A. et al. prepared Magnéli-phase titanium suboxides $TinO_{2n-1}$ and boron-doped ultrananocrystalline diamond and IrO_2-Ta_2O_5 mixed metal oxides for perfluorooctane sulfonate (PFOS) oxidation. It was observed that PFOS oxidation rates were higher for titanium suboxides $TinO_{2n-1}$ and consumed lower energy than boron-doped ultrananocrystalline diamond and IrO_2-Ta_2O_5 mixed metal oxides [102].

5.3.3 Biomolecule Detection

Biomolecules such as carbohydrates, lipids, proteins, nucleic acids are of great importance to human health. Any abnormal changes in these can lead to serious life-threatening diseases such as scurvy, leukemia, and Parkinson's disease [103]. Electrochemical sensors based on metal oxide nanomaterials can successfully be used to detect chemical or biological species. Different morphologies of metal oxide nanomaterials have been explored in studying electrochemical sensing of biological molecules. Balram et al. [104] developed a successful novel electrochemical sensor by using flower-shaped ZnO nanoparticles for sensing DA with LOD value of 0.04 μM and a sensitivity of 4.38 μA $\mu M^{-1} cm^{-2}$. Similarly, CuO leaf-like nanostructure and nano-rice-like structure also exhibit good sensitivity for DA detection. The set concertation range was 10 to 60 mM with LOD value of 50 μM and sensitivity value of 0.1975 mA mM^{-1} [105]. Urhan et al. [106] developed Cu_2O/CuO/electrochemically rGO (Cu_xO/ERGO) nanocomposite for detecting DA in the presence of ascorbic acid (AA). The modified electrode showed high electrocatalytic activity and high sensitivity. The linear range was 0.1–400 μM with LOD value of 12 nM and sensitivity value of 122.4 mA $cm^{-2 \mu M^{-1}}$. Much research showed that ZnO/rGO nanocomposite showed high sensitivity and selectivity for DA even in the presence of other analytes such as AA, uric acid (UA) [107,108]. Swamy et al. [109] reported that the CuO- and NiO-modified electrodes successfully separated anodic signals of DA, AA, and tryptophan (Tyr) in their ternary mixture. It has been proved that nanoparticles of CuO exhibit a good electrocatalytic nature than NiO nanoparticles. Coupling of metal oxide nanomaterials with rGO shows the synergistic effect, which facilitated the better electrochemical sensing of DA. Cai et al. [110] prepared various morphologies of iron oxide nanoparticles (Fe_2O_3 NPs), such as cubic, orthorhombic, and discal for the electrochemical sensing of DA and UA. The discal Fe_2O_3 NPs coupled with GO resulted in better electrocatalytic sensing of DA and UA. The linear range for DA was set from 0.02 to 10 μM with LOD value of 3.2 nM, and for UA, the linear range was set from 10 to 100 μM with LOD value of 2.5 nM. Furthermore, the effects of doping have also been explored by many researchers. For example, Suresh et al. [111] prepared Ni-doped V_2O_5 nanoparticles by thermal decomposition method, which exhibited high selectivity and sensitivity for DA. The prepared nanomaterial showed a good response for DA detection. The linear concentration range for DA was from 6.6 to 96.4 μM with LOD value of 28 nM and the sensitivity of 132 nA μM^{-1}. Therefore, metal oxide nanomaterials can successfully be used in designing electrochemical sensors for biomolecule detection with a low LOD, better selectivity, sensitivity, and good reproducibility.

5.3.4 Gas Pollutant Detection

The exponential growth in industrialization, construction, and mining has taken a toll on air quality, resulting in air pollution. Air pollution is the presence of toxic gases such as NH_3, CO, nitric oxide

(NO), NO_2, hydrogen sulfide (H_2S), sulfur oxide (SO_2), volatile organic compounds (VOC) in the environment, which has negative effects on human health. It also has a detrimental effect on the environment with acceleration in climate change, acid rain, and changes in soil quality that affect plant growth and groundwater quality. For monitoring air quality, metal oxide-based electrochemical sensors have some potential results. Some metal oxide gas sensors have already made their place in the market and are commercially available for monitoring the presence of toxic gases at trace levels. The gas-sensing mechanism of metal oxide nanomaterials involves recognition system and transducer. The recognition system involves recognizing the analyte gas molecule, which results in electronic or chemical change at the surface of metal oxide nanomaterial. The transducer's function is to convert the aforementioned surface phenomenon into an electrical signal. Metal oxide nanomaterials have oxygen molecules adsorbed on their surface. These adsorbed oxygen molecules extract electrons from the conduction band of semiconducting metal oxides and trap the electrons by forming oxide ions. This produces band bending, resulting in an electron-depleted region called as depletion layer. When these metal oxides interact with analyte gas molecules, there is a change in the depletion layer [112–115]. The increase or decrease in the depletion layer or conductivity depends on the majority charge carrier in the metal oxide (*n*-type or *p*-type) and also on the nature of the analyte gas molecules (oxidizing or reducing) [115].

In the case of *n*-type metal oxide, the electrons are removed from the conduction band surface by the adsorbed oxygen molecules, resulting in the formation of the negatively charged oxide ion O_2^-. Consequently, this results in the formation of the depletion layer and potential barrier, which decreases the conductivity or increases the resistance of the oxide layer, due to loss of electrons. Herein, the analyte gas plays an important role in the detection mechanism. When the adsorbed analyte is a reducing gas (donor) such as H_2, NH_3, H_2S, it will result in the release of electrons into the depletion layer at the metal oxide surface. This will lower the potential barrier level and therefore increase the conductivity (or decrease the resistance) of the oxide layer. On the other hand, if the analyte gas is an oxidizing gas (acceptor) such as NO, O_3, Cl_2, CH_3COCH_3, further loss of electrons causes an increase in the potential barrier and widening of the depletion layer and, consequently, decreases the conductivity or increases the resistance of the oxide layer [113–115].

Similarly, in the case of *p*-type metal oxides, the adsorption of reducing gases will decrease the conductivity of the oxide layer due to the release of the electrons by the reducing gas molecules, which will recombine with the holes in the *p*-type metal oxide depletion layer. This decreases the *p*-type charge carrier concentration, hence leading to a decrease in conductivity. If the *p*-type metal oxide is exposed to oxidizing gases, the electrons are captured from the oxide depletion layer, and therefore, *p*-type charge carrier concentration increases, resulting in an increase in conductivity or a decrease in resistance [113–115]:

$$O_{2(ads)} + e^- \leftrightarrow O^-_{2(ads)} (< 100°C)$$

$$O^-_{2(ads)} + e^- \leftrightarrow 2O^-_{(ads)} (100 - 300°C)$$

Mani et al. [116] investigated undoped and cobalt (Co)-doped nanostructured ZnO thin films for NH_3 sensing. The NH_3 vapor sensing mechanism can be explained as follows:

$$O_2 + e^-_{(MO\ surface)} \leftrightarrow O^-_{2\ (MO\ surface)}$$

$$4NH_3 + O^-_{2(MO\ surface)} \leftrightarrow 2N_2 + 6H_2O + 6e^-$$

They showed that Co-doped ZnO thin films were better candidates than undoped ZnO for sensing NH_3 at room temperature.

Tesfamichael et al. [117] demonstrated that W-doped ZnO thin films showed better electrochemical activity toward NO_2 sensing for a concentration range from 5 to 10 ppm at an operating temperature of 150°C. Furthermore, Z.S. Hosseini et al. [118] showed that vertically aligned ZnO nanorods with flower-like structures have a better gas response (S) of 581 toward 5 ppm and good sensitivity at a temperature of 250°C for better detection of H_2S. Wu et al. [119] synthesized highly porous CuO–ZnO nanocomposite for the detection of carbon monoxide. The sensing electrode was made of ZnO nanotetrapod-supported CuO nanoparticles and exhibited sensitivity toward 25 ppm carbon monoxide in air at 500°C.

Clearly, it has been established that different morphologies of nanomaterials or altering their electronic or physical properties by doping or forming hybrid or composite structures with novel metals, carbonaceous materials, and metal oxides are significantly modifying gas-sensing properties of metal oxides and paving the way for newer and better gas-sensing material.

5.4 CONCLUSIONS AND FUTURE OUTLOOKS

The development of nanoscale electrode materials for the designing of robust electrochemical sensor platforms has gained significant importance. This chapter provides readers with a broader outlook on using metal oxide as an electrochemical sensor in this vastly broadening multidisciplinary research. Through this chapter, an attempt has been made to provide a brief summary to the reader of the different electrochemical and physiochemical aspects of metal oxide nanomaterials and recent advances in designing electrochemical sensors for various applications. Undoubtedly, metal oxide nanomaterials with tunable catalytic, electrochemical, and structural properties are the potential candidates to enlarge the commercial application of electrochemical devices. However, the main disadvantages of using pure metal oxide nanomaterials in sensors are poor conductivity, agglomeration of nanoscale materials, low catalytic activity, and difficulty in bulk production of nanomaterials. These difficulties can be overcome by modification, coupling, and doping of metal oxides with other metal oxides or carbonaceous materials (graphene, CNTs, carbon nanofibers) or with metal ions (Cu, Ag, Mn, etc.). This not only enhances its physiochemical properties but also improves the selectivity and sensitivity toward various target analytes. All the applications that are discussed in this chapter show that doped or composite materials yield better or enhanced sensing results. Another challenge with metal oxide nanomaterials is less stability as they are prone to agglomeration, resulting in compromised sensitivity, resulting in more reproducibility of results. Most of the discussed results are simply shown as proof of the concept and performed in laboratories. Furthermore, most reported metal oxide sensors are rarely tested on real-life samples, such as wastewater, which shows considerable analytical complexity. Even though an exhausting amount of research has been done on metal oxide-based electrochemical sensors, there are still many element oxides that remain unexplored such as non-transition metal oxides, lanthanides. The success in the designing of metal oxide electrochemical sensors is worth appreciation; however, there is still a wide scope for upgradation in the future. The most important prerequisite parameters required in successfully transitioning the electrochemical sensors from laboratory to commercialization are selectivity, sensitivity, LOD, limit of quantification of analyte molecules, and reproducibility of results. The substantial challenges that continue to persist in designing commercially viable metal oxide electrochemical sensors are listed as follows: (1) electrochemical sensor should display a low LOD, that is, the lowest concentration of the analyte that can be reliably detected as most of the pollutants are harmful to humans or environment even in the trace amount; (2) it must also have high selectivity and sensitivity of analytes in real and complex samples like blood, industrial wastewater; and (3) reproducibility and stability of electrochemical sensor in complex real sample as most of the nanomaterials tend to deteriorate or agglomerate or leach out in real/complex samples. They are also affected by environmental conditions such as pH, temperature, humidity. Nevertheless, for the future commercialization of metal oxide electrochemical sensors, more efforts are needed to translate laboratory results into commercial applications.

REFERENCES

1. Maduraiveeran G. Bionanomaterial-based electrochemical biosensing platforms for biomedical applications. *Analytical Methods* 2020; 12: 1688–1701. DOI: 10.1039/D0AY00171F
2. Murthy H.A., Wagassa A.N., et al. Functionalized metal and metal oxide nanomaterial-based electrochemical sensors. In C.M. Hussain and J.G. Manjunatha (Eds.) *Functionalized Nanomaterial-Based Electrochemical Sensors.* Woodhead Publishing, Cambridge, 2022; pp. 369–392. DOI: 10.1016/B978-0-12-823788-5.00001-6
3. Xiao F., Wang L., Duan, H. Nanomaterial based electrochemical sensors for in vitro detection of small molecule metabolites. *Biotechnology Advances* 2016; 34(3): 234–249. DOI: 10.1016/j.biotechadv.2016.01.006
4. Bard A.J., Faulkner L.R., White H.S. *Electrochemical Methods: Fundamentals and Applications.* John Wiley & Sons, Hoboken, NJ, 2022.
5. Eliaz N., Eliezer G. *Physical Electrochemistry: Fundamentals, Techniques, and Applications.* John Wiley & Sons, Hoboken, NJ, 2019.
6. Baig N., et al. Recent trends in nanomaterial-modified electrodes for electroanalytical applications. *Trends in Analytical Chemistry* 2019; 111: 47–61. DOI: 10.1016/j.trac.2018.11.044. 10.1016
7. Karimi-Maleh H. et al. A novel electrochemical epinine sensor using amplified CuO nanoparticles and an-hexyl-3-methylimidazolium hexafluorophosphate electrode. *New Journal of Chemistry* 2019; 43: 2362–2367. DOI: 10.1039/C8NJ05581E.
8. Ungureanu, C. et al. Metallic and metal oxides nanoparticles for sensing food pathogens—An overview of recent findings and future prospects. *Materials* 2022; 15: 5374. DOI: 10.3390/ma15155374
9. Kucherenko I.S., Soldatkin O.O., et al. Advances in nanomaterial application in enzyme-based electrochemical biosensors: A review. *Nanoscale Advances* 2019; 1(12): 4560–4577. DOI: 10.1039/C9NA00491B
10. Lim W.Q., Gao Z. Metal oxide nanoparticles in electroanalysis. *Electroanalysis* 2015; 27: 1–18. DOI: 10.1002/elan.201500024
11. Wang S., Wang Z., Zha Z. Metal nanoparticles or metal oxide nanoparticles, an efficient and promising family of novel heterogeneous catalysts in organic synthesis. *Dalton Transactions* 2009; 21: 9363–9373. DOI: 10.1039/B913539A
12. Zhang H., Chen G., Bahnemann D. Photoelectrocatalytic materials for environmental applications. *Journal of Materials Chemistry* 2009; 19: 5089–5121. DOI: 10.1039/B821991E
13. Sun Y., Liu S., et al. Metal Oxide nanostructures and their gas sensing properties: A review. *Sensors* 2012; 12: 2610–2631.DOI: 10.3390/s120302610
14. Hajipour M.J., Fromm K.M., et al. Antibacterial properties of nanoparticles. *Trends Biotechnology* 2012; 30: 499–511. DOI: 10.1016/j.tibtech.2012.06.004
15. Poizot P., Laruelle S., et al. Nano-sized transition-metal oxides as negative-electrode materials for lithium-ion batteries. *Nature* 2000; 407: 496–499. DOI: 10.1038/35035045
16. Jose R., Thavasi V. Metal oxides for dye-sensitized solar cells. *Journal of the American Ceramic Society* 2009; 92: 289–301. DOI: 10.1111/j.1551–2916.2008.02870.x
17. Martin L., Chu Y., Ramesh R. Advances in the growth and characterization of magnetic, ferroelectric, and multiferroic oxide thin films. *Materials Science and Engineering R* 2010; 68: 89–133. DOI: 10.1016/j.mser.2010.03.001
18. Rakhi R., Chen W., et al. High performance supercapacitors using metal oxide anchored graphene nanosheet electrodes. *Journal of Materials Chemistry* 2011; 21: 16197–16204. DOI: 10.1039/C1JM12963E
19. Jiang J., Li Y., et al. Recent advances in metal oxide-based electrode architecture design for electrochemical energy storage. *Advanced Materials* 2012; 24: 5166–5180. DOI: 10.1002/adma.201202146
20. Qiu Y., Pan Z., et al. Current progress in developing metal oxide nanoarrays-based photoanodes for photoelectrochemical water splitting. *Science Bulletin* 2019; 64(18): 1348–1380. DOI: 10.1016/j.scib.2019.07.017
21. Rodríguez J.A., Fernández Garcia M. *Synthesis, Properties, and Applications of Oxide Nanomaterials.* Wiley-Interscience: Hoboken, NJ, 2007.
22. George J.M., Antony A., Mathew B. Metal oxide nanoparticles in electrochemical sensing and biosensing: A review. *Microchimica Acta* 2018; 185(7): 1–26. DOI: 10.1007/s00604-018-2894-3
23. Li Y., Yang X., et al. One-dimensional metal oxide nanotubes, nanowires, nanoribbons, and nanorods: Synthesis, characterizations, properties and applications. *Critical Reviews in Solid State and Materials Sciences* 2012; 37: 1–74. DOI: 10.1080/10408436.2011.606512

24. Hornyak G.L., Moore J.J., et al. *Fundamentals of Nanotechnology.* CRC Press, Taylor & Francis Group: Boca Raton, FL, 2008.
25. Zhang W., Fan G., et al. Interfacial engineering of hierarchical transition metal oxide heterostructures for highly sensitive sensing of hydrogen peroxide. *Small* 2018; 14(19): 1703713. DOI: 10.1002/smll.201703713
26. Dakshayini B.S., Reddy K.R., et al. Role of conducting polymer and metal oxide-based hybrids for applications in ampereometric sensors and biosensors. *Microchemical Journal* 2019; 147: 7–24. DOI: 10.1016/j.microc.2019.02.061
27. Kumar S., Bukkitgar S.D., et al. Electrochemical sensors and biosensors based on graphene functionalized with metal oxide nanostructures for healthcare applications. *ChemistrySelect* 2019; 4(18): 5322–5337. DOI: 10.1002/slct.201803871
28. Fazio E., Spadaro S., et al. Metal-oxide based nanomaterials: Synthesis, characterization and their applications in electrical and electrochemical sensors. *Sensors* 2021; 21(7): 2494. DOI: 10.3390/s21072494
29. Kamau G.N. (1988). Surface preparation of glassy carbon electrodes. *Analytica Chimica Acta* 1988; 207: 1–16. DOI: 10.1016/s0003-2670(00)80777-1
30. Wang F., Hu S. Electrochemical sensors based on metal and semiconductor nanoparticles. *Microchimica Acta* 2009; 165: 1–22. DOI: 10.1007/s00604-009-0136-4.
31. Siregar J., Septiani N.L.W., et al. Review- A pollutant gas sensor based on Fe_3O_4 nanostructures: A review. *Journal of Electrochemical Society* 2021; 168: 027510. DOI: 10.1149/1945-7111/abd928
32. Liu Y., Wu S., et al. Interface Co-assembly synthesis of magnetic Fe_3O_4@mesoporous carbon for efficient electrochemical detection of Hg(II) and Pb(II). *Advanced Materials Interfaces* 2022; 10: 2201631. Version of Record online: DOI: 10.1002/admi.202201631
33. Xu Z., Fan X., et al. A sensitive electrochemical sensor for simultaneous voltammetric sensing of cadmium and lead based on Fe_3O_4/multiwalled carbon nanotube/laser scribed graphene composites functionalized with chitosan modified electrode. *Materials Chemistry and Physics* 2019; 238: 121877. DOI: 10.1016/j.matchemphys.2019.121877
34. Li W.J., Yao X.Z., et al. Fe_3O_4 with novel nanoplate-stacked structure: Surfactant-free hydrothermal synthesis and application in detection of heavy metal ions. *Journal of Electroanalytical Chemistry* 2015; 749: 75–82. DOI: 10.1016/j.jelechem.2015.04.038
35. Li B.Q., Nie F., et al. An electrochemical sensor for sensitive determination of nitrites based on Ag-Fe_3O_4-graphene oxide magnetic nanocomposites. *Chemical Papers* 2015; 69: 911–920. DOI: 10.1515/chempap-2015-0099
36. Singh A.K., Singh M., et al. Electrochemical preparation of Fe_3O_4/MWCNT-polyaniline nanocomposite film for development of urea biosensor and its application in milk sample. *Journal of Food Measureme nt and Characterization* 2020; 14: 163–175. DOI: 10.1007/s11694-019-00278-2
37. Anshori I., Kepakisan K.A.A., et al. Facile synthesis of graphene oxide/Fe_3O_4 nanocomposite for electrochemical sensing on determination of dopamine. *Nanocomposites* 2022; 8: 155–166. DOI: 10.1080/20550324.2022.2090050.
38. Shen M., Li W., et al. A ratiometric electrochemical sensor for bisphenol A detection based on Ag@Fe_3O_4-rGO composite. *Microchemical Journal* 2023; 186: 108315. DOI: 10.1016/j.microc.2022.108315
39. Su Y., Zheng X., et al. Mn-Fe_3O_4 nanoparticles anchored on the urushiol functionalized 3D-graphene for the electrochemical detection of 4-nitrophenol. *Journal of Hazardous Materials* 2021; 409: 124926. DOI: 10.1016/j.jhazmat.2020.124926
40. Nehru R., Dong C.D., et al. Cobalt-doped Fe_3O_4 nanospheres deposited on graphene oxide as electrode materials for electrochemical sensing of the antibiotic drug. *ACS Applied Materials and Interfaces* 2021; 4(7): 6768–6777. DOI: 10.1021/acsanm.1c00826.
41. Kaplan S., Karatekin R.S., et al. A novel Ni–Fe_3O_4@s-rGO/GCE electrode for electrochemical detection of H_2O_2. *Materials Chemistry and Physics* 2023; 294: 127051. DOI: 10.1016/j.matchemphys.2022.127051
42. Sun X., Guo S., et al. Dumbbell-like PtPd-Fe_3O_4 nanoparticles for enhanced electrochemical detection of H_2O_2. *Nano Letters* 2012; 12: 4859–4863. DOI: 10.1021/nl302358e
43. Lee S., Oh J., Kim D., Piao Y. (2016) A sensitive electrochemical sensor using an iron oxide/graphene composite for the simultaneous detection of heavy metal ions. *Talanta* 2016; 160: 528–536. DOI: 10.1016/j.talanta.2016.07.034
44. Chauhan N., Pundir C.S., et al. (2011) An amperometric biosensor based on acetylcholinesterase immobilized onto iron oxide nanoparticles/multi-walled carbon nanotubes modified gold electrode for measurement of organophosphorus insecticides. *Analytica Chimica Acta* 2011; 701: 66–74. DOI: 10.1016/j.aca.2011.06.014

45. Zhou Q., Fang Z., et al. Applications of TiO_2 nanotube arrays in environmental and energy fields: A review. *Microporous and Mesoporous Materials* 2015; 202: 22–35. DOI: 10.1016/j.micromeso.2014.09.040
46. Mun K.S., Alvarez S.D., et al. A stable, label-free optical interferometric biosensor based on TiO_2 nanotube arrays. *ACS Nano* 2010; 4: 2070–2076. DOI: 10.1021/nn901312f
47. Lin S., et al. Highly selective and sensitive glucose sensors based on organic electrochemical transistors using TiO_2 nanotube arrays-based gate electrodes. *Sensors and Actuators, B: Chemical*, 2015; 208: 457–463. DOI: 10.1016/j.snb.2014.11.038
48. Mathew S., John B.K., et al. Metal-doped titanium dioxide for environmental remediation, hydrogen evolution and sensing: A review. *ChemistrySelect* 2021; 6 (45): 12742–12751. DOI: 10.1002/slct.202103577
49. Zhang Q., Zhang K., et al. CuO nanostructures: Synthesis, characterization, growth mechanisms, fundamental properties, and applications. *Progress in Materials Science* 2014; 60: 208–337. DOI: 10.1016/j.pmatsci.2013.09.003
50. Yang C., Xiao, F., et al. (2015). 3D flower- and 2D sheet-like CuO nanostructures: Microwave-assisted synthesis and application in gas sensors. *Sensors and Actuators, B: Chemical* 2015; 207: 177–185. DOI: 10.1016/j.snb.2014.10.063
51. Daniela C., Rachele M., et al. (2015). A novel nonenzymatic amperometric Hydrogen peroxide sensor based on CuO@Cu_2O nanowires embedded into poly(vinyl alcohol). *Talanta* 2016; 147: 124–131. DOI: 10.1016/j.talanta.2015.09.038
52. Wu N., Zhao M., et al. Porous CuO–ZnO nanocomposite for sensing electrode of high-temperature CO solid-state electrochemical sensor. *Nanotechnology* 2005; 16: 2878. DOI 10.1088/0957-4484/16/12/024
53. Fang H., Li S., et al. Enhanced qawaz NO_2 gas sensing performance by hierarchical CuO–Co_3O_4 spheres. *Sensors and Actuators, B: Chemical* 2022; 352: 131068. DOI: 10.1016/j.snb.2021.131068
54. Umar A., Ibrahim A.A., et al. Fabrication and characterization of CuO nanoplates based sensor device for ethanol gas sensing application. *Chemical Physics Letters* 2020; 763: 138204. DOI: 10.1016/j.cplett.2020.138204
55. Bhuvaneshwari S., Gopalakrishnan N., et al. (2019). Selective ammonia sensor based on copper oxide/reduced graphene oxide nanocomposite. *Journal of Alloys and Compounds* 2019; 788: 422–428. DOI: 10.1016/j.jallcom.2019.02.245
56. Yang M., He J., et al. Copper oxide nanoparticle sensors for hydrogen cyanide detection: Unprecedented selectivity and sensitivity. *Sensors and Actuators, B: Chemical* 2011; 155: 692–698. DOI: 10.1016/j.snb.2011.01.031
57. Wang Z., Xiao Y., et al. Humidity-sensing properties of urchinlike CuO nanostructures modified by reduced graphene oxide. *ACS Applied Materials and Interfaces* 2014; 6(6): 3888–3895. DOI: 10.1021/am404858z
58. Li Y.Y., Kang, P., et al. Ag nanoparticles anchored onto porous CuO nanobelts for the ultrasensitive electrochemical detection of dopamine in human serum. *Sensors and Actuators, B: Chemical* 2021; 327: 128878. DOI: 10.1016/j.snb.2020.128878
59. Doğan H., Urhan B.K., et al. Simultaneous electrochemical detection of ascorbic acid and dopamine on Cu_2O/CuO/electrochemically reduced graphene oxide (Cu_xO/ERGO)-nanocomposite-modified electrode. *Microchemical Journal* 2019; 150: 104157. DOI: 10.1016/j.microc.2019.104157
60. Gu C., Wang Q., et al. Ultrasensitive non-enzymatic pesticide electrochemical sensor based on HKUST-1-derived copper oxide @ mesoporous carbon composite. *Sensors and Actuators, B: Chemical* 2020; 305(15): 127478. DOI: 10.1016/j.snb.2019.127478
61. Zhang G., Pan P., et al. Rapid synthesis of cypress-like CuO nanomaterials and CuO/MWCNTs composites for ultra-high sensitivity electrochemical sensing of nitrite. *Microchemical Journal* 2020; 159: 105439. DOI: 10.1016/j.microc.2020.105439
62. Bai B.C., Lee Y.S., Im J.S. Activated carbon fibers for toxic gas removal based on electrical investigation: Mechanistic study of p-type/n-type junction structures. *Scientific Reports* 2019; 9: 14458. DOI: 10.1038/s41598-019-50707-x
63. Zhang J., Ma J., Zhang S., et al. (2015) A highly sensitive nonenzymatic glucose sensor based on CuO nanoparticles decorated carbon spheres. *Sensors and Actuators, B: Chemical* 2015; 211: 385–391. DOI: 10.1016/j.snb.2015.01.100
64. Vajedi F., Dehghani H. (2019). The characterization of TiO_2-reduced graphene oxide nanocomposites and their performance in electrochemical determination for removing heavy metals ions of cadmium(II), lead(II) and copper(II). *Materials Science and Engineering B* 2019; 243: 189–198. DOI: 10.1016/j.mseb.2019.04.009
65. Beitollahi H., Tajik S., et al. Recent advances in ZnO nanostruture based electrochemical sensors and biosensors. *Journal of Materials Chemistry B* 2020; 8: 5826–5844. DOI: 10.1039/D0TB00569J

66. Franke M.E. et al. Metal and metal oxide nanoparticles in chemiresistors: Does the nanoscale matter? *Small* 2006; 2(3): 301–301. DOI: 10.1002/smll.200500261
67. Shetti N.P. et al. ZnO-based nanostructured electrodes for electrochemical sensors and biosensors in biomedical applications. *Biosensors and Bioelectronics* 2019; 141: 111417. DOI: 10.1016/j.bios.2019.111417.
68. Nikos V.M. Liquid-phase synthesis of nanoparticles: Particle size distribution dynamics and control. *Chemical Engineering Science* 2005; 60: 4749–4770. DOI: 10.1016/j.ces.2005.04.012
69. Ahmed M., Zhao R., et al. Morphology evolution of ZnO by controlling solvent and electrochemical sensing of hexagonal nanotablets toward amines. *Chinese Chemical Letters* 2020; 31(8): 2091–2094. DOI: 10.1016/j.cclet.2020.01.014
70. Zhao M., Li Z., et al. Employing the interfacial barrier of P-rGO/ZnO microspheres for improving the electrochemical sensing performance to dopamine. *Sensors and Actuators, B: Chemical* 2020; 309: 127757. DOI: 10.1016/j.snb.2020.127757
71. Yan Y., Li B., et al. Vanadium based materials as electrode materials for high performance supercapacitors. *Journal of Power Sources* 2016; 329: 148–169. DOI: 10.1016/j.jpowsour.2016.08.039
72. Schneider K., Lubecka M., Czapla A. V_2O_5 thin films for gas sensor applications. *Sensors and Actuators, B: Chemical* 2016; 236(29): 970–977. DOI: 10.1016/j.snb.2016.04.059
73. Yan W., Hu M., et al. Room temperature gas sensing properties of porous silicon/V_2O_5 nanorods composite. *Applied Surface Science* 2015; 346: 216–222. DOI: 10.1016/j.apsusc.2015.01.020
74. Li Y., Huang Z., Rong S. A vanadium oxide nanotube-based nitric oxide gas sensor. *Sensors and Materials* 2006; 18: 241–249.
75. Modafferi V., Trocino S., et al. Electrospun V_2O_5 composite fibers: Synthesis, characterization and ammonia sensing properties. *Thin Solid Films* 2013; 548: 689–694. DOI: 10.1016/j.tsf.2013.03.137
76. Jin W., Yan S.L., et al. Enhancement of ethanol gas sensing response based on ordered V_2O_5 nanowire microyarns. *Sensors and Actuators, B: Chemical* 2015; 206: 284–290. DOI: 10.1016/j.snb.2014.09.064
77. Shen S., Zhang Y., et al. Heavy metal analysis in Tai lake water sample based on vanadium oxide-polypropylene carbonate modified electrode. *International Journal of Electrochemical Science* 2016; 11: 5669–5678. DOI: 10.20964/2016.07.49
78. Tchounwou P.B., Yedjou C.G., et al. Heavy Metal Toxicity and the Environment. In: Luch, A. (eds) *Molecular, Clinical and Environmental Toxicology. Experientia Supplementum*, vol 101. Springer, Basel. DOI: 10.1007/978-3-7643-8340-4_6
79. Waheed A., Mansha M., Ullah N. Nanomaterials-based electrochemical detection of heavy metals in water: Current status, challenges and future direction. *Trends in Analytical Chemistry* 2018; 105: 37–51. DOI: 10.1016/j.trac.2018.04.012
80. Deshmukh S., Kandasamy G., et al. Terephthalic acid capped iron oxide nanoparticles for sensitive electrochemical detection of heavy metal ions in water. *Journal of Electroanalytical Chemistry* 2017; 788: 91–98. DOI: 10.1016/j.jelechem.2017.01.064
81. Yantasee W., Hongsirikarn K., et al. Direct detection of Pb in urine and Cd, Pb, Cu, and Ag in natural waters using electrochemical sensors immobilized with DMSA functionalized magnetic nanoparticles. *Analyst* 2008; 133(3): 348–355. DOI: 10.1039/B711199A
82. Narouei F.H., Kirk K.A., Andreescu S. Electrochemical quantification of lead adsorption on TiO_2 Nanoparticles. *Electroanalysis* 2021; 33(1): 188–196. DOI: 10.1002/elan.202060152
83. Liu J., Zhu G., et al. Fabrication of electrospun ZnO nanofiber-modified electrode for the determination of trace Cd(II). *Sensors and Actuators, B: Chemical* 2016; 234: 84–91. DOI: 10.1016/j.snb.2016.04.073
84. Belcovici A., Muresan L., et al. Zinc oxide nanostructured platform for electrochemical detection of heavy metals. *Electroanalysis* 2023; 35: 202200395. DOI: 10.1002/elan.202200395
85. Mao A., Li H., et al. Determination of mercury using a glassy carbon electrode modified with nano TiO_2 and multi-walled carbon nanotubes composites dispersed in a novel cationic surfactant. *Journal of Electroanalytical Chemistry* 2015; 751: 23–29. DOI: 10.1016/j.jelechem.2015.04.034
86. Xu Q., Cheng X.L., et al. Fe-doping induced surface Fe^{2+}/Fe^{3+} cycle and activated redox-inert TiO_2 for enhanced Hg(II) electrochemical sensing: An efficient strategy to strengthen the redox activity. *Analytica Chimica Acta* 2022; 1232: 340472. DOI: 10.1016/j.aca.2022.340472
87. Karthik R., Thambidurai S., et al. Synthesis of cobalt doped ZnO/reduced graphene oxide nanorods as active material for heavy metal ions sensor and antibacterial activity. *Journal of Alloys and Compounds* 2017; 715: 254–265. DOI: 10.1016/j.jallcom.2017.04.298
88. Dong Y., Zhou M., Zhang L. Designed 3D porous core-shell vanadium oxide microspheres for the simultaneous electrochemical sensing of toxic metal ions. *Journal of Alloys and Compounds* 2020; 827: 154357. DOI: 10.1016/j.jallcom.2020.154357

89. Wang X., Lu, X., et al. Development of biosensor technologies for analysis of environmental contaminants. *Trends in Environmental Analytical Chemistry* 2014; 2: 25–32. DOI: 10.1016/j.teac.2014.04.001
90. Govindhan M., Adhikari B.R., et al. Nanomaterials-based electrochemical detection of chemical contaminants. *RSC Advances* 2014; 4(109): 63741–63760. DOI: 10.1039/C4RA10399H
91. Shadlaghani A., Farzaneh M., et al. Direct electrochemical detection of glutamate, acetylcholine, choline, and adenosine using non-enzymatic electrodes. *Sensors* 2019; 19: 447–492. DOI: 10.3390/s19030447
92. Khairy M., Ayoub H.A., Banks C.E. Non-enzymatic electrochemical platform for parathion pesticide sensing based on nanometer-sized nickel oxide modified screen-printed electrodes. *Food Chemistry* 2018; 255: 104–111. DOI: 10.1016/j.foodchem.2018.02.004
93. Jamil A., Solangi A.R., et al. Nonenzymatic electrochemical detection of 2,4,6-Trichlorophenol using CuO/Nafion/GCE: A practical sensor for environmental toxicants. *Langmuir* 2021; 37: 3214–3222. DOI: 10.1021/acs.langmuir.1c00165
94. Nurdin N., Maulidiyah M., et al. Nanocomposite design of graphene modified TiO_2 for electrochemical sensing in phenol detection. *Korean Journal of Chemical Engineering* 2022; 39: 209–215. DOI: 10.1007/s11814-021-0938-6
95. Anu Prathap M.U., Sun S., et al. A novel non-enzymatic lindane sensor based on CuO–MnO_2 hierarchical nano-microstructures for enhanced sensitivity. *Chemical Communications* 2015; 51(21): 4376–4379. DOI: 10.1039/C5CC00024F
96. Tian X., Liu L., et al. Nonenzymatic electrochemical sensor based on CuO-TiO_2 for sensitive and selective detection of methyl parathion pesticide in ground water. *Sensors and Actuators, B: Chemical* 2018; 256: 135–142. DOI: 10.1016/j.snb.2017.10.066
97. Zhao Y., Song X., et al. A facile route to the synthesis copper oxide/reduced graphene oxide nanocomposites and electrochemical detection of catechol organic pollutant. *CrystEngComm* 2012; 14(20): 6710–6719. DOI: 10.1039/c2ce25509j
98. Huo D., Li Q., et al. A highly efficient organophosphorus pesticides sensor based on CuO nanowires–SWCNTs hybrid nanocomposite. *Sensors and Actuators, B: Chemical* 2014; 199: 410–417. DOI: 10.1016/j.snb.2014.04.016
99. Sangamithirai D., Ramanathan S. Electrochemical sensing platform for the detection of nitroaromatics using g-C_3N_4/V_2O_5 nanocomposites modified glassy carbon electrode. *Electrochimica Acta* 2022; 434: 141308. DOI: 10.1016/j.electacta.2022.141308
100. Rajesh K., Santhanalakshmi J. Design and development of graphene intercalated V_2O_5 nanosheets based electrochemical sensors for effective determination of potentially hazardous 3,5–Dichlorophenol. *Materials Chemistry and Physics* 2017; 199: 497–507. DOI: 10.1016/j.matchemphys.2017.07.022
101. Hassan, M.H., Khan, R., Andreescu, S., Advances in electrochemical detection methods formeasuring contaminants of emerging concerns. *Electrochemical Science Advances* 2022; 2: e2100184. DOI: 10.1002/elsa.202100184
102. Fenti A. et al. Performance testing of mesh anodes for in situ electrochemical oxidation of PFAS. *Chemical Engineering Journal Advances* 2022; 9: 100205. DOI: 10.1016/j.ceja.2021.100205
103. Mahmoud L., Edward H.S., Shana O.K. Electrochemical methods for the analysis of clinically relevant biomolecules. *Chemical Reviews* 2016; 116: 9001–9090. DOI: 10.1021/acs.chemrev.6b00220
104. Balram D., Lian K.Y., Sebastian N. A novel electrochemical sensor based on flower shaped zinc oxide nanoparticles for the efficient detection of dopamine. *International Journal of Electrochemical Science* 2018; 13: 1542–1555. DOI: 10.20964/2018.02.06
105. Yadav P., Manivannan S., et al. Electrochemical properties of highly sensitive and selective CuO nanostructures based neurotransmitter dopamine sensor. *Electroanalysis* 2017; 29: 1–9. DOI: 10.1002/elan.201700195
106. Dogana H.O., Urhan B.K., et al. Simultaneous electrochemical detection of ascorbic acid and dopamine on Cu_2O/CuO/electrochemically reduced graphene oxide (CuxO/ERGO)-nanocomposite-modified electrode. *Microchemical Journal* 2019; 150: 104157. DOI: 10.1016/j.microc.2019.104157
107. Zhang X., Zhang Y.C., Ma L.X. One-pot facile fabrication of graphene-zinc oxide composite and its enhanced sensitivity for simultaneous electrochemical detection of ascorbic acid, dopamine and uric acid. *Sensors and Actuators, B: Chemical* 2016; 227: 488–496. DOI: 10.1016/j.snb.2015.12.073
108. Zhihua L., Xucheng Z., et al. A novel sensor for determination of dopamine in meat based on ZnO decorated reduced graphene oxide composite. *Innovative Food Science and Emerging Technologies* 2015; 31: 196–203. DOI: 10.1016/j.ifset.2015.06.011

109. Swamy B.K., Shiprath K., et al. Electrochemical detection of dopamine and tyrosine using metal oxide (MO, M1/4Cu and Ni) modified graphite electrode: A comparative study. *Biointerface Research in Applied Chemistry* 2020; 10: 6460–6473. DOI: 10.33263/BRIAC105.64606473
110. Cai Z., Ye Y., et al. Morphology–dependent electrochemical sensing properties of iron oxide–graphene oxide nanohybrids for dopamine and uric acid. *Nanomaterials* 2019; 9: 835. DOI: 10.3390/nano9060835
111. Suresh R., Giribabu K., et al. New electrochemical sensor based on Ni-doped V_2O_5 nanoplates modified glassy carbon electrode for selective determination of dopamine at nanomolar level. *Sensors and Actuators, B: Chemical* 2014; 202: 440–447. DOI: 10.1016/j.snb.2014.05.095
112. Wetchakun K., Samerjai T. et al. Semiconducting metal oxides as sensors for environmentally hazardous gases. *Sensors and Actuators, B: Chemical* 2011; 160: 580–591. DOI: 10.1016/j.snb.2011.08.032
113. Fazio E., Spadaro S., et al. Metal-oxide based nanomaterials: Synthesis, characterization and their applications in electrical and electrochemical sensors. *Sensors* 2021; 21: 2494–2526. DOI: 10.3390/s21072494
114. Dey A. Semiconductor metal oxide gas sensors: A review. *Materials Science and Engineering B* 2018; 229: 206–217. DOI: 10.1016/j.mseb.2017.12.036
115. Isaac N.A., Pikaar I., Biskos G. Metal oxide semiconducting nanomaterials for air quality gas sensors: Operating principles, performance, and synthesis techniques. *Microchimica Acta* 2022; 189: 196. DOI: 10.1007/s00604-022-05254-0
116. Mani G.K., Rayappan J.B.B. A highly selective and wide range ammonia sensor—Nanostructured ZnO:Co thin film. *Materials Science and Engineering B* 2015; 191: 41–50. DOI: 10.1016/j.mseb.2014.10.007
117. Tesfamichael T., Cetin C., et al. The effect of pressure and W-doping on the properties of ZnO thin films for NO_2 gas sensing. *Applied Surface Science* 2015; 357: 728–734. DOI: 10.1016/j.mseb.2014.10.007
118. Hosseini Z.S., Zad A.I., Mortezaali A. Room temperature H_2S gas sensor based on rather aligned ZnO nanorods with flower-like structures. *Sensors and Actuators, B: Chemical* 2015; 207: 865–871. DOI: 10.1016/j.snb.2014.10.085
119. Wu N., Zhao M., et al. Porous CuO–ZnO nanocomposite for sensing electrode of high-temperature CO solid-state electrochemical sensor. *Nanotechnology* 2005; 16(12): 2878–2881. DOI: 10.1088/0957-4484/16/12/024

6 Metal Nanomaterials for Electrochemical Sensing

Avinash Singh, Ankita Ojha, and Vijay Bahadur

6.1 INTRODUCTION

Nanoscience and nanotechnology play a crucial role in the improvement and revolutionization of various fields such as medicine, transportation, textiles, security, energy, environment. Materials exhibit a drastic change in their properties upon reduction in size in the nanoscale regime due to an increase in the surface area and quantum confinement effect (Ngumbi et al., 2018; Blackman and Genov, 2018). The optical, electrical, and other properties of these nanomaterials depend on their size and vary significantly with size. For example, the Ag nanoparticles (NPs) have size- and shape-dependent color and surface plasmon resonance (SPR) band position (Penga et al., 2010). The surface-to-volume ratio (*S*/*V* ratio) for a spherical-shaped NP is inversely proportional to its size (3/*r*), and hence, the reduction in size increases the *S*/*V* ratio. Nanomaterials have more atoms on their surface in comparison with their bulk counterpart, and these surface atoms are more reactive due to the unbalanced force acting on them, which enhances the reactivity of the NPs. Due to their exceptional properties in the nanorange, metal-based NPs have tremendous applications in various fields such as catalysis, medicine, energy harvesting, energy storage, agriculture, sensing (Shnoudeh et al., 2019). Metal nanomaterials especially Au, Ag, Pt, and Cu NPs are widely used in the electrochemical sensing of biomolecules such as glucose, proteins, DNA, and other molecules and ions. Sensors are mainly devices that can detect a particular analyte by showing a response in the presence of an analyte. This response can be in the form of color, fluorescence, electrical signal, etc. The output signal for an electrochemical sensor is an electrical signal that is produced as a result of electron/charge transfer between the electrode and the analyte.

The device consists of electrodes and electrolytes. The electrode acts as a surface at which the redox processes take place. The inherent drawbacks of conventional electrochemical sensors are their selectivity, limited sensitivity, and level of detection. However, due to certain limitations such as suitable redox potential windows, less surface area, and less affinity between the electrode surface and analyte, the electrode surfaces need to be modified/functionalized with suitable metal nanomaterials. Metal nanomaterials are more redox active, provide a larger surface area, and have a better affinity with the analyte molecules or ions. They work as an electron conduction pathway between the electrode surface and the analyte and hence improve sensory activity. Various researchers and scientists have developed metal NP-based electrochemical sensors to achieve improved selectivity, higher sensitivity, and low level of detection (Zhong et al., 2005; Zhu et al., 2015; Si et al., 2013; Patella et al., 2017). Electrode modified with metal nanomaterials offers efficient electron/charge transfer from analytes to the electrode due to the morphology, reactive surface, and selectivity toward the analyte. Different analytes including H_2O_2, glucose, protein, DNA, and other biomolecules and metal ions exhibit better interaction with the metal nanomaterials (Figure 6.1). The electrochemical sensors possess other advantages such as portability, robustness, quick response time, repeatability, and reduction of volume of sample (Si et al., 2013). Based on the modification of the electrode surface with the redox-active metal nanomaterials, metal NP-based electrochemical sensors can be classified into two key groups: (1) nonenzymatic sensors: In these sensors, metal nanomaterials or functionalized metal nanomaterials are used as sensors; (2) enzymatic sensors:

DOI: 10.1201/9781032654874-6

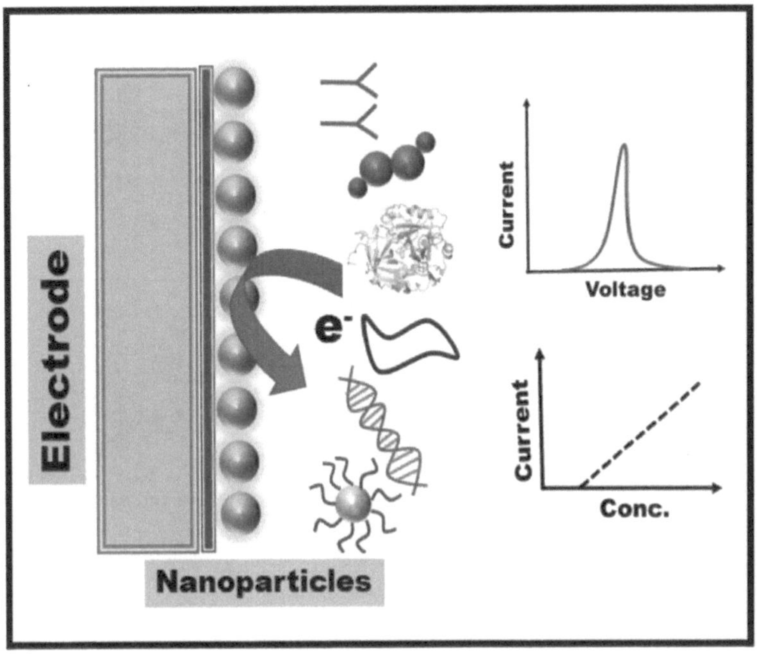

FIGURE 6.1 Scheme for the surface modification of electrode with metal nanoparticle and electrochemical sensing of various analytes.

In these sensors, metal nanomaterials modified with enzymes are used as sensors. Enzymatic sensors have better sensitivity and stability, and they are known as third-generation biosensors.

This chapter aims to discuss the various routes involved in the synthesis of metal nanomaterials, various methods of characterization, followed by their applications as an electrochemical sensor for different analytes, the limitations involved in the electrochemical sensors, and plausible solutions. This effort aims to give a precise, succinct explanation of the principle and practical approach of electrochemical sensing using metal nanomaterials. Such efforts are significant to readers who want to understand the concept of electrochemical sensing and are interested in the development of better sensing alternatives.

6.2 SYNTHESIS OF METAL NANOMATERIALS

There are two general methods for synthesizing different types of NPs including metal nanomaterials:

i. Physical or top-down method
ii. Chemical or bottom-up method

In the physical approach to the synthesis of NPs, the synthesis of NPs is achieved by mechanically reducing the size of the bulk material. For instance, Fe NPs can be synthesized by ball milling a piece of iron. There are several ways of top-down approaches to the synthesis of NPs, e.g., thermal evaporation, mechanical ball milling, laser ablation. The top-down approach does not provide sufficient strategies to control the shape, size, and morphology of the NPs. Moreover, the top-down method is not an energy-saving approach and is often high cost.

The bottom-up approach is preferred by chemists due to its versatility. It offers better control over the shape, size, and morphology of the NPs and is more energy-efficient and low cost. Further, it is easy to do surface modifications of the NPs using suitable ligands for sensing and other applications. In the bottom-up approach, a precursor of metals (usually a metal salt) is taken in a suitable

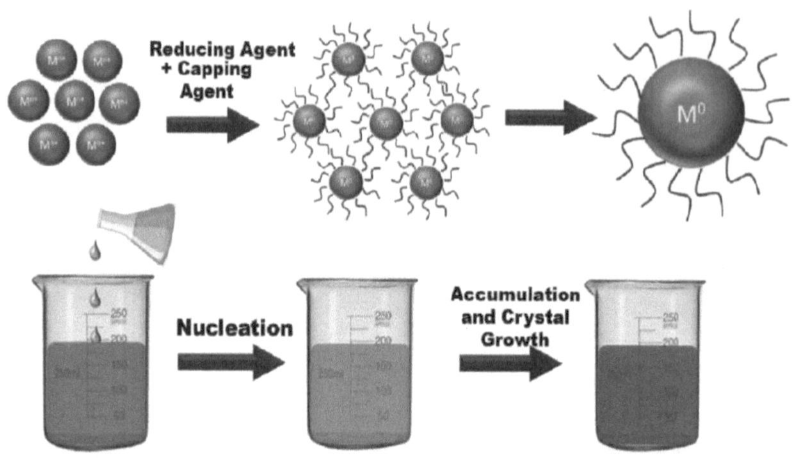

FIGURE 6.2 Scheme for the synthesis of metal nanoparticle with reducing and capping agents.

solvent (aqueous or organic) and some reducing and/or capping agents are used. The role of the reducing agent is to reduce the metal ion in a zero/low oxidation state so that the nucleation of metal nanomaterials can take place. After the nucleation, the crystal growth starts and the role of a capping agent comes into play. The capping agents control the growth of the particle in the nanoregime (Figure 6.2). The shape, size, and morphology of the NPs depend on the nature of the metal ion, the reducing agents, and the capping agents. There are many bottom-up approaches to synthesis. Hot-injection, hydrothermal/solvothermal, microwave-assisted, sonochemical, electrochemical, radiation-chemical, and photochemical methods are some widely studied methods for the synthesis of metal nanomaterials. These methods differ in generating the reducing conditions and the reaction conditions. In general, some widely used reducing agents are citric acid, ascorbic acid, sodium borohydride, sodium citrate, hydrazine hydrate, glycerol, thioglycerol, etc. (Jamkhande et al., 2019). The molecules with functional groups having an affinity toward metals such as –OH, –COOH, –SH, –NH can be used as capping agents. Molecules having a chain or branched structure, cavity, or micellar aggregation are more suitable for capping (Sau and Rogach, 2010). Metal nanomaterials can also be synthesized using biological methods, which is a kind of bottom-up approach. In biological methods, plant extract, enzymes, and microorganisms are used as reducing and capping agents (Song and Kim, 2009; Pantidos and Horsfall, 2014; Baghayeri et al., 2017). The biological methods are inexpensive, safe, and environment-friendly (Iravani, 2011; Shah et al., 2015). The hydrothermal method is used for the synthesis of various metal nanomaterials such as Ni, Cu, and Ag using glycerol as a capping agent (Kim et al., 2014).

Metal nanomaterials such as Au, Ag, Ru, Fe, Pt are used for electrochemical sensing. Wang et al. have synthesized Au NPs through the chemical reduction method where carbonized resins were used as reducing and capping agents. The synthesized Au NPs were employed sfor the electrochemical detection of Cu^{2+} ions using ratiometric methods. The ratiometric method involves the measurement of the signal of interest concerning another signal which is proportionately correlated with the signal of interest (Wang et al., 2019). A thermally reduced graphene oxide (TrGNO) was used as a scaffold to improve the charge transfer rate for better electrochemical performance in the work. Zhang et al. (2018) have synthesized Pt NPs in the hollow carbon sphere and applied them in the nonenzymatic sensing of H_2O_2 and glucose. In that study, ultra-small Pt NPs of 3.13 nm were synthesized and encapsulated with a hollow carbon sphere using one-step synthesis. Andreescu and Luck (2008) have synthesized protein-encapsulated Au NPs for the detection of glucose. Au NPs were synthesized using iso-ascorbic acid as the reducing agent. In another study, Ag NPs synthesized using the leaf of *Salvia centifolia* as a capping and reducing agent have been used for the electrochemical detection of nitrite ions, as catalysts, and as antibacterial agents (Baghayeri et al., 2017). Barman et al. (2017)

have synthesized Cu NPs in which ascorbic acid was used as a reducing agent and polyvinyl pyrrolidone (PVP) was used as a capping agent. Further, this Cu NP-modified Au electrode assembly was used for the electrochemical detection of p-nitrophenol with a detection limit of 0.34 nM. Ayaz et al. (2022) have synthesized Cu NPs and electrodeposited them on the surface of graphite pencil electrode and used this assembly for the detection of glucose in the flow injection analysis. Patella et al. (2017) have synthesized Pd nanowires using single-step galvanic displacement deposition and used them as an electrochemical sensor of H_2O_2. Zhong et al. (2005) have fabricated a glucose biosensor based on Au NPs self-assembled with polymer onto the gold substrate. A colloidal solution of Au NPs was synthesized where $HAuCl_4$ was used as Au precursor and sodium citrate was used as a reducing and capping agent.

6.3 CHARACTERIZATION TECHNIQUES FOR METAL NANOMATERIALS

Characterization of NPs is crucial after their synthesis. NPs of suitable metal with the desired shape, size, crystal structure, morphology, elemental composition, and surface characteristics play a significant role in achieving the objective of their synthesis. There are various techniques that are used in the thorough characterization of NPs. We cannot have a thorough characterization of these NPs using a single technique. Hence, all these techniques are complementary to each other. Table 6.1 provides a summary of various characterization techniques and information received from these techniques, and some of them are discussed in brief.

6.3.1 Optical Spectroscopy

Optical spectroscopy is a very essential tool for the characterization of various molecular assemblies including nanostructures. Molecules have characteristics of absorption and emission band based on the nature of the functional group, ligands, and the extent of conjugation. Nanosized materials due to their reduced size and shape have a reduced density of state (DoS), which leads to size- and shape-dependent absorption and emission spectra. The DoS is defined as the measure

TABLE 6.1
Summary of Characterization Techniques for Metal Nanomaterials

Sl. No.	Characterization Techniques	Information Provided
1.	UV-vis spectroscopy	Optical properties, i.e., band gap, surface plasmon, resonance band, particle size, concentration, agglomeration state, the preliminary idea about the shape of nanoparticles
2.	Pl. spectroscopy	Optical properties and structural feature correlation such as defects (surface as well as deep), size, composition
3.	DLS	Hydrodynamic size, agglomeration of particles
4.	BET	Surface area
5.	FTIR	Surface composition, ligand binding
6.	ICP-MS	Elemental composition, size, size distribution, concentration
7.	TEM	Size and shape of nanoparticles, aggregation state
8.	HRTEM	In addition to all the information provided by the conventional TEM crystal structure of the single particle, crystal defects, monocrystalline, polycrystalline, and amorphous
9.	SEM-HRSEM, SEM-EDX	Morphology, dispersion of nanoparticles in the matrix
10.	XRD	Crystal structure, composition, grain size
11.	XPS	Electronic structure, elemental composition, oxidation state, ligand binding

of the number of states per unit of energy. A discreteness in the DoS is observed with the decrease in shape and size in the nanoregime, which is otherwise continuous (Pederson, 2006). This leads to size- and shape-dependent absorption bands in the NPs. When a metal NP interacts with electromagnetic radiation, the surface electron cloud of the NP polarizes and starts oscillating with a certain frequency which depends on the size of the NP. When this oscillating frequency matches the frequency of electromagnetic radiation, the radiation gets absorbed. This phenomenon is called SPR, and the absorbed frequency is called the SPR band. The SPR band is the characteristic of the metal NP and depends on the shape, size, and morphology of the NPs. We can calculate the particle size (nm) of Au NPs in the range of 35–100 nm using the SPR band position of the absorption spectra by using the following equation (Haiss et al., 2007):

$$d = \frac{\frac{\lambda_{SPR} - \lambda_0}{L_1}}{L_2} \tag{6.1}$$

where λ_{SPR} is the position of SPR in the absorption spectrum, $\lambda_0=512\,nm$, $L_1=6.53$, and $L_2=0.0216$.

Researchers have studied the binding and signaling of gold NPs for the sensing of glucose (Adreescu and Luck, 2008). In that study, the SPR band of Au NPs was found to be at 523 nm. The SPR band of 20–50-nm-sized Ag NPs was found to be 430 nm (Song and Kim, 2009). In another study of one-pot synthesis of Ag NPs for biomedical applications, the researchers have reported the SPR band of Ag NPs at 410 nm (Patil et al., 2012). The SPR band of copper NPs synthesized using laser ablation in liquid media was found to be ~600 nm (Liu et al., 2015). The SPR band does not need to fall in the UV-visible region, but it may fall in the near-IR region. It depends on shape, size, and material. The concentration of the NPs in the solution can be calculated using Lambert–Beer's law by measuring the absorbance. Generally, the other metal nanomaterials except for Au and Ag such as copper, iron, ruthenium, and selenium do not give an intense colored solution, which is the characteristic of SPR.

The change in the absorption spectra (bathochromic, hypsochromic, hyperchromic, or hypochromic) of the metal nanomaterials is due to variations in shape, size, and interactions with the analytes (DNA, protein, antibodies, glucose, and other biomolecules).

6.3.2 Transmission Electron Microscopy (TEM)

After the invention of TEM by Max Knoll and Ernst Ruska in Germany in 1931, the field of nanoscience and nanotechnology received tremendous encouragement from researchers and scientists all over the world. The principle of TEM involves the transmission of a high electron beam through the specimen after its magnification and focus by the objective lens. The representative sample of the metal nanomaterials is deposited on the supporting films or grid. The TEM grid is generally made of copper, and carbon is coated on it. It is a flat disc-shaped structure with mesh or different-shaped holes to support the sample specimen. The TEM provides a picture of a specimen having a particle size of 10–100 nm with high resolution and contrast. In addition to the shape and size of the NPs, the high-resolution TEM (HRTEM) provides information about crystal structure, including whether the material is amorphous, crystalline, polycrystalline, monocrystalline, or a single crystal. It can also detect imperfection in the crystal structure of the NPs. Both the TEM and the HRTEM are complementary techniques, and together, they provide important information about the NPs. Further, the selected area electron diffraction (SAED) pattern is used to get the image of different planes of the crystal, crystal orientation, lattice constant, and defects in the lattice. M. Sanchez-Dominguez et al. proposed a novel method to synthesize metal and metal oxide NPs in microemulsion with the support of TEM and HRTEM (Sanchez-Dominguez et al., 2009). Burst et al. (1995) have synthesized 2 nm size Au NPs and studied the reactions of functionalized NPs using the TEM and SAED. Various other researchers have characterized their metal nanomaterials using the TEM and HRTEM techniques (Kanaras et al., 2002; Song and Kim, 2008; Wang et al., 2019).

6.3.3 Scanning Electron Microscopy (SEM)

SEM is the sister technique of TEM, which provides information about the shape and morphology of the NPs. The basic difference between SEM and TEM lies in the direction of the electron beam after interaction with the sample. In SEM, the image is created after the detection of a reflected electron from the surface, while in TEM as discussed in the earlier section, the image is created based on the focusing of the transmitted electrons. Essentially, the SEM images provide information about the surface morphology, topography, and elemental compositions of the NPs. It also provides the essential characterization of the deposition of metal nanomaterials on the surface of electrodes for electrochemical sensing (Baron et al., 2007; Patil et al., 2012; Barman et. al., 2017; Cao et al., 2021). Polsky et al. (2006) have synthesized nucleic acid-functionalized Pt NPs for the detection of DNA.

6.3.4 X-ray Diffraction

X-rays are high-energy radiations (0.1–100 keV) having wavelength of around 0.01–10 nm. A material upon irradiation with X-rays may cause various optical phenomena such as diffraction, scattering, interference, refraction, reflection, absorption, and/or emission. Based on these different phenomena, different characterization techniques for material characterizations have been developed. X-ray diffraction in which the diffraction phenomenon is involved is a powerful tool to obtain information about the materials such as their crystal structure, particle size, lattice strain.

The particle size of the NP can be determined by the very well-known Scherrer formula:

$$D = \frac{0.9\ \lambda}{\beta \cos\theta} \tag{6.2}$$

where D is the particle size in nm, λ is the wavelength of characteristic X-ray in nm, β is the Full Width at Half Maximum (FWHM) of the diffraction peak in the X-Ray diffraction (XRD) profile, and θ is the Bragg angle of diffraction.

Various researchers have used the X-ray diffraction technique for the characterization of metal nanomaterials. For Ag NPs, the 2θ value ranges from 20° to 80° with intense peaks at 38.11° and 70.57° (Gnanadesigan, et al., 2012).

6.3.5 X-ray Photoelectron Spectroscopy (XPS)

XPS is another X-ray-based analytical technique that provides information about the electronic structure, oxidation state, elemental composition, and ligand binding on the surface of the metal nanomaterials. The basic principle behind XPS is the photoelectric effect. When a sample is irradiated with an X-ray, the core electrons of the atoms of a sample are ejected, which is known as the photoemission process. The binding energy of core electrons of each surface atom is its characteristic binding energy, which can be considered to be its ionization energy (not exactly but similar to it). If the kinetic energy of the emitted electrons is E_k, the binding energy E_b can be calculated by the following equation, which is popularly known as Einstein's formula:

$$E_b = h\nu - E_k - \varphi \tag{6.3}$$

where $h\nu$ is the energy of the X-ray photon and φ is the work function of the spectrometer. The binding energy of the core electron of the surface atoms depends on its elements, oxidation state, and the surrounding environment of the surface atoms (ligand binding). Several researchers have used XPS as the confirmatory technique to get the above-mentioned information (Phung et al., 2003; Tunc and Suzer, 2005; Song and Kim, 2008; Shard, 2012; Belsey et al., 2015; Maiti et al., 2016).

6.3.6 Fourier Transform Infrared Spectroscopy (FTIR)

FTIR is one of the most important techniques in validating surface passivation/ligand binding with metal nanomaterials. Several functional group molecules are used for the surface passivation of NPs based on their application, level of sensitivity, and limit of detection. The FTIR provides the evidence of binding of glucose, protein, DNA, nucleic acid, and other biomolecules with metal nanomaterials. The frequency stretching or bending of different functional groups present in these biomolecules (–OH, –COOH, –SH, –C=O, –NH, etc.) changes upon binding with NPs due to changes in polarity, H-bonding, or strength of the bond (Burst et al., 1995; Devraj et al., 2013; Sharma et al., 2014; Tunc, 2014; Pandey and Pandey, 2015; Baghayeri et al., 2017; Singh et al., 2018). These are some major characterization techniques for metal nanomaterials. Apart from these techniques, some other techniques mentioned in Table 6.1 are also important for the characterization of metal nanomaterials.

6.4 APPLICATION OF METAL NANOMATERIALS IN ELECTROCHEMICAL SENSING

The application of metal nanomaterials in the field of electrochemical sensing has been a matter of great investigation, especially in environmental cleaning and pollution control. Their unique electrocatalytic behavior, low cost of synthesis as compared to noble metal-based catalysts, and higher selectivity make them a material of high potential in the field of electrochemical sensing. The high tunability of these metal nanomaterials and easily tailored electrocatalytic activity by varying the shape and size make them a promising material in sensing applications (Aralekallu and Sannegowda, 2021). The electrochemical sensors that have been applied mainly rely on three methods: potentiometry (measuring the potential difference change between two electrodes of the cell), amperometry (measuring the change in current due to ions present in the electrochemical solution), and conductometry (measuring the electrical conductivity due to chemical reaction). In the voltammetry method, molecules are oxidized or reduced on the surface of the working electrode. This method is highly selective as each molecule has its specified redox potential (Saleh et al., 2019). Here are some of the applications of these metal nanomaterials that have been studied in the past decade in the field of electrochemical sensing.

6.4.1 Volatile Compounds and Toxic Gases

Metal nanomaterials hold great utility in the detection of volatile compounds or toxic gases. An electrochemical gas sensor uses an electrode setup of two or three electrodes with a standardized atmosphere. The gaseous molecules or volatile molecules drift toward the working electrode surface, and the reaction occurring at the surface of the electrode results in the generation of electric current. Noble metallic nanomaterials such as Au, Ag, Pt, Hg, and Pd, as well as their bimetallic alloys, are well known for their unique physical, chemical, and electrochemical properties. Endogenous detection of H_2S gas by differential pulse voltammetry (DPV) using Au@Ag NPs has been reported with high selectivity for H_2S even in the presence of interfering substances such as NaBr, KI. The Au@Ag system showed a stable electrochemical signal, and the gas was detected by a redox reaction between the gas and Ag present in core shells (Zhao et al., 2018). Gutes et al. used graphene decorated with metal nanomaterials of Au, Pt, and Pd NPs for the detection of H_2S gas. Au and Pt had a higher density than Pd, and it was observed that Au-deposited graphene showed great sensitivity toward this gas, but recovery was found to be tough due to the strong Au–S bond.

6.4.2 Metal ion and Anion Detection

Huang et al. developed an electrochemical sensor out of Au NPs for the detection of As^{3+} with a detection limit of 32.5 pM and sensitivity of 16.15 μA μM^{-1}. Ethylenediaminetetraacetic acid (EDTA) was applied to eliminate the interferences due to other metal ions. Au-modified electrode surfaces have been successfully implied for the detection of Hg with detection up to 1 μM L^{-1}. Au was adsorbed on the surface of indium tin oxide (ITO) through electrostatic adhesion, and Hg was stripped from the surface of ITO preferably. The low concentration of Au enhanced the sensitivity by providing nucleation sites for the deposit of Hg (Ratner and Mandler, 2015). Graphene–Au-coated electrochemical sensors have been designed for the detection of Hg^{2+} with a detection limit of 0.001 aM. In this system, single-stranded DNA has been applied for higher selectivity and sensitivity (Zhang et al., 2015). Cu-loaded multi-walled carbon nanotubes (MWCNT)-reduced graphene Oxide (rGO) nanocomposite electrochemical sensors have been prepared for the simultaneous detection of nitrite and nitrate ions. The fabricated sensor showed a wide operative concentration range between 0.1 and 75 μM, and the detection limit was found to be 30 nM (for nitrite) and 20 nM (for nitrate ions). This sensor can easily be used for the detection of these anions in water sources as well as food samples such as sausages, salami, and cheese (Bagheri et al., 2017). A novel label-free electrochemical sensor was designed for highly selective and extremely sensitive detection of Hg^{2+}, where Hg^{2+} was captured by thymine through coordination and then Au NPs were synthesized using $HAuCl_4$-NH_2OH-mediated catalytic reactions. In this method, the unnecessary labeling of the DNA probe was skipped using a high thymine-rich DNA. This method was found to be extremely selective with very high sensitivity (Tang et al., 2014). Pd NPs embedded into porous activated carbon derived from waste biomass feedstock used on glassy carbon electrodes (GCEs) were used for the electrochemical detection of bivalent metal ions of Cd, Pb, Cu, and Hg simultaneously with a detection limit of 41, 50, 66, and 54 nm, respectively. This sensor holds the additional advantage of real-time sampling and has been used for metal ion detection in milk samples (Veerakumar et al., 2016). Cu-metal organic framework (MOF) decorated with Au NPs electrodeposited on GCE through the potentiostat method has been applied in the detection of nitrite. This system showed a synergetic catalytic effect for nitrite oxidation. In this Au@Cu-MOF electrode system, Cu-MOF had a large surface area and high porosity which prevented aggregation of Au NPs and enhanced nitrite adsorption, while Au NPs contributed to high conductivity and increased catalytic activity. The electrochemical sensor showed high sensitivity and selectivity with a detection limit of 82 nm (Chen et al., 2019). Ag NPs have been used in the detection of metal ions such as Cd^{2+}, Cr^{6+}, Cu^{2+}, Hg^{2+}, and Sb^{3+} because of their cheaper cost of synthesis yet the uniqueness in their physical and chemical properties, which increase their optical, chemical, and catalytic activity (Sawan et al., 2020).

6.4.3 Organic Compound Sensing

A refreshable photo-electrochemical sensor has been designed for the detection of bisphenol A (BPA) using Au NPs modified over a C-doped TiO_2 nanotube array. It showed great electrochemical activity and excellent sensitivity of about 2.8 μA $\mu M^{-1} cm^{-2}$, and the detection limit for it lies around 6.2 nM (Maduraiveeran and Jin, 2017). An Ag-based amperometric sensor has been developed with a silicate core shell for the detection of nitroaromatic compounds with a detection limit of 5 nM. This sensor was highly stable and recyclable due to the immobility of metal nanomaterials in the amino-functionalized silicate framework (Maduraiveeran and Ramaraj, 2009). Another Ag NP-loaded electrochemical sensor was developed by Kariuki et al. (2016). In this system, Ag was loaded into a poly(amic) acid polymer matrix. This sensor had a detection limit of 1.68 μM and a sensitivity of 7.88 μA μM^{-1} with limited interference from other compounds with the same functional groups. Polyphenols are a class of hydrocarbons that are aromatic and are one of the major contributors to environmental pollution. Au NPs and tyrosinase-modified eggshell membrane-coated GCE electrodes have been applied in the detection of polyphenols by Datta et al. (2017). The eggshell membrane contains a high amount of fiber, and hence, Au NPs and tyrosinase bind easily to its surface. The binding of enzymes on the surface electrode changes the porosity and electrode smoothness. The presence of Au NPs enhances the electron transport at the surface of the electrode because of their higher conductivity. The enzymes present on the surface of the electrode increase the current response at the anodic peak from polyphenols (Datta et al., 2017). Pt-Pd nanocubes synthesized through the hydrothermal method dispersed over graphene nanoribbons have been used in the electrochemical detection of trinitrotoluene with a detection limit of 0.8 ppb (Zhang et al., 2015). Cyclodextrin-Pt NPs on graphene sheets have been used in the electrochemical detection of naphthol isomers with a linear response range of 0.8–220 nM for naphthol and increased sensitivity (Zhu et al., 2012).

6.4.4 Biomolecule Sensing

With the increasing health awareness in public, recent time demands a point-of-care testing method for biomolecules that affect human health. The electrochemical sensors that are applied in the sensing of biomolecules or biological systems are referred to as biosensors. Metal-based biosensors have been applied in the detection of biomarkers for cancer such as carcinoembryonic antigen (CEA) and cancer antigen, herceptin-2, and other biomolecules such as dopamine, hormones, glucose, and blood cells (Islam et al., 2020). Nanoscale Au-based electrodes have been successfully applied in certain sensor devices for the detection of biomolecules. Au NPs create a highly stable and quite convenient bridging between carriers and biomolecules. 3D-printed tubes for the simultaneous detection of glucose and peroxide have been developed containing two working electrodes. The first working electrode was loaded with colloidal Pt NPs, and the second one was modified with Au-horse radish peroxidase (HRP). The Pt electrode detected glucose, while the Au-modified electrode detected peroxides (Narayanan and Slaughter, 2019). Two electrochemical sensors with Ag NPs coated on Zn-metal-organic frameworks (MOFs) with different morphologies (2D and 3D) were designed to develop a self-supporting sensor used in the detection of hydrogen peroxide. The 2D-Ag-Zn-MOFs exhibited better electrocatalytic behavior as compared to their 3D counterparts because of higher dispersion and enhanced stability along with larger surface area and increased electrical conductivity. The 2D feature showed a detection limit as low as 1.67 μM and a linear range from 5 to 70 μM. This sensor has been utilized in the real-time detection of hydrogen peroxide in living cells in response to drug stimulation (Chen et al., 2022). Another peroxide sensor was developed using 15 nm Ag and 8 nm Au NP-derived nanocomposite of ZnO nanoflowers. It was again found that the sensitivity and stability of Ag-based nanocomposite were higher, and it showed great perspective for the development of a nonenzymatic biosensor for peroxide detections (Hussain et al., 2016). Pt NP-doped MWCNT-fabricated electrodes on a silicate matrix have been used for the detection of glucose. In this, –the NH_2 group was exploited because of its high sensitivity toward metal ions. A linear range of 1–25 mM and response time of <15 seconds have been reported for the presence of glucose with sensitivity four times that of the regular MWCNT-based electrochemical sensors (Yang et al., 2006). Au-Ag alloy-coated MWCNTs have been used as volatile biomarkers for gastric cancer cells. These biosensors are ultrasensitive for MGC-803 and GES-1 cancer cells. These fabricated sensors could detect 3-octanone (detection range=0–0.0025% v/v and detection limit=0.3 ppb) and butanone (detection range=0–0.055% v/v and detection limit=0.5 ppb) with great selectivity (Zhang et al., 2014). Nucleic acid-functionalized Pt NPs were used as a catalytic label in the detection of DNA hybridization and recognition of protein or aptamer. Aptamer-functionalized Pt NPs are associated with the thrombin aptamer–thrombin complex and help in the detection of thrombin, and the sensitivity is around 10^{-9} M (Polsky et al., 2006). Pd NPs supported on porous carbon aerogel were synthesized through a microwave-assisted method and were found to be excellent in sensing dopamine and melatonin simultaneously with a detection limit of 0.0026 and 0.0071 μM, respectively (Rajkumar et al., 2017). Noble metal NPs such as Pt and Au-modified electrochemical sensors have been applied in the determination of ascorbic acid even at lower overpotential and higher specificity. The Pt-based electrochemical sensor provides excess sites for the adsorption of hydrogen and is further oxidized followed by reversible or dissociative adsorption at the side chain. Au NPs are highly biocompatible, have multiple surface functionalities, and are hence greatly applied in biomedical applications (Dhara and Debiprasad, 2019).

6.4.4.1 Pathogen Sensing

Ag NPs were employed for the detection of bacteria (*Salmonella typhi*) using an antibody-based electrochemical technique. Anodic stripping voltammetry (ASV) was employed for the selective detection of *Salmonella typhi* with a sensitivity of 10 cfu mL^{-1} (Kumar et al., 2020). Negatively charged lipopolysaccharide from the outer surface of certain Gram-negative bacteria serve as a useful biomarker for the detection of these bacteria. Positively charged chitosan-stabilized Ag NPs have been applied as sensitive electrochemical biosensors for the detection of lipopolysaccharides. Mainly, the bacterium *Escherichia coli* has been detected using this system with a detection range of 0.001–100 ng mL^{-1} (Imran et al., 2020). COVID-19, a pandemic causing severe acute respiratory disease that hit the world in recent years, has also witnessed the power of metal NP-based sensing devices. New-age sensing devices have successfully helped people in the detection of COVID-19 infections, which has hit the world recently (Islam et al., 2020). Thiolated probes were immobilized on the electrode surface made up of Au-Fe_3O_4 NPs. The method of detection was through DPV, and calixarene-8 was employed for the improved electrochemical activity. The method was found to be highly effective and much better than the commonly applied real-time polymerase chain reaction (RT-PCR) method. The sensor could be easily used with a smartphone with a plug-and-play mode of operation (Zhao et al., 2021).

6.5 LIMITATIONS AND PLAUSIBLE SOLUTIONS

Apart from the various advantages of metal NP-based electrochemical sensors, there are certain limitations of these electrochemical sensors. The first limitation is the stability of the metal nanomaterials under various conditions to increase the self-life span. The metal nanomaterials due to higher surface energy get aggregated in due course of time. The aggregation of these NPs affects their reactivity and hence their sensing efficiency (Aralekallu and Sannegowda, 2021). In this regard, a suitable capping agent should be selected which is stable and has a stronger binding affinity with the metals. The NPs deposited over the surface of the electrode should be stable for a wide temperature and pH range. Another big challenge in the application of metallic NPs as electrochemical sensors is their extra sensitivity to environmental conditions such as the ionic strength of their medium, exposure to UV radiation, and thermal instability, which needs to be taken care of during their application. Attaching suitable ligands to these metal nanomaterials can help to increase their stability and activity. The addition of stabilizing agents or forming nanocomposites is a highly effective methodology for introducing stability in NPs of metal oxides. In addition to this, the other limitations of these metal NP-based electrochemical sensors are (1) low selectivity, (2) low sensitivity, (3) cost, and (4) biocompatibility. Metal NPs can interact with many different compounds, which can reduce the sensor's selectivity. For applications where it is critical to be able to specifically detect a target molecule, this can be an issue. Here, again a suitable capping molecule having a very specific binding capability can enhance the selectivity of the electrochemical sensor. The size and surface area of metal NP-based electrochemical sensors frequently limit their sensitivity. Greater S/V ratios in smaller NPs can increase sensitivity. The sensitivity of smaller NPs can be decreased by the fact that they are also more likely to agglomerate. The stability of NPs in a small size regime and hence the increased S/V ratio tend to increase the sensitivity of the electrochemical cell. Some NPs like Ag NPs are toxic beyond a certain concentration (Khan et al., 2019). However, their toxicity can be reduced by adopting suitable capping molecules which possess biocompatibility. Lastly, a hindrance to the widespread usage of metal NPs is their high cost. A hassle-free and more efficient methodology for the synthesis of better-quality metal NPs can promote its widespread use.

6.6 FUTURE PERSPECTIVES AND CONCLUSION

The inherent advantages of the metal NP-based electrochemical sensor over other electrochemical sensors such as better sensitivity, lower detection limits, less time consumption, and cost-effectiveness prove to be the driving force for researchers and scientists to have more insight into these electrochemical sensors. This chapter provides a brief of the synthesis methods, characterizing tools, and application of metal-based NPs in sensing different analytes. Compared to other analysis tools such as spectroscopy and chromatography, electrochemical sensing using metal nanomaterials provides a quick, easier, and robust way of analysis. With the advancement in material science and nanotechnology, it is believed that new advances are to be achieved in the development of more useful, reliable, time-efficient, and robust electrochemical sensors with an even lower limit of detection. With the development of new multifunctional materials based on metal nanomaterials, improved sensitivity, selectivity, and reproducibility will be achieved. Although significant advancement has been achieved in the design and development of electrochemical sensors (Pardo et al., 2021; Ding et al., 2021; Tarapoulouzi et al., 2022), it is believed that designing binary, ternary, and other hybrid metal-based nanomaterials of superstructures and morphology will be helpful in the development of ultrasensitive sensing devices that can simultaneously monitor various biomolecules such as protein, DNA, RNA, glucose which will be used for practical applications.

REFERENCES

Andreescu S. & Luck L.A. (2008). Studies of the binding and signaling of surface-immobilized periplasmic glucose receptors on gold nanoparticles: A glucose biosensor application. *Analytical Biochemistry*, 375, 282.

Aralekallu, S. & Sannegowda, L.K. (2021). Metal nanoparticles for electrochemical sensing applications. In C.M. Hussain and S.K. Kailasa (Eds.) *Handbook of Nanomaterials for Sensing Applications, Micro & Nano Technologies*, Elsevier, Amsterdam, The Netherlands, pp. 589–629.

Ayaz, S., Karakaya, S., Emir, G., Usakligil, N., Dilgin, D.J. & Dilgin, Y. (2022). Flow-injection amperometric determination of glucose using nickel oxide-cobalt (II, III) oxide and nickel oxide-copper nanoparticle modified pencil graphite electrodes. *Analytical Letters*, 55, 2046–2057.

Baghayeri, M., Mahdavi B., Abadi, Z.H. & Farhadi, S. (2017). Green synthesis of silver nanoparticles using water extract of Salvia leriifolia: Antibacterial studies and applications as catalysts in the electrochemical detection of nitrite. *Applied Organometallic Chemistry*, 2017, 4057.

Bagheri, H., Hajian, A., Rezaei, M. & Shirzadmehr, A. (2017). Composite of Cu metal nanoparticles-multiwall carbon nanotubes-reduced graphene oxide as a novel & high-performance platform of the electrochemical sensor for simultaneous determination of nitrite & nitrate. *Journal of Hazardous Materials*, 324, 762–772.

Barman K., Changmai B. & Jasimuddin, S.K. (2017). Electrochemical detection of para-nitrophenol using copper metal nanoparticles modified gold electrode. *Electroanalysis*, 29, 1.

Baron, R., Sljukic, B., Salter, C., Crossley, A. & Compton, R.G. (2007). Development of an electrochemical sensor nanoarray for hydrazine detection using a combinatorial approach. *Electroanalysis*, 10, 162.

Belsey, N.A., Shard, A.G. & Milleni, C. (2015). Analysis of protein coatings on gold nanoparticles by XPS and liquid-based particle sizing techniques. *Biointerfases*, 10, 019012.

Blackman, III G.N. & Genov, D.A. (2018). Bounds on quantum confinement effects in metal nanoparticles. *Physical Review B*, 97, 115440.

Burst, M., Fink, J., Bethell, D., Schiffrin, D.J. & Kiely, C. (1995). Synthesis and reactions of functionalised gold nanoparticles. *Journal of the Chemical Society, Chemical Communications*, 1995, 1655.

Cao, L., Zhang, W., Lu, S., Guo, C., Wang, P., Zhang, D. & Ma, W. (2021). A label-free electrochemical immunosensor for CEA detection on a novel signal amplification platform of Cu_2S/Pd/CuO nanocomposites. *Frontiers in Bioengineering and Biotechnology*, 9, 767717.

Chen, S., Xie, Y., Guo, X. & Sun, D. (2022). Self-supporting electrochemical sensors for monitoring of cell-released H_2O_2 based on metal nanoparticle/MOF nanozymes. *Microchemical Journal*, 181, 107715.

Datta, S., Kanjilal, B. & Sarkar, P. (2017). Electrochemical sensor for detection of polyphenols in tea & wine with differential pulse voltammetry & electrochemical impedance spectroscopy utilizing tyrosinase & gold nanoparticles decorated biomembrane. *Journal of The Electrochemical Society*, 164(4), B118–B126.

Devaraj, P., Kumari, P., Aarti, C. & Renganathan, A., (2013). Synthesis and characterization of silver nanoparticles using cannonball leaves and their cytotoxic activity against MCF-7 cell line. *Journal of Nanotechnology*, 2013, 598328.

Dhara, K. & Debiprosad, R.M. (2019). Review on nanomaterials-enabled electrochemical sensors for ascorbic acid detection. *Analytical Biochemistry*, 586, 113415.

Ding R., Cheong Y. H., Ahamed A. & Lisak G. (2021). Heavy metals detection with paper-based electrochemical sensors. *Analytical Chemistry*, 93, 4, 1880–1888.
Gnanadesigan, M., Anand, M., Ravikumar, S., Maruthupandy, M., Ali, M.S., Vijayakumar, V. & Kumaraguru, A.K. (2012). Antibacterial potential of biosynthesized silver nanoparticles using Avicennia marina mangrove plant. *Applied Nanoscience*, 2, 143.
Haiss, W., Thanh, N.T.K., Aveyard, J. & Fernig, D.G. (2007). Determination of size and concentration of gold nanoparticles from UV-vis spectra. *Analytical Chemistry*, 79, 4215.
Hussain, M., Sun, H., Karim, S. et al. (2016). Noble metal nanoparticle-functionalized ZnO nanoflowers for photocatalytic degradation of RhB dye & electrochemical sensing of hydrogen peroxide. *Journal of Nanoparticle Research*, 18, 95.
Khan, I., Khalid, S. & Khan, I. (2019). Nanoparticles: Properties, applications and toxicities. *Arabian Journal of Chemistry*, 12, 908–931.
Imran, M., Ehrhardt, C.J., Bertino, M.F., Shah, M.R. & Yadavalli, V.K. (2020). Chitosan stabilized silver nanoparticles for the electrochemical detection of lipopolysaccharide: A facile biosensing approach for gram-negative bacteria. *Micromachines*, 11(4), 413.
Iravani, S. (2011). Green synthesis of metal nanoparticles using plants. *Green chemistry*, *13*(10), 2638-2650.
Islam, T., Hasan, M.M., Awal, A., Nurunnabi, M. & Ahammad, A.J.S. (2020). Metal nanoparticles for electrochemical sensing: Progress & challenges in the clinical transition of point-of-care testing. *Molecules*, 25, 5787.
Jamkhande, P.G., Ghule, N.W., Bamer, A. H. & Kalaskar, M.G. (2019). Metal nanoparticles synthesis: An overview on methods of preparation, advantages and disadvantages, and applications. *Journal of Drug Delivery Science and Technology*, 53, 101174.
Kanaras, A.G., Kamounah, F.S., Schaumburg, K., Kiely, C.J. & Burst, M. (2002). Thioalkylated tetraethylene glycol: A new ligand for water soluble monolayer protected gold clusters. *Chemical Communications*, 2002, 2294.
Kariuki, V.M., Fasih-Ahmad, S.A., Osonga, F.J. & Sadik, O.A. (2016). An electrochemical sensor for nitrobenzene using pi-conjugated polymer-embedded nanosilver. *Analyst*, 141, 2259–2269.
Kim, M., Son, W., Ahn, K.H., Kim, D.S., Lee, H. & Lee, Y. (2014). Hydrothermal synthesis of metal nanoparticles using glycerol as a reducing agent. *Journal of Supercritical Fluid*, 90, 53–59.
Kumar, V., Chopra, A., Bisht, B. & Bhalla, V. (2020). Colorimetric & electrochemical detection of pathogens in water using silver ions as a unique probe. *Scientific Reports*, 10(1), 11986.
Liu, P., Wang, H., Li, X., Rui, M. & Zeng, H. (2015). Localized surface plasmon resonance of cu nanoparticles by laser ablation in liquid media. *RSC Advances*, 5, 79738–79745. doi: 10.1039/C5RA14933A.
Maduraiveeran, G. & Jin, W. (2017). Nanomaterials based electrochemical sensor & biosensor platforms for environmental applications. *Trends in Environmental Analytical Chemistry*, 13, 10–23.
Maduraiveeran, G. & Ramaraj, R. (2009). Potential sensing platform of silver nanoparticles embedded in functionalized silicate shell for nitroaromatic compounds. *Analytical Chemistry*, 81, 7552–7560.
Maiti, N., Thomas, S., Debnath, A. & Kapoor, S. (2016). Raman and XPS study on the interaction of taurine with silver nanoparticles. *RSC Advances*, 6, 56406.
Narayanan, J.S. & Slaughter, G. (2019). Towards a dual in-line electrochemical biosensor for the determination of glucose & hydrogen peroxide. *Bioelectrochemistry*, 128, 56–65.
Ngumbi, P.K., Mugo, S.W. & Ngaruiya, J.M. (2018). Determination of gold nanoparticles sizes via surface plasmon resonance. *IOSR Journal of Applied Chemistry (IOSR-JAC)*, 11, 25.
Pandey, P.C. & Pandey, D. (2015). Novel synthesis of Nickel-iron hexacyanoferrate nanoparticles and its application in electrochemical sensing. *Journal of Electroanalytical Chemistry*, 763, 63–70. doi: 10.1016/j.jelechem.2015.12.048.
Pantidos, N. & Horsfall, L.E. (2014). Biological synthesis of metallic nanoparticles by bacteria, fungi and plants. *Journal of Nanomedicine & Nanotechnology*, 5, 1000233.
Patella, B., Inguanta, R., Piazza, S. & Sunseri, C. (2017). A Nanostructured sensor of hydrogen peroxide. *Sensors and Actuators B Chemical*, 245, 44–54.
Patil, R.S., Kokate M.R., Jambhale, C.L., Pawar, S.M., Han, S.H. & Kolekar, S.S. (2012). One-pot synthesis of PVA-capped silver nanoparticles their characterization and biomedical application. *Advances in Natural Sciences: Nanoscience and Nanotechnology*, 3, 015013.
Pardo, D.D., Zapata Z. & Briceño-Gamba N.O. (2021). Electrochemical sensors for the detection of heavy metals in wastewater. *ECS Meeting Abstracts*, MA2021-01(63), 1663.
Pedersen, K. (2006). *Quantum Size Effects in Nanostructures, Organic and Inorganic Nanostructures.*
Penga, S., McMahona, J.M., Schatzb, G.C., Graya, S.K. & Sun, Y. (2010). Reversing the size-dependence of surface plasmon resonances. *PNAS*, 107, 14530.

Phung, X., Groza, J., Stach, E.A., Williams, L.N. & Ritchey, S.B. (2003). Surface characterization of metal nanoparticles. *Materials and Engineering A*, 359, 261.

Polsky, R., Gill, R., Kaganovsky, L. & Willner, I. (2006). Nucleic acid-functionalized Pt nanoparticles: Catalytic labels for the amplified electrochemical detection of biomolecules. *Analytical Chemistry*, 78(7), 2268–2271.

Rajkumar, C., Veerakumar, P., Chen, S.-M., Thirumalraj, B. & Liu, S.B. (2017). Facile & novel synthesis of palladium nanoparticles supported on a carbon aerogel for ultrasensitive electrochemical sensing of biomolecules. *Nanoscale*, 9(19), 6486–6496.

Ratner, N. & Mandler, D. (2015). Electrochemical detection of low concentrations of mercury in water using gold nanoparticles. *Analytical Chemistry*, 87, 5148–5155.

Saleh, T.A., Fadillah, G. & Saputra, O.A. (2019). Nanoparticles as components of electrochemical sensing platforms for the detection of petroleum pollutants: A review. *TrAC Trends in Analytical Chemistry*, 118, 194–206.

Sanchez-Dominguez, M., Boutonett, M. & Solans, C. (2009). A novel approach to metal and metal oxide nanoparticle synthesis: The oil-in-water microemulsion reaction method. *Journal of Nanoparticle Research*, 11, 1823–1829.

Sau, T.K. & Rogach, A.L. (2010). Nonspherical noble metal nanoparticles: Colloid-chemical synthesis and morphology control. *Advanced Materials*, 22, 1781.

Sawan, S., Maalouf, R., Errachid, A. & Jaffrezic-Renault, N. (2020). Metal and metal oxide nanoparticles in the voltammetric detection of heavy metals: A review. *TrAC Trends in Analytical Chemistry*, 131, 116014.

Shah, M., Fawcett, D., Sharma, S., Tripathy, S.K. & Poinern, G.E.J. (2015). Green synthesis of metallic nanoparticles via biological entities. *Materials*, 8, 7278.

Shard, A.G. (2012). A straightforward method for interpreting XPS data from core–shell nanoparticles. *Journal of Physical Chemistry C*, 116, 16806.

Sharma, G., Sharma, A.R., Bhavesh, R., Park, J., Ganbold, B., Nam, J. & Lee S. (2014). Biomolecule-mediated synthesis of selenium nanoparticles using dried vitis vinifera (raisin) extract. *Molecules*, 19, 2761.

Shnoudeh, A.J., Hamad, I., Abdo, R.W., Qadumii, L., Jaber, A.Y., Surchi, H.S. & Alkelany, S.Z. (2019). Chapter 15, Synthesis, characterization, & applications of metal nanoparticles. *Biomaterials & Bionanotechnology*, 9, 527.

Si, P., Huang, Y., Wanga, T. & Ma, J. (2013). Nanomaterials for electrochemical non-enzymatic glucose biosensors. *RSC Advances*, 3, 3487.

Singh, A., Guleria, A., Kunwar, A., Neogy, S. & Rath, M.C. (2018). Highly facile and rapid one-pot synthetic protocol for the formation of Se nanoparticles at ambient conditions with controlled phase and morphology: Role of starch and cytotoxic studies. *Materials Research Express*, 6, 015029.

Song, J. Y., & Kim, B. S. (2008). Biological synthesis of bimetallic Au/Ag nanoparticles using Persimmon (Diopyros kaki) leaf extract. *Korean Journal of Chemical Engineering*, *25*, 808-811.

Song, J.Y. & Kim, B.S. (2009). Rapid biological synthesis of silver nanoparticles using plant leaf extracts. *Bioprocess and Biosystems Engineering*, 32, 79.

Tang, S., Tong, P., Lu, W., Chen, J., Yan, Z. & Zhang, L. (2014). A novel label-free electrochemical sensor for Hg^{2+} based on the catalytic formation of metal nanoparticle. *Biosensors & Bioelectronics*, 59, 1–5.

Tarapoulouzi, M., Ortone, V. & Cinti, S. (2022). Heavy metals detection at chemometrics-powered electrochemical (bio)sensors. *Talanta*, 244, 123410.

Tunc, I. (2014). The effect of the presence of Ag nanoparticles on the photocatalytic degradation of oxalic acid adsorbed on TiO_2 nanoparticles monitored by ATR-FTIR. *Materials Chemistry and Physics*, 144, 444.

Tunc, I. & Suzer, S. (2005). XPS characterization of Au (Core)/SiO_2 (Shell) nanoparticles. *Journal of Physical Chemistry Letters*, 109, 7597.

Veerakumar, P., Veeramani, V., Chen, S.-M., Madhu, R. & Liu, S.-B. (2016). Palladium nanoparticle incorporated porous activated carbon: Electrochemical detection of toxic metal ions. *ACS Applied Materials & Interfaces*, 8(2), 1319–1326.

Wang, X., Liu, G., Qi, Y., Yuan, Y., Gao, J., Luo, X. & Yang, T. (2019). Embedded Au nanoparticles-based ratiometric electrochemical sensing strategy for sensitive and reliable detection of copper ions. *Analytical Chemistry*, 91(18), 12006–12013.

Yang, M., Yang, Y., Liu, Y., Shen, G. & Yu, R. (2006). Platinum nanoparticles-doped sol–gel/carbon nanotubes composite electrochemical sensors & biosensors. *Biosensors & Bioelectronics*, 21(7), 1125–1131.

Zhang, Y., Gao, G., Liu, H., Fu, H., Fan, J., Wang, K., et al. (2014). Identification of volatile biomarkers of gastric cancer cells & ultrasensitive electrochemical detection based on sensing interface of Au-Ag alloy coated MWCNTs. *Theranostics*, 4(2), 154–162.

Zhang, R., Sun, C.-L., Lu, Y.-J. & Chen, W. (2015). Graphene nanoribbon-supported PtPd concave nanocubes for electrochemical detection of TNT with high sensitivity & selectivity. *Analytical Chemistry*, 87(24), 12262–12269.

Zhang, Y., Zeng, G.M., Tang, L., Chen, J., Zhu, Y., He, X.X. & He, Y. (2015). Electrochemical sensor based on electrodeposited graphene-Au modified electrode & nanoAu carrier amplified signal strategy for attomolar mercury detection. *Analytical Chemistry*, 87, 989–996.

Zhang, C., Zhang, R., Gao, X., Cheng, C., Lin, H., Li, X. & Chen, W. (2018). Small naked Pt nanoparticles confined in mesoporous shell of hollow carbon spheres for high-performance nonenzymatic sensing of H_2O_2 and glucose. *ACS Omega*, 3, 96.

Zhao, Y., Yang, Y., Cui, L., Zheng, F. & Song, Q. (2018). Electroactive Au@Ag nanoparticles driven electrochemical sensor for endogenous H_2S detection. *Biosensors & Bioelectronics*, 117, 53–59.

Zhao, H., Liu, F., Xie, W., Zhou, T.C., OuYang, J., Jin, L., Li, H., Zhao, C.Y., Zhang, L., Wei, J., et al. (2021). Ultrasensitive supers sandwich-type electrochemical sensor for SARS-CoV-2 from the infected COVID-19 patients using a smartphone. *Sensors & Actuators B Chemistry*, 327, 128899.

Zhong, X., Yuan, R., Chai, Y., Liu, Y., Dai, J. & Tang, D. (2005). Glucose biosensor based on self-assembled gold nanoparticles and double-layer 2D-network (3-mercaptopropyl)-trimethoxysilane polymer onto gold substrate. *Sensors and Actuators B Chemical*, 104, 191.

Zhu, G., Gai, P., Wu, L., Zhang, J., Zhang, X. & Chen, J. (2012). β-cyclodextrin-platinum nanoparticles/graphene nanohybrids: Enhanced sensitivity for electrochemical detection of naphthol isomers. *Chemistry - An Asian Journal*, 7(4), 732–737.

Zhu, C., Yang, G., Li, H., Du, D. & Lin, Y. (2015). Electrochemical sensors & biosensors based on nanomaterials & nanostructures. *Analytical Biochemistry*, 87(1), 230–249.

7 Two-dimensional Materials for Electrochemical Sensing

Ashish Kumar Bhui and Sanvedna Shukla

7.1 INTRODUCTION

Two-dimensional (2D) materials offer extraordinary physicochemical properties because of their high aspect ratio, distinctive surface chemistry, and quantum-size effect (Zhang et al., 2015; Xu et al., 2013). A broad library of 2D materials has grown since the discovery of graphene including transition metal dichalcogenides (TMDs) (Muller et al., 2008; Geim et al., 2007; Novoselov et al., 2004; Huang et al., 2011), hexagonal boron nitride (h-BN) (Weng et al. 2016; Lin et al., 2010; Chen et al., 2018), graphitic carbon nitride (Zhang et al., 2015), black phosphorus (BP) (Liu et al. 2015), layered double hydroxides (LDHs) (Ma et al. 2015), silicene (Tao et al. 2015), 2D metal oxides/sulfides (Ma et al., 2015), transition metal carbides, and carbonitrides (MXenes) 2D metals (Huang et al., 2011), and 2D polymers (Wang et al., 2016; Hu et al., 2015). Due to their anisotropic structure, interesting characteristics, and prospective catalytic applications, 2D metallic nanomaterials have garnered significant scientific attention (Huang et al., 2011; Zhao et al., 2015), surface-enhanced Raman scattering (SERS) (Viarbitskaya et al., 2013; Huang et al., 2011; Xiong et al., 2005; Yin et al., 2012), sensing (Gao et al., 2012), bioimaging (Homan et al., 2012; Bao et al., 2013), photothermal therapy (Tang et al., 2014; Huang et al., 2011), solar cells (Yao et al., 2014), etc.

Modern nanoscience pioneered metal nanomaterials. Faraday synthesized Au nanoparticles chemically in 1857 (Faraday, 1857). Since the electron microscope can observe nanostructures, controlled production of colloidal Au nanoparticles has been studied extensively (Turkevich et al., 1951). In the meanwhile, controlled production of various metal nanostructures and their intriguing features has been extensively studied. Synthesizing metal nanostructures with regulated sizes (Campbell et al., 2002; Corma et al., 2013; Qian et al. 2012), shapes/facets, and compositions as well as architectures has progressed (Chen et al., 2014; Zhang et al., 2015; Xia et al., 2017).

The unique sheet-like structure known as graphene, which was first discovered in 2004, has several intriguing features. Due to their exceptional electrical, mechanical, and optical features, which render them appropriate for a broad spectrum of practical applications, 2D materials are receiving growing interest in the current times. Researchers worldwide are attracted to 2D materials due to their exclusive properties and promising practical applications. Over a decade, several 2D materials were isolated. The development of 2D materials in carbides and nitrides has been the subject of research. With charge and heat transfer restricted to a single plane, 2D nanostructured materials exhibit unique physical characteristics.

Scientists refrained from assuming the existence of 2D materials in nature due to the presence of environmental instability. While 700 stable 2D materials exist theoretically, their synthesis is lacking. Before graphene, germanene and silicene's electrical and atomic structures were studied in 1994. Boehm called single-layer carbon foils "ene" in 1962. Steel, dry lubricants, and brake linings use graphite. Kroto developed fullerene carbon allotropes in 1985. Single-dimensional carbon nanotubes are incredibly tiny and cylindrical. Carbon nanotubes, a nanomaterial with exceptional strength and electrical conductivity, are valuable in many applications. Nanotechnology is interested in one-dimensional carbon nanotubes for use in electronics, energy storage, and other sophisticated technologies. Iijima discovered nanotubes in 1991. According to Gupta et al. (2015), these discoveries have spurred research on one-dimensional nanoribbons and graphite isolation from 2D

 DOI: 10.1201/9781032654874-7

crystals. Fullerenes, graphite, and nanotube articles followed. Novoselov isolated monolayer graphene in 2004. Material technology has advanced here. This implies that graphene is made from carbon-based graphite. Graphene is atom-thick. This is a one-atom-thick layer of closely packed carbon atoms with sp^2 hybridization. Mas-Ballesté (2011) states that graphite layers around the atom minimize inter-planar stresses and speed layer separation. In 2004, a Manchester group led by Andre Geim separated graphite layers, as Novoselov et al. reported in 2005. Novoselov et al. used scotch tape to peel graphene sheets and arrange them on a silicon wafer covered with SiO_2. The user discusses a wonderful substance. It is thinner than paper, stronger than diamond, and conducts electricity better than copper.

Graphene, the first carbon-based 2D material, shows the technological promise of stable, single- and few-atom-thick layers. Choi (2017) discovered TMDs as the first and most studied 2D materials. Electrical qualities and translucent and flexible nature are its distinct properties. Transistors, light detectors, memory devices, photovoltaic devices, Li-ion batteries, and H_2 evolution catalysis use semiconductor TMDs. TMDs have two chalcogens between the metal transition layers, unlike graphene (Lv, 2015). S, Te, Se, and others are chalcogens; Mo, Nb, and W are transition metals. TMDs have strong intra-plane and weaker interlayer non-covalent bonds. Exfoliation isolates TMD 2D structures. Li intercalation, sticky tape exfoliation, and chemical/solvent-aided isolation isolate materials (Zeng, 2012).

7.2 CLASSIFICATION OF 2D MATERIALS

All 2D materials have layers of atoms bound together by forces. These items fall under three categories:

i. 2D graphene family materials
ii. 2D oxides
iii. 2D chalcogenides

Carbon's unique features have long fascinated researchers. Material science advances in nanoscale carbon allotropes, including graphite, fullerenes, and Bucky balls, have boosted this interest. Quantum confinement fundamentals recover these structures' optical, chemical, and bio-interfacing features.

7.2.1 Two-Dimensional Graphene Family Materials

These are further classified under the graphene family as follows:

a. *Graphene*: Graphene is the thinnest and strongest known with a single-atom thickness and 2D hexagonal honeycomb structure. A few synthesis processes include the production of 96% transparent graphene (Wassei & Kaner, 2010) with a high degree of heat and electricity conducting feature. Graphene-based materials (GBMs) are used in sensors, energy, coating, and more due to their exceptional qualities.
b. *Graphene Oxide*: Graphene has various advantages, but producing pure graphene is exigent due to the instability of the single atomic layers. A powerful oxidant is required for graphene synthesis, which oxidizes graphite to render it hydrophilic. This facilitates and stabilizes graphene exfoliation. In aqueous media, this solution may be sonicated to form a single or some layers of graphene oxide (GO) dispersible in any medium. Oxidizing species alter the properties of GO. Because of layer-disturbed sp^2 bonding, GO is an electric insulator. GO is reduced to reestablish its hexagonal form and properties.
c. *Fluorographene*: Fluorinating graphene yields fluorographene (FG) (Robinson et al., 2010), developed to overcome the hydrophobicity and structural defects of graphene.

d. *Hexagonal Boron Nitride (h-BN)*: h-BN is characterized by a single-layered ultra-flat structure isomorphic with graphene, having excellent electrical–optical capabilities, structural uniqueness, thermal stability, and chemical behavior. Metal chalcogenides, GO sheets, and h-BN may form composites. Compared to graphene, hexagonal BN (h-BN) contains boron and nitrogen atoms connected in a 2D plane through sp^2 hybridization forming a honeycomb hexagonal structure (Song et al., 2010). Their unique features, such as enormous thermal, chemical, and mechanical durability, along with near-transparent feature, low dielectric constant, and broad band gap, make this inorganic material indigenous with a broad range of applications ranging from aerospace to medical (Chen et al., 2022; He et al., 2019). In general, h-BN might be created using several top-down and bottom-up procedures (Li et al., 2018) with exfoliation being the most extensively used way to get a monolayered atomic plane with exceptional chemical stability (Gautam et al., 2021).

e. *Boron–Carbon Nitride (BCN)*: BCN is a newer carbon composite with the carbon–boron nitride hybrid having adjustable band gaps (0–5.5 eV). It possesses characteristic optical absorptivity, electrochemical activity, and cathodoluminescence (Wu et al., 2021). Bonds between B, C, and N affect BCN's structural configuration (Lei et al., 2022). Doping improves various properties such as energy storage, capacitors, photocatalytic processes, and gas storage devices. Stimulation techniques indicate that the lowest energy atoms prefer B–N and C–C bonds for the stake of stability. The BCN phase separates into single-layered BN and graphene if B–C, C–N, B–B, and N–N bonds are common. BCN may be 0 dimension, 1 dimension, or 2 dimension (Figure 7.1).

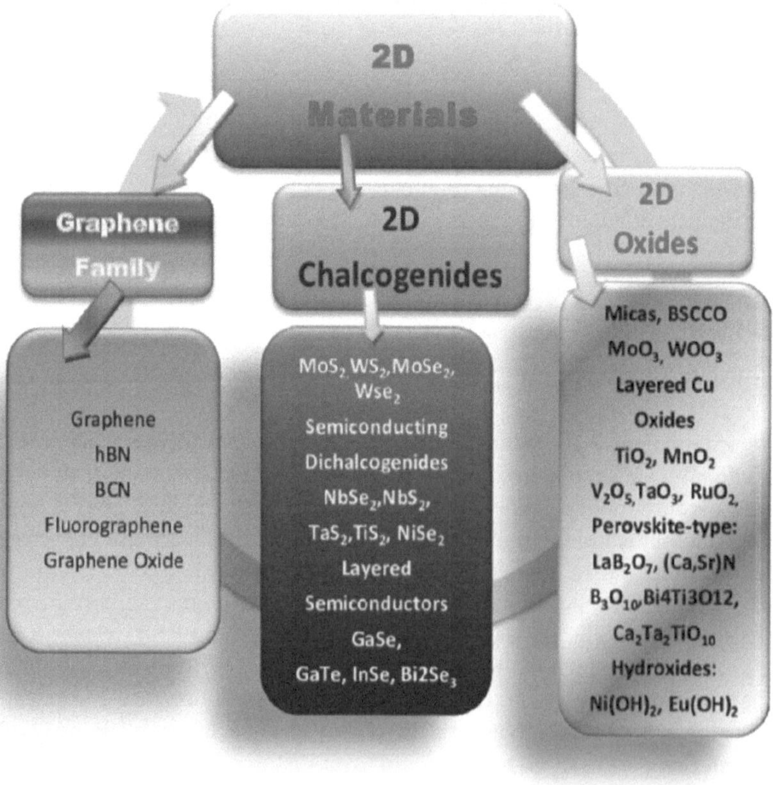

FIGURE 7.1 Family of 2D materials.

7.2.1.1 Synthesis Processes for Graphene

This includes top-down techniques with liquid-phase exfoliation (LPE), mechanical exfoliation (ME), and electrochemical exfoliation (EE) to convert graphite into microscopic flakes of graphene. Exfoliated graphene has several advantages, like its solubility, liquid processability, and low production cost, but it also has some disadvantages, including an increased number of flaws, lower conductivity, and optical opacity.

Chemical vapor deposition (CVD) and epitaxial growth, on the other hand, are the two most used bottom-up techniques (Prekodravac et al., 2021). These approaches result in monolayer graphene with a highly regulated chemical composition and crystalline structure, as well as increased conductivity and optical transparency (Mas-Ballest et al., 2011).

7.2.1.2 Electronic and Optical Property of Graphene

With a s^2p^4 electron distribution, graphene possesses a zero-overlap conduction and valence band structure, leaving 4e entirely for chemical bonding. However, in the case of sheets, there is sp^2 hybridization utilizing three of the four electrons, leaving just one extremely active electron (e) delocalized, left below and above the planar structure, resulting in super-high conductivity displayed by graphene or GBMs.

At ambient temperature, graphene has an exceptionally high electrical mobility of 250,000 $cm^2 Vs^{-1}$ (Cui et al., 2015) allowing it to behave like photons. It also has minimal inherent electrical noise, which explains its remarkable sensitivity to even little changes in the chemical, electrical, or mechanical characteristics of the material. It has strong electrical conductivity with low resistance (lower than Ag wires at ambient temperature). Because electrons and holes behave like quasi-particles (also known as massless Dirac fermions), graphene exhibits the quantum Hall effect (Geim & Novoselov, 2007), which is the quantization of graphene energy into discrete levels (Landau levels) due to quantum confinement in sheet (Geim & Novoselov, 2007).

Graphene has about 98% transmittance (Wassei & Kaner, 2010). While white light absorption depends mainly on the fine structure constant in contrast to material qualities being one-atom thick, graphene absorbs 2.3% white light. Graphene is the stellar material for mode-locking fiber lasers with a universal dynamic conduction value $G=e2/4$ (±2%–3%) over the visible spectrum.

7.2.1.3 Two-Dimensional Metal Oxides

2D oxides are metal–oxygen sheets, which mainly include metal oxides of titanium, cobalt, copper, and tungsten (Gupta et al., 2015). Iron oxide is a novel 2D oxide with intrinsic electrical characteristics due to oxygen ions (Rasmussen & Thygesen, 2015).

Material synthesis can also be done via ion layer adsorption and reactions, lasers, or molten salts (Gao et al., 2020). These materials have outstanding electronic characteristics and are employed in batteries, transistors, and supercapacitors (Yin & Peng et al., 2019). 2D oxides' high electron mobility makes them catalytic and photocatalytic (Haque et al., 2017). These can detect nitrogen dioxide, acetone, ethanol, formaldehyde, etc. (Dral & ten Elshof, 2018). Biosensing applications employ manganese dioxide, zinc oxide, and copper oxide. These may be used for biosensing glucose, hydrogen peroxide, DNA, and proteins, including BSA (Shavanova et al., 2016).

7.2.1.4 Two-Dimensional Chalcogenides

2D chalcogenide behaves like the other 2D materials, with a transition metal (M) that covalently binds two chalcogens (X) and weak van der Waal forces holding these layers together (Lv et al., 2015; Zhou et al., 2021). TMDs' bulk-to-monolayer crystal structure changes (Yu et al., 2021) (Figure 7.2). These materials' crystal structures determine their interesting characteristics.

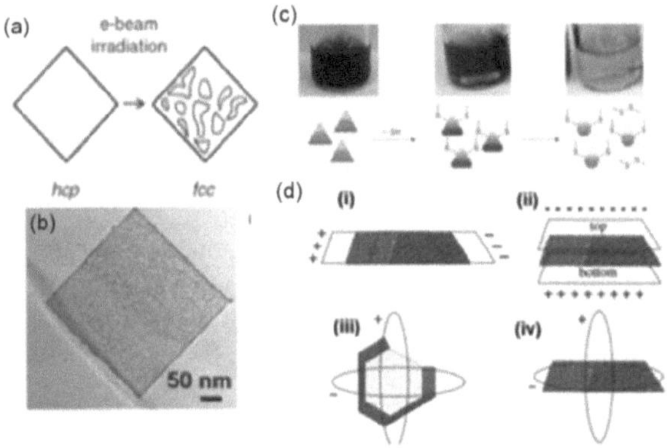

FIGURE 7.2 (a) Schematic representation of an Au square sheet's e-beam-induced crystal phase change from hcp to fcc. (b) TEM picture of the resulting porous Au square sheet after 20 seconds of e-beam irradiation. (Adapted from Huan et al., 2011). (c) Photographs and schematic representation of the shape distortion of Ag nanoprisms when exposed to thiol-terminated molecules. Reprinted with permission from Liu et al. 2013. (d) Diagram of the four Plasmon resonance modes involved in Ag nanodisks. They are the dipolar in-plane mode, the dipolar out-of-plane mode, and the matching quadrupolar modes, in that order. (Adapted from Brioude et al. 2005).

They are optically, electrochemically, and thermos-electronically active. When TMD energy band gaps cover the visible and near-infrared (NIR) light ranges, they are utilized in various applications such as LEDs, transistors, and lasers (Zheng et al., 2022). TMD electronic characteristics change with band gap modulation (Jung et al., 2014). These have also been intensively researched as electrical devices such as field-effect transistors (FETs) and energy storage devices (Huang et al., 2011). TMDs are semiconductors containing transition metals belonging to group VI elements represented by the formula MX^2, where M is a transition metal (e.g., Mo, W, Ti, Nb) and X might be S, Se, or Te (Wu et al., 2018). M atoms are arranged in one plane with two X planes surrounding it through a covalent linkage. Strong charge mobility, layer-dependent photonics, and sensing to optoelectronics were among the unique chemical and physical characteristics of TMDs (Xu et al., 2022; Martínez et al. 2021). Optoelectronic devices integrate optics and electronics (Lee et al., 2017). These are employed in energy generation, storage, catalysis, transistors, solar cells, and photodetectors (Choi et al., 2017; Cui, 2015). These materials generate and control heat (Zhang & Zhang, 2017). TMDs, like graphene, may exhibit band gaps, light absorption, and emission (Chowdhury et al., 2020). TMDs' present uses range from creation using ME, LPE, and CVD deposition (Kang et al., 2020). Ion intercalation during the exfoliation process allows the crystalline structure of the 2D material to change from 2H to 1T in TMDs, especially molybdenum disulfide (MoS_2) (Kang et al., 2020).

BP was originally synthesized in 1914; however, it received widespread attention during the latter part of the last decade (Li et al., 2014; Xu et al., 2019). It is a layered semiconductor material in which van der Waals interactions stack individual atomic layers together. To generate a puckered honeycomb structure, each phosphorus atom is covalently connected with three nearby phosphorus atoms (Castellanos-Gomez et al., 2014). BP is a promising material for various applications due to its direct band gap, characteristic wrinkled structure, high hole mobility, and good mechanical, electrical, and optical properties.

However, its low chemical stability and rapid chemical breakdown in ambient circumstances limit its usage in many sectors. Recent uses of BP include energy storage devices (Sui et al., 2021), photocatalysis (Zhao et al., 2021), sensing, and biological applications such as CVD (Li et al., 2021;

Cui et al., 2015; Abbas et al., 2015). LPE with CVD is the most used technique for creating a thin-film BP (Akbarnejad et al., 2022; Zeng et al., 2021).

Transition metal carbides and/or carbonitrides called MXenes initially made their scientific debut in 2011 (Naguib et al., 2012). MXenes are formulated as M_{n+1} AX_n, where M stands for transition metal (Sc, Ti, Mo, Nb, V, Zr, Cr, or Hf) and A represents In. Sn, Si, or Al and X stand for C and/or N, and n may be 1, 2, or 3. MXenes may be ideal candidates for multiple uses in different fields ranging from energy storage to photocatalysis and biomedicine due to their diversity in morphological and chemical structures, hydrophilic nature, and better electrical conductivity (Mashtalir et al., 2013). Exfoliation of the A layer from the MAX phase followed by selective etching produces MXenes (Verger et al. 2019): etching concentrated HF acid solutions is normally utilized (Naguib et al., 2012). MXenes have hexagonal symmetry structurally, which comes from the lattice structure of the parent MAX phase. There are three ways to prepare MXenes: one with a single M element (Ti_2C and Mo_2C); another with one or more M elements randomly arranged (Ti, V_2C_2 and Cr, V_3C_2); and lastly double-ordered M elements, like Mo_2TiC_2 and $Mo_2Ti_2C_3$, where either a single M element layer or two layers of a single M element are positioned between layers of a second M element. Accordingly, all of these MXenes are represented as M_2X, M_3X_2, and M_4X_3, which are derivations of their respective MAX phases, M_2AX, M_3AX_2, and M_4AX_3, respectively (Khazaei et al., 2017).

Being ceramic, they are stable both chemically and structurally. They can be found structurally in both mono-layer and multilayer forms, with extremely high interlayer spacing—even larger than that found in graphite—and controllable layer thickness. The material must be utilized as a cathode in energy conversion applications so that this open space between the layers is capable of ion intercalation, which implies that ions of various sizes may be inserted into this area. As a result, the partly delaminated layered structure allows not only ion intercalation but also the production of electric double layers (EDLs), which aid in charge storage in capacitors. van der Waals attraction and hydrogen bonding are the major factors in the interaction between layers of MXenes. The presence of functional groups as terminations opens up numerous possibilities for designing desirable surface properties and aids in regulating their electrochemical, thermoelectric, and dielectric characteristics. The otherwise metallic MXenes are transformed into semiconductors such as Sc_2CF_2, Sc_2CO_2, and Ti_2CO_2 by surface terminations. Additionally, the MXenes' transition elements (M) are quite active. The same may be made from a variety of transition metals, each of which can be customized for certain purposes (Tang et al., 2018). The direct functional theory is the fact that few MXenes have in-plane elastic constants above 500 GPa, which is even higher than the stiffness value of routinely used structural steel, which is 400 GPa (Kurtoglu et al., 2012). The MXenes must have excellent structural, mechanical, electrochemical, ion mobility, and electronic transport characteristics to be used as an electrode material. The "M" atoms are the source of their electrical origin. Surface-functionalized MXenes often exhibit a semiconducting nature, whereas all non-functionalized MXenes exhibit a metallic nature. The vast surface area of 2D materials generally available to interact with the analyte allows for greater sensitivity and extremely low detection limits. Additionally, the active spots on their surface encourage interaction with the target species or the immobilization of additional recognition elements, such as metallic nanoparticles or receptors, which can increase the device's sensing effectiveness. Another property of graphene and other 2D materials that enhances sensitivity is conductivity: electron mobility in the 2D plane provides extraordinary electronic properties. This property is especially useful in electrically transduced sensors (Meng et al., 2019). Sensitivity in 2D materials occurs by translating changes in the matter's physical qualities into electrosensitive, optical, or mechanical outputs. Sensing technology in some 2D materials is shown in Table 7.1. Sensitivity adjustments in 2D thus rely on morphological intercalation species, single or multiple layers, and externally switched potential bias, which are examples of such properties (Dai et al., 2022).

TABLE 7.1
Sensing Technology in 2D Materials

Material	Tuning Approach	Mechanism[a]	Target	Limit of Detection	Sensitivity	References
MXene/Prussian blue ($Ti_3C_2T_x$)	Non-covalent	ECL	Lactate and glucose	0.67×10^{-6} M	11.4 μA $mmC^{-1}cm^{-2}$	Lei et al., (2019)
MXene/PVA hydrogel ($Ti_3C_2T_x$)	Non-covalent	ECL	Facial expression	–	25 (GF)	Zhang et al., (2022)
MXene/PDMS (Ti_3C_2)	Non-covalent	PI	Human motion	–	7400 (GF)	Kedambaimoole et al., (2020)
MXene ($Ti_3C_2T_x$)	NA	E	VOC	50 ppb	<ppm	Kim et al., (2018)
MXene/PVA ($Ti_3C_2T_x$)	Non-covalent	E	VOC	50 ppb	0.10 ppm^{-1}	Yuan et al., (2018)
YAuNPs/MXene-PDDA ($Ti_3C_2T_x$)	Non-covalent	ECL	Nitrite	0.059 μM	250 μA $mM^{-1}cm^{-2}$	Wang et al., (2021)
MXene (Ti_3C_2)	NA	OP	Hg^{2+}	42.5 ppm	0.05–50 mM	Lu et al., (2021)
MXene ($Ti_3C_2T_x$)	NA	OP	Ag	0.615 μM		Wang et al., (2020)
Graphene-pyrene	Non-covalent	E	Tumor necrosis factor	476×10^{-15}M 608×10^{-15}M	–	Hao et al., (2021)
			Interleukin	611×10^{-15} M		
			Interferon			
Graphene	NA	E	Brain stimulation	116.07–174.10 μC cm^{-2}	–	Park et al., (2018)
Graphene–MoS_2	Non-covalent	PI	Human motion	–	6.06 kPa^{-1}	Kim et al., (2018)
rGO	NA	PI	Human motion	–	25.1 kPa^{-1}	Pang et al., (2018)
rGO/cellulose fibers	Non-covalent	PI	Human motion	8 μm	–8.9 (GF)	Yin et al., (2019)
Carboxyphenyl-G	Covalent	ECL	Ovalbumin	0.9 pg mL^{-1}	–	Eissa et al., (2015)
rGO–aryl fluoride	Covalent	ECL	Ammonia	20 ppm	63%	Khurshid et al., (2018)
rGO–ferrocenylaniline	Covalent	ECL	Tumor necrosis	pg mL^{-1}	–	Qi et al., (2016)
Pt-loaded/h-BN	Non-covalent	E	Propane	2%	20,000 ppm	Harley et al., (2016)

(Continued)

TABLE 7.1 (*Continued*)
Sensing Technology in 2D Materials

Material	Tuning Approach	Mechanism[a]	Target	Limit of Detection	Sensitivity	References
h-BN	NA	E	NO_2	20 ppb	1.645 ppm^{-1}	He et al., (2019)
$MoTe_2$	NA	E	Ketone	0.2 ppm	100 ppm	Wu et al. (2018)
WS_2	NA	E	NH_3	1 ppm	10 ppm	Li et al., (2017)
MoS_2	NA	E	Pressure and temperature	–	~51 kΩ $°C^{-1}$	Chen et al., (2018)
MoS_2/h-BN	Non-covalent	E	NO_X	6 ppb	6 ppb	Ali et al., (2020)
MoS_2	NA	E	NO_2	0.15 ppb	1 ppb	Tabata et al., (2021)
MoS_2	NA	OP	Methanol	2.7 ppm	–	Zhao et al., (2018)
MoS_2/Au	Non-covalent	OP	Antigen	0.5 g mL^{-1}	–	Su et al., (2019)
MoS_2/rGO	Non-covalent	ECL	H_2O_2	0.19 μM	101.70 μA $mM^{-1}cm^{-2}$	Yang et al., (2021)
MoS_2/rGO	Non-covalent	ECL	NO_2	0.17 μM	0.46 μA $μM^{-1}cm^{-2}$	Hu et al., (2018)
BP/PEI-PEG	Non-covalent	F	CO_2	200 ppm	–	Liu et al., (2022)
BP	Non-covalent	E	NO_2	20 ppb	20% for 20 ppb	Cui et al., (2015)
BP	NA	E	NO_2	5 ppb	5 ppb	Abbas et al., (2015)
BP	NA	E	NH_3	80 ppb	10 ppm	Cho et al., (2015)
BP-AuNPs	Non-covalent	ECL	Patulin	0.03×10^{-3}M	–	Xu et al., (2019)

Source: Reproduced with permission from Manuel (2022).

ECL, electrochemical; E, electrical; OP, optical; PI, piezoelectric; F, flexible.

7.3 STABILITY OF TWO-DIMENSIONAL NANOMATERIALS

Controlling growth kinetics has produced diverse 2D metal nanostructures; however, stability concerns during characterization and application are unavoidable. Due to its high surface energy and unsaturated surface atom ratio, the 2D shape is thermodynamically unstable for metal materials. With the availability of adequate energy, all metal nanostructures will become isotropic to minimize the surface-to-volume ratio (Liz-Marzan et al., 2017). Heat, high-energy electrons, or chemical assaults from oxidizers or other severe chemical conditions might provide external energy (Huang et al., 2011; Tang et al., 2009; Lee et al., 2014; Kan et al., 2006). Thus, ultrathin Au square sheets synthesized on the GO surface were stable in solution at ambient conditions, but high-energy electron beam irradiation destroyed them within a few seconds, causing crystal phase transformation and nanosheet pores, and Au nanoplates fragmented at 450°C during thermal stability testing (Kan et al., 2006). XRD spectra at varied annealing temperatures showed that edges with (110) facets melted and broke first, whereas {111} surfaces remained stable. Localized corrosion from big Au nanoplates' {111} surfaces was obtained (Ma et al., 2015). Dodecylbenzenesulfonate and $AuCl_4^-$ anions caused this counter-Ostwald ripening mechanism. At normal temperatures, chemically active species like halide ions circle the sharp edges of metal nanoprisms (Tang et al., 2009). Kelly et al. (2013) observed that thiol-terminated poly(ethylene glycol) (PEG-SH) at 0.1 mm changed Ag nanoprisms from triangular to disklike. When PEG-SH is raised to 1.0 mm, Ag nanodisks can deform into irregular Ag nanoparticles (AgNPs). In subsequent stability and degradation research, thiols, amines, halide ions, and high temperatures caused an irreversible breakdown of Ag, Au, and Ag@Au nanoprisms (Lee et al., 2014). Meanwhile, 2D metal nanostructure stability countermeasures have been devised. Inert protective coatings or alloys with another metal have been shown to assist 2D metal nanostructures in retaining their forms throughout the chemical and physical postprocessing (Sun et al., 2009; Xue et al., 2007; Huang et al., 2011; Xiong et al., 2005).

Ag@Pd core–shell nanoplates have better photochemical stability than pure-Ag nanoplates (Xue et al., 2007). Yin et al. established (2012) that epitaxial Au coating increased Ag nanoplate resilience against H_2O_2 oxidation. Xue et al. (2017) found that edge-Au-coated Ag nanoprisms were stable after chemical etching. Nonmetals can also shield nonactive metals. Mirkin et al. (2007) stated that silica-coated Ag nanoprisms were relatively more stable (Figure 7.2).

7.4 MECHANISMS OF TRANSDUCTION SENSING IN TWO-DIMENSIONAL MATERIALS

Several features can account for the exceptionally high-sensitivity 2D material-based strain sensors. Exfoliated 2D materials have unusual piezoelectric (PI) properties with high gauge factors (GFs) because they are made up of layers with tiny gaps between them. Devices made of 2D materials have a very high sensitivity as a result, enabling the detection of even the smallest mechanical stresses (Yan et al., 2021). Covalent or non-covalent surface modification or functionalization techniques, i.e., chemical processes, are typically used to enhance the sensing performance of 2D materials (Chia et al., 2021; Li et al., 2017; Xu et al., 2022).

A sensor is a machine or system that identifies, monitors, and responds to various inputs and converts them into forms that may be perceived. Instead, sensors use a unique approach to turn these differences into a marker to track the variable of the device. The elementary ideas governing the common transduction procedures of 2D materials used for sensing are as follows:

a. Electrochemical mechanism
b. Electrical mechanism
c. Piezoelectric mechanism, and
d. Optical mechanism

7.4.1 Electrochemical Mechanism of Transduction

Electrochemical sensors utilize electrodes to transduce analytes. Electrochemical sensors are common analytic platforms because of their accuracy, specificity, sensitivity, quick response, and easy setup (Islam et al., 2020). Electrochemical sensors detect analyte–electrode responses. They provide quantitative electric impulses from chemical concentrations. Electrochemical sensors use amperometry, potentiometry, and conductometry. To maximize the quantitative assessment of an electrochemical signal, an operational electrode made of 2D materials is therefore advantageous (e.g., electron transfer of a specific analyte, oxidation, or reduction) (Lee and Suh, 2019). External mediators are required to provide an electrochemical signal proportionate to analytes' concentration if they are unable to initiate redox reactions (Chowdhury et al., 2022).

Prominent among electrochemical sensors are voltammetric sensors and amperometric sensors. Voltammetric sensors assess current based on potential fluctuation during electroactive analyte or mediator redox processes (Scholz et al., 2015). Voltammetric sensors detect tiny, bio-relevant chemical compounds including uric acid, ascorbic acid, and dopamine to monitor human health (Lee et al., 2019; He et al., 2010). Measurements of amperometry are made at a constant potential. The analyte concentration is determined by continually recording the faradaic current created by the electroactive molecule's redox reactions. Health, environmental, food, and agricultural monitoring employs amperometric sensors made from 2D materials (Kudr et al., 2019).

7.4.2 Electrical Mechanism

The most exciting 2D electrical sensors are FETs, which have a high sensitivity, long lifespan, rapid detection, and simplicity of usage (Kajale et al., 2021; Vargas-Bernal et al., 2019). Due to their high electron and hole mobilities, 2D materials are used in FETs as transport channels to drive charges injected from the electrode and collected in the semiconductor. Ambipolar FETs conduct both positive and negative species. Semiconductive 2D materials and adsorbed molecules transfer charges when exposed to various species. Interaction affects resistance (Zhang et al., 2021). Compared to chemiresistor devices, FETs are more flexible, reliable, low energetic, and recover faster (Hu et al., 2021). 2D FETs and electrical sensors for gas, optical, protein, mechanical, ion, and bioelectric sensing are proof-of-concept devices (Dai et al., 2022). Electrical transduction also uses conductometric data (Hashtroudi et al., 2020; Wang et al., 2021). Conductometric sensors use analytes' capacity to change electrical current between electrodes. Gas detection at ambient temperature relies on 2D hybrid nanomaterials including BP, graphene, and layered TMDs (Hashtroudi et al., 2020; Wang et al., 2021). These materials function effectively due to their substantial surface areas, active sites, room-temperature conductivity, and tunable optical properties.

7.4.3 Piezoelectric Mechanism

High-sensitivity PI sensors are used in many applications. Geometric considerations and the resistivity effect affect PI sensor resistance under stress. The average cross-sectional area, length, and resistivity of a semiconductor determine its tension resistance. External inputs are converted into electrical impulses through instruments, which typically calculate the sensitivity of strain sensor via a parameter known as GF. The ratio of normalized resistance change to applied strain signifies the sensitivity. Due to its piezoelectrical features, such as a linear resistance change with strain, graphene is commonly used in PI sensors. Structural deformation, overconnection, and tunneling are the causes of graphene piezoresistivity (Irani et al., 2022). TMDs also exhibit in-plane and vertical piezoelectricity (d33). TMDs with a size of a few centimeters, good crystallinity, and good repeatability are hard to make, which makes it hard to use them in large-area integrated, flexible sensor arrays (Zheng et al., 2022). Many PI sensors use graphene in 3D scaffolds, polymer microstructures, or (micro)fibers (Dominguez-Alfaro et al., 2021).

7.4.4 Optical Mechanism

Simple and inexpensive optical sensors measure light's wavelength, intensity, frequency, light's polarizations, and scattering. Unlike electrical sensors, optical sensors are chemically inert, electromagnetically insensitive, and capable of remote sensing. Optical sensors can identify single molecules. However, the 2D materials' inimitable optical characteristics, better conductivity, remarkable electron mobility, and huge surface-to-volume ratio make them ideal for optical sensors (Lei et al., 2022). Dopants or metallic nanostructures might boost the intended signal (Duan et al., 2021). Biomedicine, environmental monitoring, and food analysis are progressively adopting 2D optical sensors. *Surface-plasmon resonance (SPR)*, *SERS*, and fluorescence resonance energy transfer *(FRET)* are the primary optical sensor transduction methods (Lei et al., 2022).

7.4.4.1 Surface-Plasmon Resonance

Applications of plasmonic sensors for environmental monitoring apart from conventional biochemical studies have grown tremendously. Plasmonic sensors function by oscillating electromagnetic waves and conducting electron density on a material surface in response to a particular stimulus such as photons, electrons, or phonons. Graphene and TMDs are exploited for SPR sensing due to their enticing physicochemical features (Zhou et al., 2020). Few SPR sensors have directly used 2D materials' plasmonic effect (Zhou et al., 2020; Ge et al. (2017). 2D nanomaterials' poor visible and NIR absorption limits their direct use as plasmonic sensors. Thus, plasmonic sensors typically use Au or Ag instead of 2D nanomaterials.

7.4.4.2 Surface-Enhanced Raman Spectrum (SERS)

Raman spectroscopy belongs to the class of typical non-destructive analytical technique for molecular structure identification and fingerprinting. Raman scattering's weak signals at low analyte concentrations limit its sensing use. SERS in the modern era has generated interest because it is extremely sensitive to the detection of several analyte traces along with its capability to undergo ultrasensitive detection of multifold analyte traces (up to 1010) (Wang et al., 2017). Electromagnetic and chemical enhancements drive SERS. Interactions of dipole–dipole between the substrate and the analyte give rise to CM (Ji et al., 2011). Localized SPR (LSPR) is the underlying cause of EM. SERS substrates are 2D materials because of their inexpensive cost, facile production, excellent optical characteristics, and biocompatibility (Chen, 2020). 2D materials are an intriguing substitute for noble metals, serving as cutting-edge materials for SERSs. Noble metal, however efficient, has limitations such as strong metal-adsorbate interaction, lower stability over time, high-cost factor, catalytic effect, and photobleaching property.

SERS is a highly accurate molecule-detecting technology that captures the Raman scattering signal amplification caused by the molecule absorbed by the SERS substrate (Xue et al., 2007). SERS has been found effective in numerous sectors, including explosive detection, cancer detection, and food quality analysis, as a potent detection instrument with single-molecule precision (Banaei et al., 2017). Noble metal nanoparticles, particularly Ag and Au, have been widely used as the most efficient SERS substrates (Zoric et al., 2011). Notably, the spectrum ranges and enhancement factors in SERS have been demonstrated to be highly dependent on the shape of the metal substrates (Orendorff et al., 2006). 2D metal nanostructures with well-defined morphologies, such as nanoprisms, nanoplates, and nanosheets, might serve as ideal high-activity SERS substrates. Sharp corners and edges provide considerably amplified local electric fields, which can intensify Raman scattering signals as Raman hot spots. Xia et al. revealed the shape dependency of SERS substrate in the detection of 4-mercaptopyridine as an example (Xiong, 2005). They established that the strength of the SERS signal on triangular and hexagonal Pd nanoplate substrates was about four and three times that of Pd cuboctahedra, respectively. Assembling the edges of Ag nanoplates sharper can make the SERS signal much better at identifying 1,4-benzenedithiol molecules (Yang et al., 2014). Compared to edges that are straight or curved, edges and corners that are rough or sharp on

Ag nanoplates with wavy edges may give off much stronger SERS signals of 1,4-benzenedithiol. In particular, the peak at 1564 cm for Ag nanoplates with wavy edges was found to be 6.5 and 13 times greater than that of nanoplates with straight and curved edges. Ru-capped columns and Ru triangular nanoplates generated greater SERS signals than Ru nanoparticles (Yin et al., 2012). Meanwhile, because sharp corners and edges in 2D metal nanomaterials might be unstable, heterostructures or alloys based on 2D metal nanomaterials have been created to improve their stability as SERS substrates (Huang & Li, 2011). Zheng and Huang (2011) demonstrated that hexagonal Pd@Ag core–shell nanoplates have high photothermal stability by keeping an eye on the Raman signals of 4-pyridinethiol on Pd@Ag nanoplates and Ag nanoprisms while they were exposed to NIR light (Huang et al., 2011). After 40 minutes of laser irradiation, the SERS signals on Pd@Ag kept more than 90% of their original strength, while the signal on Ag nanoprisms had already dropped by 60%. This shows that core–shell nanoplates are a better NIR–SERS substrate than pure-Ag nanostructures. Jin et al. demonstrated high repeatability and stability of GO-supported Ag nanoplates as the SERS substrate in another paper (Hou et al., 2015).

Raman hot spots can be made from sharp edges and corners of isolated 2D metal nanostructures, as well as from properly arranged 2D metal nanostructures with adjustable gaps between them. This includes assemblies of 2D metal nanomaterials and 2D metal nanostructure arrays. Yu et al. (2008) investigated the theoretical and practical effects of Ag nanoplate dimer gap distance on SERS activity. The calculated local electromagnetic field distribution showed that a strong plasmon coupling effect caused a large resonance amplification when the distance between two Ag nanoplates was only 2 nm. The SERS signal of adenine molecules that were stuck to an Ag nanoplate film substrate was many orders of magnitude higher than that of AgNPs that were more spread out. Another more perfect approach for studying the influence of internanostructure distance on SERS characteristics is to use lithography to create a succession of Au nanodisk arrays of varying diameters (Yu et al., 2008). Because of the larger electromagnetic coupling effect between the Au nanodisks, the smaller diameter Au nanodisk arrays displayed a stronger SERS signal of 4-mercaptopyridine. Also, Han et al. made different SERS performances that depended on the wavelength of excitation oppositely by changing the orientation mode of Ag nanoprism assemblies (Bae et al., 2008). They consider that the various ways that 4-aminobenzenethiol sticks to various surfaces of exposed crystals are what causes the selective SERS amplification (Figure 7.3).

7.4.4.3 Fluorescence Resonance Energy Transfer (FRET)

Fluorescence sensors are portable, sensitive, and inexpensive. The photoluminescent 2D materials are employed as quenchers or fluorophores in fluorescence-based chemical sensors (Neema et al., 2020). FRET uses fluorescent donors and light-absorbing acceptors to transfer energy non-radiatively. When the donor emission and absorption spectra match (<10 nm) (Jones, 2019), 2D materials have fair quenching properties (Tian et al., 2017). TMD (Yan et al., 2017) acceptors has been used in FRET sensors for single-molecule spectroscopy, biochemical analysis, cellular imaging, and environmental monitoring (Zhou et al 2020).

7.5 TWO-DIMENSIONAL MATERIALS AND SENSING APPLICATIONS

Out of the available physical and chemical sensors, physical sensors convert material changes into electro-resistive optical or mechanical outputs. These alterations depend on the number of layers, the intercalation species, and the externally switched potential bias (Dai et al., 2020). Identifying and quantifying chemical species are crucial. For environmental protection, safety, and medical diagnosis, monitoring and ultra-sensitive detection of these compounds have been widely investigated (Anichini, 2018). 2D materials may sense chemical species and interact with analytes, comparable to previous methods. Detection of chemical species and trapping may be done through chemical functionalization, natural flaws, and surface doping of 2D materials.

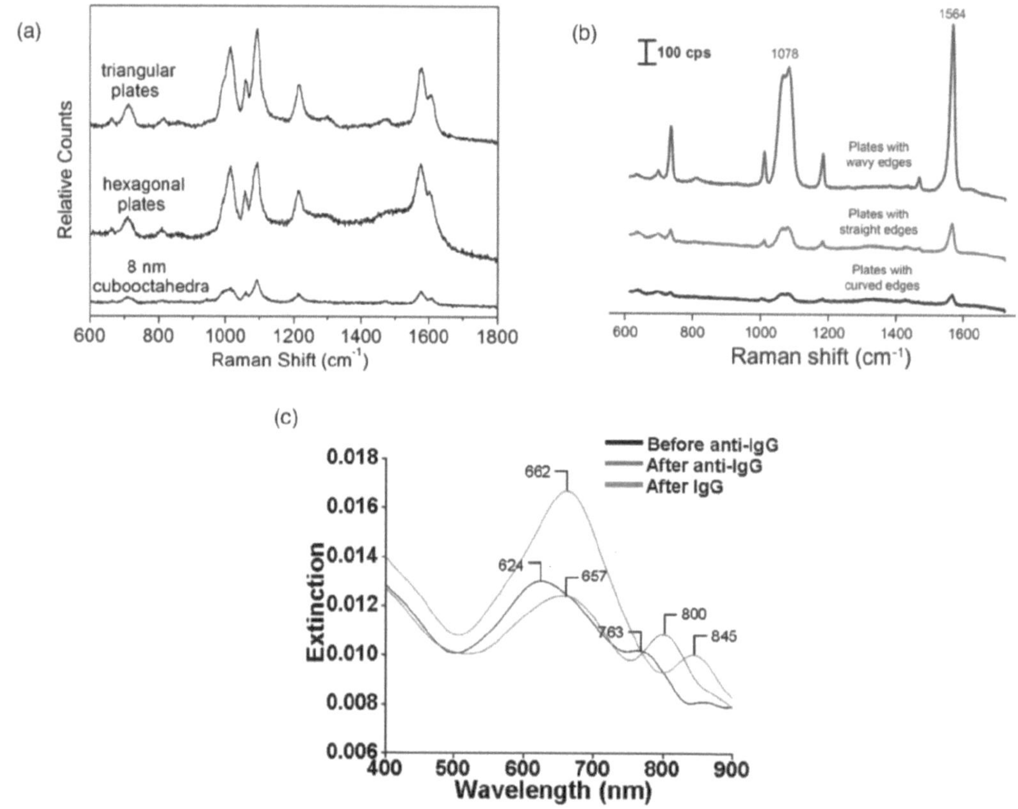

FIGURE 7.3 (a) Raman spectra of 4-mercaptopyridine molecules adsorbed on Pd triangular, hexagonal, and 8 nm cuboctahedra. Reproduced from (Xiong et al. 2005) (b) A comparison of SERS spectra obtained from aqueous solutions of 1,4-benzenedithiol-adsorbed Ag nanoplates with straight, curved, and wavy edges. Reproduced from Yang et al. 2014 (c) LSPR spectra of an Au nanoplates-coated glass substrate before (black), after (red), and after functionalization with human anti-IgG (green). Reproduced from Beeram et al. 2010.

7.5.1 Electrosensitivity

The most popular use for 2D materials with electroresistivity is gas detection, which makes use of chemiresistor and FETs (Chia et al., 2021). Chemiresistor resistance and FET drain current variation during gaseous analyte adsorption underpin their operation. With bare 2D material and gaseous analyte, the interaction is weak, but the chemical composition of the material can engineer it. MXene and g-CN have terminal oxygen groups and amines, permitting chemical alterations and hydrogen-bonding interactions. h-BN or BP nanosheets are less stable to oxygen and light (Dae et al., 2020). H_2, NH_3, NO_x, and SO_2 are the most prevalent gas-sensing analytes. Gas sensor studies often include theoretical predictions of gaseous analyte–pristine material interactions (Chia et al., 2021). TMDs have limited sensitivity to nonpolar molecules; however, dopant insertion, which is usually metals, changes their electronic structures and increases sensor performance (Chia et al., 2021).

Many chemiresistive 2D material sensors exist. Zhou et al. used density functional theory (DFT) to examine hazardous gas molecules' interactions with WS_2 monolayers. Band structure calculations found that monolayer WS_2's valence and conduction bands are unaffected by H_2, NH_3, and CO. When adsorbed on WS_2 monolayers, O_2, NO, and NO_2's lowest unoccupied orbitals are trapped around the Fermi level (Zhou et al., 2015). Numerous authors used n- and p-type WS_2 materials for proton-rich analytes including water and ammonia in addition to NO_2. Adsorption and/or solvation-assisted interaction of these molecules onto the material's surface may enhance acidity and catalyze reactions (Jha et al., 2019). Matatagui and colleagues created a simple, inexpensive WS_2 nanoplatelets device on paper via abrasion-induced deposition. The device detected room-temperature NO_2 with great sensitivity and selectivity (Matatagui et al., 2022). Drop casting of liquid-exfoliated WS_2 nanosheets on interdigitated electrodes (IDEs) created chemoresistive humidity sensors. Exfoliated WS_2's affinity for protons suggested n-type performance. The response of sensor was between 11.9% and 80% RH. The sensor responds quickly to RH changes. The range from 13 s (11.9%) to 17 s (80%) was the reaction time (Jha et al., 2019). In another study, the fabrication of p-type nanoflake was done through ball milling of commercial bulk WS_2 powder (Li et al., 2016). IDE having 200-μm gap electrodes is produced through dip coating. The sensor is relatively more sensitive with high response/recovery times for ammonia concentrations from 1 to 10 ppm at room temperature (Li et al., 2017). Other approaches develop material "in situ" on the apparatus. O'Brien et al. sulfurized tungsten trioxide (WO_3) with H_2S plasma to synthesize WS_2 at low temperatures. WO_3-sputter-dependent interdigitated WS_2 electrodes were made. Room-temperature sensitivity was 1.4 ppm NH_3 in nitrogen (Li et al., 2017). An affordable hydrothermal technique and calcination procedure produced ultra-thin WS_2 nanosheets having very low thickness of 5 nm, similar to CVD. WS_2 nanosheets produced a wall-like structure. Resistive gas sensors had an appreciable degree of sensitivity ranging from 9.3% to 0.1 ppm NO_2 gas at 25°C and were quite stable under low and moderate humidity (Xu et al., 2018).

N-methyl-2-pyrrolidone (NMP) was used by Guo et al. (2019) to exfoliate WSe_2 into nanosheets. An outstanding selectivity was demonstrated by a chemiresistive sensor, for low NO_2 detection concentration (50 ppb), with an exceptionally high degree of sensitivity and consistent long-term stability for a considerable period of 8 weeks. The authors suggested a technique based on charge transfer between NO_2 molecules and WSe_2 nanosheets, as well as NO_2 gas physisorption. The WSe_2 nanosheets' electrophilic NO_2 surface molecules increase the concentration of holes and lower sensor resistance by removing electrons from the valence band (Guo et al., 2019).

A chemical sensor with exceptional selectivity for trimethylamine (TEA) was a monolayer of MoS_2. Transient surface physisorption influences MoS_2 monolayer conductance. TEA and carbon nanotube gas sensors were compared. MoS_2-based devices outperformed tube-based devices in selectivity in a series of observations (Cho et al., 2015). CVD-layered MoS_2 is obtained from MoO_3 and sulfur powder. IDEs with 100-μm spacing were placed on a sapphire substrate. NO_2 and NH_3 transient resistance responses were examined within a range of 1.2–50 ppm. The resistance for the NO_2 gas mode showed positive sensitivity at ambient temperature and 100°C. In order to p-dope the Fermi level and bring it closer to the valence-band edge, NO_2 served as an electron acceptor. In order to get the Fermi level of MoS_2 near the conduction-band edge, NH_3 was used as an electron donor (n-doping), resulting in negative sensitivity at both temperatures, and the material surface was saturated at 20 ppm regardless of operation temperature. Finally, DFT validated the charge-transfer process between gas molecules and MoS_2 (Cho et al., 2015).

Zhao et al. (2018) revealed a convenient yet effective technique for synthesizing extensive TMD layers on flexible substrates. CVD using molybdenum carbonyl ($Mo(CO)_6$) and H_2S produced MoS_2 directly on SiO_2 and PI substrates at 200°C. Raman spectroscopy shows that low-temperature-grown crystals have a layered structure with high homogeneity, stoichiometry, and control layers.

MoS_2 on soft substrates detected ammonia and NO_2. The 2D MoS_2-based flexible gas sensor responded 300% to NO_2 and 5% to NH_3 at 500 ppm analyte. The 2D nature of the flaw or basal plane affects gas molecule adsorption. A flexible gas sensor containing MoS_2 with a bending radius of 7 mm was tested for durability after 2000, 4000, and 6000 bending cycles (Zhao et al., 2018). Cho et al. (2015) employed CVD to produce MoS_2 nanosheets vertically and horizontally, improving the material's sensing capabilities over the exposed region. They tested MoS_2 nanosheet NO_2 sensing capabilities using the basal plane, edge, and mixed electrode designs. Edge-exposed arrangement increased sensitivity fivefold and showed area-in-contact dependency (Cho et al., 2015). Cone-shaped (CS)–MoS_2 bilayers were created by depositing 2-nm-thick MoO_3 layers on a 2″ three-dimensional

(3D) cone-patterned sapphire substrate (CPSS) followed by CVD sulfurization. The sensor detected NO at 200% ppb levels. The 30% exposed area increase in this construction increased sensing at ambient temperature (Chen et al., 2019).

Single- and few-layer $MoSe_2$ nanosheets at room temperature were mechanically exfoliated using bulk crystals onto pre-cleaned 300 nm SiO_2/Si substrates at room temperature. Electron-beam lithography was used to construct $MoSe_2$ gas sensor devices that responded nonlinearly to ammonia concentrations from 50 to 500 ppm (Late et al., 2014). Other work synthesized ultrasonication-assisted liquid exfoliation $MoSe_2$ chemiresistive devices. Centrifugation speed controlled nanoflakes' size. The exfoliated material was drop cast in an IDE to detect selective H_2S gas at 200°C. The device's response ranged from 15.87% to 53.04% for H_2S values of 50 ppb to 4.45 ppm in ambient air and from 7.13% to 19.87% for 500 ppb to 4.45 ppm in synthetic air (Jha et al., 2019). Choi et al. (2017) presented a niobium-doped $MoSe_2$ nanomaterial for NO_2 gas sensing. Plasma-incorporated Nb was compared to $MoSe_2$. Doping control optimized gas sensing responsiveness. The sensor detected 3 ppm NO_2 best after the initial doping ($MoSe_2$: Nb-1C). Nb improved the surface area-to-volume ratio, making it more receptive compared to pure $MoSe_2$.

The amount of metallic niobium diselenide ($NbSe_2$) clusters boosted by Nb dopants reduced the sensor's efficiency, where these clusters were non-sensitive to gaseous molecules (Choi et al., 2017). In $MoTe_2$, a novel multilayer TMD, Zhang and co-workers (2021) exploited TMDs' NO_2 chemisorption and photoresponsivity to improve sensor recovery. Under 254-nm UV irradiation, exfoliated material has a detection limit of 252 ppt. The recovery rate rose from 10% to 40% in dark conditions to 100% under UV irradiation within 160 seconds at various NO_2 concentrations. Finally, the p-type $MoTe_2$ sensor detected NO_2 in ambient air well, suggesting its practical potential (Wu et al., 2018). TMDs behaved similarly when approached analogously (Tabata et al., 2021).

A tridimensional h-BN architecture was synthesized with boron trioxide (B_2O_3) and h-BN for sensing humidity and LPG (Gautam et al., 2016). The material was tridimensionally porous in nature after preparation and had the capability to show sensation via chemisorptions just like other 2D materials. h-BN's resistivity ranged from 1.47 to 0.29 M/%RH for low (10%–40%) and high (40%–90%) humidity, respectively (Gautam et al., 2016; Goel et al., 2021). A low-temperature solvothermal technique produced h-BN with a practically transparent shape and a thickness of ~1. 36 nm (four atomic layers). Charge transfer from oxygen molecule chemical absorption/desorption on the surface of sensor generated fluctuation in conductance. It can measure up to 100 ppm at various operating temperatures. At temperatures of 240°C, 260°C, and 300°C, ethanol sensitivity was 1.72, 2.9, and 4.5, respectively, with high recovery and response of the equipment recovered (Lin et al., 2016). MXenes are good chemical sensors because of their good conductivity and the large number of functional groups on their surface, unlike pristine graphene, which possesses an inert surface and low uptake of chemical species, or MoS_2, which has a high selectivity but a lower signal-to-noise ratio (Kim et al., 2018). They created a VOC-selective MXene chemical sensor. Material conductivity improves sensor LOD. The surface's functional groups enhance VOC selectivity and interaction. The authors evaluate one form of MXene ($Ti_3C_2T_x$) and emphasize the broad diversity of this family of compounds' enhancement potential. Yuan and colleagues also electrospun MXene to great effect. After 1000 cycles, the sensors maintained performance, indicating industrial potential (Yuan et al., 2018).

7.5.2 Optical Sensor

Lu et al. used SERS to identify nitroaromatic chemicals using GO coated with AgNPs. Rhodamine 6G, methyl violet, and methylene blue were identified as fluorophores. GO-AgNP gave strong peaks. Results revealed a significant improvement over pristine GO. These findings demonstrated the AgNP's signal-boosting action (Lu et al., 2016). Electrostatic interactions self-assembled poly(N-vinyl-2-pyrrolidone)-stabilized gold nanorods (AuNR) with GO. The GO–AuNR materials raised the Raman signal for crystal violet and neutral red aromatic dyes, proving that cationic and anionic aromatic dyes may be molecular probes (Hu et al., 2015). Highly enriched 1T phase MoS_2 up to 94.5% was presented through an exfoliation approach (Criado et al., 2019), which displayed low-detection SERS in crystal violet and rhodamine 6G. Coherently, the same material exfoliated 1T phase MoS_2 modified with a monoclonal antibody was employed as a capture probe for α-fetoprotein (AFP) at highly detectable levels (1 pg mL^{-1} to 10 ng mL^{-1}) with LOD less than 0.03 pg. They added silver-coated gold nanocubes to functionalized MoS_2 to enhance immunosensor "hot spots" (Silvestri et al., 2021). Su et al. (2014) used gold nanoparticles to detect rhodamine 6G on MoS_2 nanosheets. The signal was amplified by the mixing process. A maximum optical signal requires functionalization control and accuracy. Particle aggregation maximized hot spots, enhancing SERS (Su et al., 2014). To detect methylene blue, TMDs and Ti_3C_2-functionalized platinum, gold, and silver nanoparticles were synthesized during SERS. The functionalized 2D material enhanced the analyte signal in all TMDs and shifted peak locations. Silver, gold, and palladium nanoparticle-decorated Ti_3C_2 had enhancement factors of 1.50, 1.17, and 9.61104, respectively (Satheeshkumar et al., 2016).

7.5.3 Colorimetric Sensors

Colorimetric assays using 2D material sensing systems detect metal ions easily such as Hg^{2+} in aqueous solutions in the presence of peroxidase (Zhang et al., 2017). 2D rGO/PEI/Pd nanohybrid was synthesized for detection where mercury ions stimulate oxidation in the nanohybrids rGO/PEI/Pd, yielding 3,3′,5,5′-tetramethylbenzidine (TMB) having visible blue color. Nanohybrids detected Hg^{2+} in ddH_2O at 0.39 nM and wastewater at ~1 nM (Zhang et al., 2017). Wang and colleagues demonstrated a simple method to activate optical signals using MXenes to reduce Ag^+ to AgNPs. Selective NPs stay on the material's surface. Besides a UV-vis spectrometer, a smartphone can detect it (Wang et al., 2020).

7.6 BIOMEDICAL APPLICATIONS OF 2-DIMENSIONAL NANOMATERIALS

7.6.1 Application of Graphene

Gene therapy is a revolutionary treatment for hereditary problems. Gene delivery uses functionalized graphene extensively. A graphene gene delivery vector modified with polyamidoamine (PAMAM) dendrimer and oleic acid was biocompatible and efficient (Liu et al., 2014). In cell culture assays, graphene–oleate–PAMAM hybrids had 13 times the gene transfection capability of ultrasonicated graphene. Graphene-based nanomaterials also promote cell necrosis. Few-layer graphene (FLG) kills monocytes without harming other immune cells (Russier et al., 2017). FLG is used to treat myeloid monocytic leukemia. FLG also outperformed etoposide in specificity and toxicity. Tissue engineering is exploring graphene's unique features. Tissue engineering uses cells, biochemical agents, and designed materials to restore, maintain, or improve tissue or organ function (Cheng et al., 2017). Lu et al. (2016) created a graphene hydrogel (MGH) membrane by face-to-face stacking multiple chemically converted graphene (CCG) sheets for guided bone regeneration (GBR). CCG-based membranes maintained bone space, promoted early osteogenesis, accelerated mineralization, and improved bone quality and regeneration. MGH membranes remained in place 8 weeks after implantation, suggesting their surface texture adhered to bone tissues. This study showed CCG's GBR therapy and biological tissue engineering potential. Another study produced a graphene-based SPR biosensor to identify biomolecules (Zeng et al., 2015) and developed a graphene–gold metasurface biosensor that directly detects unlabeled ssDNA via pi-stacking. The graphene sensing substrate, made from pure graphene layers, has superior optical and electrical characteristics, a high electron transfer rate, and optical transparency, enhancing plasmons. Graphene gold biosensing structure detected ssDNA at 1×10^{-18} mol/L, i.e. 1 aM, greater than SPR biosensor. As seen above, graphene is crucial to multisensor platform design for illness detection (Li et al., 2023).

7.6.2 Application of Graphene Oxide (GO)

GO has good water dispersion, biocompatibility, and drug molecule affinity because of its primitive graphite structure, oxygen-containing functional groups, and hydrophobic and hydrophilic characteristics. These features make GO nanosheets popular drug carriers. Photothermal treatment (PTT) agents frequently employ GO due to their excellent NIR absorption. Yin et al., (2017) created a doxorubicin hydrochloride (DOX)-loaded hyaluronic acid-decorated GO nanosheet (HSG). In vivo and in vitro investigations showed that this nanosystem may give photothermal-controlled selective fast drug delivery and photo-chemo-synergetic treatment. Tissue engineering uses GO, a common 2D nanomaterial (Ruan et al., 2016). GO was covalently cross-linked with carboxymethyl cellulose (CMC) to make biocompatible scaffolds. Porous GO-CMC scaffolds have greater modulus and hardness than CMC scaffolds. GO sheets' p–p stacking interactions increased GO-CMC scaffolds' osteoinductivity. Systematic in vitro and in vivo tests showed that these GO-CMC scaffolds repaired rats' skull defects, advancing regenerative medicine and tissue engineering. Advanced biosensors can also benefit from GO's unique and flexible features. Cheeveewattanagul et al. (2017) created a useful immunosensing platform using GO-coated nanopaper (GONAP). GONAP's photoluminescence-quenching ability quenched Ab-QDs' photoluminescence. These antibodies might form antigen–Ab–QD complexes and be desorbed from GONAP, restoring photoluminescence. A very sensitive immunosensor technology with ready-to-use immune sensing qualities is almost ubiquitous. Human serum analysis was successful using the test. In summary, GONAP allows immunosensing for automated diagnosis.

7.6.3 Application of Reduced Graphene Oxide (RGO)

RGO is produced by reducing GO to eliminate most of its oxygenated functional groups, resulting in carbon nanosheet structural flaws. rGO's electrical conductivity comes from its restored graphitic network (Cheng et al., 2017). rGO may also be constructed via non-covalent interactions (van der Waals forces and p–p stacking) to adsorb drugs or biomolecules. Drug delivery, PTT, tissue engineering, and biosensing employ rGO because of its biocompatibility and electrical conductivity (Hwang et al., 2023).

Drug delivery research has focused on rGO due to its high specific surface area and electrochemical characteristics. He et al. (2017) created an electrochemically triggered drug delivery system using DOX-loaded rGO thin films and p–p stacking and hydrophobic interaction. At pH 9, DOX loaded 98% efficiently onto rGO. Another study created a cancer-treating rGO–PTT agent. Yan et al. (2019) created multifunctional rGO-based nanosheets with indoleamine 2, 3-dioxygenase (IDOi) that activated PTT to eliminate primary tumors under NIR laser irradiation. rGO nanosheets releasing IDOi at tumor locations may influence tryptophan catabolism, boosting the anticancer immune response. IDOi/rGO nanosheets synergistically induced systemic antitumor immunity and directly killed tumor cells under laser irradiation in vitro and in vivo. rGO also works in tissue engineering. Huo et al. (2017) created rGO-enzyme-coated tissue-engineered blood vessels (TEBVs) to prevent platelet aggregation by carrying two enzymatic catalysts, apyrase and 5-nucleotidase (5-NT), enabling cascade reactions. These rGO–enzyme complexes might convert ADP to AMP and thereafter to adenosine, preventing ADP-induced platelet aggregation and thrombosis.

Due to its availability and electrical conductivity, RGO-based biosensors have been widely researched. Using rGO/Rh nanoparticles, a sensitive electrochemical enzyme biosensor for 17b-estradiol detection was created (Povedano et al., 2017). 17b-estradiol is an emerging pollutant because it harms animal and human endocrine systems. This rGO-based biosensor detected 17b-estradiol at 0.54 pmol L^{-1} and 25.7 A (mol L^{-1}) cm, diagnosing endocrine function. These findings suggest that rGO nanomaterial-based biosensors can quickly and accurately detect hormones in urine samples.

7.6.4 Biomedical 2-Dimensional Silicate Clays

Silicate clays are versatile and have grown in popularity in recent years. Crystalline clays like smectite are prevalent. An octahedral alumina sheet is placed between two tetrahedral silica sheets (Chrzanowski et al., 2013). Clays have aqueous stability, high drug loading, shear thinning, etc. Clays' layered structure gives their surfaces a permanent negative charge and their edges a positive charge. The pharmaceutical sector uses clays for drug delivery, tissue engineering, and biosensing due to these benefits (Chang et al., 2010). Silicate clays' superior biological safety has spurred gene or drug delivery research during the past decade (Lvov et al., 2016). They have additional great delivery system qualities than biocompatibility. Silicate clays adsorb nucleic acid and medicinal molecules with structural stability, preventing biological molecules from deterioration or destruction in harsh environments. Silicate clays, unlike other vectors, are safe and effective therapeutic carriers that avoid significant immunological and toxicological reactions (Khlibsuwan et al., 2018). Jesus et al. (2018) created a clay-based nanocomposite to administer sodium diclofenac (SDCF) using sodium montmorillonite (MMT) in a hybrid hydrogel of siloxane-poly (ethylene oxide). This hybrid hydrogel nanocomposite may sustain SDCF release, treating arthritis, sprains, gout, migraine, and postoperative pain long term.

Tissue engineering research generated silicate nanomaterials (Kerativitayanan et al., 2015). Scientists created a nanocomposite hydrogel of laponite nanosilicates and gelatin methacrylate (GelMA) to regulate the fast release of growth factor-rich stem cell secretome (Waters et al., 2016). The nanocomposite hydrogel loaded with secretome promoted angiogenesis and heart protection, enhancing the therapeutic effectiveness of stem cell-derived secretome for cardiac tissue repair and regeneration. Laponites (XLG) can be added to isolated hydroxypropylmethylcellulose (Si-HPMC) hydrogel for tissue engineering (Boyer et al., 2018). The composite hydrogel developed a hybrid interpenetrating network, increasing its mechanical characteristics. The composite hydrogel's physicochemical properties and in vitro cytocompatibility were assessed. The composite hydrogel increased cell survival and clustering. Boyer et al. (2018) concluded from these preliminary results that Si-HPMC/XLG hydrogel-containing chondrocytes might be used in cartilage tissue engineering. Silicate clays, 2D natural clays that activate blood platelets to coagulate blood, are used in tissue engineering for hemostasis. Gaharwar et al. (2014) synthesized nanocomposite hydrogels made of silicate nanoplatelets and gelatin to increase blood coagulation in vitro and in vivo. Silicate nanoplatelets added to gelatin improved physiological stability, injectability, hemostatic efficacy, and in vitro blood clotting time by 77%. These nanocomposites were biocompatible and promoted hemostasis in deadly liver lacerations in vivo experiments. These silicate-based gelatin nanocomposites might be employed as injectable hemostatic agents to treat urgent wounds outside the hospital.

7.7 CONCLUSION

The inherent features of 2D materials and their use in biomedicine have been examined in this study. The synthesis of 2D materials using various methods and their properties have been discussed. Many graphene-based compounds, including functionalized GO, and reduced GO have been examined. About the most modern biosensing technology, including biosensors, the mechanics are explored. One of the most well-known 2D materials is graphene. A zero-band gap semiconductor, linear-like at the Dirac point relativistic-like charge velocity, ultra-high charge mobility, transparency, large surface area, non-toxicity, proximity-induced ability, high tensile strength, and high thermal conductivity are just a few of graphene's outstanding characteristics and qualities. The mechanisms and methods used in the engineering of biosensing platforms are explored. This provides a wide range of opportunities and is essential for sensor and biosensor applications that make use of a big surface area. These 2D new materials will be improved upon and optimized for bioreceptor specificity shortly. These materials may be used and incorporated into various sensor and biosensor systems, providing ultra-high sensitivity and potentially solving problems like early cancer detection.

REFERENCES

Abbas, A. N.; Liu, B.; Chen, L.; Ma, Y.; Cong, S.; Aroonyadet, N.; Köpf, M.; Nilges, T.; Zhou, C. Black phosphorus gas sensors. *ACS Nano* 2015, 9(5), 5618–5624. https://doi.org/10.1021/acsnano.5b01961.

Akbarnejad, E.; Salehi, F.; Mohajerzadeh, S. Facile and rapid exfoliation of black phosphorus assisted by acetic acid. *Journal of Materials Science: Materials in Electronics* 2022, 33(5), 2499–2508. https://doi.org/10.1007/s10854-021-07457-2.

Ali, A.; Koybasi, O.; Xing, W.; Wright, D. N.; Varandani, D.; Taniguchi, T.; Watanabe, K.; Mehta, B. R.; Belle, B. D. Single digit parts-per-billion NO_x detection using MoS_2/hBN transistors. *Sensors Actuators A Physical* 2020, 315, 112247. https://doi.org/10.1016/j.sna.2020.112247.

Anichini, C.; Czepa, W.; Pakulski, D.; Aliprandi, A.; Ciesielski, A.; Samorì, P. Chemical sensing with 2D materials. *Chemical Society Reviews* 2018, 47(13), 4860–4908. https://doi.org/10.1039/C8CS00417J.

Bae, Y.; Kim, N. H.; Kim, M.; Lee, K. Y.; Han, S. W. Anisotropic assembly of Ag nanoprisms. *Journal of the American Chemical Society* 2008, 130, 5432–5433.

Banaei, N.; Foley, A.; Houghton, J. M.; Sun, Y.; Kim, B. Multiplex detection of pancreatic cancer biomarkers using a SERS based immunoassay. *Nanotechnology* 2017, 28, 455101.

Bao, C.; Beziere, N.; del Pino, P.; Pelaz, B.; Estrada, G.; Tian, F.; Ntziachristos, V.; de la Fuente, J. M.; Cui, D. Nanoprisms: Gold nanoprisms as optoacoustic signal nanoamplifiers for in vivo bioimaging of gastrointestinal cancers. *Small* 2013, 9, 67.

Boyer, C.; Figueiredo, L.; Pace, R. et al., Laponite nanoparticle-associated silated hydroxypropylmethyl cellulose as an injectable reinforced interpenetrating network hydrogel for cartilage tissue engineering. *Acta Biomaterialia* 2018, 65, 112–122.

Campbell, C. T.; Parker, S. C.; Starr, D. E. The effect of size- dependent nanoparticle energetics on catalyst sintering. *Science* 2002, 298, 811–814.

Castellanos-Gomez, A.; Vicarelli, L.; Prada, E.; Island, J. O.; Narasimha-Acharya, K. L.; Blanter, S. I.; Groenendijk, D. J.; Buscema, M.; Steele, G. A.; Alvarez, J. V.; Zandbergen, H. W.; Palacios, J. J.; van der Zant, H. S. J. Isolation and characterization of few-layer black phosphorus. *2D Materials* 2014, 1(2), 025001. https://doi.org/10.1088/2053-1583/1/2/025001.

Chang, C.; Spreeuwel, A.; Zhang, C. PEG/clay nanocomposite hydrogel: A mechanically robust tissue engineering scaffold. *Soft Matter* 2010, 6, 5157–5164.

Cheeveewattanagul, N.; Morales-Narváez, E.; Bergua, J. F. Straightforward immunosensing platform based on graphene oxide-decorated nanopaper: A highly sensitive and fast biosensing approach. *Advanced Functional Materials* 2017, 27, 1702741.

Chen, C.; Kang, Y.; Huo, Z.; Zhu, Z.; Huang, W.; Xin, H. L.; Snyder, J. D.; Li, D.; Herron, J. A.; Mavrikakis, M. et al., Highly crystalline multimetallic nanoframes with three-dimensional electrocatalytic surfaces. *Science* 2014, 343, 1339–1343.

Chen, M.; Liu, D.; Du, X.; Lo, K. H.; Wang, S.; Zhou, B.; Pan, H. 2D materials: Excellent substrates for surface-enhanced Raman scattering (SERS) in chemical sensing and biosensing. *TrAC Trends in Analytical Chemistry* 2020, 130, 115983. https://doi.org/10.1016/j.trac.2020.115983.

Chen, X.; Lin, S.; Zhang, H. Screening of single-atom catalysts sandwiched by boron nitride sheet and graphene for oxygen reduction and oxygen evolution. *Renewable Energy* 2022, 189, 502–509. https://doi.org/10.1016/j.renene.2022.03.003.

Chen, X.; Park, Y. J.; Kang, M.; Kang, S. K.; Koo, J.; Shinde, S. M.; Shin, J.; Jeon, S.; Park, G.; Yan, Y.; MacEwan, M. R.; Ray, W. Z.; Lee, K. M.; Rogers, J. A.; Ahn, J. H. CVD grown monolayer MoS_2 in bioabsorbable electronics and biosensors. *Nature Communications* 2018, 9(1), 1690. https://doi.org/10.1038/s41467-018-03956-9.

Chen, Y. Z.; Wang, S. W.; Yang, C. C.; Chung, C. H.; Wang, Y. C.; Huang Chen, S. W.; Chen, C. W.; Su, T. Y.; Lin, H. N.; Kuo, H. C.; Chueh, Y. L. An indoor light-activated 3D cone-shaped MoS_2 bilayer-based NO gas sensor with PPb-level detection at room-temperature. *Nanoscale* 2019, 11(21), 10410–10419. https://doi.org/10.1039/C8NR10157D.

Cheng, C.; Li, S.; Thomas, A. Functional graphene nanomaterials based architectures: Biointeractions, fabrications, and emerging biological applications. *Chemical Reviews* 2017, 117, 1826–1914.

Chia, H. L.; Mayorga-Martinez, C. C.; Pumera, M. Doping and decorating 2D materials for biosensing: Benefits and drawbacks. *Advanced Functional Materials* 2021, 31(46), 2102555. https://doi.org/10.1002/adfm.202102555.

Cho, S. Y.; Kim, S. J.; Lee, Y.; Kim, J. S.; Jung, W. B.; Yoo, H. W.; Kim, J.; Jung, H. T. Highly enhanced gas adsorption properties in vertically aligned MoS_2 layers. *ACS Nano* 2015, 9(9), 9314–9321. https://doi.org/10.1021/acsnano.5b04504.

Choi, S. Y.; Kim, Y.; Chung, H. S.; Kim, A. R.; Kwon, J. D.; Park, J.; Kim, Y. L.; Kwon, S. H.; Hahm, M. G.; Cho, B. Effect of Nb doping on chemical sensing performance of two-dimensional layered $MoSe_2$. *ACS Applied Materials & Interfaces* 2017, 9(4), 3817–3823. https://doi.org/10.1021/acsami.6b14551.

Chowdhury, T.; Sadler, E. C.; Kempa, T. J. Progress and prospects in transition-metal dichalcogenide research beyond 2D. *Chemical Reviews* 2020, 120(22), 2563–12591. https://doi.org/10.1021/acs.chemrev.0c00505.

Chowdhury A.; Biswas, S.; Singh, T.; Chandra, A. Redox mediator induced electrochemical reactions at the electrode-electrolyte interface: Making sodium- ion supercapacitors a competitive technology. *Electrochemical Science Advances* 2022, 2(1), e2100030. https://doi.org/10.1002/elsa.202100030.

Chrzanowski, W.; Kim, S. Y.; AbouNeel, E. A. Biomedical applications of clay. *Australian Journal of Chemistry* 2013, 66, 1315–1322.

Corma, A.; Concepcion, P.; Boronat, M.; Sabater, M. J.; Navas, J.; Yacaman, M. J.; Larios, E.; Posadas, A.; Lopez-Quintela, M. A.; Buceta, D. et al., Exceptional oxidation activity with size-controlled supported gold clusters of low atomicity. *Nature Chemistry* 2013, 5, 775–781.

Cui, S.; Pu, H.; Wells, S. A.; Wen, Z.; Mao, S.; Chang, J.; Hersam, M. C.; Chen, J. Ultrahigh sensitivity and layer-dependent sensing performance of phosphorene-based gas sensors. *Nature Communications* 2015, 6(1), 8632. https://doi.org/10.1038/ncomms9632.

Dai, C.; Liu, Y.; Wei, D. Two-dimensional field-effect transistor sensors: The road toward commercialization. *Chemical Reviews* 2022, 122(11), 10319–10392. https://doi.org/10.1021/acs.chemrev.1c00924.

Dai, J.; Ogbeide, O.; Macadam, N.; Sun, Q.; Yu, W.; Li, Y.; Su, B. L.; Hasan, T.; Huang, X.; Huang, W. Printed gas sensors. *Chemical Society Reviews* 2020, 49(6), 1756–1789. https://doi.org/10.1039/C9CS00459A.

Dominguez-Alfaro, A.; Gómez, I. J.; Alegret, N.; Mecerreyes, D.; Prato, M. 2D and 3D immobilization of carbon nanomaterials into PEDOT via electropolymerization of a functional Bis-EDOT monomer. *Polymers* 2021, 13(3), 436. https://doi.org/10.3390/polym13030436.

Dong, L. M.; Ye, C.; Zheng, L. L.; Gao, Z. F.; Xia, F. Two-dimensional metal carbides and nitrides (MXenes): Preparation, property, and applications in cancer therapy. *Nanophotonics* 2020, 9(8), 2125–2145. https://doi.org/10.1515/nanoph-2019–05500.

Dral, A. P.; ten Elshof, J. E. 2D metal oxide nanoflakes for sensing applications: Review and perspective. *Sensors and Actuators B: Chemical* 2018, 272, 369–392. https://doi.org/10.1016/j.snb.2018.05.157.

Duan, Q.; Liu, Y.; Chang, S.; Chen, H.; Chen, J. H. Surface plasmonic sensors: Sensing mechanism and recent applications. *Sensors* 2021, 21(16), 5262. https://doi.org/10.3390/s21165262.

Eissa, S.; Jimenez, G. C.; Mahvash, F.; Guermoune, A.; Tlili, C.; Szkopek, T.; Zourob, M.; Siaj, M. Functionalized CVD monolayer graphene for label-free impedimetric biosensing. *Nano Research* 2015, 8(5), 1698–1709. https://doi.org/10.1007/s12274-014-0671-0.

Faraday, M. The Bakerian lecture: Experimental relations of gold (and other metals) to light. *Philosophical transactions of the Royal Society of London. Series B, Biological Sciences* 1857, 147, 145–181.

Gaharwar, A. K.; Avery, R. K.; Assmann, A. Shear-thinning nanocomposite hydrogels for the treatment of hemorrhage. *ACS Nano* 2014, 8, 9833–9842.

Gao, C.; Lu, Z.; Liu, Y.; Zhang, Q.; Chi, M.; Cheng, Q.; Yin, Y. Highly stable silver nanoplates for surface plasmon resonance biosensing. *Angewandte Chemie International Edition* 2012, 51, 5629–5633.

Gao, J.; He, P.; Yang, T.; Wang, X.; Zhou, L.; He, Q.; Jia, L.; Deng, H.; Zhang, H.; Jia, B.; He, X. Short rod-like Ni-MOF anchored on graphene oxide nanosheets: A promising volta for highly sensitive determination of pchloronitrobenzene. *Journal of Electroanalytical Chemistry* 2020, 861, 113954. https://doi.org/10.1016/j.jelechem.2020.113954.

Gautam, C.; Tiwary, C. S.; Machado, L. D.; Jose, S.; Ozden, S.; Biradar, S.; Galvao, D. S.; Sonker, R. K.; Yadav, B. C.; Vajtai, R.; Ajayan, P. M. Synthesis and porous h-BN 3D architectures for effective humidity and gas sensors. *RSC Advances* 2016, 6(91), 87888–87896. https://doi.org/10.1039/C6RA18833H.

Gautam, C.; Chelliah, S. Methods of hexagonal boron nitride exfoliation and its functionalization: Covalent and non-covalent approaches. *RSC Advances* 2021, 11(50), 31284–31327. https://doi.org/10.1039/D1RA05727H.

Ge, Y.; Chen, S.; Xu, Y.; He, Z.; Liang, Z.; Chen, Y.; Song, Y.; Fan, D.; Zhang, K.; Zhang, H. Few-layer selenium-doped black phosphorus: Synthesis, nonlinear optical properties and ultrafast photonics applications. *Journal of Materials Chemistry C* 2017, 5(25), 6129–6135. https://doi.org/10.1039/C7TC01267E.

Geim, A. K.; Novoselov, K. S. The rise of graphene. *Nature Materials* 2007, 6, 183–191.

Goel, N.; Kumar, M. Recent advances in ultrathin 2D hexagonal boron nitride based gas sensors. *Journal of Materials Chemistry C* 2021, 9(5), 1537–1549. https://doi.org/10.1039/D0TC05855F.

Guo, R.; Han, Y.; Su, C.; Chen, X.; Zeng, M.; Hu, N.; Su, Y.; Zhou, Z.; Wei, H.; Yang, Z. Ultrasensitive room temperature NO_2 sensors based on liquid phase exfoliated WSe_2 nanosheets. *Sensors & Actuators, B: Chemical* 2019, 300, 127013. https://doi.org/10.1016/j.snb.2019.127013.

Gupta, A.; Sakthivel, T.; Seal, S. Recent development in 2D materials beyond graphene. *Progress in Materials Science* 2015, 73, 44–126. https://doi.org/10.1016/j.pmatsci.2015.02.002.

Hao, Z.; Luo, Y.; Huang, C.; Wang, Z.; Song, G.; Pan, Y.; Zhao, X.; Liu, S. An intelligent graphene-based biosensing device for cytokine storm syndrome biomarkers detection in human biofluids. *Small* 2021, 17(29), 2101508. https://doi.org/10.1002/smll.202101508.

Harley-Trochimczyk, A.; Pham, T.; Chang, J.; Chen, E.; Worsley, M. A.; Zettl, A.; Mickelson, W.; Maboudian, R. Platinum nanoparticle loading of boron nitride aerogel and its use as a novel material for low-power catalytic gas sensing. *Advanced Functional Materials* 2016, 26(3), 433–439. https://doi.org/10.1002/adfm.201503605.

Hashtroudi, H.; Mackinnon, I. D. R.; Shafiei, M. Emerging 2D hybrid nanomaterials: Towards enhanced sensitive and selective conductometric gas sensors at room temperature. *Journal of Materials Chemistry C*, 2020, 8(38), 13108–13126. https://doi.org/10.1039/D0TC01968B.

Haque, M.; Masum, B.; Rakibu, H.; Suhanur, R. Synthesis of ZnO nanoparticles by two different methods & comparison of their structural, antibacterial, photocatalytic and optical properties. *Nano Express* 2020, 1(1), 1007. https://doi.org/10.1088/2632–959X/ab7a43.

He, B.; Zhao, Q.; Zeng, Z.; Wang, X.; Han, S. Effect of hydrothermal reaction time and calcination temperature on properties of Au@CeO_2 core–shell catalyst for CO oxidation at low temperature. *Journal of Materials Science* 2015, 50, 6339–6348. https://doi.org/10.1007/s10853-015-9181-z.

He, L.; Sarkar, S.; Barras, A. Electrochemically stimulated drug release from flexible electrodes coated electrophoretically with doxorubicin loaded reduced graphene oxide. *Chemistry Communications* 2017, 53, 4022–4025.

He, Q.; Sudibya, H. G.; Yin, Z.; Wu, S.; Li, H.; Boey, F.; Huang, W.; Chen, P.; Zhang, H. Centimeter-long and large-scale micropatterns of reduced graphene oxide films: Fabrication and sensing applications. *ACS Nano* 2010, 4(6), 3201–3208. https://doi.org/10.1021/nn100780v.

He, Y.; Li, D.; Gao, W.; Yin, H.; Chen, F.; Sun, Y. High-performance NO_2 sensors based on spontaneously functionalized hexagonal boron nitride nanosheets via chemical exfoliation. *Nanoscale* 2019, 11(45), 21909–21916. https://doi.org/10.1039/C9NR07153A.

Homan, K. A.; Souza, M.; Truby, R.; Luke, G. P.; Green, C.; Vreeland, E.; Emelianov, S. Silver Nanoplate contrast agents for in vivo molecular photoacoustic imaging. *ACS Nano* 2012, 6, 641–650.

Hou, H.; Wang, P.; Zhang, J.; Li, C.; Jin, Y. Graphene oxide - Supported Ag nanoplates as LSPR tunable and reproducible substrates for SERS applications with optimized sensitivity. *ACS Applied Materials & Interfaces* 2015, 7, 18038–18045.

Hu, H.; Zhou, J.; Kong, Q.; Li, C. Two-dimensional Au nanocrystals: Shape/size controlling synthesis, morphologies, and applications. *Particle & Particle Systems Characterization* 2015, 32, 796–808.

Hu, W.; Sheng, Z.; Hou, X.; Chen, H.; Zhang, Z.; Zhang, D. W.; Zhou, P. Ambipolar 2D semiconductors and emerging device applications. *Small Methods* 2021, 5(1), 2000837, https://doi.org/10.1002/smtd.202000837.

Huang, X.; Li, S.; Huang, Y.; Wu, S.; Zhou, X.; Li, S.; Gan, C. L.; Boey, F.; Mirkin, C. A.; Zhang, H. Synthesis of hexagonal close- packed gold nanostructures. *Nature Communications* 2011, 2, 292.

Huang, X.; Tang, S.; Liu, B.; Ren, B.; Zheng, N. Enhancing the photothermal stability of plasmonic metal nanoplates by a core-shell architecture. *Advanced Materials* 2011, 23, 3420–3425.

Huo, D.; Liu, G.; Li, Y. Construction of antithrombotic tissue-engineered blood vessel via reduced graphene oxide based dual-enzyme biomimetic cascade. *ACS Nano* 2017, 11, 10964–10973.

Hwang, H.; Kim, S.; Lee, S.; Lee, B. Reconfigurable single-layer graphene radio frequency antenna device capable of changing resonant frequency. *Nanomaterials* 2023, 13, 1203. https://doi.org/10.3390/nano13071203.

Iijima, S. Helical microtubules of graphitic carbon. *Nature* 1991, 354, 56–58. https://doi.org/10.1038/354056a0.

Irani, F. S.; Shafaghi, A. H.; Tasdelen, M. C.; Delipinar, T.; Kaya, C. E.; Yapici, G. G.; Yapici, M. K. Graphene as a piezoresistive material in strain sensing applications. *Micromachines* 2022, 13, 119. https://doi.org/10.3390/mi13010119.

Islam, M. N.; Channon, R. B. Electrochemical sensors, in: Ladame, S.; Chang, J. Y. H. (Eds.) *Bioengineering Innovative Solutions for Cancer*, Academic Press, New York, 2020, 47–71, https://doi.org/10.1016/B978-0-12–813886-1.00004-8.

Jesus, C. R.; Molina, E. F.; Pulcinelli, S. H. Highly controlled diffusion drug release from ureasil-poly (ethylene oxide)-Na±-montmorillonite hybrid hydrogel nanocomposites. *ACS Applied Materials & Interfaces* 2018, 10, 19059–19068.

Jha, R. K.; D'Costa, J. V.; Sakhuja, N.; Bhat, N. $MoSe_2$ nanoflakes based chemiresistive sensors for ppb-level hydrogen sulfide gas detection. *Sensors & Actuators, B: Chemical* 2019, 297, 126687. https://doi.org/10.1016/j.snb.2019.126687.

Ji, W.; Xue, X.; Ruan, W.; Wang, C.; Ji, N.; Chen, L.; Li, Z.; Song, W., Zhao, B.; Lombardi, J. R. Scanned chemical enhancement of surface-enhanced Raman scattering using a charge-transfer complex. *Chemistry Communications* 2011, 47(8), 2426–2428. https://doi.org/10.1039/C0CC03697H.

Jones, G. A.; Bradshaw, D. S. Resonance energy transfer: From fundamental theory to recent applications. *Frontiers of Physics* 2019, 7, 100.

Kajale, S. N.; Yadav, S.; Cai, Y.; Joy, B.; Sarkar, D. 2D material based field effect transistors and nanoelectromechanical systems for sensing applications. *iScience* 2021, 24, 12. https://doi.org/10.1016/j.isci.2021.103513.

Kan, C. X.; Wang, G. H.; Zhu, X. G.; Li, C. C.; Cao, B. Q. Structure and thermal stability of gold nanoplates. *Applied Physics Letters* 2006, 88, 071904.

Kang, K.; Chen, S.; Yang, E. H. Synthesis of transition metal dichalcogenides, in: Yang, D.; Datta, J.; Ding, G. H. (Eds.) *Synthesis, Modeling, and Characterization of 2D Materials, and Their Heterostructures*, Elsevier, Amsterdam, The Netherlands, 2020, 247–264, https://doi.org/10.1016/B978-0-12-818475-2.00012-X.

Kedambaimoole, V.; Kumar, N.; Shirhatti, V.; Nuthalapati, S.; Sen, P.; Nayak, M. M.; Rajanna, K.; Kumar, S. Laser-induced direct patterning of free-standing Ti_3C_2–MXene films for skin conformal tattoo sensors. *ACS Sensors* 2020, 5(7), 2086–2095. https://doi.org/10.1021/acssensors.0c00647.

Kerativitayanan, P.; Carrow, J. K.; Gaharwar, A. K. Nanomaterials for engineering stem cell responses. *Advanced Healthcare Materials* 2015, 4, 1600–1627.

Khazaei, M.; Ranjbar, A.; Arai, M.; Sasaki, T.; Yunoki, S. Electronic properties and applications of MXenes: A theoretical review. *Journal of Materials Chemistry C* 2017, 5(10), 2488–2503.

Khlibsuwan, R.; Siepmann, F.; Siepmann, J. Chitosan-clay nanocomposite microparticles for controlled drug delivery: Effects of the MAS content and TPP crosslinking. *Journal of Materials Chemistry B* 2018, 6, 6544–6549.

Khurshid, F.; Jeyavelan, M.; Takahashi, K.; Leo Hudson, M. S.; Nagarajan, S. Aryl fluoride functionalized graphene oxides for excellent room temperature ammonia sensitivity/selectivity. *RSC Advances* 2018, 8(36), 20440–20449. https://doi.org/10.1039/C8RA01818A.

Kim, J.; Mondal, S.; Min, B. K.; Choi, C. G. Highly sensitive and flexible strain–pressure sensors with cracked paddy-shaped MoS_2/Graphene Foam/Ecoflex hybrid nanostructures. *ACS Applied Materials & Interfaces* 2018, 10(42), 36377–36384. https://doi.org/10.1021/acsami.8b11233.

Kudr, J.; Adam, V.; Zitka, O. Fabrication of graphene/molybdenum disulfide composites and their usage as actuators for electrochemical sensors and biosensors. *Molecules* 2019, 24(18), 3374.

Kurtoglu, M.; Naguib, M.; Gogotsi, Y.; Barsoum, M. W. First principles study of two dimensional early transition metal carbides. *MRS Communications* 2012, 2(04), 133–137.

Late, D. J.; Doneux, T.; Bougouma, M. Single-layer $MoSe_2$ based NH_3 gas sensor. *Applied Physics Letters* 2014, 105(23), 233103. https://doi.org/10.1063/1.4903358.

Lee, C. W.; Suh, J. M.; Jang, H. W. Chemical sensors based on two-dimensional (2D) materials for selective detection of ions and molecules in liquid. *Frontiers in Chemistry* 2019, 7, 178.

Lee, K. E.; Hesketh, A. V.; Kelly, T. L. Chemical stability and degradation mechanisms of triangular Ag, Ag@Au, and Au nanoprisms. *Physical Chemistry Chemical Physics* 2014, 16, 12407–12414.

Lei, Y.; Zhao, W.; Zhang, Y.; Jiang, Q.; He, J. H.; Baeumner, A. J.; Wolfbeis, O. S.; Wang, Z. L.; Salama, K. N.; Alshareef, H. N. A MXene-based wearable biosensor system for high-performance in vitro perspiration analysis. *Small* 2019, 15(19), 1901190. https://doi.org/10.1002/smll.201901190.

Lei, Z. L.; Guo, B. 2D material-based optical biosensor: Status and prospect. *Advanced Science* 2022, 9(4), 2102924. https://doi.org/10.1002/advs.202102924.

Li, D.; Muller, M. B.; Gilje, S.; Kaner, R. B.; Wallace, G. G. Processable aqueous dispersions of graphene nanosheets. *Nature Nanotechnology* 2008, 3, 101–105.

Li, K., Ji, Q., Liang, H. L. Biomedical application of 2D nanomaterials in neuroscience. *Journal of Nanobiotechnology* 2023, 21, 181. https://doi.org/10.1186/s12951-023-01920-4.

Li, Q.; Wu, J. T.; Liu, Y.; Qi, X. M.; Jin, H. G; Yang, C.; Liu, J.; Li, G. L.; He, Q. G. Recent advances in black phosphorus-based electrochemical sensors: A review. *Analytica Chimica Acta* 2021, 1170, 338480. https://doi.org/10.1016/j.aca. 2021.338480.

Li, X.; Li, X.; Li, Z.; Wang, J.; Zhang, J. WS_2 nanoflakes based selective ammonia sensors at room temperature. *Sensors & Actuators, B: Chemical* 2017, 240, 273–277. https://doi.org/10.1016/j.snb.2016.08.163.

Lin, L.; Liu, T.; Zhang, Y.; Sun, R.; Zeng, W.; Wang, Z. Synthesis of boron nitride nanosheets with a few atomic layers and their gas-sensing performance. *Ceramics International* 2016, 42(1), 971–975. https://doi.org/10.1016/j.ceramint.2015.08.109.

Lin, Y.; Williams, T. V.; Connell, J. W. Soluble, exfoliated hexagonal boron nitride nanosheets. *Journal of Physical Chemistry Letters* 2010, 1, 277–283.

Ling, T.; Wang, J. J.; Zhang, H.; Song, S. T.; Zhou, Y. Z.; Zhao, J.; Du, X. W. Freestanding ultrathin metallic nanosheets: Materials, synthesis, and applications. *Advanced Materials* 2015, 27, 5396–5402.

Liu, H.; Du, Y.; Deng, Y.; Ye, P. D. Semiconducting black phosphorus: Synthesis, transport properties and electronic applications. *Chemical Society Reviews* 2015, 44, 2732–2743.

Liu, L.; Sang, S.; Han, D.; Liu, Z.; Han, X.; Li, D.; Chen, Y.; Liu, D.; Liu, X.; Yang, K.; Cheng, Y. PEI/PEG functionalized Black Phosphorus prepared by a one-pot method for a wide detection range CO_2 gas sensor. *Sensors & Actuators, B: Chemical* 2022, 369, 132303. https://doi.org/10.1016/j.snb.2022.132303.

Liu, L.; Burnyeat, C. A.; Lepsenyi, R. S.; Nwabuko, I. O.; Kelly, T. L. Mechanism of shape evolution in Ag nanoprisms stabilized by thiol-terminated poly(ethylene glycol): An in situ kinetic study. *Chemistry of Materials* 2013, 25, 4206–4214.

Liu, X.; Ma, D.; Tang, H. Polyamidoamine dendrimer and oleic acid functionalized graphene as biocompatible and efficient gene delivery vectors. *ACS Applied Materials & Interfaces* 2014, 6, 8173–8183.

Liz-Marzán, L. M.; Grzelczak, M. Growing anisotropic crystals at the nanoscale. *Science* 2017, 356, 1120–1121.

Lu, J.; Cheng, C.; He, Y. S. Multilayered graphene hydrogel membranes for guided bone regeneration. *Advanced Materials* 2016, 28, 4025–4031.

Lu, L.; Han, X.; Lin, J.; Zhang, Y.; Qiu, M.; Chen, Y.; Li, M.; Tang, D. Ultrasensitive fluorometric biosensor based on Ti_3C_2 MXenes with Hg^{2+}-triggered exonuclease III-assisted recycling amplification. *Analyst* 2021, 146(8), 2664–2669. https://doi.org/10.1039/D1AN00178G.

Lv, R. Transition metal dichalcogenides and beyond: Synthesis, properties, and applications of single- and few-layer nanosheets. *Accounts of Chemical Research* 2015, 48(1), 56–64. https://doi.org/10.1021/ar5002846.

Lvov, Y.; Wang, W.; Zhang, L. Halloysite clay nanotubes for loading and sustained release of functional compounds. *Advanced Materials* 2016, 28, 1227–1250.

Ma, R.; Sasaki, T. Two-dimensional oxide and hydroxide nanosheets: Controllable high-quality exfoliation, molecular assembly, and exploration of functionality. *Accounts of Chemical Research* 2015, 48, 136–143.

Manuel, V. S.; Dominguez-Alfaro, A.; Alegret, N.; Silvestri, A.; Gomez, I. J. 2D Materials towards sensing technology: From fundamentals to applications. *Sensing and Bio-Sensing Research* 2022, 38, 100540. https://doi.org/10.1016/j.sbsr.2022.100540.

Martínez-Perinan, E.; García-Mendiola, T.; Enebral-Romero, E.; del Caño, R.; Vera-Hidalgo, M.; Vázquez Sulleiro, M.; Navío, C.; Pariente, F.; Pérez, E. M.; Lorenzo, E. A MoS_2 platform and thionine-carbon nanodots for sensitive and selective detection of pathogens. *Biosensors & Bioelectronics* 2021, 189, 113375. https://doi.org/10.1016/j.bios.2021.113375.

Mas-Ballesté, R.; Gómez-Navarro, C.; Gómez-Herrero, J.; Zamora, F. 2D materials: To graphene and beyond. *Nanoscale* 2011, 3(1), 20–30. https://doi.org/10.1039/C0NR00323A.

Mashtalir, O.; Naguib, M.; Mochalin, V. N.; Dall'Agnese, Y.; Heon, M.; Barsoum, M. W.; Gogotsi, Y. Intercalation and delamination of layered carbides and carbonitrides. *Nature Communications* 2013, 4(1), 1716. https://doi.org/10.1038/ncomms2664.

Matatagui, D.; Cruz, C.; Carrascoso, F.; Al-Enizi, A. M.; Nafady, A.; Castellanos- Gomez, A.; Horrillo, M. D. Eco-friendly disposable WS_2 paper sensor for sub-ppm NO_2 detection at room temperature. *Nanomaterials* 2022, 12(7), 1213. https://doi.org/10.3390/nano12071213.

Meng, Z.; Stolz, R. M.; Mendecki, L.; Mirica, K. A. Electrically-transduced chemical sensors based on two-dimensional nanomaterials. *Chemical Reviews* 2019, 119(1), 478–598. https://doi.org/10.1021/acs.chemrev.8b00311.

Naguib, M.; Mashtalir, O.; Carle, J.; Presser, V.; Lu, J.; Hultman, L.; Gogotsi, Y.; Barsoum, M. W. Two-dimensional transition metal carbides. *ACS Nano* 2012, 6(2), 1322–1331. https://doi.org/10.1021/nn204153h.

Neema, P. M.; Tomy, A. M.; Cyriac, J. Chemical sensor platforms based on fluorescence resonance energy transfer (FRET) and 2D materials. *TrAC Trends in Analytical Chemistry* 2020, 124, 115797. https://doi.org/10.1016/j.trac.2019.115797.

Novoselov, K. S. Electric field effect in atomically thin carbon films. *American Association for the Advancement of Science* 2004, 306(5696), 666–669. https://doi.org/10.1126/science.1102896.

Orendorff, C. J.; Gearheart, L.; Jana, N. R.; Murphy, C. J. Aspect ratio dependence on surface enhanced Raman scattering using silver and gold nanorod substrates. *Physical Chemistry Chemical Physics* 2006, 8, 165–170.

Pang, Y.; Zhang, K.; Yang, Z.; Jiang, S.; Ju, Z.; Li, Y.; Wang, X.; Wang, D.; Jian, M.; Zhang, Y.; Liang, R.; Tian, H.; Yang, Y.; Ren, T. L. Epidermis microstructure inspired graphene pressure sensor with random distributed spinosum for high sensitivity and large linearity. *ACS Nano* 2018, 12(3), 2346–2354. https://doi.org/10.1021/acsnano.7b07613.

Park, D. W.; Ness, J. P.; Brodnick, S. K.; Esquibel, C.; Novello, J.; Atry, F.; Baek, D. H.; Kim, H.; Bong, J.; Swanson, K. I.; Suminski, A. J.; Otto, K. J.; Pashaie, R.; Williams, J. C.; Ma, Z. Electrical neural stimulation and simultaneous in vivo monitoring with transparent graphene electrode arrays implanted in GCaMP6f mice. *ACS Nano* 2018, 12(1), 148–157. https://doi.org/10.1021/acsnano.7b04321.

Povedano, E.; Cincotto, F. H.; Parrado, C. Decoration of reduced graphene oxide with rhodium nanoparticles for the design of a sensitive electrochemical enzyme biosensor for 17b-estradiol. *Biosensors & Bioelectronics* 2017, 89, 343–351.

Prekodravac, J. R.; Kepić, D. P.; Colmenares, J. C., Giannakoudakis, D. A.; Jovanović, S. P. A comprehensive review on selected graphene synthesis methods: From electrochemical exfoliation through rapid thermal annealing towards biomass pyrolysis. *Journal of Materials Chemistry C* 2021, 21, 6722–6748. https://doi.org/10.1039/D1TC01316E.

Qi, M.; Zhang, Y.; Cao, C.; Zhang, M.; Liu, S.; Liu, G. Decoration of reduced graphene oxide nanosheets with aryldiazonium salts and gold nanoparticles toward a label-free amperometric immunosensor for detecting cytokine tumor necrosis factor-α in live cells. *Analytical Chemistry* 2016, 88(19), 9614–9621. https://doi.org/10.1021/acs.analchem.6b02353.

Qian, H.; Zhu, M.; Wu, Z.; Jin, R. Quantum sized gold nanoclusters with atomic precision. *Accounts of Chemical Research* 2012, 45, 1470–1479.

Robinson, J. T.; Burgess, J. S.; Junkermeier, C. E.; Badescu, S. C.; Reinecke, T. L.; Perkins, F. K.; Zalalutdniov, M. K.; Baldwin, J. W.; Culbertson, J. C.; Sheehan, P. E. Properties of fluorinated graphene films. *Nano Letters* 2010, 10(8), 3001–3005.

Ruan, J.; Wang, X.; Yu, Z. Enhanced physiochemical and mechanical performance of chitosan-grafted graphene oxide for superior osteoinductivity. *Advanced Functional Materials* 2016, 26, 1085–1097.

Russier, J; Leon, V; Orecchioni, M. Few-layer graphene kills selectively tumor cells from myelomonocytic leukemia patients. *Angewandte Chemie International Edition* 2017, 56, 3014–3019.

Satheeshkumar, E.; Makaryan, T.; Melikyan, A.; Minassian, H.; Gogotsi, Y.; Yoshimura, M. One-step solution processing of Ag, Au and Pd@MXene hybrids for SERS. *Scientific Reports* 2016, 6(1), 32049. https://doi.org/10.1038/srep32049.

Scholz, F. Voltammetric techniques of analysis: The essentials. *ChemTexts* 2015, 1(4), 17. https://doi.org/10.1007/s40828-015-0016-y.

Shavanova, K.; Bakakina, Y.; Burkova, I.; Shtepliuk, I.; Viter, R.; Ubelis, A.; Beni, V.; Starodub, N.; Yakimova, R.; Khranovskyy, V. Application of 2D non-graphene materials and 2D oxide nanostructures for biosensing technology. *Sensors* 2016, 16, 223. https://doi.org/10.3390/s16020223.

Shu, Y.; Li, B.; Chen, J.; Xu, Q.; Pang, H.; Hu, X. Facile synthesis of ultrathin nickel–cobalt phosphate 2D nanosheets with enhanced electrocatalytic activity for glucose oxidation. *ACS Applied Materials & Interfaces* 2018, 10, 2360–2367. https://doi.org/10.1021/acsami.7b17005.

Song, L.; Ci, L; Lu, H; Sorokin, P. B.; Jin, C.; Ni, J.; Kvashnin, A. G.; Kvashnin, D. G.; Lou, J.; Yakobson, B. I.; Ajayan, P. M. Large scale growth and characterization of atomic hexagonal boron nitride layers. *Nano Letters* 2010, 10(8), 3209–3215. https://doi.org/10.1021/nl1022139.

Su, S.; Li, J.; Yao, Y.; Sun, Q.; Zhao, Q.; Wang, F.; Li, Q.; Liu, X.; Wang, L. Colorimetric analysis of carcinoembryonic antigen using highly catalytic gold nanoparticles decorated MoS_2 nanocomposites. *ACS Applied Bio Materials* 2019, 2(1), 292–298. https://doi.org/10.1021/acsabm.8b00598.

Su, S.; Zhang, C.; Yuwen, L.; Chao, J.; Zuo, X.; Liu, X.; Song, C.; Fan, C.; Wang, L. Creating SERS hot spots on MoS_2 nanosheets with in situ grown gold nanoparticles. *ACS Applied Materials & Interfaces* 2014, 6(21), 18735–18741. https://doi.org/10.1021/am5043092.

Sui, Y.; Zhou, J.; Wang, X.; Wu, L.; Zhong, S.; Li, Y. Recent advances in black-phosphorus-based materials for electrochemical energy storage. *Materials Today* 2021, 42, 117–136. https://doi.org/10.1016/j.mattod.2020.09.005.

Sun, Y.; Lei, C. Synthesis of out-of-substrate Au–Ag nanoplates with enhanced stability for catalysis. *Angewandte Chemie International Edition* 2009, 48, 6824–6827.

Tabata, H.; Matsuyama, H.; Goto, T.; Kubo, O.; Katayama, M. Visible-light-activated response originating from carrier-mobility modulation of NO_2 gas sensors based on MoS_2 monolayers. *ACS Nano* 2021, 15(2), 2542–2553. https://doi.org/10.1021/acsnano.0c06996.

Tang, B.; Xu, S.; An, J.; Zhao, B.; Xu, W.; Lombardi, J. R. Kinetic effects of halide ions on the morphological evolution of silver nanoplates. *Physical Chemistry Chemical Physics* 2009, 11, 10286–10292.

Tang, H.; Hu, Q.; Zheng, M., Chi, Y.; Qin, X.; Pang, H.; Xu, Q. Mxene–2D layered electrode materials for energy storage. *Progress in Natural Science: Materials International* 2018, 28(2), 133–147.

Tao, L.; Cinquanta, E.; Chiappe, D.; Grazianetti, C.; Fanciulli, M.; Dubey, M.; Molle, A.; Akinwande, D. Silicene field-effect transistors operating at room temperature. *Nature Nanotechnology* 2015, 10, 227–231.

Tian, F.; Lyu, J.; Shi, J.; Yang, M. Graphene and graphene-like two-denominational materials based fluorescence resonance energy transfer (FRET) assays for biological applications. *Biosensors & Bioelectronics* 2017, 89, 123–135. https://doi.org/10.1016/j.bios.2016.06.046.

Turkevich, J.; Stevenson, P. C.; Hillier, J. A study of the nucleation and growth processes in the synthesis of colloidal gold. *Discussions of the Faraday Society* 1951, 11, 55–75.

Vargas-Bernal, R. Electrical properties of two-dimensional materials used in gas sensors. *Sensors* 2019, 19(6), 1295. https://doi.org/10.3390/s19061295.

Verger, L.; Xu, C.; Natu, V.; Cheng, H. M.; Ren, W.; Barsoum, M. W. Overview of the synthesis of MXenes and other ultrathin 2D transition metal carbides and nitrides. *Current Opinion in Solid State & Materials Science* 2019, 23(3), 149–163. https://doi.org/10.1016/j.cossms.2019.02.001.

Viarbitskaya, S.; Teulle, A.; Marty, R.; Sharma, J.; Girard, C.; Arbouet, A.; Dujardin, E. Tailoring and imaging the plasmonic local density of states in crystalline nanoprisms. *Nature Materials* 2013, 12, 426–432.

Wang, L.; Xiong, Q.; Xiao, F.; Duan, H. 2D nanomaterials based electrochemical biosensors for cancer diagnosis. *Biosensors & Bioelectronics* 2017, 89, 136–151. https://doi.org/10.1016/j.bios.2016.06.011.

Wang, Y.; Wang, S.; Dong, N.; Kang, W.; Li, K.; Nie, Z. Titanium carbide mxenes mediated in situ reduction allows label-free and visualized nanoplasmonic sensing of silver ions. *Analytical Chemistry* 2020, 92(6), 4623–4629. https://doi.org/10.1021/acs.analchem.0c00164.

Wang, Y.; Zeng, Z.; Qiao, J.; Dong, S.; Liang, Q.; Shao, S. Ultrasensitive determination of nitrite based on electrochemical platform of AuNPs deposited on PDDA modified MXene nanosheets. *Talanta* 2021, 221, 121605. https://doi.org/10.1016/j.talanta.2020.121605.

Wassei, J. K.; Kaner, R. B. Graphene, a promising transparent conductor. *Materials Today* 2010, 13(3), 52–59.

Waters, R.; Pacelli, S.; Maloney, R. Stem cell secretome-rich nanoclay hydrogel: A dual action therapy for cardiovascular regeneration. *Nanoscale* 2016, 8, 7371–7376.

Weng, Q.; Wang, X.; Wang, X.; Bando, Y.; Golberg, D. Functionalized hexagonal boron nitride nanomaterials: Emerging properties and applications. *Chemical Society Reviews* 2016, 45, 3989–4012.

Wu, E.; Xie, Y.; Yuan, B.; Hao, D.; An, C.; Zhang, H.; Wu, S.; Hu, X.; Liu, J.; Zhang, D. Specific and highly sensitive detection of ketone compounds based on p-Type $MoTe_2$ under ultraviolet illumination. *ACS Applied Materials & Interfaces* 2018, 10(41), 35664–35669. https://doi.org/10.1021/acsami.8b14142.

Xia, Y.; Yang, X. Toward cost-effective and sustainable use of precious metals in heterogeneous catalysts. *Accounts of Chemical Research* 2017, 50, 450–454.

Xiong, Y.; McLellan, J. M.; Chen, J.; Yin, Y.; Li, Z. -Y.; Xia, Y. Kinetically controlled synthesis of triangular and hexagonal nanoplates of palladium and their SPR/SERS properties. *Journal of the American Chemical Society* 2005, 127, 17118–17127.

Xu, T.; Liu, Y.; Pei, Y.; Chen, Y.; Jiang, Z.; Shi, Z.; Xu, J.; Wu, D.; Tian, Y.; Li, X. The ultra-high NO_2 response of ultra-thin WS_2 nanosheets synthesized by hydrothermal and calcination processes. *Sensors & Actuato rs, B: Chemical* 2018, 259, 789–796. https://doi.org/10.1016/j.snb.2017.12.070.

Xu, X.; Yao, W.; Xiao, D.; Heinz, T. F. Spin and pseudospins in layered transition metal dichalcogenides. *Nature Physics* 2014, 10(5), 343–350. https://doi.org/10.1038/nphys2942.

Xu, Y.; Shi, Z.; Shi, X.; Zhang, K.; Zhang, H. Recent progress in black phosphorus and black-phosphorus-analogue materials: Properties, synthesis and applications. *Nanoscale* 2019, 11(31), 14491–14527. https://doi.org/10.1039/C9NR04348A.

Xu, Y.; Xie, X.; Zhang, R.; Yuan, W. Functionalization of black phosphorus nanosheets via self-polymerization of dopamine and subsequent secondary reactions for gas sensing investigation. *Sensors & Actuators, B: Chemical* 2022, 372, 132670. https://doi.org/10.1016/j.snb.2022.132670.

Xue, C.; Chen, X.; Hurst, S. J.; Mirkin, C. A. Self-assembled monolayer mediated silica coating of silver triangular nanoprisms. *Advanced Materials* 2007, 19, 4071–4074.

Xue, C.; Mirkin, C. A. pH-switchable silver nanoprism growth pathways. *Angewandte Chemie International Edition* 2007, 46, 2036–2038.

Yan, L.; Shi, H.; Sui, X.; Deng, Z.; Gao, L. MoS_2-DNA and MoS_2 based sensors. *RSC Advances* 2017, 7(38), 23573–23582. https://doi.org/10.1039/C7RA02649H.

Yan, M.; Liu, Y.; Zhu, X. Nanoscale reduced graphene oxide-mediated photothermal therapy together with IDO inhibition and PD-L1 blockade synergistically promote antitumor immunity. *ACS Applied Materials & Interfaces* 2019, 11, 1876–1885.

Yan, W.; Fuh, H. R.; Lv, Y.; Chen, K. Q.; Tsai, T. Y.; Wu, Y. R.; Shieh, T. H.; Hung, K. M.; Li, J.; Zhang, D.; Coileáin, C. O.; Arora, S. K.; Wang, Z.; Jiang, Z.; Chang, C. R.; Wu, H. C. Giant gauge factor of Van der Waals material based strain sensors. *Nature Communications* 2021, 12(1), 2018. https://doi.org/10.1038/s41467-021–22316-8.

Yang, H.; Zhou, J.; Bao, J.; Ma, Y.; Zhou, J.; Shen, C.; Luo, H.; Yang, M.; Hou, C.; Huo, D. A simple hydrothermal one-step synthesis of 3D-MoS_2/rGO for the construction of sensitive enzyme-free hydrogen peroxide sensor. *Microchemical Journal* 2021, 162, 105746. https://doi.org/10.1016/j.microc.2020.105746.

Yang, Y.; Zhong, X. L.; Zhang, Q.; Blackstad, L. G.; Fu, Z. W.; Li, Z. Y.; Qin, D. The role of etching in the formation of Ag nanoplates with straight, curved and wavy edges and comparison of their SERS properties. *Small* 2014, 10, 1430–1437.

Yin, A. X.; Liu, W. C.; Ke, J.; Zhu, W.; Gu, J.; Zhang, Y. W.; Yan, C. H. Ru nanocrystals with shape-dependent surface-enhanced Raman spectra and catalytic properties: Controlled synthesis and DFT calculations. *Journal of the American Chemical Society* 2012, 134, 20479–20489.

Yin, F.; Li, X.; Peng, H.; Li, F.; Yang, K.; Yuan, W. A highly sensitive, multifunctional, and wearable mechanical sensor based on RGO/synergetic fiber bundles for monitoring human actions and physiological signals. *Sensors & Actuators, B: Chemical* 2019, 285, 179–185. https://doi.org/10.1016/j.snb.2019.01.063.

Yin, T.; Liu, J.; Zhao, Z. Redox sensitive hyaluronic acid-decorated graphene oxide for photothermally controlled tumor-cytoplasm-selective rapid drug delivery. *Advanced Functional Materials* 2017, 27, 1604620.

Yu, Q.; Guan, P.; Qin, D.; Golden, G.; Wallace, P. M. Inverted size-dependence of surface-enhanced raman scattering on gold nanohole and nanodisk arrays. *Nano Letters* 2008, 8, 1923–1928.

Yuan, W.; Yang, K.; Peng, H.; Li, F.; Yin, F. A flexible VOCs sensor based on a 3D Mxene framework with a high sensing performance. *Journal of Materials Chemistry A* 2018, 6(37), 18116–18124. https://doi.org/10.1039/C8TA06928J.

Zeng, S.; Sreekanth, K. V.; Shang, J. Graphene-gold metasurface architectures for ultrasensitive plasmonic biosensing. *Advanced Materials* 2015, 27, 6163–6169.

Zeng, Y.; Guo, Z. Synthesis and stabilization of black phosphorus and phosphorene: Recent progress and perspectives. *iScience* 2021, 24, 10. https://doi.org/10.1016/j.isci.2021.103116.

Zeng, Z. An effective method for the fabrication of few-layer-thick inorganic nanosheets. *Angewandte Chemie - International Edition* 2018, 51(36), 9052–9056. https://doi.org/10.1002/anie.201204208.

Zhang, J.; Chen, Y.; Wang, X. Two-dimensional covalent carbon nitride nanosheets: Synthesis, functionalization, and applications. *Energy & Environmental Science* 2015, 8, 3092–3108.

Zhang, S.; Zhang, D.; Zhang, X.; Shang, D.; Xue, Z.; Shan, D.; Lu, X. Ultratrace naked-eye colorimetric detection of Hg2+ in wastewater and serum utilizing mercury-stimulated peroxidase mimetic activity of reduced graphene oxide-PEI-Pd nanohybrids. *Analytical Chemistry* 2017, 89(6), 3538–3544. https://doi.org/10.1021/acs.analchem.6b04805.

Zhang, Y. Z.; Lee, K. H.; Anjum, D. H.; Sougrat, R.; Jiang, Q.; Kim, H.; Alshareef, H. N. MXenes stretch hydrogel sensor performance to new limits. *Science Advances* 2022, 4(6), eaat0098. https://doi.org/10.1126/sciadv.aat0098.

Zhao, G. Q.; Hu, J.; Long, X.; Zou, J.; Yu, J. G.; Jiao, F. P. A critical review on black phosphorus-based photocatalytic CO_2 reduction application. *Small* 2021, 17(49), 2102155. https://doi.org/10.1002/smll.202102155.

Zhao, Y.; Song, J. G.; Ryu, G. H.; Ko, K. Y.; Woo, W. J.; Kim, Y.; Kim, D.; Lim, J. H.; Lee, S.; Lee, Z.; Park, J.; Kim, H. Low-temperature synthesis of 2D MoS_2 on a plastic substrate for a flexible gas sensor. *Nanoscale* 2018, 10(19), 9338–9345. https://doi.org/10.1039/C8NR00108A.

Zheng, L.; Wang, X.; Jiang, H.; Xu, M.; Huang, W.; Liu, Z. Recent progress of flexible electronics by 2D transition metal dichalcogenides. *Nano Research* 2022, 15(3), 2413–2432. https://doi.org/10.1007/s12274-021-3779-z.

Zhou, C.; Yang, W.; Zhu, H. Mechanism of charge transfer and its impacts on Fermi-level pinning for gas molecules adsorbed on monolayer WS_2. *Journal of Chemical Physics* 2015, 142(21), 214704. https://doi.org/10.1063/1.4922049.

Zhou, J.; Chen, J.; Ge, Y.; Shao, Y. Two-dimensional nanomaterials for Förster resonance energy transfer–based sensing applications. *Nanophotonics* 2020, 9(7), 1855–1875. https://doi.org/10.1515/nanoph-2020–0065.

Zoric, I.; Zach, M.; Kasemo, B.; Langhammer, C. Gold, platinum, and aluminum nanodisk plasmons: Material independence, subradiance, and damping mechanisms. *ACS Nano* 2011, 5, 2535–2546.

8 Recent Advancement in Electrochemical Biosensors
The Role of Conducting Polymer Nanohybrid Materials

Habibun Nabi Muhammad Ekramul Mahmud, Nor Kartini Abu Bakar, Noman Nawaz, and Waqar Ahmed

8.1 INTRODUCTION

Sensors were developed in response to interest in developing and exploiting analytical devices for measuring, monitoring, and detecting particular chemical reactions and species. Sensors are analytical devices consisting of active sensing materials and signal transducers. Sensors are typically divided into two groups: chemical sensors and biosensors. This classification is rooted in their distinct sensing abilities. Biosensors, for instance, are capable of identifying biological elements like antibodies, cells, enzymes, tissues, and genetic materials such as DNA and RNA, while chemical sensors convert chemical interactions or concentrations of various species into quantifiable signals (Aydemir, Malmström, & Travas-Sejdic, 2016; Moon et al., 2018). Biosensors are devices encompassing biochemical sensing components that generate selective and sensitive quantitative signals. When a target analyte interacts with biomolecules such as antibodies, proteins, DNA, and enzymes, physicochemical changes occur. These alterations are subsequently converted into measurable electrical signals using a transducer, where the magnitude of the signal is directly linked to the concentration of the analyte. Multiple applications of biosensors have been reported in multiple disciplines including food safety, point of care (POC), environmental monitoring, and drug discovery. Biosensors are classified as colorimetric, piezoelectric, mechanical, electrochemical, and optical.

In terms of composition, biosensors are categorized into the following two groups: (1) based on transducer type such as optical, colorimetric, and electrochemical transducers, or (2) based on type of biorecognition elements such as affinity and biocatalytic biosensors. A biocatalytic sensor functions by immobilizing tissue slices, cells, and enzymes which recognize a target molecule and produce electroactive signals. On the contrary, affinity sensors rely on a distinct binding interaction between the target analyte and immobilized biochemical elements, such as aptamers or antibodies (Wang, 2006). Electrochemical transducers are most frequently used for the biorecognition of most bioanalytes. Their performance can be assessed based on factors including reproducibility, sensitivity, and susceptibility to interference.

8.2 ELECTROCHEMICAL BIOSENSORS

The interest among researchers in electrochemical biosensors has grown significantly owing to their advantageous properties including simplicity in design, minimal cost, and efficient identification of a wide range of molecules including glucose, neurotransmitters, microbial pathogens, DNA, RNA, antibodies, and amino acids. Electrochemical biosensors are at the forefront of research, especially for POC analysis, due to their effectiveness and low detection limits (Liu et al., 2022; Maduraiveeran, 2022; Ning et al., 2022). An electrochemical cell with two or three electrodes

 DOI: 10.1201/9781032654874-8

serves as the basic instrument for producing electrochemical biosensors. A three-electrode system typically comprises a counter/auxiliary electrode, a reference electrode (RE), and a working electrode (WE). A graphite rod or platinum wire generally serves as the counter electrode; Ag/AgCl is usually used as the RE; and chemically stable solid conductive substances, such as gold, platinum, indium tin oxide (ITO), and glassy carbon (GC) serve as the WE. In comparison, a two-electrode setup involves the reference and WEs only (Moon et al., 2018; Wang, 2001). Depending on how measurements are performed, electrochemical transducers can be categorized into three groups: (1) potential difference (potentiometry), (2) current (amperometric and voltammetric), and (3) impedance [electrochemical impedance spectroscopy (EIS)]. Among these, biosensors that involve current measurements are the most commonly employed.

In potentiometric sensors, the concentration of analyte is evaluated by measuring the change in potential difference between WE and RE while generating negligible current. Potentiometric sensors monitor charge accumulation under equilibrium conditions at zero current due to selective binding at the electrode surface. In addition to being non-invasive, inexpensive, and easy to miniaturize, potentiometric sensors can be used for real-time measurements. The sensors based on current measurements are characterized by measuring the current in response to the applied potential to a WE electrode in comparison to the applied potential to a RE electrode. Electrochemical oxidation or reduction leads to the generation of current through electrolysis on the electrode surface. The rate of electron transfer at the electrode surface and the rate at which molecules are transported in mass to the electrode, however, are both necessary for this redox process to occur.

A range of voltammetric approaches are available, such as square wave, differential pulse, cyclic, linear sweep, and stripping voltammetry. The voltammetric technique involves scanning a potential across a predefined range, and the resulting current response is typically measured as a plateau or peak that correlates directly with the concentration of the analyte in the sampled substance. The amperometric approach centers on keeping a steady potential at the WE in relation to the RE. The resulting changes in current due to electrochemical oxidation or reduction are then directly observed over time. Overall, amperometric biosensors offer exceptional sensitivity and selectivity as they rely on maintaining a consistent and specific potential for each target analyte. As a result, the potential interference from other electroactive substances on their performance can be constrained. EIS is a powerful and robust approach applied to monitor the interfacial characteristics associated with biorecognition events that usually occur at the substrate surface. This includes substrate–enzyme interaction, whole-cell detection, and antibody–antigen biorecognition. Furthermore, a small-amplitude sinusoidal alternating current (AC) excitation signal from EIS is used to monitor the capacitances and resistances of various materials. An impedance spectrum can be produced by changing the frequency over a wide range. Using Nyquist plot, the impedance technique can measure charge transfer resistance (R_{ct}) at higher frequencies and mass transfer at lower frequencies; therefore, affinity-type biosensors are primarily equipped with impedimetric detection equipment. Bode plots are valuable for determining the capacitance of electrochemical systems by illustrating the relationship between frequency and phase angle (Magar, Hassan, & Mulchandani, 2021; Moon et al., 2018). EIS biosensors are experiencing a substantial rise in usage, attributed to their swift response, user-friendly operation, miniaturization potential, and seamless integration with lab-on-a-chip technologies. This integration allows for cost-effective and real-time measurement of exceedingly low concentrations.

The discovery and use of numerous novel materials in recent decades, including carbon-based materials [graphene, graphene oxide (GO), reduced graphene oxide (rGO), carbon nanotubes, and multiwalled carbon nanotubes (MWCNTS)], inorganic nanoparticles (metal or metal oxide), and conducting polymers (CPs), has facilitated the development of electrodes for electrochemical biosensors (Ahmadi & Ahour, 2020; Hwang et al., 2020). It's intriguing to observe that CPs possess unique properties compared to other materials because of their chain conformation modifications and π peculiar orbital structure. When CPs are integrated into electrochemical biosensors, they exhibit excellent selectivity and sensitivity to specific bioanalytes as well as rapid electrical signals (Naseri, Fotouhi, & Ehsani, 2018; Zhang et al., 2020). The convenience of adjusting CP characteristics is also expected to facilitate the development of electrochemical biosensors based on CPs. Modifying CP properties can easily be achieved by coupling or functionalizing CP monomers with various functional groups (Aydemir, Malmström, & Travas-Sejdic, 2016), which may result in substantial enhancement in CP electrical stability and electronic properties. Additionally, it has been shown that CP-integrated biosensors perform very differently depending on certain significant parameters, such as shape, size, structural design, morphology, and electrochemical conductivity (Park, Lee, & Kwon, 2016). To enhance the utilization of CPs in electrochemical biosensors, various schemes have been employed to boost the electrochemical stability, flexibility, selectivity, sensitivity, and reproducibility of the measurements. Aside from incorporating functional groups, the strategies have been primarily focused on designing CPs nanocomposites, including CP nanowires, nanospheres, nanotubes, as well as forming hybrid nanoparticles, hybrid nanocomposites, and nanohydrogels by combining them with other functional materials.

8.3 CONDUCTING POLYMERS

The term conductive polymers (CP) refers to an emerging class of materials that integrate the properties of semiconductors and organic polymers, also known as smart polymers. CPs have similar optical and electrical properties to metals, whereas polymers retain mechanical properties. Owing to their flexibility and inherent properties, CPs have dominated all fields of applications for more than three decades. The higher surface area of CPs makes them extremely advantageous for numerous types of applications, considering that natural polymers do not contain such properties and do not belong to the class of nanoparticles, which are the most explored and sought-after components by researchers. Conductive polymers are semiconductors in their purest form, which can be doped to become highly conductive materials (Kaur, Kaur, & Kaur, 2021; Shahabuddin et al., 2021). The conductivity of CPs depends exclusively on the delocalized π-electrons. In doping, the oxidation and reduction states of the π-conjugated chains are modified to form p- or n-doped CPs. The oxidized or reduced forms of CPs are produced depending on the doping techniques and chemical additives used. Mainly, polymers exist as either cation or anion radical or di-cation or di-anion species (Lu et al., 2011). In electrochemical doping, doping and dedoping of ions occur in a reversible state in an electrolyte comprising of supporting salts. The movement of electrons takes place from CPs to the electrode when a positive potential is scanned. In contrast, CPs obtain electrons from the electrode when a negative potential is applied. It is possible for the electrolytes or ions in aqueous solutions to move into or out of the CPs to compensate for charges formed during the electrochemical oxidation and reduction event. The removal or addition of ions significantly alters CP volume, making them promising materials for biosensing applications.

The conductivity of these doped CPs can be tailored by modifying the dopant characteristics, adjusting the extent of doping, manipulating the structure of the polymer chain, or by blending with additional polymers. The CPs are organic compounds such as PANI, PPy, PTh, polyacetylene (PA), poly(phenylenevinylene) (PPV), polyfuran (PF), and poly(para-phenylene) (PPP), etc. PANI, PPy, poly(3,4-ethylenedioxythiophene) PEDOT, and PTh stand out among them for having some unmatched properties that are carefully considered for a range of applications. Table 8.1 shows the formula, synthesis method, conductivity, and applications of some CPs.

8.4 SYNTHESIS OF CPs

Although gas phase techniques like plasma polymerization and chemical vapor deposition (CVD) can be used to prepare CPs (Liu, Goeckner, & Walker, 2017; Park et al., 2016), however, in most cases, CPs are synthesized through electrochemical oxidative polymerization or chemical processes. For example, non-oxidative polymerization methods such as dehydrobrominative or polycondensation Grignard metathesis are sometimes possible for the synthesis of CPs (Tamba et al., 2013). CPs can be prepared conventionally (Malinauskas, 2001) by using redox enzymes to generate hydrogen

TABLE 8.1
Formula, Conductivities, Synthesis Method, and Applications of Some Important Conducting Polymers

Sr. No.	Conducting Polymers	Formula	Conductivity (S cm^{-m})	Synthesis Method	Applications	References
1	Polypyrrole	$(C_4H_5N)_n$	10^2–7.5×10^3	Photochemical, chemical, and electrochemical method	Applied in the development of electromagnetic tapes, microelectronic devices, supercapacitors, antistatic coatings, electrochemical sensors, electrode-based detectors, screens, etc.	Tat'yana and Efimov (1997)
2	Polyaniline	$(C_6H_5NH_2)_n$	30–200	Metathesis, electrochemical, chemical method	PANI has numerous applications across various disciplines, including energy storage devices, corrosion-resistant coatings, electromagnetic screens, flexible electrodes, conductive fibers, supercapacitors, fabrication of electrochemical biosensors, etc.	Kaur, Kaur, and Kaur (2021)
3	Poly (3,4 ethylenedioxythiophene)	$(C_{12}H_8S_2O_4)_n$	300	Chemical polymerization, emulsion polymerization	Due to its conductive nature, PEDOT proves valuable in diverse applications such as energy conversion, solar cells, transparent electronic devices, capacitors, and electrochemical detection	Gueye et al. (2020)
4	Polythiophene	$(C_4H_4S)_n$		Electrochemical, chemical method	Polythiophene and its derivatives can be used in magneto-optical devices, biosensors, thermoelectric devices, water purification devices	Husain, Ahmad, and Mohammad (2020)

peroxide which initiates the polymerizing reaction. This is done more rarely at the liquid/air interface under mild conditions (Ramanavičius, Kaušaitė, & Ramanavičienė, 2005) and under harsh oxidative polymerization conditions by employing the oxidizing agents such as $K_2S_2O_8$, $K_2Cr_2O_7$, $FeCl_3$, KIO_3, and $KMnO_4$ (Erdem, Karakışla, & Sacak, 2004). The most used method, however, is oxidative electrochemical polymerization, because it offers a superior degree of control over polymer deposition (Heinze, Frontana-Uribe, & Ludwigs, 2010). Typically, electrochemical polymerization occurs in an electrochemical cell with three electrodes. The process of electropolymerization may be attained either by applying a constant current to initiate the polymerization in the case of galvanostatic method, by providing constant potential to initiate the polymerization via potentiostatic technique, or by applying the potential linearly which is scanned at high and low potential through potentiodynamic techniques such as cyclic voltammetry. The galvanostatic approach is usually regarded as the most effective technique since it allows monitoring of the changes in the potential with time that reflect the conductivity as the conducting polymer film grows. While the potentiostatic approach provides ease in controlling polymer film thickness via Faraday's law, the potentiodynamic approach results in more adherent and homogeneous polymer films.

CPs have been extensively studied in the field of biomedical and bioanalytical science (Nair, Mishra, & Kumar, 2019; Park, Jung, & Chang, 2019), drug delivery (Inal et al., 2017), cell culture (Dong, Ma, & Guo, 2020; Zarrintaj et al., 2018), and tissue engineering (Boehler, Oberueber, & Asplund, 2019), owing to their intrinsic properties and biocompatibility (He et al., 2019; Humpolíček et al., 2018). Additionally, CPs are one of the most sought-after materials used in the construction of electrochemical biosensors because of their electrical properties, which enable them to convert biological data into electrical signals. In addition, coupling the functional groups to modify the CPs is very simple which allows the probability to maximize their capabilities to quantify and detect biomolecules or to enhance the interaction between the functionalized polymer and bioanalytes.

8.5 CP NANOHYBRID-BASED BIOSENSORS

In modern analytical science, the emergence of CPs is a landmark, and they are commonly used in electroanalytic applications as signal-enhancing components. Nevertheless, CPs have inadequate electrical conductivity and mechanical strength. Therefore, numerous novel strategies have been explored to enhance the properties, such as mechanical strength and electronic conductivity of polymers, which involve the integration of nanoparticles and conducting fillers. Hybrid materials are unavoidably categorized as nanocomposites that are formed by specifically combining two or more nano-dimensional objects under optimized conditions. Nanocomposites play a pivotal role in enhancing and optimizing various scientific and technological applications. The properties of these composites can be tuned by combining filler and matrix in suitable proportions.

Conducting substrates were fabricated by integrating conductive fillers including carbon-based additives, metal particles, and CPs to non-conductive substrates (Nada, Eckstein Andicsová, & Mosnáček, 2022). Here, CPs may offer the surface area for the matrix, and nanoparticles may serve as a filler (Mansour, Poncin-Epaillard, & Debarnot, 2019). When CPs are hybridized in the vicinity of nanoparticles, organic-inorganic conducting polymer nanohybrid materials are obtained. These nanohybrid materials provide enhanced biosensing platforms with excellent electrochemical properties due to their exceptional characteristics. CPs, which contain an organic component, may also be polymerized using one or more of the materials listed below:

- Carbon nanomaterials: graphene, fullerene, carbon nanotubes,
- Inorganic compounds: halides, oxides, and
- Metal nanoparticles: palladium, silicon, silver, platinum, gold.

Conductive polymers containing sulfur (S) or nitrogen (N) atoms within their repeating units show great potential as materials for constructing metal-polymer nanohybrids. For instance, PTh and

PEDOT contain an S atom in their heterocyclic rings, and PANI and PPy contain an N atom in their heterocyclic rings. Therefore, Au, Pd, and Ag nanoparticles might be easily synthesized onto the surface of PEDOT, PTh, PANI, and PPy nanohybrids without the need for any reducing or dispersing agents (Qin et al., 2011; Tseng et al., 2005; Zhang et al., 2006). Several essential factors related to sensing geometry design should be taken into account when developing high-performance sensing devices, including (1) facile and portable geometry, (2) enlarged surface areas, and (3) specific selectivity. Based on such fundamental requirements, organic-inorganic nanohybrid materials based on CPs are one of the most efficient conducting platforms for electrochemical biosensor applications. Adding or adsorbing pristine metals or metal oxide nanoparticles to conducting polymer nanohybrid materials enhances biomolecule interaction and target selectivity.

8.6 DETECTION OF BIOMOLECULES

The creation of portable, precise, and rapid diagnostic tools for the early detection of fatal diseases is one of the main objectives of developing biochemical sensing platforms. A number of biosensing nanomaterials have recently been developed, including metal transistors, carbon nanotubes, graphene, and nanohybrids. Specifically, CP nanohybrid-based materials have been investigated for the detection of directly or indirectly associated bioanalytes, such as nucleic acid (DNA or RNA), dopamine (DA), and glucose owing to their biocompatibility, affinity for bioanalytes, and electrical conductivity. Electrochemical biosensors based on CPs nanohybrid materials used from 2018 to 2022 have been reviewed for the detection of nucleic acids, DA, and glucose.

8.6.1 Nucleic Acid Detection

Protein information in the human body can be affected by nucleic acids such as DNA or RNA. Due to these developments, advanced biosensing platforms are increasingly used to analyze DNA in applications like clinical diagnosis, drug discovery, and infectious diseases. Electrochemical biosensors incorporating conjugated/biomaterial nanohybrid geometries are among the most attractive DNA sensing technologies.

Electrochemical DNA biosensors demonstrate fast and direct electric signals toward target bioanalytes. This enables straightforward sensing designs without sophisticated signal transduction equipment or costly analyzers. Furthermore, compared to pristine transducers, DNA sensors based on hybridization exhibited more selective and sensitive sensing performances. Especially, high-performance DNA biosensors were developed due to the superior electrical characteristics of DNA detectors based on conducting polymer nanohybrids (Park et al., 2014). In the field of nanotechnology, DNA biosensors are one of the most significant tools for identifying pathogens (as shown in Figure 8.1). In the past, numerous methods of detecting DNA have been developed, such as piezoelectric, magnetic, electrochemical, and optical methods. Among these, electrochemical biosensors are regarded as a potential tool since they are highly sensitive, affordable, and portable.

Tran et al. constructed an electrochemical biosensor for the identification of pathogenic *E. coli*-specific genes. The MWCNTs/MnO_2/PANI nanowires were used as a biosensing material. On the surface of the platinum microelectrode (Pt), the nanostructure was created using a newly developed combined method (chemical/electrochemical synthesis), synthesis of MWCNTs/MnO_2 by chemical method, and synthesis of PANI nanowires by electrochemical technique. The fabricated biosensor has an estimated limit of detection (LOD) of 4.42×10^{-13} M and can recognize the target gene with a relatively linear range from 5 pM to 500 nM (Tran et al., 2021). The research highlighted the substantial potential of utilizing electrochemical DNA biosensors incorporating PANI-based nanowires. These biosensors offer a promising platform for creating electrochemical DNA sensors with exceptional selectivity and sensitivity, particularly for identifying pathogenic agents.

Nguyet and co-authors developed a simple label-free DNA biosensor consisting of cerium oxide nanorods functionalized with polypyrrole nanoparticles (PPy NPs) (CeO_2/NRs-PPy/NPs) substrate

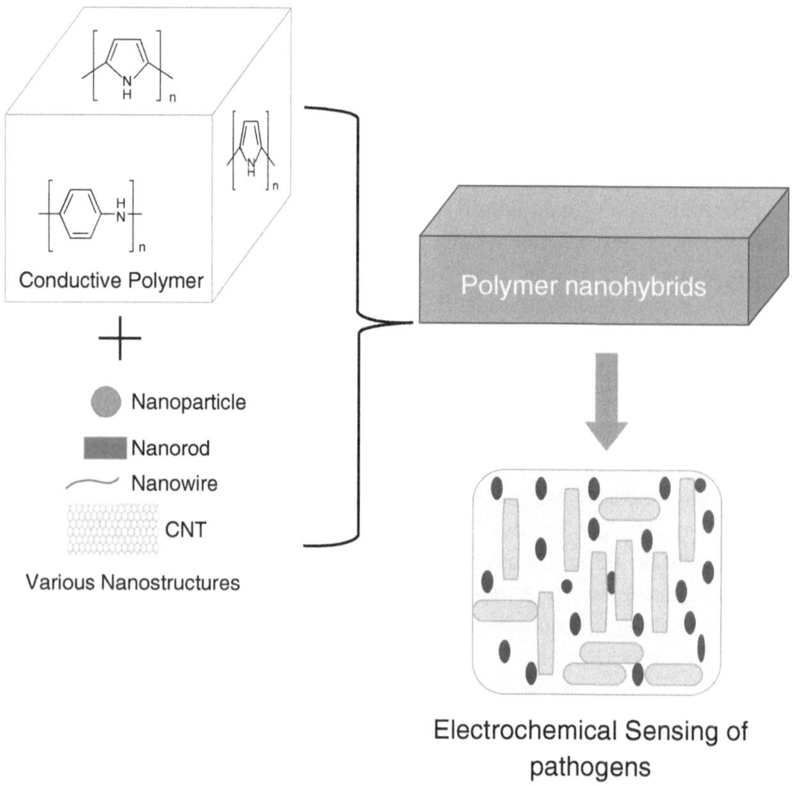

FIGURE 8.1 Electrochemical biosensing based on polymer nanohybrids for the recognition of pathogens (Kavitha et al., 2022).

for recognition of *Salmonella* DNA sequence (Nguyet et al., 2019). The biosensor was constructed using a hydrothermal synthesis technique, wherein CeO_2 nanorods were grown on the surface of an electrode. Subsequently, in situ chemical oxidative polymerization of PPy onto the cerium oxide nanorods was conducted, leading to the development of a modified electrode termed CeO_2/NR-PPy/NPs. The ssDNA sequences were affixed onto the modified microelectrode surfaces through covalent bonding. The research findings demonstrated that the developed biosensor showed a consistent linear response in the range of 1.0×10^{-9}mol L^{-1} to 1.0×10^{-6}mol L^{-1} with a sensitivity of 14.7×10^{6} Ωmol L^{-1} cm^{-1}. The DNA biosensor exhibited low LOD and limit of quantification (LOQ) with the values of 2.86×10^{-7}mol L^{-1} and 9.56×10^{-7}mol L^{-1}, respectively.

Shoaie et al. successfully created a novel electrochemical biosensor that employs DNA for fast and accurate *E. coli* identification, using a dual signal amplification strategy (Shoaie, Forouzandeh, & Omidfar, 2018). Cyclic voltammetry was used to prepare the PANI and gold nanoparticles (AuNPs) onto an electrode. In the subsequent step, avidin was attached to PANI-AuNP nanocomposite present on the surface of the modified electrode. Following this, the DNA probe was attached to the PANI-AuNP-avidin-modified electrode through an avidin–biotin interaction. The cyclic voltammetry electrochemical response was recorded at a scan rate of 50 mV s^{-1} and a potential of −0.1 V (vs. Ag/AgCl). The biosensor demonstrates the ability to detect *E. coli* in the range of 4×10^{6} to 4 CFU without the need for DNA amplification. Furthermore, the assay exhibited exceptional specificity toward other pathogens. It is noteworthy that the developed genosensor holds substantial potential for the fast and effective identification of *E. coli* infections.

Among the several CP nanoparticle (NP) hybrids, PANI nanocomposites were investigated in detail. For example, the electrochemical biosensing capability of carbon dot (CD)/zinc oxide nanorod/PANI nanohybrids deposited onto electrode has been explored for the biosensing of

E. coli O157:H7 present in H_2O samples (Pangajam, Theyagarajan, & Dinakaran, 2020). To prepare CDs/ZnO/PANI, the authors prepared zinc oxide nanorods using sol-gel approach, while CD was synthesized from honey, and then, the nanocomposite was deposited onto screen-printed carbon electrode (SPCE). *E. coli* O157:H7 DNA was efficiently identified using the CD/ZnO/PANI/SPCE-modified electrode up to a limit of 1.3×10^{-18} M. The fabricated electrochemical-based biosensor also showed excellent selectivity.

DNA electrochemical biosensors are also used for POC testing. Recently Fani et al. fabricated a label-free and rapid DNA electrochemical biosensor for the recognition of human T-lymphotropic virus type 1 (HTLV-1) oligonucleotide (Fani et al., 2021). The nanostructured material based on PPy, rGO, and AuNPs (rGO/PPy/(l-Cys)/AuNPs) was synthesized using an eco-friendly electropolymerization method. The constructed DNA biosensor showed good linearity between 0.1 fM and 100 µM with an LOD of 20 attomolar. The designed biosensor demonstrated exceptional reproducibility, repeatability, sensitivity, and selectivity.

A label-free microRNA (miRNA) biosensor has been fabricated consisting of (PPy)/silver nanofoam (AgNF) pyrrolidinyl peptide nucleic acid (acpcPNA)/PPy on gold electrode (Kangkamano et al., 2018). AgNF was electropolymerized as a redox marker onto an Au electrode surface. To immobilize the peptide nucleic acid (PNA) probes, a conducting polymer PPy layer was electrodeposited on the surface and functionalized with AgNF. The designed biosensor recognized miRNA-21, an abnormally regulated gene found in nearly all malignancies. Cyclic voltammetry (CV) was used to observe the response by monitoring the current of the AgNF redox reaction before and after hybridization. Optimal experimental settings exhibit a proportional relationship between the electrochemical response and miRNA-21 concentrations between 0.20 fM and 1.0 nM, indicating a low detection limit of 0.20 fM. The developed biosensor exhibited exceptional specificity, accurately differentiating between complementary, non-complementary, single-base, and double-base mismatched targets. The biosensor also showed excellent recovery (81%–119%). The findings showed that the biosensor has a high potential for investigation without RNA extraction and amplification. This suggests that it might be exploited clinically for cancer diagnosis and prognosis.

An electrochemical biosensor for the identification of Zika virus genomic RNA was developed by Moço et al. and tested on actual samples collected from infected individuals. Polytyramine, which serves as a CP, and electrochemically rGO are used to construct modified graphite electrodes for Zika virus biosensing (Moço et al., 2019).

The COVID-19 N-gene has recently been detected using a newly created peptide and electrodeposited PANI nanowire antifouling electrochemical biosensor (Song et al., 2021). To detect the nucleic acid of COVID-19, a biotin-labeled marker designed for the N-gene (nucleocapsid phosphoprotein) was immobilized onto PANI nanowires coated with a peptide. This immobilization was achieved through the use of a biotin/streptavidin affinity platform, resulting in the creation of an electrochemical sensing interface that is highly antifouling and sensitive. The antifouling biosensor exhibited a linear range between 10^{-14} and 10^{-9} M and showed very low LOD of 3.5 fM. The outstanding functionality of the biosensor is ascribed to the elevated electrochemical conductivity of PANI, which was selected as the biosensor material. Tailored peptides provided remarkable antifouling properties, enabling reliable recognition in complicated systems. These key characteristics are crucial components for fast and decentralized clinical monitoring.

More recently, Avelino and co-authors developed a novel label-free electrochemical biosensor for the identification of breakpoint cluster region/Abelson proto-oncogene (BCR/ABL) fusion oncogene, permitting an early identification of patients suspected of leukemia (Avelino et al., 2022). A nanohybrid composite was used as a biosensing platform based on chitosan and zinc oxide NPs (CS-ZnO/NPs) deposited on a PPy film. DNA sequences were covalently attached, which enable biomolecular detection. LOD and LOQ were estimated to be 1.34 and 4.08 fM, respectively, and sensitivity was reported in the range of 34.03 µA fM^{-1} cm^2 for the ABL/BCR fusion gene. The biosensing platform presented high selectivity, specificity, and repeatability with a relative standard deviation (RSD) of 4.21%. Furthermore, a linearity correlation was noticed between 138.80 aM to 13.88 pM.

TABLE 8.2
Electrochemical Biosensors Designed for the Identification of Nucleic Acids

Sr. No.	Analyte	Material	Electrochemical Method	LOD	References
1	*E. coli* Gene	$MWCNTs/MnO_2$/ PANI nanowires	EIS	4.42×10^{-13} M	Tran et al. (2021)
2	*Salmonella* DNA	CeO_2/NRs-PPy/NPs	EIS	2.86×10^{-7} mol L^{-1}	Nguyet et al. (2019)
3	*E. coli* Gene	PANI-AuNP-avidin	CV	NA	Shoaie, Forouzandeh, and Omidfar (2018)
4	*E. coli* O157:H7 DNA	CDs/ZnO/PANI	DPV	1.3×10^{-18} M	Pangajam, Theyagarajan, and Dinakaran (2020)
5	HTLV-1 oligonucleotide	rGO/PPy/(l-Cys)/ AuNPs	DPV	20 attomolar	Fani et al. (2021)
6	miRNA-21	acpcPNA/PPy/AgNF	CV	0.20 fM	Kangkamano et al. (2018)
7	Genomic RNA	rGO/polytyramine	DPV	NA	Moço et al. (2019)
8	COVID-19 N-gene	PANI nanowires/ peptide	DPV	3.5 fM	Song et al. (2021)
9	BCR/ABL fusion oncogene	PPy/Chit-ZnONP	CV and EIS	1.34 fM	Avelino et al. (2022)

Furthermore, the biosensor's user-friendly operation and quick response time make it highly suitable for early cancer diagnosis. The developed nanohybrid device offers a viable substitute for the identification of the fusion gene, presenting a promising approach to cancer diagnostics. Table 8.2 presents electrochemical biosensors based on CP nanostructures for the identification of some nucleic acids (RNA or DNA).

8.6.2 Dopamine (DA) Detection

Arvid Carlsson discovered DA in the 1950s. DA is essential for maintaining both mental and physical health, as it affects the circulatory, central nervous, hormonal, and renal systems (Gowrishankar, Hahn, & Blakely, 2014). Research has provided evidence that DA in the frontal lobe is essential for memory, attention, and problem-solving, as it regulates the transmission of information to other brain regions. DA is also known as the "reward chemical" due to its release during successful reward events such as having sex, eating, and other stimulating activities. Irregular DA levels have been linked with various disorders, including eating disorders, insomnia disorders, anxiety, schizophrenia, and Parkinson's disease (Camardese et al., 2014). Early detection of abnormal DA levels is of the utmost importance. As a result, there is a growing focus on developing excellent sensing approaches for DA molecule detection. Uge and co-authors designed an electrochemical biosensor for the identification of DA in medicinal samples. Multiwalled carbon nanotubes were electropolymerized onto glassy carbon electrode (GCE); then, this modified electrode was functionalized with ceric oxide-poly(3,4-ethylenedioxythiophene) (CeO_2/PEDOT), and the final composite consisting of MWCNTs/CeO_2/PEDOT was obtained. The modified GCE demonstrated a more efficient electro-catalytic response to DA oxidation in comparison with bare GCE (Üğe, Zeybek, & Zeybek, 2018). Under optimal circumstances, the fabricated biosensor revealed two linearity ranges of 0.10–10 μmol L^{-1} and 40–400 μmol L^{-1} as well as offers an LOD of 0.03 μmol L^{-1}.

DA is determined using an electrochemical biosensor that is resistant to biofouling in complicated biological media (Song et al., 2019). The biosensor consists of PEDOT which was doped with the ionic liquid (IL). The PEDOT/IL-modified GCE has shown the capability to accurately measure DA in the availability of high protein concentrations while effectively preventing biofouling in native serum conditions. The LOD for DA was as low as 33 nM, covering linear ranges from 0.2 to

328 μM (at 0.2 V vs. SCE). The modified PEDOT/IL/GCE exhibits good stability, excellent conductivity, and a porous microstructure. Accurate DA measurement in samples of human urine can be achieved with the developed biosensor.

It has been demonstrated that linear sweep voltammetry (LSV) is more useful than cyclic voltammetry to electropolymerize the AuNPs over PANI-GCE-modified electrodes for the detection of DA molecules (Mahalakshmi & Sridevi, 2021). At neutral pH, the modified GCE demonstrated higher electro-catalytic activity, which is desirable for biosensing biological molecules, such as DA despite interferences from structurally similar molecules. The DA oxidation previously reported in the literature took place between 0.270V and 0.361V for PANI-modified electrode, but the range was reduced to 0.168 V in the present work which was synthesized by electropolymerzation technique. For dopamine-uric acid (DA-UA) and dopamine-ascorbic acid (DA-AA) mixture, the peak-to-peak separation is 345 and 108 mV, respectively, demonstrating an overall linear range of 20–100 μM with an LOD of 16 μM for DA. The enhanced electro-catalytic performance is ascribed to the combined effect of the PANI film and the electropolymerized gold NPs.

Ratlam et al. presented two types of DA biosensors, i.e., fluorescent and electrochemical biosensors consisting of CPs and quantum dot nanocomposite (Ratlam, Phanichphant, & Sriwichai, 2020). Fluorine-doped tin oxide (FTO) glass substrate was employed to fabricate the electrospun nanofiber film of PANI/carbon quantum dots (CQDs) for electrochemical-based biosensors. CV and amperometry were carried out to analyze the electro-catalytic activity of the deposited PANI/CQDs-modified FTO toward identification of DA at neutral pH. Composites prepared from PANI and CQDs can serve as potential DA biosensors for real-time analysis in the future.

Adhikari and co-authors reported a rod-like PPy-Ag nanostructure that was deposited on GCE to detect nanomolar DA concentration employing sodium cholate as a soft template (Adhikari et al., 2020). Electrochemical measurement of DA was performed on the deposited nanostructure using LSV. The PPy-Ag-nanostructure-coated electrode exhibited enhanced sensitivity of 8.22 mA μM^{-1} for DA with an LOD of 0.00005 μM.

CP polythionine (PTh) films have been reportedly synthesized by electropolymerization technique with single-step and double-step methods (PTh-S and PTh-D, respectively) onto GCE surface for the identification of DA with varying sensitivity (Rahman & Lee, 2021). PTh films with identical chemical structures are analyzed spectroscopically and electrochemically. However, their sensitivity to DA detection varies substantially depending on polymerization conditions. Minor changes to electropolymerization parameters significantly affected the electro-catalytic performance of polythionine films for DA oxidation. The study also explored the mechanism of PTh formation and its response to interfering substances. The PTh created using the double-step method displayed significantly greater surface charge compared to PTh-S, resulting in improved sensitivity for DA recognition up to an order of magnitude relative to PTh-S.

Minta et al. synthesized a nanocomposite using a hydrothermal process based on PANI and thermally reduced graphene oxides (TRGOs) (PANI/TRGO) which was prepared at 400°C and 700°C for electrochemical biosensing of DA (Minta et al., 2022). The distribution of PANI and optimal oxygen content are crucial factors for DA biosensing. It has been found that PANI/TRGO outperformed other active materials in terms of sensitivity, achieving an LOD of 430 nM. This composite exhibited exceptional sensitivity (6.7 μA μM^{-1}) and remained stable even after 30 days of testing.

A promising approach was investigated for crosslinking different CPs including poly-aminoindole (PIn-X-NH_2), PPy, and PANI using PEDOT:PSS as a conductive dopant. This strategy enables simultaneous detection of hydrogen peroxide and DA (Yang et al., 2022). The formation of hydrogels relies on the electrostatic interaction between positively charged polymer chains (such as PPy, PANI, and PIn-X-NH_2, where X=4,5,6,7) and negatively charged groups (SO_3^-) in PEDOT:PSS. The DA biosensor exhibited a direct correspondence within the range of 0.02–1288.74 μM, revealing a sensitivity of 0.2102 μA $cm^{-2}\mu M^{-1}$. The estimated LOD for DA was determined to be 6.7 nM (S/N=3).

TABLE 8.3
Electrochemical-Based Biosensors for the Detection of Dopamine

Sr. No.	Analyte	Material	Electrochemical Method	LOD	References
1	Dopamine	MWCNTs/CeO_2/PEDOT	CV and DPV	0.03 μmol L^{-1}	Üğe, Zeybek, and Zeybek (2018)
2	Dopamine	PEDOT/IL	CV	33 nM	Song et al. (2019)
3	Dopamine	PANI/AuNPs	CV	16 μM	Mahalakshmi and Sridevi (2021)
4	Dopamine	PANI/CQDs	CV and amperometry	0.1013 μM	Ratlam, Phanichphant, and Sriwichai (2020)
5	Dopamine	PPy-Ag nanostructure	LSV	0.00005 μM	Adhikari et al. (2020)
6	Dopamine	PTH-S and PTH-D films	DPV	NA	Rahman and Lee (2021)
7	Dopamine	PANI/TRGO	CV and EIS	430 nM	Minta et al. (2022)
8	Dopamine	AuNPs/PPy/PEDOT:PSS	Amperometry	6.7 nM	Yang et al. (2022)
9	Dopamine	AuNPs/PEDOT/CNTs	CV and DPV	0.283 μM	Chen et al. (2022)

Chen et al. developed a nanohybrid biosensing platform through a simple one-pot synthesis method. The platform was based on a material composed of AuNPs incorporated into a PEDOT/CNTs matrix (Chen et al., 2022). The nanohybrid material was subsequently applied as a coating on a bare GCE to amplify the detection capability of DA in the presence of elevated concentrations of uric acid and ascorbic acid. Under optimum conditions, the developed biosensor demonstrated a linear response within the range of 9.14–29.704 μM, with a sensitivity of 1.557 μA μM^{-1} and an LOD of 0.283 μM. The developed biosensor demonstrated good average recovery when applied to real samples and exhibited excellent stability and reproducibility. Table 8.3 shows electrochemical biosensors based on CP nanocomposites for DA detection.

8.6.3 Glucose Detection

The construction of glucose biosensors is crucial for addressing diabetes mellitus, a condition that is regarded as a global public health concern that raises the risk of renal failure, blindness, heart disease, surgical and wound infections (Labib, Sargent, & Kelley, 2016; Wang & Lee, 2015). Glucose levels in the blood, sweat, tears, and saliva can be monitored quickly, accurately, and simply with glucose biosensors due to their accuracy, simplicity, and high speed (Menezes et al., 2015; Shu et al., 2014). Glucose sensors are typically categorized into two types: enzyme free and enzyme based. In enzyme-based sensors, glucose level is monitored by interactions with one of three enzymes: glucose-1-dehydrogenase (GDH), glucose oxidase (GO_x), and hexokinase (Price, 2003). Glucose biosensors generally employed for self-monitoring of blood glucose (SMBG) typically consist of two enzyme families: GDH and GO_x. These enzymes vary in their redox potential, turnover rate, cofactors, and glucose selectivity. Several clinical laboratories use the hexokinase enzymes as a standard method to measure glucose using spectrophotometry (Heller & Feldman, 2008). GO_x is considered a reference enzyme for the creation of biosensors because of its high selectivity for glucose. GO_x offers advantages such as easy availability; affordability; and tolerance to high pH, ionic strength, and extreme temperatures. These features allow for relaxed manufacturing conditions and user-friendly storage provisions in glucose biosensors for lay users (Bankar et al., 2009). A non-enzymatic glucose biosensor uses direct electrochemical oxidation to measure glucose levels. In recent years, non-enzymatic and enzyme-based glucose biosensors have been developed with multiple sensing platforms, including carbon nanotubes, enzymes, molecularly imprinted polymers, and CPs.

Highly selective, binder-free, and efficient, non-enzymatic electrochemical biosensor have been reported by Bilal and coworkers for glucose monitoring based on PANI/CuNi nanostructures. The PANI/CuNi nanohybrids were synthesized by combining PANI solution with various NPs loading ratio (1:0.25, 1:0.33, 1:0.5 and 1:1) (Bilal & Ullah, 2018). The PANI/CuNi-functionalized electrode showed a linear range (LR) of 5.6 mM (R^2 = 0.992), low LOD of 0.2 μM, and excellent sensitivity of 1030 μA mM^{-1} cm^{-2}. The developed biosensor also possesses high stability and efficient recognition of glucose in real blood samples.

Batool et al. reported a nanocomposite synthesized by decorating the carboxy-functionalized graphene oxide (fGO) with magnetic NPs and coupling with the CP PANI (Batool et al., 2019). An enzyme-free glucose sensor was obtained by depositing the resulting nanocomposite on a substrate. The charge transfer capability of PANI, along with the large surface area of Fe_3O_4 and fGO, contribute to a robust electrochemical response (typically at 0.18 V vs Ag/AgCl). However, this response is attenuated in the presence of glucose. An extremely selective technique was designed to determine blood glucose concentrations based on the observed reduction in current. This method offers a reliable and accurate means of measuring glucose levels. The fabricated glucose biosensor response was found linear in the range of 0.05 μM to 5 mM, with a lower LOD of 0.01 μM.

A glucose biosensor was developed by modifying an SPCE with a combination of MWCNT, gold NPs, rGO, and PANI (Maity, Minitha, & Rajendra Kumar, 2019). To achieve a glucose biosensor with both high selectivity and sensitivity, GO_x was immobilized onto the electrode surface. Using cyclic voltammetry, GO_x coating on modified SPCE was monitored electrochemically at different stages. The modified SPCE demonstrated a 13.43-fold increase in reduction current in comparison to the bare SPCE when the working potential was lowest. The glucose biosensor exhibited a wide linear range (1^{-10}mM), low LOD (64 μM), high stability, excellent reproducibility (90.23%, n=7), and excellent sensitivity (246 μA $cm^{-2}mM^{-1}$). An amperometric technique was also employed to validate the biosensor for glucose detection in serum samples from human subjects. Designed nanocomposite-based SPCE can be used to develop efficient glucose biosensors as well as a platform for a wide range of biochemical sensors.

Zahed and co-authors fabricated a flexible glucose-based biosensor. The biosensor platform consists of novel materials such as PEDOT and polystyrene sulfonate (PEDOT:PSS); porous and stable laser-induced graphene (LIG) with 3-dimensional (3D) properties was also decorated on the resulting PEDOT:PSS for glucose detection (Zahed et al., 2020). To enhance electrode robustness and provide uniform electrical conductivity, PEDOT and polystyrene sulfonate were spray-coated onto the LIG. The constructed biosensor presented an outstanding amperometric response to glucose. It showed a low LOD of 3 μM and an excellent sensitivity of 247.3 μA mM^{-1} cm^{-2} with a wide linear range of 10 μM to 9.2 mM. The feasibility of the biosensor was ultimately demonstrated through sweat samples gathered during physical exercise.

Two types of CP-based composites were investigated by German et al. for glucose detection. Polypyrrole, short-chain PANI-based nanostructures, glucose oxidase, and gold NPs (6 nm) were immobilized on the surface of a graphite rod followed by GO_x deposition (German, Ramanaviciene, & Ramanavicius, 2021). The enhanced enzymatic-based biosensors consisting of PPy-AuNPs/GR-GO_x/GO_x and PANI-AuNPs/GR-GO_x/GO_x electrodes were characterized by wider linear response (up to 16.5 mmol L^{-1}), low LOD (0.071 and 0.070 mmol L^{-1}), suitable repeatability (RSD 5.89% and 4.67%), good stability, and excellent sensitivity (55.4 and 65.4 μA mM^{-1} cm^{-2}). The PANI-AuNPs/GR-GO_x/GO_x electrode has shown outstanding resistance to interference from substances such as uric acid and ascorbic acid. Additionally, successful application of this electrode in real-sample analysis has been demonstrated for glucose monitoring in serum samples.

A glucose biosensor was created for amperometric glucose detection, employing reduced rGO, PANI, glucose oxidase, and Nafion as integral components. rGO was prepared using a novel method involving thermal reduction in the presence of malonic acid (Popov et al., 2021). A GR electrode was served as the WE of the biosensor, which was modified with an rGO, GO_x, Nafion and dispersion of PANI. A 1:10 ratio between PANI and rGO was found optimum in the dispersion applied for GO_x attachment. The designed biosensor demonstrated a broad linear range of detection spanning from 0.5 to 50 mM, with an LOD of 0.089 mM. The biosensor also shows remarkable stability, reproducibility, and excellent selectivity. Thus, the constructed biosensor can be used to determine glucose levels in human serum. An rGO and PANI nanostructure dispersion could be used to develop electrochemical glucose biosensors.

Glucose electrochemical biosensor was reported using a molecularly imprinted polymers strategy, in which a hierarchical $Ti_2Nb_{10}O_{29}$-based N-rich carbon conductive-deposited TNO nanostructure (TNO/NC) was synthesized (Karaman et al., 2022). Initially, the TNO/NC structure was prepared through PPy-chemical vapor deposition technique with minimal waste production. Subsequently, the bare GCE was modified with synthesized material by dropping TNO/NC particles onto GCE surfaces via an infrared heat lamp. Using cyclic voltammetry, an imprinted glucose electrochemical biosensor was fabricated by utilizing 25.0 mM glucose as a template and 75.0 mM pyrrole as a monomer. The electrochemical polymerization process took place at potentials ranging from +0.20 to +1.20 V versus Ag/AgCl. The results showed that sensitive, stable, and selective electrochemical response were correlated with various glucose concentrations, and the sensitivity of imprinted biosensor for glucose identification was calculated to be 18.93 $\mu A\ \mu M^{-1}\ cm^{-2}$ ($R^2=0.99$) at +0.30 V with low LOD of 1.0×10^{-6} M. Therefore, it is conceivable that the developed imprinted glucose biosensor may be employed in a variety of discipline, such as food safety and public health.

Kawakami et al. presented a glucose biosensor relying on AuNPs functionalized with pyrrole (Kawakami et al., 2022). The PPy was synthesized by electrochemical technique containing 0.1 M Py monomer and 0.05 M polymerization in 0.05 M PBS. A solution containing 0.1 mM $KAuCl_4$ was applied directly onto the PPy-deposited electrode to produce the gold NP decorations. The catalytic electrode was prepared without enzymes. The size and amount of AuNPs was shown to be dependent on the amount of $KAuCl_4$ mixture poured on the PPy-modified electrode. When the AuNPs decorated with PPy were employed as a catalytic electrode, the electro-catalytic activity in glucose oxidation was confirmed. The performance was revealed to be regulated by the volume of $KAuCl_4$ solution used to prepare the electrode.

A glucose electrochemical biosensor was designed composed of PEDOT:4-sulfocalixarene (PEDOT:SCX)/MXene followed by affixing GO_x onto a modified electrode (Murugan et al., 2022). A redox peak of flavin adenine dinucleotide (FAD)-GO_x was detected at a potential of −0.435 V on a modified electrode, which suggested direct electron movement between the electrode surface and the enzyme. The modified electrode exhibited a linear range from 0.5 to 8 mM for glucose concentrations. Finally, the innovative biosensor was effectively employed to recognize glucose in typical fruit juice samples, yielding acceptable recovery rates. Table 8.4 illustrates electrochemical biosensors based on CP nanohybrids for glucose detection.

8.7 CONCLUSION AND FUTURE PERSPECTIVE

Electrochemical biosensors are a potential approach for multiple biomolecule recognition. Recently, the majority of the work has been aimed at developing novel sensing/detecting probe materials to enhance biosensor performance, including selectivity, sensitivity, and biocompatibility. Of those developed nanomaterials, CPs and their hybrids were extensively employed in biosensor surface design. In addition to enhanced sensitivity and selectivity, these nanohybrids also exhibit superior mechanical stability. Numerous CPs, such as PPy, PANI, PTh, and PEDOT, offer many advantages owing to their electrochemical redox efficiency and charge transport properties, which correspond to the π-electrons delocalization over the CP backbone. In this chapter, the latest advancements in the fabrication of electrochemical biosensors based on conductive-nanohybrids for the detection of the most critical bioanalytes such as nucleic acid (DNA and RNA), DA, and glucose are covered in depth.

Researchers have made significant advances in combining CPs with carbon nanomaterials (CNTs, MWCNTs, graphene, GO, rGO), quantum dots, metal, and metal oxide NPs to catalyze

TABLE 8.4
Electrochemical-Based Biosensors for the Identification of Glucose

Sr. No.	Analyte	Material	Electrochemical Method	LOD	References
1	Glucose	PANI/CuNi nanostructures	Chronoamperometry	0.2 µM	Bilal and Ullah (2018)
2	Glucose	fGO-Fe_3O_4 NP-PANI	CV	0.01 µM	Batool et al. (2019)
3	Glucose oxidase	AuNP/PANI/rGO/ NH_2-MWCNTs	Amperometry	64 µM	Maity, Minitha, and Rajendra Kumar (2019)
4	Glucose	PEDOT:PSS/LIG	Amperometry	3 µM	Zahed et al. (2020)
5	Glucose	PPy-AuNPs and PANI-AuNPs	Amperometry	0.071 and 0.070 mmol L^{-1}	German, Ramanaviciene, and Ramanavicius (2021)
6	Glucose oxidase	rGO/GO_x/PANI/Nafion	Amperometry	0.089 mM	Popov et al. (2021)
7	Glucose	TNO@NC/PPy	CV and DPV	1.0×10^{-6} M	Karaman et al. (2022)
8	Glucose	AuNPs/PPy	CV	NA	Kawakami et al. (2022)
9	Glucose oxidase	PEDOT:SCX/MXene/GO_x	CV	22.5 µM	Murugan et al. (2022)

reactions with target analytes with powerful synergistic effects. While these methods often enhance LOD and sensitivity, other problems may occur, such as nonspecific adsorption of other substances and fouling. These challenges could potentially complicate the process of commercializing the developed electrochemical biosensors. Attaining a comprehensive understanding of the electrochemical processes taking place at the interface between the electrode and electrolyte remains a central challenge in customizing sensing devices. It is imperative to note that although most developed electrochemical biosensors have been analyzed using real samples, there is still a sizable gap that exists between laboratory testing and the commercialization of these sensing platforms. Effective collaboration between industry and diverse research groups is essential for the progress and commercialization of electrochemical-based sensors and biosensors designed to detect biomolecules, pharmaceutical drugs, and various other substances. Apart from selectivity and sensitivity, it is crucial to comprehensively assess the cost-effectiveness and robustness of electrochemical sensors and biosensors.

REFERENCES

Adhikari, A., De, S., Rana, D., Nath, J., Ghosh, D., Dutta, K., & Chattopadhyay, D. (2020). Selective sensing of dopamine by sodium cholate tailored polypyrrole-silver nanocomposite. *Synthetic Metals*, 260, 116296.

Ahmadi, M., & Ahour, F. (2020). An electrochemical biosensor based on a graphene oxide modified pencil graphite electrode for direct detection and discrimination of double-stranded DNA sequences. *Analytical Methods*, 12(37), 4541–4550.

Avelino, K. Y., Oliveira, L. S., Santos, M. R., Lucena-Silva, N., Andrade, C. A., & Oliveira, M. D. (2022). Electrochemical DNA biosensor for chronic myelocytic leukemia based on hybrid nanostructure. *Bioelectrochemistry*, 147, 108176.

Aydemir, N., Malmström, J., & Travas-Sejdic, J. (2016). Conducting polymer based electrochemical biosensors. *Physical Chemistry Chemical Physics*, 18(12), 8264–8277.

Bankar, S. B., Bule, M. V., Singhal, R. S., & Ananthanarayan, L. (2009). Glucose oxidase—an overview. *Biotechnology advances*, 27(4), 489–501.

Batool, R., Akhtar, M. A., Hayat, A., Han, D., Niu, L., Ahmad, M. A., & Nawaz, M. H. (2019). A nanocomposite prepared from magnetite nanoparticles, polyaniline and carboxy-modified graphene oxide for non-enzymatic sensing of glucose. *Microchimica Acta*, 186(5), 1–10.

Bilal, S., & Ullah, W. (2018). Polyaniline@ CuNi nanocomposite: A highly selective, stable and efficient electrode material for binder free non-enzymatic glucose sensor. *Electrochimica Acta*, 284, 382–391.

Boehler, C., Oberueber, F., & Asplund, M. (2019). Tuning drug delivery from conducting polymer films for accurately controlled release of charged molecules. *Journal of Controlled Release*, 304, 173–180.

Camardese, G., Di Giuda, D., Di Nicola, M., Cocciolillo, F., Giordano, A., Janiri, L., & Guglielmo, R. (2014). Imaging studies on dopamine transporter and depression: A review of literature and suggestions for future research. *Journal of Psychiatric Research*, 51, 7–18.

Chen, S., Chen, W., Wang, Y., Wang, X., Ding, Y., Zhao, D., & Liu, J. (2022). Facile one-pot method of AuNPs/PEDOT/CNT composites for simultaneous detection of dopamine with a high concentration of ascorbic acid and uric acid. *RSC Advances*, 12(24), 15038–15045.

Dong, R., Ma, P. X., & Guo, B. (2020). Conductive biomaterials for muscle tissue engineering. *Biomaterials*, 229, 119584.

Erdem, E., Karakışla, M., & Sacak, M. (2004). The chemical synthesis of conductive polyaniline doped with dicarboxylic acids. *European Polymer Journal*, 40(4), 785–791.

Fani, M., Rezayi, M., Pourianfar, H. R., Meshkat, Z., Makvandi, M., Gholami, M., & Rezaee, S. A. (2021). Rapid and label-free electrochemical DNA biosensor based on a facile one-step electrochemical synthesis of rGO–PPy–(l-Cys)–AuNPs nanocomposite for the HTLV-1 oligonucleotide detection. *Biotechnology and Applied Biochemistry*, 68(3), 626–635.

German, N., Ramanaviciene, A., & Ramanavicius, A. (2021). Dispersed conducting polymer nanocomposites with glucose oxidase and gold nanoparticles for the design of enzymatic glucose biosensors. *Polymers*, 13(13), 2173.

Gowrishankar, R., Hahn, M. K., & Blakely, R. D. (2014). Good riddance to dopamine: Roles for the dopamine transporter in synaptic function and dopamine-associated brain disorders. *Neurochemistry International*, 73, 42–48.

Gueye, M. N., Carella, A., Faure-Vincent, J., Demadrille, R., & Simonato, J.-P. (2020). Progress in understanding structure and transport properties of PEDOT-based materials: A critical review. *Progress in Materials Science*, 108, 100616.

He, H., Zhang, L., Guan, X., Cheng, H., Liu, X., Yu, S., & Ouyang, J. (2019). Biocompatible conductive polymers with high conductivity and high stretchability. *ACS Applied Materials & Interfaces*, 11(29), 26185–26193.

Heinze, J., Frontana-Uribe, B. A., & Ludwigs, S. (2010). Electrochemistry of conducting polymers persistent models and new concepts. *Chemical Reviews*, 110(8), 4724–4771.

Heller, A., & Feldman, B. (2008). Electrochemical glucose sensors and their applications in diabetes management. *Chemical Reviews*, 108(7), 2482–2505.

Humpolíček, P., Kašpárková, V., Pacherník, J., Stejskal, J., Bober, P., Capáková, Z., & Lehocký, M. (2018). The biocompatibility of polyaniline and polypyrrole: A comparative study of their cytotoxicity, embryotoxicity and impurity profile. *Materials Science and Engineering: C*, 91, 303–310.

Husain, A., Ahmad, S., & Mohammad, F. (2020). Preparation and applications of Polythiophene nanocomposites. *Journal of Computing Science and Engineering (JESC)* I, 36–53.

Hwang, M. T., Heiranian, M., Kim, Y., You, S., Leem, J., Taqieddin, A., & van der Zande, A. M. (2020). Ultrasensitive detection of nucleic acids using deformed graphene channel field effect biosensors. *Nature Communications*, 11(1), 1–11.

Inal, S., Hama, A., Ferro, M., Pitsalidis, C., Oziat, J., Iandolo, D., & Marchat, D. (2017). Conducting polymer scaffolds for hosting and monitoring 3D cell culture. *Advanced Biosystems*, 1(6), 1700052.

Kangkamano, T., Numnuam, A., Limbut, W., Kanatharana, P., Vilaivan, T., & Thavarungkul, P. (2018). Pyrrolidinyl PNA polypyrrole/silver nanofoam electrode as a novel label-free electrochemical miRNA-21 biosensor. *Biosensors and Bioelectronics*, 102, 217–225.

Karaman, C., Karaman, O., Atar, N., & Yola, M. L. (2022). A molecularly imprinted electrochemical biosensor based on hierarchical $Ti_2Nb_{10}O_{29}$ (TNO) for glucose detection. *Microchimica Acta*, 189(1), 1–11.

Kaur, G., Kaur, A., & Kaur, H. (2021). Review on nanomaterials/conducting polymer based nanocomposites for the development of biosensors and electrochemical sensors. *Polymer-Plastics Technology and Materials*, 60(5), 504–521.

Kavitha, N., Elavarasan, M., Ramachandran, R., Uthayakumar, S., Chandramohan, A., & Dinakaran, K. (2022). Polymer nanohybrid composites as conductive platform for the electrochemical sensing of pathogens. *Current Research in Green and Sustainable Chemistry*, 5, 100316.

Kawakami, H., Ito, Y., Chien, Y.-A., Chen, C.-Y., Chiu, W.-T., Chakraborty, P., & Chang, T.-F. M. (2022). Development of polypyrrole/nano-gold composite for non-enzymatic glucose sensors. *Micro and Nano Engineering*, 14, 100109.

Labib, M., Sargent, E. H., & Kelley, S. O. (2016). Electrochemical methods for the analysis of clinically relevant biomolecules. *Chemical Reviews*, 116(16), 9001–9090.

Liu, C., Goeckner, M. J., & Walker, A. V. (2017). Plasma polymerization of poly (3, 4-ethylenedioxyethene) films: The influence of plasma gas phase chemistry. *Journal of Vacuum Science & Technology A: Vacuum, Surfaces, and Films*, 35(2), 021302.

Liu, X., Bu, S., Feng, J., Wei, H., Wang, Z., Li, X., & Wan, J. (2022). Electrochemical biosensor for detecting pathogenic bacteria based on a hybridization chain reaction and CRISPR-Cas12a. *Analytical and Bioanalytical Chemistry*, 414(2), 1073–1080.

Lu, X., Zhang, W., Wang, C., Wen, T.-C., & Wei, Y. (2011). One-dimensional conducting polymer nanocomposites: Synthesis, properties and applications. *Progress in Polymer Science*, 36(5), 671–712.

Maduraiveeran, G. (2022). Metal nanocomposites based electrochemical sensor platform for few emerging biomarkers. *Current Analytical Chemistry*, 18(5), 509–517.

Magar, H. S., Hassan, R. Y., & Mulchandani, A. (2021). Electrochemical impedance spectroscopy (EIS): Principles, construction, and biosensing applications. *Sensors*, 21(19), 6578.

Mahalakshmi, S., & Sridevi, V. (2021). In situ electrodeposited gold nanoparticles on polyaniline-modified electrode surface for the detection of dopamine in presence of ascorbic acid and uric acid. *Electrocatalysis*, 12(4), 415–435.

Maity, D., Minitha, C., & Rajendra Kumar, R. T. (2019). Glucose oxidase immobilized amine terminated multiwall carbon nanotubes/reduced graphene oxide/polyaniline/gold nanoparticles modified screen-printed carbon electrode for highly sensitive amperometric glucose detection. *Materials Science and Engineering: C*, 105, 110075.

Malinauskas, A. (2001). Chemical deposition of conducting polymers. *Polymer*, 42(9), 3957–3972.

Mansour, A., Poncin-Epaillard, F., & Debarnot, D. (2019). Affinity and distribution of silver nanoparticles within plasma polymer matrices. *Journal of Materials Science*, 54(19), 12972–12987.

Menezes, F. G., Neves, A. C., Lima, D. F. D., Lourenço, S. D., Silva, L. C. D., & de Lima, K. M. (2015). Bioorganic concepts involved in the determination of glucose, cholesterol and triglycerides in plasma using the enzymatic colorimetric method. *Química Nova*, 38, 588–594.

Minta, D., González, Z., Melendi-Espina, S., & Gryglewicz, G. (2022). Highly efficient and stable PANI/TRGO nanocomposites as active materials for electrochemical detection of dopamine. *Surfaces and Interfaces*, 28, 101606.

Moço, A. C., Guedes, P. H., Flauzino, J. M., da Silva, H. S., Vieira, J. G., Castro, A. C., & Madurro, J. M. (2019). Electrochemical detection of zika virus in biological samples: A step for diagnosis point-of-care. *Electroanalysis*, 31(8), 1580–1587.

Moon, J.-M., Thapliyal, N., Hussain, K. K., Goyal, R. N., & Shim, Y.-B. (2018). Conducting polymer-based electrochemical biosensors for neurotransmitters: A review. *Biosensors and Bioelectronics*, 102, 540–552.

Murugan, P., Annamalai, J., Atchudan, R., Govindasamy, M., Nallaswamy, D., Ganapathy, D., & Sundramoorthy, A. K. (2022). Electrochemical sensing of glucose using glucose oxidase/PEDOT: 4-sulfocalix [4] arene/MXene composite modified electrode. *Micromachines*, 13(2), 304.

Nada, A. A., Eckstein Andicsová, A., & Mosnáček, J. (2022). Irreversible and self-healing electrically conductive hydrogels made of bio-based polymers. *International Journal of Molecular Sciences*, 23(2), 842.

Nair, S. S., Mishra, S. K., & Kumar, D. (2019). Recent progress in conductive polymeric materials for biomedical applications. *Polymers for Advanced Technologies*, 30(12), 2932–2953.

Naseri, M., Fotouhi, L., & Ehsani, A. (2018). Recent Progress in the development of conducting polymer-based nanocomposites for electrochemical biosensors applications: A mini-review. *The Chemical Record*, 18(6), 599–618.

Nguyet, N. T., Van Thu, V., Lan, H., Trung, T., Le, A.-T., Pham, V.-H., & Tam, P. D. (2019). Simple label-free DNA sensor based on CeO_2 nanorods decorated with ppy nanoparticles. *Journal of Electronic Materials*, 48(10), 6231–6239.

Ning, G., Wang, H., Fu, M., Liu, J., Sun, Y., Lu, H., & Wang, H. (2022). Dual signals electrochemical biosensor for point-of-care testing of amino acids enantiomers. *Electroanalysis*, 34(2), 316–325.

Pangajam, A., Theyagarajan, K., & Dinakaran, K. (2020). Highly sensitive electrochemical detection of E. coli O_{157}: H_7 using conductive carbon dot/ZnO nanorod/PANI composite electrode. *Sensing and Bio-Sensing Research*, 29, 100317.

Park, C. S., Lee, C., & Kwon, O. S. (2016). Conducting polymer based nanobiosensors. *Polymers*, 8(7), 249.

Park, C.-S., Kim, D. H., Shin, B. J., & Tae, H.-S. (2016). Synthesis and characterization of nanofibrous polyaniline thin film prepared by novel atmospheric pressure plasma polymerization technique. *Materials*, 9(1), 39.

Park, S. J., Kwon, O. S., Lee, J. E., Jang, J., & Yoon, H. (2014). Conducting polymer-based nanohybrid transducers: A potential route to high sensitivity and selectivity sensors. *Sensors*, 14(2), 3604–3630.

Park, Y., Jung, J., & Chang, M. (2019). Research progress on conducting polymer-based biomedical applications. *Applied Sciences*, 9(6), 1070.

Popov, A., Aukstakojyte, R., Gaidukevic, J., Lisyte, V., Kausaite-Minkstimiene, A., Barkauskas, J., & Ramanaviciene, A. (2021). Reduced graphene oxide and polyaniline nanofibers nanocomposite for the development of an amperometric glucose biosensor. *Sensors*, 21(3), 948.

Price, C. P. (2003). Point-of-care testing in diabetes mellitus. *Clinical Chemistry and Laboratory Medicine (CCLM)*, 41(9), 1213–1219.

Qin, X., Lu, W., Luo, Y., Chang, G., & Sun, X. (2011). Preparation of Ag nanoparticle-decorated polypyrrole colloids and their application for H_2O_2 detection. *Electrochemistry Communications*, 13(8), 785–787.

Rahman, M. M., & Lee, J.-J. (2021). Sensitivity control of dopamine detection by conducting poly (thionine). *Electrochemistry Communications*, 125, 107005.

Ramanavičius, A., Kaušaitė, A., & Ramanavičienė, A. (2005). Polypyrrole-coated glucose oxidase nanoparticles for biosensor design. *Sensors and Actuators B: Chemical*, 111, 532–539.

Ratlam, C., Phanichphant, S., & Sriwichai, S. (2020). Development of dopamine biosensor based on polyaniline/carbon quantum dots composite. *Journal of Polymer Research*, 27(7), 1–12.

Shahabuddin, S., Mazlan, N. A., Baharin, S. N. A., & Sambasevam, K. P. (Eds.) (2021). Introduction to conducting polymers. In *Advances in Hybrid Conducting Polymer Technology* (pp. 1–18): Springer, Berlin, Heidelberg.

Shoaie, N., Forouzandeh, M., & Omidfar, K. (2018). Voltammetric determination of the Escherichia coli DNA using a screen-printed carbon electrode modified with polyaniline and gold nanoparticles. *Microchimica Acta*, 185(4), 1–9.

Shu, H., Cao, L., Chang, G., He, H., Zhang, Y., & He, Y. (2014). Direct electrodeposition of gold nanostructures onto glassy carbon electrodes for non-enzymatic detection of glucose. *Electrochimica Acta*, 132, 524–532.

Song, Z., Ma, Y., Chen, M., Ambrosi, A., Ding, C., & Luo, X. (2021). Electrochemical biosensor with enhanced antifouling capability for COVID-19 nucleic acid detection in complex biological media. *Analytical Chemistry*, 93(14), 5963–5971.

Song, Z., Sheng, G., Cui, Y., Li, M., Song, Z., Ding, C., & Luo, X. (2019). Low fouling electrochemical sensing in complex biological media by using the ionic liquid-doped conducting polymer PEDOT: Application to voltammetric determination of dopamine. *Microchimica Acta*, 186(4), 1–9.

Tamba, S., Fuji, K., Meguro, H., Okamoto, S., Tendo, T., Komobuchi, R., & Mori, A. (2013). Synthesis of high-molecular-weight head-to-tail-type poly (3-substituted-thiophene) s by cross-coupling polycondensation with [CpNiCl (NHC)] as a catalyst. *Chemistry Letters*, 42(3), 281–283.

Tat'yana, V. V., & Efimov, O. N. (1997). Polypyrrole: A conducting polymer; its synthesis, properties and applications. *Russian Chemical Reviews*, 66(5), 443.

Tran, L. T., Tran, H. V., Tran, T., Nguyen, N. T., Van Bui, D., Tran, P. Q., & Van Chu, T. (2021). A highly sensitive electrochemical DNA sensor based on nanostructured electrode of multi-walled carbon nanotubes/manganese dioxide nano-flowers-like/polyaniline nanowires nanocomposite. *Journal of the Electrochemical Society*, 168(5), 057518.

Tseng, R. J., Huang, J., Ouyang, J., Kaner, R. B., & Yang, Y. (2005). Polyaniline nanofiber/gold nanoparticle nonvolatile memory. *Nano Letters*, 5(6), 1077–1080.

Üğe, A., Zeybek, D. K., & Zeybek, B. (2018). An electrochemical sensor for sensitive detection of dopamine based on MWCNTs/CeO_2-PEDOT composite. *Journal of Electroanalytical Chemistry*, 813, 134–142.

Wang, H.-C., & Lee, A.-R. (2015). Recent developments in blood glucose sensors. *Journal of Food and Drug Analysis*, 23(2), 191–200.

Wang, J. (2001). Glucose biosensors: 40 years of advances and challenges. *Electroanalysis: An International Journal Devoted to Fundamental and Practical Aspects of Electroanalysis*, 13(12), 983–988.

Wang, J. (2006). *Analytical Electrochemistry*, Third edition. John Wiley & Sons, Inc., Hoboken, NJ.

Yang, T., Xu, C., Liu, C., Ye, Y., Sun, Z., Wang, B., & Luo, Z. (2022). Conductive polymer hydrogels cross-linked by electrostatic interaction with PEDOT: PSS dopant for bioelectronics application. *Chemical Engineering Journal*, 429, 132430.

Zahed, M. A., Barman, S. C., Das, P. S., Sharifuzzaman, M., Yoon, H. S., Yoon, S. H., & Park, J. Y. (2020). Highly flexible and conductive poly (3, 4-ethylene dioxythiophene)-poly (styrene sulfonate) anchored 3-dimensional porous graphene network-based electrochemical biosensor for glucose and pH detection in human perspiration. *Biosensors and Bioelectronics*, 160, 112220.

Zarrintaj, P., Bakhshandeh, B., Saeb, M. R., Sefat, F., Rezaeian, I., Ganjali, M. R., & Mozafari, M. (2018). Oligoaniline-based conductive biomaterials for tissue engineering. *Acta Biomaterialia*, 72, 16–34.

Zhang, X., Lee, J.-S., Lee, G. S., Cha, D.-K., Kim, M. J., Yang, D. J., & Manohar, S. K. (2006). Chemical synthesis of PEDOT nanotubes. *Macromolecules*, 39(2), 470–472.

Zhang, X., Yu, Y., Shen, J., Qi, W., & Wang, H. (2020). Design of organic/inorganic nanocomposites for ultrasensitive electrochemical detection of a cancer biomarker protein. *Talanta*, 212, 120794.

9 Emerging Materials for Gas Sensing

Aemun Khalid, Muhammad Shakeel Ahmad, and Bushra Rashid

9.1 INTRODUCTION

The earth's atmosphere is composed of approximately nine types of gases. Gases such as methane are used for household cooking activities. Propane is used in lighters and also as a carrier gas in aerosols, shaving foams, etc. Industries utilize a plethora of gases that are utilized for the manufacturing of different products.

In metallurgical process production of iron, titanium, and vanadium, the quantity of dissolved gases such as oxygen, hydrogen, and nitrogen must be monitored to avoid deterioration of mechanical properties. Hydrogen absorbs into the molten metal causing hydrogen embrittlement; if the concentration of hydrogen is high, it can cause hydrogen blistering and hydride formation; both phenomena cause the deterioration of mechanical properties of metals, particularly steels. On the contrary, if the concentration of hydrogen is low, it will only cause embrittlement. Similarly, in the casting process, defects such as blow holes and porosity arise due to the absorption of gases in molten metal. Porosity can exist on the casting's surface or it might be trapped inside the metal, forming a pore that weakens the area around it. The main contributors to metals' gas porosity are nitrogen, oxygen, and hydrogen. The only gas that can significantly dissolve in molten aluminium and produce porosity is hydrogen. The holes for castings that weigh a few kilograms are typically between 0.02 and 0.6 mm in size. In larger casting, the pores can reach a size of up to 1.0 mm in diameter. Moreover, continuous monitoring of ammonia gas during the production of urea and monitoring of sulfur trioxide and sulfur dioxide gases during the production of sulfuric acid is of prime importance.

In medical science, measurement of blood oxygen level is the preliminary step for the treatment of diseases and surgeries. Carbon monoxide (CO) poisoning is another life-threating situation that impairs body and brain function; continuous monitoring of CO levels in blood is crucial throughout the treatment process.

Above are only a few uses of gases for commercial, industrial, and medical applications. It is important that gases must be measured and monitored. Gas sensors are pivotal to all medical, industrial, and medical applications. There are various types of sensors available for the detection of different types of gases.

Hydrogen gas is a promising candidate for alternative and clean energy. Hydrogen gas burns with a blue flame with an adiabatic flame temperature of 2045°C in the air. The burning reaction is governed by the following equation:

$$H_2 + \frac{1}{2}O_2 \rightarrow H_2O + 284.8\ \text{kJ mol}^{-1}$$

Besides, hydrogen also has promising physical properties when compared with other gases such as propane, methane, or gasoline vapors. These properties include low density, high coefficient of diffusion (0.61 cm^2s^{-1}) and buoyancy, a wide range of flammability (4%–75%), high heat of combustion (284.8 kJ mol^{-1} H_2), and low minimum ignition energy (0.017 mJ).

DOI: 10.1201/9781032654874-9

For a nuclear reactor to be safe, hydrogen concentration monitoring is crucial. Unwanted hydrogen gas can be produced during the reprocessing of plutonium, during radioiodine treatment of water, and by thermal breakdown of water molecules upon interaction with high-temperature reactor core and cladding materials (uranium oxide, zirconium). Both the Fukushima tragedy in 2011 and the Three Mile Island incident in 1979 were caused by hydrogen explosions. In coal mines, coal-dust explosions, spontaneous heating, or low-temperature coal oxidation can lead to the formation of hydrogen gas in ppm range. During the early stages of electrical transformer failure, hydrogen gas is produced; if presence of hydrogen is detected, then power failure can be mitigated. The semiconductor industry also utilizes hydrogen gas for various processes and monitoring and measurement of hydrogen is critical. In the lighting industry, hydrogen is a contaminant that must be quantified during the production of krypton, xenon, and neon [1]. Hydrogen gas molecules being small in size can leak through small cavities/cracks and their high diffusion coefficient aggrandizes their concentration; explosive limit will be attained in a short span of time.

9.2 HYDROGEN SENSORS

These facts stated in the introduction make it imperative that the quantity of hydrogen must be measured and monitored constantly. Hydrogen sensors are transducer devices; the presence of hydrogen causes changes in attributes such as physical, chemical, optical, and electrical of the sensing material in the sensor. This change in attribute/property can be converted into an electrical response and further correlated with hydrogen concentrations. Many types of hydrogen sensors are currently being used; the working principle of sensors is schematically depicted in Figure 9.1.

All of the sensors mentioned in Figure 9.1 will be discussed separately with special focus on the latest material being used in these sensors.

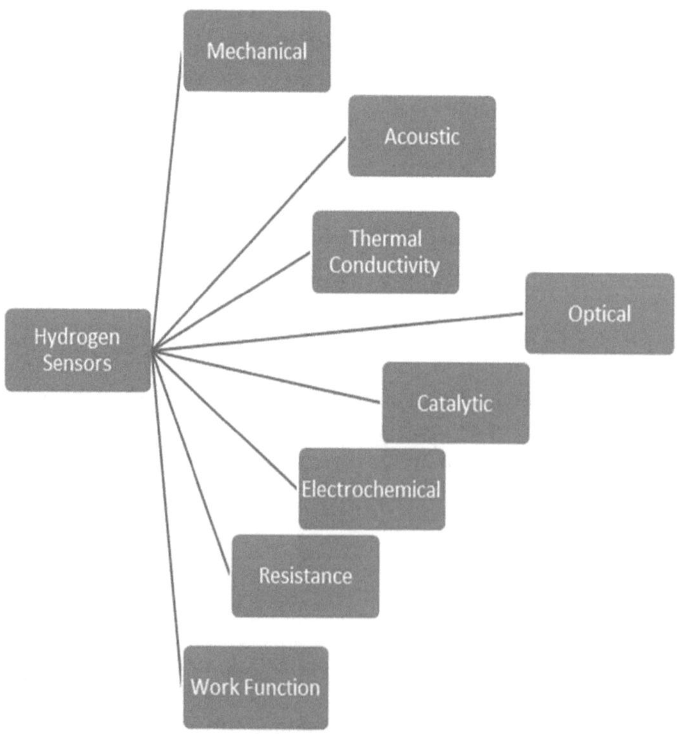

FIGURE 9.1 Types of hydrogen sensors.

9.2.1 Mechanical

Hydrogen gas is absorbed in certain metals at interstitial sites in the crystal lattice of metal causing lattice expansion and consequently changing the physical properties of metal. Platinum and palladium metals are most susceptible to hydrogen absorption in their crystal lattices. Micro electric mechanical systems (MEMS)-based microcantilever mechanical hydrogen is the most commonly used.

9.2.2 Description

Microfabricated cantilever beams consist of a substrate such as copper coated with palladium which utilizes expansion of palladium film followed by hydrogen absorption for detection. Henceforth, the presence and consequent absorption of hydrogen causes volume expansion of film; this expansion of film is hindered by the substrate coated with the film, and the stresses thus induced cause mechanical bending or curvature of the cantilever (Figure 9.2). The amount of mechanical bending/curvature can be appropriated to detect the percentage of hydrogen present in the environment. Optical, capacitive, and resistive transduction methods are used in case of the chemomechanical cantilever sensors.

The limiting factor in cantilever sensor is the adhesion of palladium film with the substrate. After repeated cycles of expansion/contraction defects such as peeling, delamination and even detachment of palladium film from the substrate can take place.

Li et al. [2] reported an innovative resistive sensor for hydrogen sensing. The sensor comprises a palladium-sputtered microcantilever and a composite piezoresistor of silver nanowires-polyimide. Detection range of 0.4%–2.0% with a correspondingly high sensitivity of 2825–47,083 was reported. However, the limitation of this type of sensor is the minimum percentage of hydrogen (0.4%) required for deflection and consequent resistance response.

Another extremely sensitive sensor consists of two opposing sets of carbon nanotube (CNT) arrays integrated between a micro-cantilever attached to an anchor electrode and a counter-electrode. On one side of the cantilever, a thin Pd coating for physical reaction/interaction with hydrogen has been applied. When the sensor is fabricated, only some percentage of the synthesized CNTs are electrically connected before the sensor is exposed to hydrogen, and other CNTs form a small gap. The length of the CNTs is not uniformly distributed and varies. The resistance of the

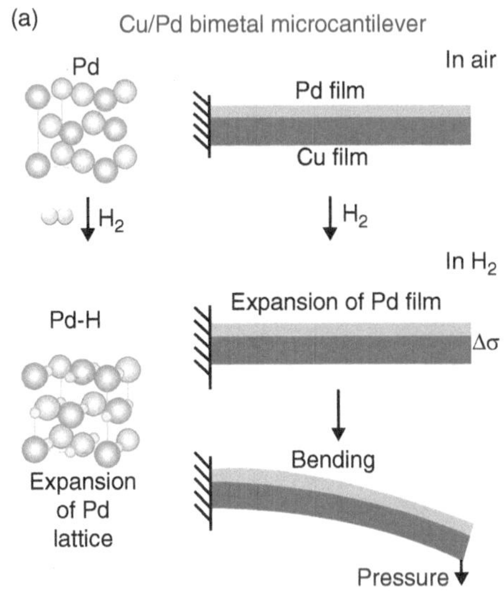

FIGURE 9.2 Cantilever bending due to hydrogen absorption in Pd film.

sensor at this state can be taken as a base resistance or **zero state**. Upon exposure to hydrogen gas, the cantilever beam bends, creating new connections between the CNTs, thus reducing the net resistance of sensor's electric circuit. The change in resistance can be quantified and related to concentration of hydrogen. Since some of the CNTs are already in connection in the zero state and there is some value of resistance that can be measured and quantified, this type of sensor reduces the minimum required concentration of hydrogen for measurable deflection in cantilever from 0.4% to 0.1% [2]. The author reported that the higher elasticity and wear resistance of CNTs makes the sensor more durable. Moreover, to prevent the delamination of thin Pd film from the cantilever, Cr was used as an interlayer. This type of sensor is more durable, has a long life, and is reliable [3].

9.2.3 Acoustic

Acoustic gas sensors function by detecting changes in the characteristics of acoustic waves caused by an absorbate, either in the bulk material or on the surface.

9.2.3.1 Surface Acoustic Wave Sensors

Surface acoustic wave sensors (SAW) commonly use piezoelectric crystals as their medium (such as quartz crystals, for example). An additional positive voltage produces sound waves, which propagate via a substrate before being converted into an electrical signal. The piezoelectric effect is the primary active effect of the SAW sensor. When designing SAW sensors, a number of considerations like relative size, sensitivity, and efficiency must be taken into account. A wireless passive surface wave sensor typically operates in the 40 MHz to several GHz signal frequency range. The general design of a sound surface wave sensor is shown in Figure 9.3, where the key components are a piezoelectric substrate, an antenna, a sensitive film, and an interdigital transducer (IDT). The sensitive layer of the sensor exhibits a frequency change by changing the speed of the acoustic surface wave.

The most commonly used substrate materials are $LiNbO_3$ and $LiTaO_3$. A hydrogen-sensitive film such as Pd, or Pt is coated between two IDTs as depicted in Figure 9.3. The sensitive film changes in mass density and electrical conductivity when it absorbs hydrogen, and this results in a detectable change in the SAW's frequency.

Langasite ($La_3Ga_5SiO_{14}$, LGS) was used as a substrate material, and indium oxide (In_2O_3, IO) and tin-doped indium oxide (ITO) were used as a hydrogen-sensitive film for monitoring of hydrogen at 350°C. In ITO, the dopant concentration was varied from 2% to 10% exhibiting an improvement in sensor sensitivity [4].

Pd/Cu nanowires [5] and Pd/Ni [6] nanowires have been reported as new SAW sensors exhibiting fast response and recovery. Graphene oxide (GO) has emerged as sensitive material; it has strong surface activity and a large surface-to-volume ratio. In addition, GO's capacity for gas adsorption is increased due to its multi-layer pore structure coupled with a variety of oxygen functional groups. The high sensitivity of GO hydrogen sensors has also been noted in numerous research.

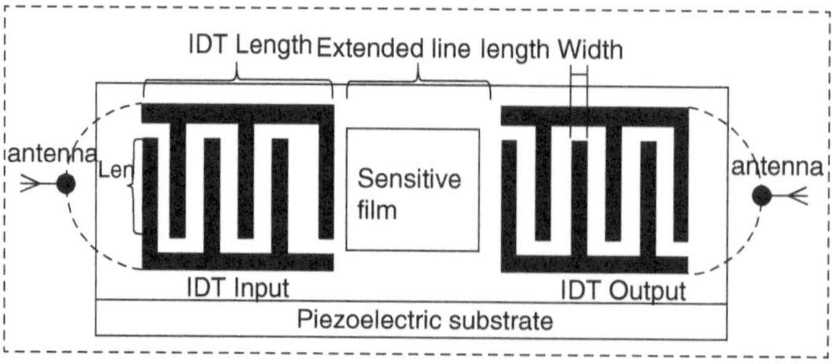

FIGURE 9.3 SAW sensor structure.

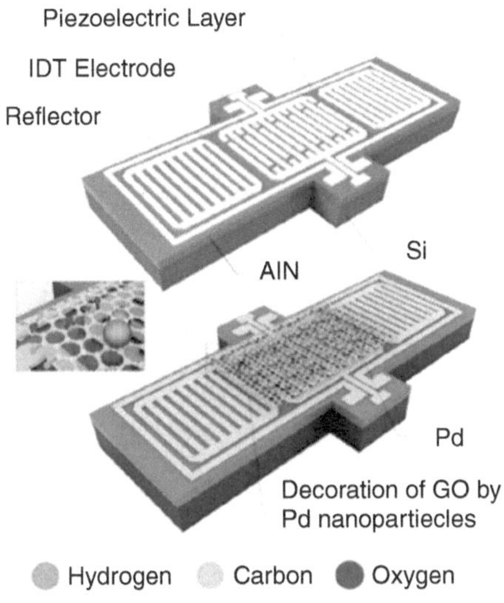

FIGURE 9.4 SAW resonator schematic with graphene oxide (GO) decorated by palladium nanoparticles.

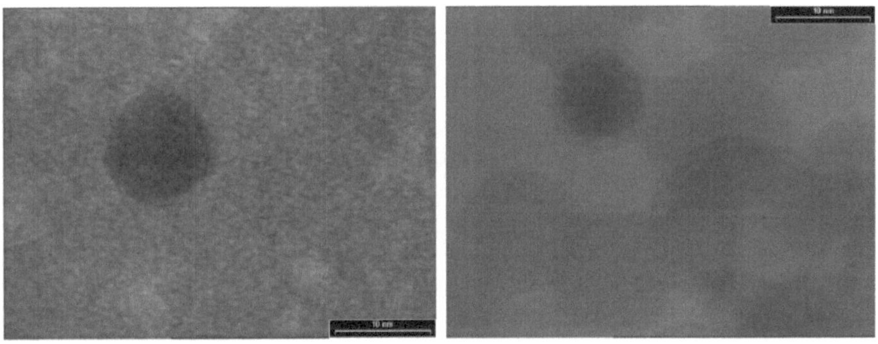

FIGURE 9.5 Images of transmission electron microscope Si/SiO_2 QDs.

The oxygenous functional groups in GO interact with a variety of different gas species, reducing the selectivity for gas sensing. GO itself is a good material to serve as a gas-detecting layer. In one study, palladium nanoparticles are distributed on GO surface Figure 9.4 [7]. This type of sensor exhibited better selectivity and reduced cross-sensitivity to other gases.

Quantum dots (QDs) dispersed in polymeric films have been used by Constantinoiu, Iz et al. [8]. Si/SiO_2 QD were produced by the laser ablation method. QDs were ultrasonically dispersed in poly(methyl methacrylate) (PMMA) (dissolved in anisole) and polydimethylsiloxane (PDMS) (dissolved in toluene). The solution was drop casted on a quartz substrate. The obtained sensor exhibited high sensitivity toward hydrogen gas (Figure 9.5).

9.2.4 Thermal Conductivity

Since hydrogen has a higher thermal conductivity than air (0.174 W m^{-1} K at 20°C), techniques based on thermal conductivity can be used to determine the amount of hydrogen in the air. Since thermal-conductivity-based hydrogen sensors don't require oxygen to function, they have a wide operating range (1%–100%), which is a major benefit. Conversely, detection of low hydrogen levels (<1%) is difficult using this technique.

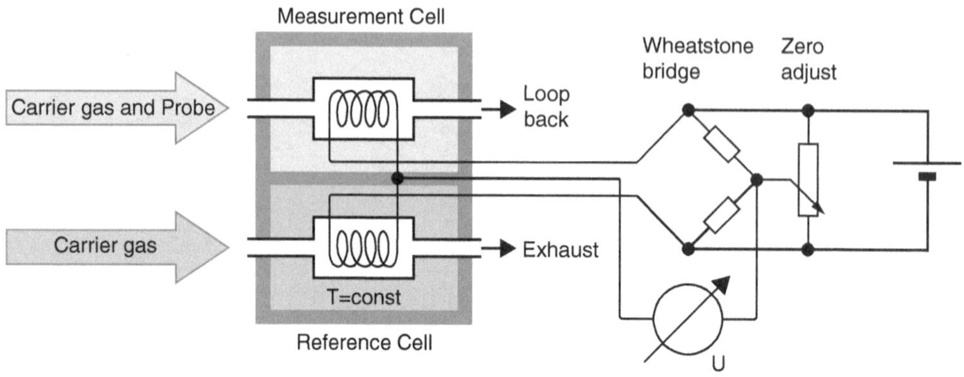

FIGURE 9.6 Thermal conductivity sensor connected in a Wheatstone bridge circuit.

The loss of heat from a hot substance to the surrounding gas governs the operation of thermal-conductivity-based sensors. There are two variations of this technique. The first variation consists of two inert resistor beads, one of which serves as a reference and the other as a detecting resistor. While the reference resistor is contained inside a chamber containing a reference gas (often air), the detecting resistor is exposed to the gas being measured. Figure 9.6 demonstrates the two beads connected in a Wheatstone bridge circuit. The rate of heat loss is identical upon exposure of resistors to the same gas, producing a zero reading. Depending on the analyte gas's relative thermal conductivity to the reference gas, the detecting resistor either loses less or more heat upon exposure to the analyte gas. This causes the temperature of the bead to either increase or decrease; thus, variation in resistance is produced; this resistance change creates an imbalance in Wheatstone bridge which can be correlated to the concentration of analyte gas.

The sensor's second type is a newer variation which does not need to employ a reference cell. The "hot" and "cold" elements have a constantly maintained preset temperature difference between them. Thermal conduction occurs through the analyte gas as heat is transmitted from "hot" to "cold" element. The amount of power needed to maintain "hot" element at a constant temperature is an intrinsic to analyte gas's thermal conductivity, which can afterward be linked to the analyte gas's concentration.

9.2.5 Optical

When some materials interact with hydrogen, their optical characteristics change, and this phenomenon can be used to detect hydrogen [1]. Among the many hydrogen-sensitive materials, palladium (Pd) is the preferred functional material. Pd films can be used in both hydrogen detection and separation devices, which are owing to their capacity to dissolve hydrogen effectively in ambient circumstances, as well as their reversible phase transition at room temperature from metal-to-metal hydride (PdH_x), which can result in considerable optical and mechanical property transformation. Normally, the hybrid Pd/PdH_x film achieves absorption equilibrium and no longer expands after a period of exposure to a specific concentration of hydrogen. Optical sensors for hydrogen detection have been noticed as a new technique because of their intrinsic safety, contrary to other techniques that have an electrical readout, thus posing a risk for an explosion due to sparking. Various methods and strategies are being utilized to improve the capabilities of optical sensors.

9.2.5.1 Description

Various detection techniques are applied to measure the change in optical properties of materials. Reflectivity, interferometry, surface plasmon resonance (SPR), evanescent field interaction, fiber Bragg grating, etc., are the most commonly used methods for the correlation of optical properties with a percentage of hydrogen. These techniques will be discussed briefly. However, the material currently being used will be discussed in detail.

Zhang et al. [9] have fabricated and discussed an optical sensor based on reflectivity measurement. The sensor was fabricated by cleaving the end of an optical fiber and a 30-nm Pd film was bonded using a UV-curable epoxy resin. When hydrogen is absorbed in the Pd film, it causes a lattice expansion due to which the reflectivity of Pd film is changed. This change was detected using Fabry–Pérot (FP) interferometer. The sensor exhibited a high sensitivity of –0.334 nm/1% concentration, a large response range (0.5%–3.5%), and a short response time (10 seconds)/recovery time (25 seconds).

Xu et al. fabricated a sensor with Pt-loaded WO_3/SiO_2 coated on polymer-filled hollow core fiber [10]. Fiber Bragg grating was employed by the sensor detection of hydrogen. The sensor exhibited a fast recovery time scale and high sensitivity in the range of 0%–4% of hydrogen gas concentration.

Hydrogen sensors based on metamaterial have also been produced. Figure 9.7a shows the construction of the sensor, the gold nanorods are encompassed in a palladium shell which is known as gold/palladium core/shell nanorod metamaterial. This type of structure exhibits such a robust response to hydrogen exposure, that the resultant modification of light transmission characteristics can easily be observed by the naked eye. Moreover, this type of sensor structure can be produced at macroscopic levels for the detection of hydrogen gas through naked eye. Exposure to 2% hydrogen produces more than 30% reflection and transmission from the metamaterial layer (Figure 9.7d). This high sensitivity is an indicator of the hyperbolic dispersion of the nanorod metamaterial and is eloquent of the physical (size) and composition (permittivity) change of the palladium shell upon exposure to hydrogen; optical properties of both the individual nanorod components and the inter-nanorod near-field interactions are changed massively; thus, this type of sensor can be used to identify explosive levels of hydrogen with the naked eye [11].

Hydrogen detection has also been demonstrated by localized surface plasmon resonance (LSPR). Gold nanoparticles (GNPs) were prepared by pulsed laser ablation (PLA). $PdCl_2$ solution was added

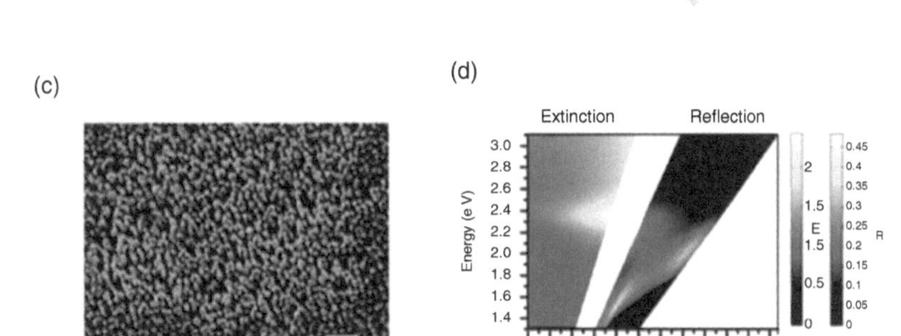

FIGURE 9.7 (a) Illustration of Au (core) Pd (shell) nanorod array structure. (b) Schematic for measurement of optical reflection and optical transmission. (c) Scanning electron microscope (SEM) image of synthesized Au-Pd nanorod array. (d) Experimental dispersion of extinction (for the angles below total internal reflection) and reflection.

to gold nanoparticles. The Pd^{+2} ions in the solution surround the gold particles leading to the formation of core-shell structure. Figure 9.8 shows X-ray diffraction (XRD) peaks of only the Au sample. The (111), (200), (220), (311) diffraction peaks are characteristic of cubic structure of Au crystallographic planes/orientations. The XRD peaks verify that GNPs are formed successfully by the laser ablation process. As stated earlier, the addition of $PdCl_2$ solution generates Pd^{+2} ions in the solution, and these ions surround GNPs forming a core-shell structure; the same can be seen in XRD pattern of AuPd solution. The strong diffraction peaks (101), (110), and (520) show the crystallographic planes of rhombohedral $PdCl_2$, whereas diffraction peaks associated with Au (111) and (220) are very weak exhibiting that palladium has surrounded GNPs effectively and core-shell structure is formed successfully [12].

These core-shell structures change their LSPR when exposed to hydrogen gas. The optical spectra change in presence of hydrogen by disappearance of absorption peak corresponding to $PdCl_2$, and reappearance of gold LSPR peak in terms of shape and position Figure 9.9.

These findings imply that following hydrogen exposure, $PdCl_2$ is chemically reduced to metallic Pd. This causes a change in the refractive index of the GNPs' surroundings, which results in shifting and variation of LSPR spectrum.

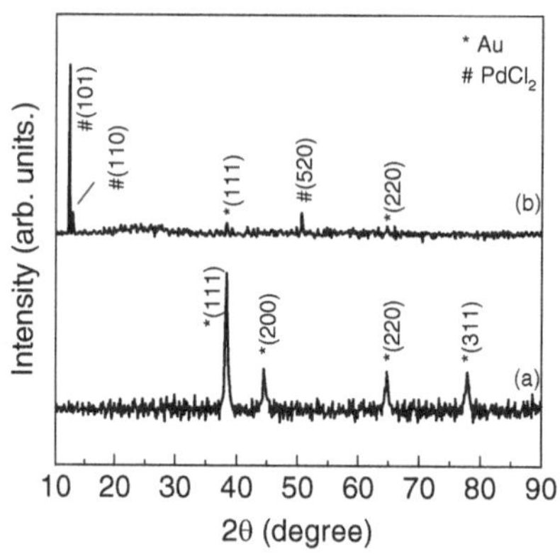

FIGURE 9.8 XRD patterns of (a) pristine GNPs (sample Au) and (b) sample Au1Pd2.

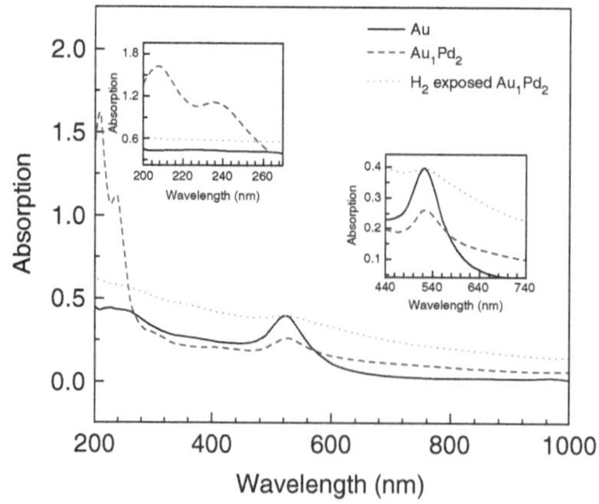

FIGURE 9.9 Spectra of optical absorption for pristine GNPs (sample Au), sample Au1Pd2 before and after 10% H_2 exposure.

9.2.6 Catalytic

As stated in the introduction section that hydrogen is a combustible gas having combustion energy of 142 kJ g^{-1}, exothermic oxidation of hydrogen generates a substantial amount of heat, which can be measured as a change in temperature and correlated with the percentage of hydrogen. This is the basis for pellistor-type sensors, which are frequently used for detection of gases. Two platinum coils that are coated in ceramic beads are used in pellistor-type sensors. The term "active bead" refers to a bead on which a hydrogen-sensitive material (such as Pd, etc.) has been coated, and the term "inactive bead" is used for the bead used as a reference. To measure the resistance imbalance within the materials, the active and inactive beads are both connected to a Wheatstone bridge. During the gas sensing procedure, an electrical current is used to heat the ceramic beads. The catalyst (Pd/Pt, etc.) can adsorb hydrogen molecules on its surface due to the high temperature. The oxygen that has already been adsorbed reacts with the hydrogen gas molecules to produce water molecules exothermically. The active bead's temperature rises as a result of this exothermic process, which also increases resistance. The imbalance caused by the active bead's change in resistance can be measured, quantified, and correlated to the amount of hydrogen gas in the atmosphere under observation.

The second type of catalytic sensors known as "thermoelectric sensors" are fabricated on the principle of Seebeck effect. The Seebeck effect is the development of a voltage difference between two locations of a conducting material when the temperature of those two sites changes (Schematic in Figure 9.10). When heat is produced by the exothermic oxidation of the hydrogen molecules at the active point (catalyst surface) causing temperature increase, the Seebeck effect can be used for hydrogen sensing applications. The increase in temperature creates voltage change governed by the Seebeck effect, indicating the presence of hydrogen gas. The variation in voltage is related to the variation in the concentration of hydrogen. The voltage signal can be described by the following equation:

$$E = s.\Delta T$$

In this equation, "s" is the Seebeck coefficient and ΔT is the temperature difference generated between the active point and reference point as a result of exothermic oxidation of hydrogen.

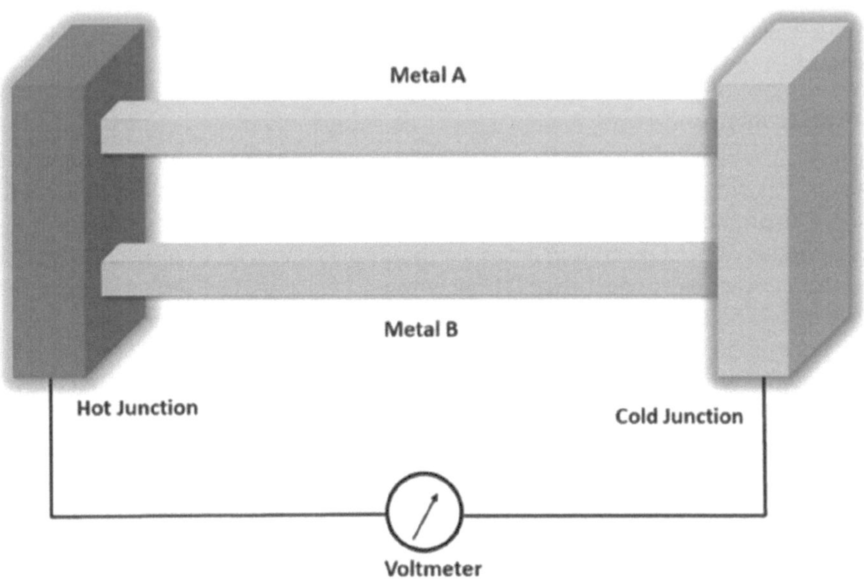

FIGURE 9.10 Schematic representation of the Seebeck effect.

Hydrogen-sensitive material is coated on active point. Metals such as Pd and Pt have high adsorption of hydrogen and serve as ideal candidates for coatings at active ends. Furthermore, employment of nanotechnology, such as the use of platinum nanoparticles (Brauns et al. [13]) as a catalyst layer for the detection of hydrogen gas, the adsorption and surface activity can be further improved. Due to its small size, this sensor has a high Seebeck coefficient. A fast response time of less than 150 ms was demonstrated by the sensor and a high response signal of 0.22 mV at a 10 ppm H_2 gas concentration level.

The inherent drawbacks of catalytic sensors prevent them from being used as reliable sensors. The selectivity of hydrogen gas is seriously hampered by the catalyst materials' sensitivity to other combustible gases. A significant drawback of contemporary sensor technology is the high amount of electricity needed to heat and cool the beads, which inhibits the ability of the sensing devices to be deployed in the field.

Pd/titanate nanotubes and Pt/titanate nanotubes have also been deployed as catalytic hydrogen sensors [14]. Compared to sensors made with TiO_2 nanoparticles or titanate nanotubes, this one had a better response. The Pd/titanate and Pt/titanate nanotube sensors' increased sensitivity may be the result of absorbed oxygen and hydrogen reacting more quickly than usual. Since titanate nanotubes have evenly dispersed Pd and Pt catalysts at a nanoscale level, the direct absorption of hydrogen on titanate nanotube further facilitates the oxidation reaction of hydrogen.

A catalytic sensor with TiO_2 and UV light-emitting diode (LED) has also been fabricated [15]. The operational temperature for the sensor was 82°C which is quite lower than previously developed sensors. This can be attributed to the direct adsorption of hydrogen on the surface coupled with enhanced activity of adsorbed oxygen which changes to superoxide radical anions facilitating the oxidation of hydrogen at a lower temperature.

Catalytic hydrogen sensors have also been fabricated using microheated platinum nanoparticle-loaded graphene aerogel [16].

9.2.7 Electrochemical

A profound knowledge of electrochemistry is imperative for understanding electrochemical sensors. Electrochemical reactions occurring at the sensing electrode cause changes in electrical properties. This change in electrical properties is detected by electrochemical sensor, whereby the magnitude of electrical change translated into the amount of sensed gas in the atmosphere. The classification of electrochemical sensors based on their working principle is amperometric, potentiometric, conductometric.

9.2.7.1 Amperometric

The three major components of an amperometric sensor are a sensing electrode, a counter electrode, and a reference electrode. An external voltage is applied to drive the reaction in one direction; potentiostat is used to maintain a constant voltage. The sensing and counter electrodes are prone to surface reactions causing them to polarize and limiting the amount of measurable reactant gas. Therefore, the reference electrode, shielded from unwanted reactions and surface activity, maintains the thermodynamic potential of the working electrode constant.

The electrodes used in amperometric electrochemical sensors are generally composed of noble metals such as gold, silver, and platinum. Platinum is often because it is a noble metal as well as a catalyst for the oxidation of hydrogen. Hydrogen gas is diffused through the inlet; oxidation takes place at the sensing electrode and its potential is changed. Oxygen is reduced chemically at the counter electrode. A flow of electrons is generated from anode to the cathode which is detected in the form of an electric current. This electric current is directly proportional to hydrogen gas concentration. Electrolyte is also elementary for completion of an electrochemical reaction; the oxidation of hydrogen produces electrons and H^+ ions (proton). The electrons travel to the cathode and are detected as electrical change; the protons must travel to the anode for the completion of the electrochemical reaction. This purpose of proton travel is served by a material known as an

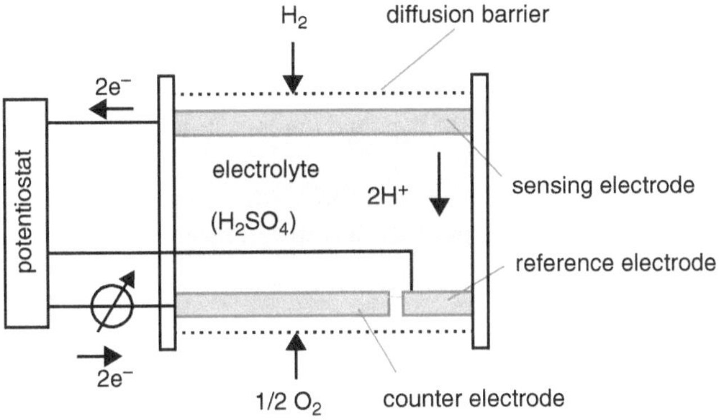

FIGURE 9.11 Schematic of 3-electrode amperometric sensor.

electrolyte, which is a proton-conducting material; sulfuric acid (H_2SO_4) is most commonly used as electrolyte. However, solid electrolytes such as Nafion®, ceramics, etc., are also being used; the solid electrolytes eliminate the problem of leakage, corrosion, volatilization associated with liquid electrolytes (Figure 9.11).

Amperometric sensors with operating temperature range as low as –20°C have been reported. However, the electrolyte for such a low temperature is selected such that it does not freeze at the operating temperature. Amperometric sensors can also be designed to operate at high temperatures >80°C.; Again, the physical properties of electrolyte must be taken into consideration so that it doesn't change its phase, physical properties or evaporates. Ceramic electrolytes can be used for operation at higher temperatures [1]. Variables such as temperature, pressure, and humidity adversely affect the sensor performance. Amperometric sensors can detect hydrogen concentrations of 5 ppm [17] up to 100% [18]. Since amperometric sensors require direct oxidation of target gas at sensing electrode, these sensors exhibit cross-sensitivity to various species such as hydrocarbon which when oxidized at the sensing electrode generates a false signal. Another disadvantage of amperometric sensors is their susceptibility to poisoning by CO and CO_2.

Amperometric hydrogen sensors based on Pt nanoparticles supporting multiwalled carbon nanotubes have been fabricated exhibiting high sensitivity, reproducibility, and long-term stability [19].

Perovskite crystallographic structure has excellent chemical and physical stability; furthermore, perovskite ceramics have a high proton conductivity, which makes them ideal for use as an electrolyte in amperometric and potentiometric sensor. Perovskite materials such as $BaCe_{0.6}Zr_{0.3}Y_{0.1}O_{3-\alpha}$ and $Sr(Ce_{0.9}Zr_{0.1})_{0.95}Yb_{0.05}O_{3-\alpha}$ [20] have been reported as an electrolyte with high sensitivity and response at 600°C.

9.2.7.2 Potentiometric

Potentiometric sensors have an electrolyte and two electrodes, identical to a two-electrode amperometric sensor. Potentiometric sensors operate at zero current, and the measured quantity is the potential difference or electromotive force caused by various hydrogen gas concentrations between the sensing electrode and the reference electrode. The hydrogen content and the potential difference between the electrodes are related by the Nernst equation given below.

$$E = \frac{RT}{2F} \ln \ln \frac{p_{H_2}^{sens}}{p_{H_2}^{ref}}$$

E: Potential difference in volts (V)
R: Gas constant 8.314 J mol^{-1} K^{-1}
T: Absolute temperature in Kelvin (K)
F: Faraday's constant 9.648×10^4 C mol^{-1}

$p_{H_2}^{sens}$: sensing electrode hydrogen pressure
$p_{H_2}^{ref}$: reference electrode hydrogen pressure

The diffusion and chemical reduction of hydrogen must achieve an equilibrium state to obtain an accurate reading.

SnO_2-ordered films were investigated for potentiometric hydrogen sensors [21]. SnO_2 films of varying pore sizes are discussed for use in potentiometric sensors. High sensitivity and selectivity were achieved.

9.2.8 Conductometric

Conductometric sensors measure and quantify the changes in electrical conductivity due to the production of H^+ ions at substrates causing changes in electrical properties.

Wu, Junbiao et al. [22] designed a conductometric sensor comprising of $NaTiO_3$ (NTO) nanoribbons decorated with Pd nanoparticles (Pd NPs) (3.5 nm). Ti_3C_2 MXene was used as a precursor in the preparation of $NaTiO_3$ nanoribbons; these Pd NPs were dispersed uniformly and served as active sites for hydrogen absorption and dissociation. The laterally paralleled structure of $NaTiO_3$ nanoribbons has abundant oxygen vacancies. These vacancies facilitate the formation of O_2^-. According to theory, the reaction between dissociated hydrogen atom and O_2^- will only occur as a result of an effective collision. Abundant oxygen vacancies of the NRs coupled with rapid dissociation of H atoms, the as-prepared Pd-NTO NRs have a fast response of 1.1 seconds to 1% H_2, which is faster than many advanced and sophisticated sensors. The sensor also has a wide detection range (0.8 ppm-10% H_2), ideal for use in hydrogen safety.

9.2.9 Resistance

9.2.9.1 Semiconducting Metal Oxide

When exposed to reducing gases, metal oxides show a change in their electrical characteristics. A change in the sensor's conductance (or resistance) is the typical mechanism for metal oxide gas sensors. As the gas concentration increases, the resistance changes reversibly as a function of gas concentration. The gas is entirely or partially oxidized by the detecting substance. The surface of sensor absorbs the gas molecules, and subsequently, gas molecules decompose or dissociate at sensor surface. Molecules of reducing gases such as CO and H_2 when absorb on the sensor's surface inject electrons into the sensing surface. Oxidizing gases like NO exhibit the opposite phenomenon i.e., these gases take electrons from the surface of the sensor. When a reducing gas comes in contact with the sensor, it reduces the resistance in the case of n-type semiconductors, whereas when an oxidizing gas increases the resistance of p-type semiconductors. The sensitivity of sensors is calculated as the ratio of conductivity in the target gas to conductivity in air. The sensitivity is not significantly changed until the attainment of a critical gas concentration and the conductivity varies greatly [23]. The functioning portion of sensors typically need to be heated. Resistive gas sensors are typically made of homogenous semiconducting oxides that can operate between 300°C and 750°C. In particular, in this temperature range, the concentration of oxygen vacancies in the bulk is in thermodynamic equilibrium with the outside oxygen pressure.

The metal-oxide film is applied on an insulating substrate material such as alumina (Al_2O_3) in between two electrodes Figure 9.12. Alumina being a ceramic material can withstand temperatures up to 1650°C; moreover, alumina has a porous morphology resulting in effective adhesion of metal oxides. The sensor is often heated by using a heating element embedded within the sensor to promote a reaction with hydrogen and removal of water by-products formed during the reaction. Semiconducting oxides, including In_2O_3, ZnO, NiO, TiO_2, NiO, FeO, SnO_2, Ga_2O_3, Sb_2O_5 Fe_2O_3, WO_3, and MoO_3, show a large change in their electrical resistance following adsorption of hydrogen gas [1].

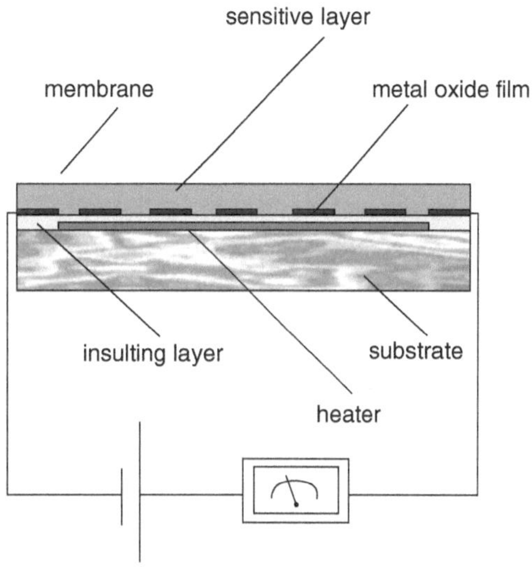

FIGURE 9.12 Schematic of metal-oxide sensor.

Reducing gases such as carbon monoxide, nitrous oxide, etc., and hydrogen-containing compounds such as alcohol, methane, and water molecules present in the environment as humidity reduce chemically and generate electrical signals. The signal thus generated leads to incorrect detection of analyte (hydrogen) gas. The stated problem can be overcome by doping metal oxides with hydrogen-sensitive catalytic metals such Pt, Au, Ag, Cu, and Pd.

9.2.9.2 Metallic Resistor

Some metals and alloys exhibit a significant change in electrical resistivity after hydrogen is absorbed. Pd is preferred for metallic resistor sensors because of extensive and selective solubility of hydrogen. Palladium hydride is formed after absorption of hydrogen. The electrical resistance of palladium hydride relative to pure palladium is high; the change in resistance can be detected and correlated with the percentage of hydrogen gas. Typically, a metal film is deposited between two electrical contacts on a substrate (like silicon) to create metallic resistor sensors. Film deposition methods such as vacuum sputtering, evaporation, electrodeposition, sputtering, and pulsed laser deposition can be used for the preparation of sensor.

Metallic resistor sensor having an arrangement of four palladium resistors in a Wheatstone bridge was fabricated. Two of the four resistors were coated with a passivation layer to account for the resistance change of palladium to increase in environment temperature. This type of sensor can operate in the temperature range of 0°C–200°C and detects hydrogen in the range of 0.5%–30%. Furthermore, the cost of this type of sensor is very low. A disadvantage of this sensor is that it is susceptible to poisoning by gases like CO, SO_2. Morphology of palladium film which is a function of film fabrication methods and film fabrication conditions plays a vital role in the sensor's response, response time, and resistance to CO poisoning.

Various materials and material shapes such as Pd-Ni thin films [24], thin palladium-coated Mg–Ni alloy film [25], rare earth metals such as samarium [26] and yttrium [27] also show change in electrical resistivity when exposed to hydrogen. These rare earth metals are often covered with a thin film of Pd to promote the dissociation of hydrogen and protect rare earth metals from oxidation. Reports indicate that discontinuous ultrathin palladium films expand creating new connections thereby reducing net electrical resistance followed by hydrogen absorptions [28]. A similar phenomenon has also been demonstrated with palladium mesowire structure [29].

9.2.10 Work Function

The minimum amount of energy required to remove an electron from a solid to infinity is known as work function. Work hydrogen detection using work function can be carried out by a three-layered structure.

- *Layer 1*: Hydrogen-sensitive metal (Pd, Pt, Au, Ag, Cu, etc.),
- *Layer 2*: Oxide (Al_2O_3, SiO_2, TiO_2, Nb_2O_5, etc.), and
- *Layer 3*: Semiconductor.

These three layers—layer 1 (metal), layer 2 (oxide), and layer 3 (semiconductor)—are stacked together in the same order; this structure is knows as metal oxide semiconductor (MOS). The hydrogen molecule absorbs onto the metal layer and diffuses until it reaches the interface of metal and oxide; since the molecules cannot further diffuse into the oxide layer, they start to accumulate at the interface between the metal and oxide layer. These accumulated hydrogen atoms polarize and form a dipole layer. The work function of the metal changes as a result of this dipole layer, and shift in energy levels at the metal–insulator interface is observed. As a result, a net voltage change ΔV is produced; this change is added to an externally applied voltage and can be measured providing the percentage of hydrogen gas in analyte gas.

The classification of work-function-based hydrogen sensor is as follows (Figure 9.13):

- Metal semiconductor (Schottky diode),
- MOS capacitor, and
- MOS field effect transistor (MOSFET).

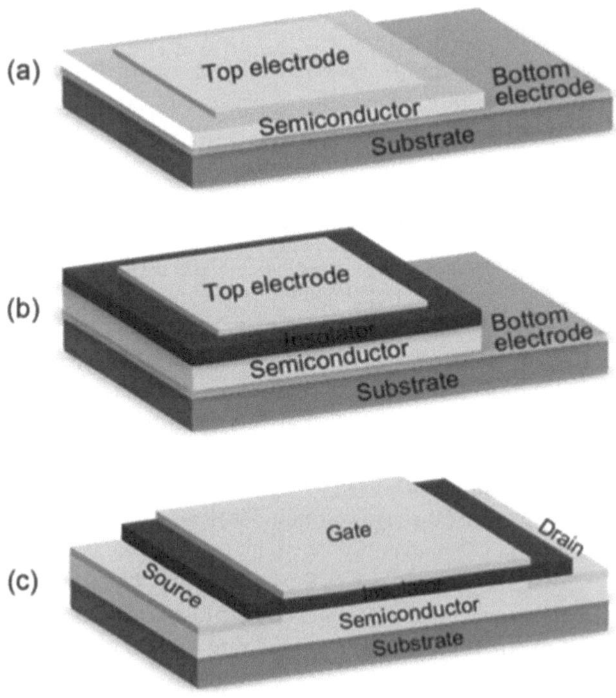

FIGURE 9.13 (a) Schematic of Schottky diode type. (b) Schematic of MOS capacitor type. (c) Schematic of MOSFET type.

9.2.10.1 Metal Semiconductor Schottky Diode

The typical construction of a Schottky diode sensor is a metal in contact with a semiconductor material. This type of sensor structure is known as metal semiconductor (MS); another variant of this type of structure is that a very thin insulating layer (generally an oxide) is placed between the metal and semiconductor. This structure is known as MOS sensor structure.

In the case of MS sensor, when a metal comes in contact with the semiconductor, the fermi level of electrons of the semiconductor is instinctively aligned and adjusted with the "fermi level" of the dominating metal. The amount by which fermi level is adjusted is equal to the difference between the work function of metal and the semiconductor. The work function is known as the Schottky barrier height. In the case of the MOS sensor, a thin insulating layer e.g., SiO_2 is placed between the metal and semiconductor. Despite the presence of an insulating layer, fermi-level adjustment will still take place.

In Schottky sensors, Pd or Pt is frequently utilized; hydrogen molecules are absorbed on the metal's surface and split into hydrogen atoms. These atoms diffuse into the metal and gather at the metal–insulator interface where they produce a dipole layer that changes the work function of the metal and changes the Schottky barrier height. When the diode is operated at constant current bias, the change in Schottky barrier height can be monitored as a current-voltage (I-V) response.

A Schottky diode based on β-Ga_2O_3-2D-flake-based Pt as shown in Figure 9.14 has been investigated [30]. The (100) plane β-Ga_2O_3 flake was formed from the side wall of a mother (201) plan by using mechanical exfoliation method Figure 9.15.

500 ppm hydrogen gas reduces the Schottky barrier height of (100) plane β-Ga_2O_3 from 1.57 to 1.37 eV. To determine the hydrogen selectivity, the current response of the β-Ga_2O_3-flake-based Schottky diode sensor to various gases was examined. Figure 9.16 shows the response of the sensor as current (A) when subjected to 100% N_2, 0.1% CO, 10% CO_2, 4% CH_4, 0.05% NO_2, NH_3 2 ppm, and 100% O_2.

A novel MOS Schottky diode consisting of Pd, NiO, and GaN has been fabricated (Figure 9.17). The simple structure of the device coupled with high sensing speed and good sensing response was demonstrated [31].

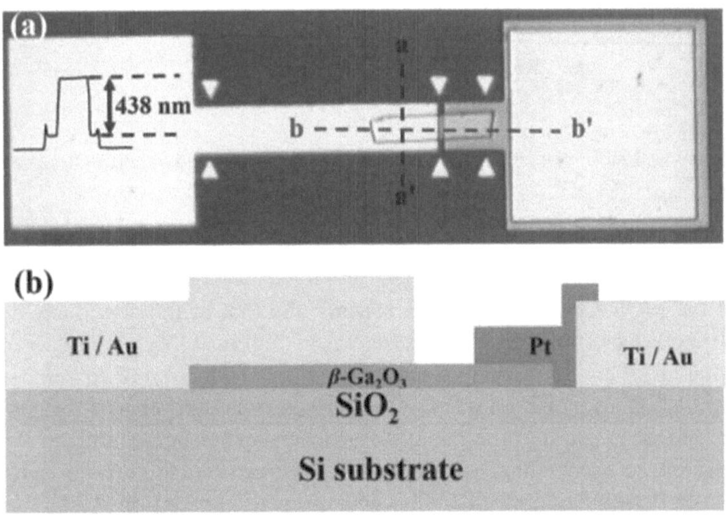

FIGURE 9.14 (a) Microscopic image of the (100) plane β-Ga_2O_3-flake-based Schottky diode. a-a′ line indicates the thickness profile of the surface profiler. (b) Cross-sectional schematic image of the sensor along b-b′ in Figure 9.14a.

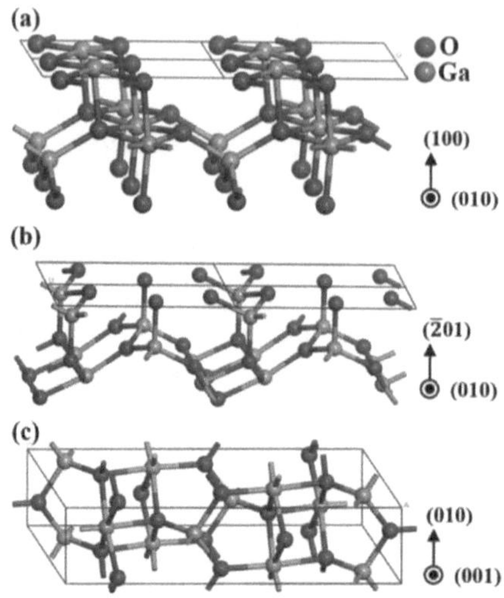

FIGURE 9.15 Schematic atomic configurations of (a) (100) plane, (b) (201), and (c) (010) plane β-Ga_2O_3. The red and gray balls denote oxygen and gallium atoms, respectively.

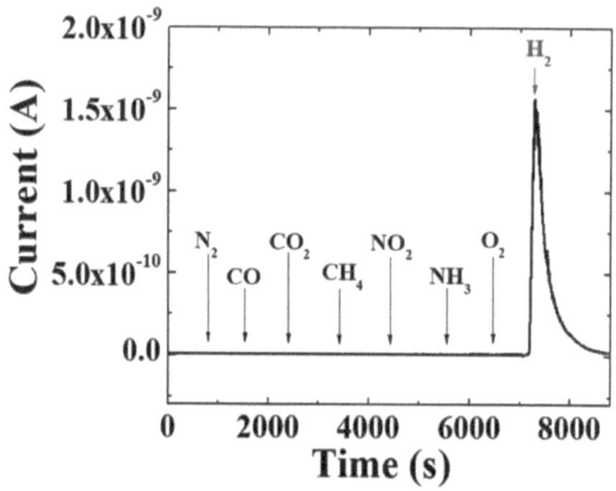

FIGURE 9.16 Response in terms of current (a) of the β-Ga_2O_3-flake-based Schottky diode to 100% N_2, 0.1% CO, 10% CO_2, 4% CH_4, 0.05% NO_2, 2 ppm NH_3, 100% O_2, and 500 ppm H_2.

9.2.10.2 MOS Capacitor

The structure of the MOS capacitor hydrogen sensors and that of the previously described MOS Schottky diode-type sensors are identical, as shown in Figure 9.13b, a thick insulating layer is placed between the metal and semiconductor to prevent the conduction of current and the creation of a charge-accumulation layer on both sides. When the sensor is exposed to hydrogen gas, hydrogen atoms dissociate and permeate into the metal/insulator causing a lateral shift in the capacitance-voltage (C-V) and conductance-voltage (G-V) plots that corresponds to the hydrogen concentration.

The performance optimization of an MOS capacitor sensor based on Pd/TiO_2/Si/Al structure was investigated by Ratan et al. [32]. The construction of this sensor is depicted in Figure 9.18; the sensor is fabricated on a p-type silicon substrate; nanostructured TiO_2 was fabricated using thermal evaporation. Hydrogen molecules diffuse through Pd thin film and interact with TiO_2; this causes a change in work function of Pd, and consequently, the capacitance and conductance of the sensor are changed. The sensor can detect H_2 gas room temperature.

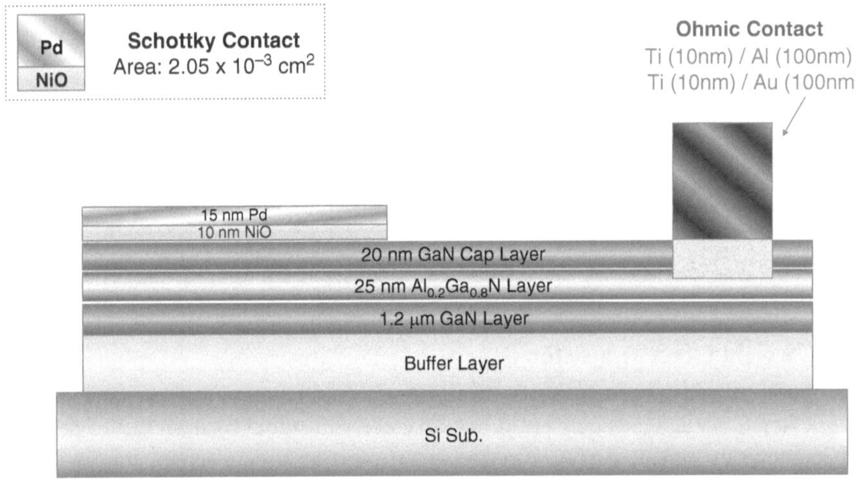

FIGURE 9.17 Representative cross-section diagram of the studied Pd/NiO/GaN MOS Schottky diode hydrogen sensor.

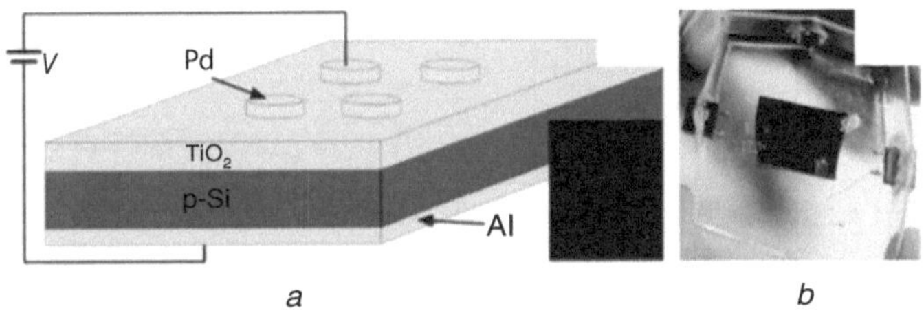

FIGURE 9.18 (a) Structure of fabricated MOS sensor and the camera image of the top surface of final film. (b) Camera image of fabricated MOS devices on the p-Si substrate packaged in a plastic case.

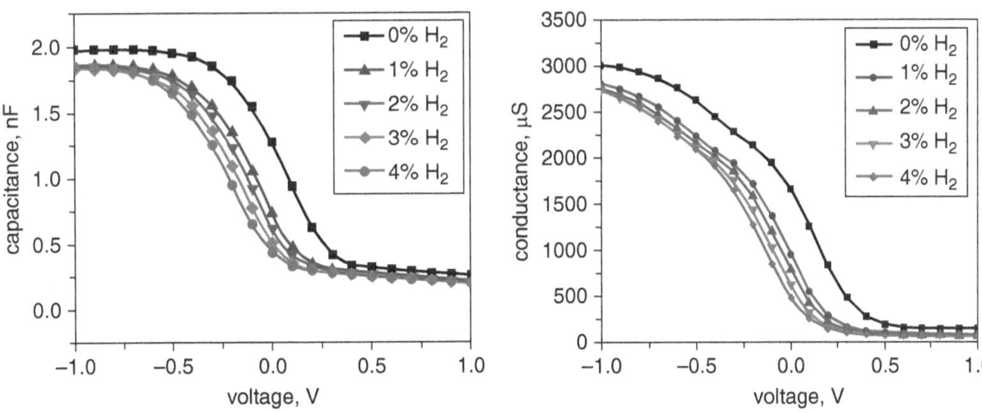

FIGURE 9.19 C–V characteristics at 1-MHz frequency for different concentrations of H_2.

Figure 9.19a and b, respectively, indicate the change in capacitance and conductance for various H_2 concentrations. Through the Pd thin layer, the exposed hydrogen gas diffuses and interacts with the TiO_2. The variation in the work function of Pd after exposure to H_2 is the primary cause of the change in capacitance and conductance.

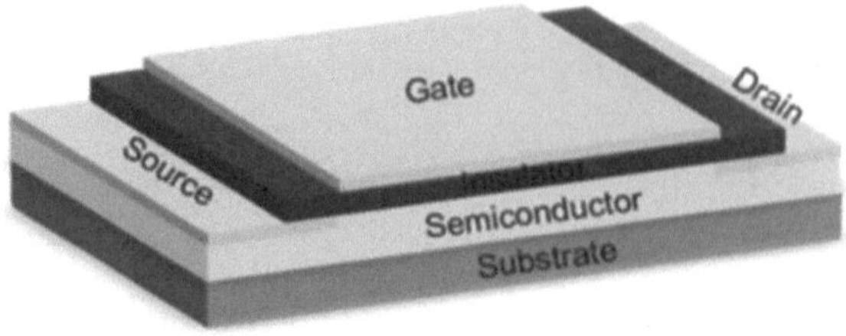

FIGURE 9.20 G–V characteristics at 1-MHz frequency for different concentrations of H_2.

9.2.10.3 MOS Field Effect Transistor

The work function of a hydrogen-sensitive catalytic material changes when exposed to hydrogen gas. This work function change is converted into an electrical signal by a MOSFET sensor.

Figure 9.20 shows the schematic construction of a typical MOSFET sensor comprising of a gate, insulator, semiconductor, and substrate.

9.2.10.3.1 Gate

The gate is composed of hydrogen-gas-sensitive metal which changes its properties upon exposure to hydrogen. Typically, platinum and palladium are used as gate materials.

9.2.10.3.2 Insulator

The function of an insulator is to prevent the flow of current between the substrate and semiconductor.

9.2.10.3.3 Semiconductor

The semiconductor layer is the primary layer and is deposited on a substrate before the insulator layer and gate layer. The two opposite ends of the semiconductor are modified by ion implantation. These ends are known as source and drain.

The main charge carriers are either holes or electrons in the semiconductor layer of a MOSFET. These charge carriers enter from the source terminal and exit through the drain.

Hydrogen is detected by MOSFET sensor; first, hydrogen is absorbed on the outer surface of metal gate; second, hydrogen atoms dissociate and penetrate through the catalytic film, and then, these hydrogen atoms form a dipole layer at the interface between the metal gate and insulator. The dipole layer of hydrogen atoms changes the work function of metal and changes the characteristic voltage of a MOSFET. This voltage change is correlated to the concentration of hydrogen gas.

9.3 CONCLUSION

The world is awaiting another revolution and hydrogen is its driving force. Hydrogen sensors will be playing a vital role in the development and sustainability of hydrogen-based economies and industries. The working principle of a hydrogen sensor dictates its accuracy, advantages, and limitations, for e.g., mechanical sensors require a minimum threshold value of hydrogen for a measurable deflection of the cantilever. This limitation can be overcome by the use of advanced materials and/or material fabrication methods such as the use of CNTs and silver nanowires. Acoustic hydrogen sensors have shown improved sensitivity by using advanced materials such as Langasite as substrate, graphene oxide, and Si/SiO_2 QD as sensing materials. The use of gold/palladium core/shell nanorod metamaterial has improved the optical response of the sensor exorbitantly and the response of sensor can be seen with the naked eye.

Similar improvements in sensor response, sensitivity, and lifetime have been made by using advanced materials. These performance improvements suggest that there is room and potential for advanced materials in hydrogen sensing.

REFERENCES

1. Hübert, T., et al., Hydrogen sensors–a review. *Sensors and Actuators B: Chemical*, 2011. 157(2): pp. 329–352.
2. Li, H., et al., Ultra-high sensitive micro-chemo-mechanical hydrogen sensor integrated by palladium-based driver and high-performance piezoresistor. *International Journal of Hydrogen Energy*, 2021. 46(1): pp. 1434–1445.
3. Kim, M.-O., et al., Highly sensitive cantilever type chemo-mechanical hydrogen sensor based on contact resistance of self-adjusted carbon nanotube arrays. *Sensors and Actuators B: Chemical*, 2014. 197: pp. 414–421.
4. Devkota, J., et al., A surface acoustic wave hydrogen sensor with tin doped indium oxide layers for intermediate temperatures. *Sensors and Actuators B: Chemical*, 2022. 354: p. 131229.
5. Wang, W., et al., Development of a Pd/Cu nanowires coated SAW hydrogen gas sensor with fast response and recovery. *Sensors and Actuators B: Chemical*, 2019. 287: pp. 157–164.
6. Wang, X., et al., Pd/Ni nanowire film coated SAW hydrogen sensor with fast response. *Sensors and Actuators B: Chemical*, 2022. 351: p. 130952.
7. Li, D., et al., A SAW hydrogen sensor based on decoration of graphene oxide by palladium nanoparticles on AlN/Si layered structure. *Journal of Micromechanics and Microengineering*, 2019. 29(4): p. 045007.
8. Constantinoiu, I. and C. Viespe, Hydrogen detection with SAW polymer/quantum dots sensitive films. *Sensors*, 2019. 19(20): p. 4481.
9. Zhang, X., et al., Optics-mechanics synergistic fiber optic sensor for hydrogen detection. *Optics Express*, 2022. 30(18): pp. 32769–32782.
10. Xu, B., et al., Hydrogen sensor based on polymer-filled hollow core fiber with Pt-loaded WO_3/SiO_2 coating. *Sensors and Actuators B: Chemical*, 2017. 245: pp. 516–523.
11. Nasir, M.E., et al., Hydrogen detected by the naked eye: Optical hydrogen gas sensors based on core/shell plasmonic nanorod metamaterials. *Advanced Materials*, 2014. 26(21): pp. 3532–3537.
12. Farnood, A., M. Ranjbar, and H. Salamati, Localized surface plasmon resonance (LSPR) detection of hydrogen gas by Pd^{2+}/Au core/shell like colloidal nanoparticles. *International Journal of Hydrogen Energy*, 2020. 45(1): pp. 1158–1169.
13. Brauns, E., et al., A fast and sensitive catalytic gas sensors for hydrogen detection based on stabilized nanoparticles as catalytic layer. *Sensors and Actuators B: Chemical*, 2014. 193: pp. 895–903.
14. Han, C.-H., et al., Synthesis of Pd or Pt/titanate nanotube and its application to catalytic type hydrogen gas sensor. *Sensors and Actuators B: Chemical*, 2007. 128(1): pp. 320–325.
15. Han, C.-H., et al., Catalytic combustion type hydrogen gas sensor using TiO_2 and UV-LED. *Sensors and Actuators B: Chemical*, 2007. 125(1): pp. 224–228.
16. Harley-Trochimczyk, A., et al., Catalytic hydrogen sensing using microheated platinum nanoparticle-loaded graphene aerogel. *Sensors and Actuators B: Chemical*, 2015. 206: pp. 399–406.
17. Ramesh, C., et al., Improved Nafion-based amperometric sensor for hydrogen in argon. *Journal of Solid State Electrochemistry*, 2008. 12: pp. 1109–1116.
18. Chao, Y., et al., Amperometric sensor for selective and stable hydrogen measurement. *Sensors and Actuators B: Chemical*, 2005. 106(2): pp. 784–790.
19. Rashid, M., et al., An amperometric hydrogen sensor based on Pt nanoparticles supported multi-wall carbon nanotubes. *Journal of Electroanalytical Chemistry*, 2019. 833: pp. 173–180.
20. Hinojo, A., et al., Evaluation of high-temperature hydrogen sensors based on $BaCe_{0.6}Zr_{0.3}Y_{0.1}O_{3-\alpha}$ and $Sr(Ce_{0.9}Zr_{0.1})_{0.95}Yb_{0.05}O_{3-\alpha}$ perovskites for industrial applications. *Sensors*, 2020. 20(24): p. 7258.
21. Zhang, H., et al., Potentiometric hydrogen sensing of ordered SnO_2 thin films. *Sensors and Actuators B: Chemical*, 2020. 321: p. 128505.
22. Wu, J., et al., Ti_3C_2 MXene-derived sodium titanate nanoribbons for conductometric hydrogen gas sensors. *Sensors and Actuators B: Chemical*, 2022. 361: p. 131693.
23. Aroutiounian, V., Metal oxide hydrogen, oxygen, and carbon monoxide sensors for hydrogen setups and cells. *International Journal of Hydrogen Energy*, 2007. 32(9): pp. 1145–1158.
24. Hughes, R. and W. Schubert, Thin films of Pd/Ni alloys for detection of high hydrogen concentrations. *Journal of Applied Physics*, 1992. 71(1): pp. 542–544.
25. Yoshimura, K., et al., New hydrogen sensor based on sputtered Mg–Ni alloy thin film. *Vacuum*, 2008. 83(3): pp. 699–702.
26. Kumar, P. and L. Malhotra, Palladium capped samarium thin films as potential hydrogen sensors. *Materials Chemistry and Physics*, 2004. 88(1): pp. 106–109.

27. DiMeo Jr, F., et al., MEMS-based hydrogen gas sensors. *Sensors and Actuators B: Chemical*, 2006. 117(1): pp. 10–16.
28. Xu, T., et al., Self-assembled monolayer-enhanced hydrogen sensing with ultrathin palladium films. *Applied Physics Letters*, 2005. 86(20): pp. 203104.
29. Favier, F., et al., Hydrogen sensors and switches from electrodeposited palladium mesowire arrays. *Science*, 2001. 293(5538): pp. 2227–2231.
30. Kim, Y., et al., ß-Ga_2O_3 flake based Schottky diode hydrogen sensor. *Sensors and Actuators B: Chemical*, 2023. 379: p. 133212.
31. Liu, I.-P., et al., Hydrogen sensing characteristics of a Pd/Nickel oxide/GaN-based Schottky diode. *International Journal of Hydrogen Energy*, 2019. 44(12): pp. 5748–5754.
32. Ratan, S., et al., Room temperature high hydrogen gas response in Pd/TiO_2/Si/Al capacitive sensor. *Micro & Nano Letters*, 2020. 15(9): pp. 632–635.

10 Mycotoxins as Food Contaminants and Their Detection Methods

Avinash Kumar Singh, Manvendra Pratap Singh, Parmanand, Vijay Bahadur, and Shatruhan Singh Rajput

10.1 INTRODUCTION

Food contamination has always been considered a significant challenge to human health. One of the main causes of these contaminations is food toxins such as mycotoxins (Bennett & Klich, 2003). Advanced methods for the identification of such lethal toxins are needed because the efforts undertaken to identify and reduce these contaminants in food grains are insufficient. Mycotoxins are naturally occurring toxic substances that come from fungi that infest food crops like corn, cereals (rice, wheat, sorghum, walnut, cotton, peanuts, barley, and oats), and spices, all of which serve as the primary sources of the essential components of foods for both humans and animals (Haschek et al., 2013). Therefore, exposure to these chemicals might have a negative impact on living things. The mycotoxins are classified into various groups by the International Agency for Research on Cancer (IARC) based on their toxicity (Smoke & Smoking, 2004). In this chapter, we have discussed four different mycotoxins named aflatoxin, Fumonisin, ochratoxin, and DON. One of the most precarious mycotoxins, aflatoxin, is a major contributor to liver cancer. It comes in more than 20 various forms, the most frequent of which is aflatoxin B1 (AFB1), which is found in corn, grains, and spices (Inan et al., 2007). Fumonisin B1 and ochratoxin A (OchA), two other toxins, are produced by the fungi *Fusarium* and *Penicillium verrocusum* and are primarily found in corn, wheat, and barley (Bennett & Klich, 2003). Fumonisin B1 is also shown to be responsible for liver and renal illness, and these poisons are found to cause DNA damage and abrupt death (Soriano et al., 2005). For the detection of these toxins, a variety of detection methods have been used. These include high-performance liquid chromatography (HPLC), gas chromatography (GC), ELISA kits, thin layer chromatography (TLC), and polymerase chain reaction (PCR) (Abramson et al., 1995). These approaches, however, frequently involve challenging procedures, making them difficult to apply for everyday tasks due to the intricacy of their apparatus. Moreover, electrochemical biosensors, which are effective in comparison to conventional approaches and have a very large detection range from hundreds of micrograms per ml to picograms per ml, are frequently utilized to get around these problems (Singh et al., 2021). Electrochemical biosensors have all been explored in this article.

10.2 MYCOTOXINS

Mycotoxins are naturally occurring hazardous chemical substances that are created as secondary metabolites by fungi and have a negative impact on all living things, including people and animals, when they are exposed to them (Bennett, 2003). Mycotoxins can be exposed by eating, inhalation, and skin absorption, which can result in bad health issues and even the demise of animals and birds. Mycotoxicosis, which is indicated by the biological system being poisoned, causes biochemical, physiological, and pathological alterations in several species (Richard, 2007). The term "mycotoxin" is made up of the words "myco" and "toxin," which refer to mold or fungi and the poison they produce,

DOI: 10.1201/9781032654874-10

respectively. There are various plant species that are the source of fungus metabolite infection, including corn, barley, maize, cottonseed, wheat, cereals, spices, oats, peanuts, soybeans, herbs, bran, gluten, milo, and rye, which are some basic elements for human and animal consumption (Haschek et al., 2013; Kumar et al., 2017). Because they acquire their nourishment from dead organic matter, which can be thought of as a saprophyte but lives as an endophyte in the living plant, fungi are aerobic organisms, and their interactions with plants are exceedingly intricate. Although certain fungi multiply before harvest and throughout storage, fungi infect the plants before harvest, in dry conditions, or after storage. Variations in temperature and moisture have a significant impact on how quickly these fungi grow and how much poison they produce. Fungi can infiltrate both old and growing seed grains in plants when the moisture content is between 22% and 25%, which is the ideal range for growth. Although most fungi are created above 30°C and below 20°C, they primarily flourish in the 20°C–30°C temperature range. Therefore, the ideal temperature range is from 0°C to 60°C (Tola & Kebede, 2016).

Mycotoxicosis is a form of naturally occurring poisoning that happens when fungi-produced natural poisons cause pathological alterations in the body. Mycotoxicosis is influenced by a number of variables, including the kind of mycotoxins; the amount and duration of exposure; the person's age, health, and sex; exposure to contaminated food due to fungi; and inhalation of toxins released by fungi, among others. Therefore, a lack of vitamins, alcohol intake, calorie restriction, the impact of infectious diseases, etc., can all increase the severity of mycotoxin poisoning (Dixon et al., 1996).

Mycotoxin classification is exceedingly difficult because of the wide range of traits associated with their various chemical structures and biosynthesis sources. However, mycotoxins can generally be divided into hepatotoxins, nephrotoxins, neurotoxins, immunotoxins, and other categories. These mycotoxins were categorized by cell biologists into general categories such as teratogens, carcinogens, mutagens, and allergies. Organic chemists have categorized them based on their chemical make-up, while biochemists have done the same for their polyketide and amino-acid-derived biosynthetic origins, and doctors have done so for diseases like St. Anthony's fire and stachybotryotoxicosis (Dixon et al., 1996; El-Nezami et al., 1998).

Mycotoxins have been recognized to be poisonous for a very long time, but it wasn't until the 1960s that the first instance of a foodborne fungus caused sickness and death in animals (Fink-Grernmels, 1999). Mycotoxins have recently been identified as an epidemic in both people and animals. In Europe, ergotism has caused thousands upon thousands of deaths over the past millennium. Aside from these, at least 100,000 people in the USSR died between 1942 and 1948 from alimentary toxic aleukia (ATA), stachybotryotoxicosis (a type of mycotoxicosis), which killed about a lakh horses in the USSR in the 1930s, and aflatoxicosis, which also caused 100,000 young turkeys to die in the UK in the 1960s (Gajęcka et al., 2013; Blount, 1961).

Aspergillus, *Penicillium*, *Fusarium*, and *Claviceps* are the four main genera into which the generation of mycotoxin from fungi has been grouped, which account for the majority of food contamination worldwide (Cancer, 1982). *Fusarium* species are harmful pathogens connected to cereals and other agricultural products that produce mycotoxin before or just after instant harvesting. Although *Aspergillus* and *Penicillium* are commensals or plant pathogens as well, these genera contribute to the development of mycotoxin in food products during the drying and storage of food grains. *Aspergillus flavus* is a unique type of pathogen that can live on plants without harming them or a storage fungus that can grow in all three of these environments and create mycotoxin (Rodricks et al., 1977). A large pathogenic organism belonging to the fourth genus, *Claviceps* is also present in many cereals and crops with ergots as their resting structure.

Numerous mycotoxins have been discovered through research; however, only a handful of them, including those in Table 10.1, have been consistently found to have negative effects on human health.

In this section, four distinct mycotoxins are discussed which are given as:

- Aflatoxin,
- Fumonisin,
- Ochratoxin, and
- DON.

TABLE 10.1
Fungal Mycotoxins by Category and Frequency of Occurrence

Fungal Species	Mycotoxins	Commodities	Disease	References
Aspergillus flavus and *parasiticus*	Aflatoxin	Corn, milo, cottonseed, cassava, peanut meal, palm meal, sesame meal, soybean meal, pineapple meal	Liver cancer	Klich et al. (2000)
Penicillium verrucosum, A. ochracoeus	Ochratoxin A	Milo, barley, wheat, oats, rye, bran, screening pellets	Anorexia, dehydration, and depression	Bayman et al. (2002)
F. verticillioides, F. proliferatum	Fumonisins	Corn, milo, barley, rye, corn gluten feed, corn gluten meal, barley bran, bran, corncob meal	Liver necrosis	Marasas et al. (2001)
Fusarium sporotrichioides	T-2 toxin	Corn, milo, corn gluten feed	Leukopenia	Beasley (2017)
F. graminearum (roseum)	Deoxynivalenol (DON)	Corn, milo, barley, wheat, rye, bran, corn gluten feed, screening pellets	Acute disease	Pietsch et al. (2014)
F. graminearum (roseum)	Zearalenone	Milo, corn, barley, corn gluten feed, corn gluten meal	Estrous behavior, vulval swelling	Bakos et al. (2013)
Claviceps spp., Neotyphodium coenophialum	Ergot alkaloids	Wheat, rye, barley, oats, and ryegrasses	Ischemic necrosis and dry gangrene	Reddy et al. (2020)
Penicillium citrinum, Penicillium viridicatum	Citrinin	Wheat, rye, barley, beans, fruits, vegetable juices, herbs, and spices	Kidney problems	Narvaez et al. (2021)

10.2.1 Aflatoxin

The aflatoxins are well-known highly toxic mycotoxins, secondary metabolites construed from polyketide pathways produced by fungal species including *Aspergillus flavus (A. flavus), A. paraciticus, A. bombycis, A. ochraceoroseus, A. nomius, and A. pseudotamari* (Payne & Brown, 1998; Sarma, 2010; Kelly et al., 1997). Among these species *A. flavus* and *A. paracitticus* are the most common fungi which are responsible for the contamination of food, while other remaining species such as *A. bombycis, ochraceoroseus, nomius, and pseudotamari* are less frequently involved in aflatoxin-producing species (Goto et al., 1996; Klich et al., 2000; Peterson et al., 2001). These toxins are a group of carcinogenic di-furanocoumarin derivatives derived from *A. flavus* and *paracitticus* spread worldwide and mostly found in subtropical climate including south-eastern United States, southern Asia, and Africa where a warm climate is conducive for fungal growth (Azziz-Baumgartner et al., 2005). Generally, the soil is a suitable place for their production but under favorable conditions invade the host plant through various pathways. There are some additional conditions such as draught, heat, and insect infection favoring the aflatoxin production in the field. The sources of these toxins are staple food items such as corn, nuts, peanuts, Brazil nut, copra, cereals (rice, wheat, sorghum, walnut, cotton, barley, and oats), spices (ginger, chili, turmeric, black pepper, and nutmeg), and decaying vegetation (Jelinek et al., 1989; Severns et al., 2003; Suárez-Bonnet et al., 2013; Mukherjee & Lakshminarasimham, 1995) and can be promoted to various critical diseases for humans and animals. However, other agricultural products like melon seeds, linseeds, cocoa beans, and sunflower seeds are rarely contaminated by aflatoxins. Temperature and humidity have a great importance in aflatoxin production. Optimal humidity for aflatoxin production is considered as 85% and interaction between temperature and water activity (a_w) plays an important role in the

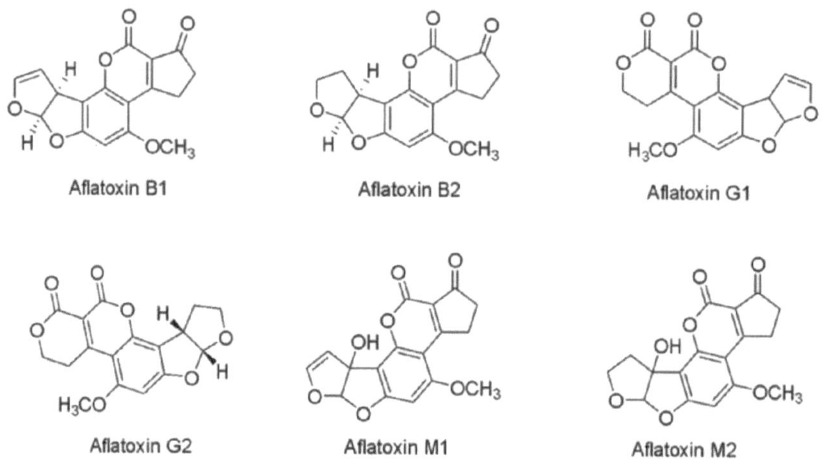

FIGURE 10.1 Chemical structures of aflatoxins.

production of aflatoxin because these interactions influence the ratio of regulatory gene (*aflR/aflS*) which is directly proportional to aflatoxin production (Schmidt-Heydt et al., 2009, 2010). The ideal temperature reported for aflatoxin production is 28°C with 0.96 a_w, while 20°C temperature with a_w 0.90 and 0.93 gives very less or no aflatoxin production; conditions and increasing temperature up to 37°C also significantly reduce the production of aflatoxin (Gallo et al., 2016).

These aflatoxins produced by saprophytic fungi have been classified into more than 20 different forms. Few of them like AFB1, AFB2, AFG1, AFG2, and other hydrolyzed forms of aflatoxin AFM1 and AFM2 are well known in the nature (Figure 10.1) (Inan et al., 2007; Giray et al., 2007; Hussain & Anwar, 2008; Kumar et al., 2017). The toxicity level of these toxins varies in the following order AFB1 > AFG1 > AFB2 > AFG2 (Jaimez et al., 2000). The designation of B and G represent the blue and green color fluorescence which is visualized in corn, peanuts, and other commodities through ultraviolet light (Fallah et al., 2011). AFB1 is the most potent and prevalent natural carcinogenic known so far which has adverse effect on human and animal health. According to WHO report, large doses of AFB1 are carcinogenic and cause liver damage or cancer (Henry et al., 1999). Aflatoxin M1 is found in milk like cows, buffalos, and humans after feed consumption of contaminated AFB1. CYP1A2 enzyme catalyzes the conversion of AFB1 to less toxic AFM1. AFM1 is estimated 10 times less toxic and carcinogenic than AFB1. The IARC) has classified AFB1 as a group 1 carcinogen and AFM1 as a group 2B carcinogen (International Agency for Research on Cancer, 1982). The US department of Agriculture (USDA) and Food and Drug Administration (FDA) set a tolerance limit of 20 μg kg^{-1} for aflatoxins in view of their toxic effects on foods, while animals have different permissible limits between 0 and 50 μg kg^{-1} (Sarma et al., 2017). The commission of European communities set the limit parameter for AFM1 toxin to be 50 μg kg^{-1} for milk. But the toxic level of aflatoxin in humans is unknown. According to a report, a laboratory worker who intentionally ingested AFB1 at 12 mg kg^{-1} body weight for 2 days developed a rash, nausea, and headache. Median lethal dose (LD_{50}) of AFB1 varies among animals of different figures or body weights. For example LD_{50} for rabbits is 0.3 mg kg^{-1} body weight, while it is 18 mg kg^{-1} body weight in rats (Food and Drug Administration, 2012).

In 1974, the first outbreak of aflatoxin was reported through hepatitis in the state of Gujarat in India and more than 100 people died due to the consumption of maize; it was heavily contaminated (Bennett & Klich, 2003; Krishnamachari et al., 1975). A major outbreak of aflatoxin exposure was recorded in Kenya in 1981 with aflatoxin intake at 38 μg kg^{-1} body weight (Ngindu et al., 1982). In 1988, in Malesia, 13 Chinese children died after eating noodles due to acute hepatic encephalopathy disease; post mortem report confirmed the exposure of aflatoxin in their body.

Monooxygenase enzyme Cytochrome P450 participates in the conversion of AFB1 to highly reactive AFB1-exo-8,9-epoxide which is capable of binding to both DNA and proteins. The reactive exo-epoxide group of AFB1 binds with guanine base moiety of DNA at the N7 position and forms trans-8,9-dihydro-8-(N7-guanyl)-9-hydoxy-AFB1 adducts, which, upon further rearrangement, form a stable adduct (Eaton & Groopman, 2013; Eaton & Ramsdell, 1992). Moreover, AFB1-DNA adducts can result in GC to TA transversions. This can be measured in tumor tissues. This specific mutation at codon 249 of the p53 tumor suppressor gene may be important in the development of primary hepatocellular carcinoma (HCC). Studies of liver cancer patients in Southeast Asia and sub-Saharan Africa, where AFB1 contamination in foods was high, have shown that a mutation in the p53 at codon 249 is associated with a G to T transversion (Dohnal et al., 2014; Raj et al., 1986).

Aflatoxicosis in animals leads to mainly liver injury, gastrointestinal dysfunction, reduced reproductivity, efficiency, anemia, reduced feed utilization, reduced immunity, and the reduction of milk and egg production. Aflatoxins are hepatotoxic in both humans and animals (Abdel-Wahhab et al., 2007; Etzel, 2002). Aflatoxin exposure mainly occurs from food-related items resulting in aflatoxicosis, which may turn from acute to chronic disease and from mild to severe, leading to cirrhosis (severe liver damage) and even liver cancer.

10.2.2 Fumonisin

Fumonisin is one of the prevalent toxic compounds among mycotoxins and is produced by species of fungi including *Fusarium verticillioide,* and many other species like *F. proliferatum*, *F. anthophilum*, *F. dlamini*, *F. napiforme*, and *Alternaria alternata* f. sp. *lycopersici* (Moss, 1998) are also responsible for fumonisin toxins. Fumonisin was first isolated by Gelderblom in 1988 and chemically characterized by Bezuidenhout in the same year (Bezuidenhout et al., 1988). *Verticilllioides* species is considered as the most worthwhile fungi throughout the world regarding contamination of corn. Maize- and wheat-based products are contaminated mainly through fumonisin (Frisvad et al., 2007) and their effects are also responsible for the contamination of other food items such as grains (rice, wheat, barley, maize, rye, oat, and millet) and grain products (tortillas, corn flask, chips) which play a crucial role in human health (Dall'Asta & Battilani, 2016; Cendoya et al., 2018). More than fifteen varieties of fumonisins have been reported so far and are categorized into four major groups A, B, C, and P (Braun & Wink, 2018; Marasas, 1996) which are as follows: FA1, FA2, FA3, FAK1; FB1, FB2, FB3, FB4; FC1, FC2, FC3, FC4; FP1, FP2, FP3, etc. (Abbas et al., 2002). "B" group fumonisin including fumonisin B, FB1, FB2, and FB3 are abundant in nature (Damiani et al., 2019) and contaminate mainly food items, out of which FB1 is the most toxic form that can co-exist with other forms of fumonisin. The IARC has suggested the cancerous effect of fumonisin B1 (FuB1 or FB1) to humans and categorized it in class 2B (IARC Working Group on the Evaluation of Carcinogenic Risks to Humans, 2002).

The optimum growing temperature for the species *F. verticillioide* and *proliferatum* is at 25°C and the maximum temperature is between 32°C and 37°C, while the minimum growing temperature is between 2°C and 2.5°C. The fungi growth requires water activity (a_w) 0.87 (Pitt & Hocking, 2009).

Maize is one of the most evident sources of occurrence of fumonisin in several countries like United States, Canada, Western Europe, and African countries. Natural occurrence of FB1 at a high rate in maize causes esophageal cancer and the associated countries are northern Italy, South Africa, China, Iran, and Southeast America. Moreover, consumption of corn and sorghum contaminated with these toxic compounds cause outbreak as foodborne diseases leading to symptoms like abdominal pain, diarrhea in some villages in India (Krishnamachari et al., 1975).

Worldwide, FB1 has been detected at mg kg^{-1} concentration mostly in maize and maize-related products, and this concentration has also been reported for human food consumption. The distribution of FB1 is present in maize-based food items in both the dry and wet milling manufactured food products such as germ, flour, bran, fiber, gluten, and steep water, but its effect is absent in starch. It shows great stability in maize and polenta, whereas it is unstable in nixtamalized maize-based food item preparation of corn or maize in alkaline solution (e.g., limewater) (Anwar et al., 2011). FB1 is less or non-toxic for animals i.e., it is absent in eggs, milk, and meat. Human exposure estimates for the USA, Canada, Switzerland, the Netherlands, and the Transkei (South Africa) ranged from 0.017 to 440 g kg^{-1} body weight per day. There were no reports available of fumonisin absorption through inhalation or dermal exposure, although fumonisins are present in *F. verticillioides* cells (mycelia, spores, and conidiophores) (Tejada-Simon et al., 1995).

The structure of fumonisins is similar to the structure of sphingosine which is a major component of sphingolipid (it is the second type of lipid found in cell membrane & nerves cells) and have the capability to inhibit the ceramide synthase (membrane protein and participates in ceramide which is backbone of lipid (Riley et al., 1998). It is found that the N-acetylated group present in the FuB1 rearranged spontaneously into O-acetylated derivative and inhibited the ceramide synthase which acts as toxicity inhibition by fumonisin.

Due to the structural similarity between FuB1 with sphinganine, FuB1 disrupts the important component of the sphingolipid metabolism which leads to several events such as transformed cell growth, cellular differentiation, and cell injury. Fumonisin is a great inhibitor for both the substrates, sphingosine and ceramide synthase, causing blockage of complex sphingolipids biosyntheses and accumulation of sphinganine. This transformation causes the toxicity of FuB1 which leads to several effects such as liver lesions (abnormal cell growth of liver), heart failure (Zhang, 1997), apoptosis (Bose et al., 1995), (cell death), pulmonary edema (fluid in lungs causes breathing problem) (Zhang, 1997), nephrotoxicity and liver cancer in rats, immunosuppression in poultry, brain hemorrhage in rabbits, and carcinogenicity (Bucci et al., 1996). Except for these effects, FuB1 shows genotoxicity which enhances the cancerous risk where the reduction of DNA repair or decreasing the DNA replication take place and DNA damage by apoptosis process through inhibiting pathways as well (Soriano et al., 2005; Abbax et al., 1993). A higher dose of AF1 causes leukoencephalomalacia in horses and rabbits (Marasas et al., 1988; Wilson et al., 1992), and FB1 is also identified as a rodent carcinogen which induces renal tumors in male rats and hepatic tumors in female mice (Gelderblom et al., 2001) (Figure 10.2).

10.2.3 Ochratoxin

Ochratoxin is a developed class of mycotoxin which is the secondary metabolite of *Aspergillus ochraceous, Aspergillus carbonarius, Aspergillus niger, Penicillium verrocusumv, Aspergillus glaucus, Aspergillus melleus,* and other species (Abarca et al., 1994; Heenan et al., 1998; Ciegler, 1972; Freire et al., 2017). The first OchA species was discovered in 1965 during the fungal metabolite of *Aspergillus carbonarius* species (Van der Merwe et al., 1965). The species *Aspergillus niger* is commercially important at the industrial level because it is used in the production of enzymes and citric acid for the consumption of humans (Heenan et al., 1998). Usually, sources of ochratoxins

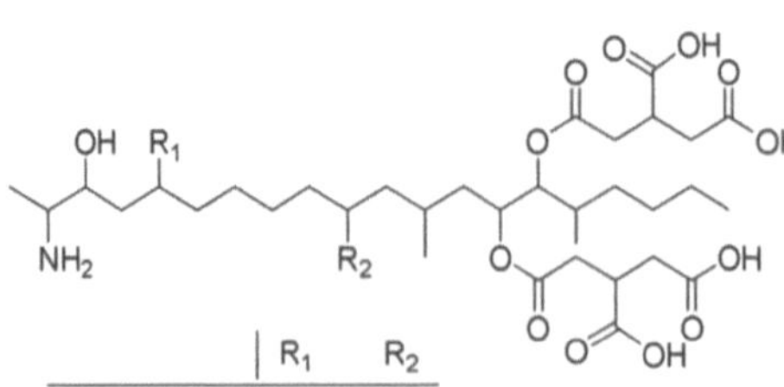

	R_1	R_2
Fumonisin B_1	OH	OH
Fumonisin B_2	OH	H
Fumonisin B_3	H	OH
Fumonisin B_4	H	H

FIGURE 10.2 Chemical structure of fumonisins.

are food commodities such as cereal grains (rye, maize, wheat, sorghum, oats, and barley), as well as coffee beans, dairy products, wine, raisins, vegetables (yam, garlic, onions, potatoes, and tomatoes), fruits (grapes, apples, peaches, pears, strawberries, oranges, watermelons, mangoes, and figs), and spices. Besides these items, other commodities like meat, eggs, pork products are also a great exposure to humans (Duarte et al., 2010; Lim et al., 2015; Sun et al., 2017; Quintela et al., 2013; Dai et al., 2014; Marin & Taranu, 2015; Pfohl-Leszkowicz & Manderville, 2007, 2012). Ochratoxin has been classified into three different forms including A, B, and C where these different forms are differentiated by different chemical structures (Pohland et al., 1992; Hope & Hope, 2012). Ochratoxin compounds contain phenylalanine group attached via amide bond with dihydroisocoumarin group. OchA is more potent and the most common toxin other than ochratoxin B and C. Structurally OchA is chlorinated while ochratoxin B is not, and ochratoxin C is ethyl ester of ochratoxin which is less common and less toxic (Kőszegi & Poór, 2016; Gallo et al., 2017; Chen et al., 2018).

Aspergillus and Penicillium are plant pathogens responsible for ochratoxin and produce mycotoxin during the drying and storage of the commodities. OchA is a common type of toxin and its growth requires several specific parameters like temperature, water activity (a_w), and low-consuming elements. Species *A. carbonarius* requires 15°C–20°C temperature with 0.95–0.98 a_w; (Mitchell et al., 2004); *A. ochraceous* need 25°C–30°C temperature with 0.98 a_w (Ramos et al., 1998), while *Penicillium viridicatum* requires 24°C temperature and 0.95–0.98 a_w feasible condition for OchA production (Northolt et al., 1979). OchA is cancerous in humans; therefore, the IARC has categorized this toxin in 2B group (Smoke & Smoking, 2004). Besides this, OchA also has neurotoxic, immunotoxic, mutagenic, and nephrotoxic properties (Pfohl-Leszkowicz et al., 2002). World health organization (WHO) set the tolerable parameter for OchA as 5 ng kg^{-1} body weight per day consumption. The LD_{50} of OchA is different for different animals: for young mice, LD_{50} is 20 mg kg^{-1}; LD_{50} for duck chicken is 3.6 mg kg^{-1}, and this dose for pig is between 1 and 6 mg kg^{-1} (Pitt, 2000).

OchA is found worldwide but mostly in the temperate climate (the region consists of extreme variation in the environment including moderate rainfall across the year, sporadic draught, mild-warm summer, and cool winter). The countries most exposed to ochratoxin are northern European countries and Balkan countries such as Serbia, Bulgaria, Yugoslavia, Romania, and Canada. In one Bulgarian study, it is reported that ochratoxin contamination in food and human serum was found in families with Balkan endemic nephropathy and urinary tract tumors than other unaffected families (Castegnaro et al., 1987).

OchA has structural and binding site similarity with phenylalanine which works as a substrate for phenylalanyl-tRNA synthetase. OchA works like an inhibitor for protein synthesis and inhibits the phenylalanyl tRNA synthetase which influences the transcription of many cellular proteins resulting in intracellular effects (Joint FAO, 1999; Creppy et al., 1996). Inhibition of mitochondrial respiration also takes place through ochratoxin and has a negative impact on adenosine triphosphate (ATP) (cellular energy) production which leads to mitochondrial dysfunction and decreases the protein synthesis. Ochratoxin A (OchA), which has a parachlorophenol part in its structure, is transformed by cytochrome P450 enzymes. The process, called oxidative dichlorination, converts OchA into a highly reactive quinoidal form. This quinoidal form can then form strong, irreversible bonds with both sulfur-containing proteins (thiol groups) and the DNA base 2'-deoxyguanosine (dG). The DNA binding results in an adduct and can disrupt normal DNA function (Tozlovanu et al., 2006). Oxidation of OchA by cytochrome p450 enzyme also forms the reactive electrophilic product ochratoxin-quinone (OTQ) (Gillman et al., 1999) which are partially detoxified and after reduction forms Ochratoxin A-hydroquinone (OTHQ) which is detected both in human and rat urine samples (Mally et al., 2004). Adduct of OchA and DNA lead to a non-genotoxic mechanism in which foramidopyrimidine- DNA-glycosylase (Fpg) enzyme participates in the recognition and repair of DNA damage suggesting that OchA causes DNA damage via oxidation (Cavin et al., 2009; Arbillaga et al., 2007; Russo et al., 2005) (Figure 10.3).

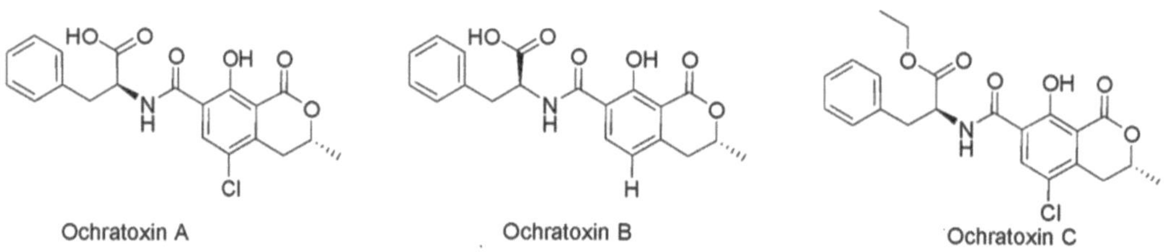

FIGURE 10.3 Chemical structure of ochratoxins.

10.2.4 Deoxynivalenol (DON)

Trichothecene mycotoxins have been recognized as major food contamination issue worldwide. This mycotoxin is categorized into four groups A, B, C, and D, where A relates to T-2 toxin and HT-2 toxin, B group stands for DON, and other C & D groups consist of roridin A and verrucarin A mycotoxins (Foroud & Eudes, 2009). The trichothecene group of mycotoxins is generally found in wheat or wheat-based products. Deoxynivalenol (DON) is a prevalent mycotoxin commonly found in wheat. Produced primarily by *Fusarium* species like *F. graminearum* and *F. culmorum*, DON-producing fungi act as plant pathogens, causing significant damage to cereal crops in the field. These infections lead to Fusarium head blight (FHB), resulting in poor yields and lower quality wheat grain (Khaneghah et al., 2018; Yazar & Omurtag, 2008) (Cirlini et al., 2013; Malbrán et al., 2014). DON is colloquially considered as "vomitoxin" due to its emetic effects in pigs (Sobrova et al., 2010). The sources of DON are mainly found in wheat, wheat-based products, barley, corn, rye, oats, sunflower seed, etc. (Bennett, 2003; Berthiller et al., 2013; Rychlik et al., 2014). The optimal growth of DON mycotoxin occurs up to 25°C and the minimum temperature is from 0.9°C to 25°C. *F. graminearum* is common species producing DON in warm climates, but due to genetic modifications, these fungi can also produce DON at lowertemperatures (Hope et al., 2005; Pitt & Hocking, 2009), while *F. culmorum* species grow at a higher temperature at 25°C. The required water activity (a_w) for DON production is 0.87 at 20°C–25°C and pH 6.5 (Sanchis & Magan, 2004). It has been previously reported that DON is a stable compound between temperatures 170°C–350°C (Generotti et al., 2015; Gärtner et al., 2008; Program, 2009). But, according to the WHO report, DON is stable at 120°C, moderately stable at 180°C, and partially stable at 210°C (FAO, 2001). DON is a polar organic compound and its chemical name is 12,13-epoxy-3α,7α,15-trihydroxytrichothec-9-en-8-on. In its molecular structure, DON contains three hydroxy groups (–OH) which is associated with its toxicity (Nagy et al., 2005). In 2010, Joint FAO/WHO Expert Committee on Food Additives (JECFA) set the acute dose parameter 8 μg kg^{-1} body weight for the DON and its derivatives (3-acetyl-DON and 15-acetyl-DON) (Authority, 2013). The toxicity of DON is lower in comparison to other groups of trichothecene, T-2, and HT-2 toxins (Sobrova et al., 2010). IARC has classified DON into group 3 category which shows that DON is non-carcinogenic in nature (International Agency for Research on Cancer, 1982). High dose exposure to these toxins, may cause several acute diseases such as nausea, diarrhea, and vomiting in animals. Pigs and other farm animals with lower doses exposure display weight loss and food refusal (Rotter, 1996). DON is also responsible for chronic diseases, and experiments on mice reported that excess consumption of DON exhibit a reduction of daily food consumption and weight loss. In humans, DON induces adverse effects such as anorectic, vomiting effects, reduction in food consumption and affects the peripheral nervous systems (Hussain & Bloom, 2013). Besides, DON may inhibit gastrointestinal toxicity, immunotoxicity, hepatotoxicity, reproductivity toxicity, neurovirulence, ribosomal toxicity etc. It is related to the synthesis of protein as well as mitochondrial electron chain function; oxidative stress, apoptosis, endoplasmic reticulum (ER) stress lipid peroxidation andmitogen-activated protein kinase (MAPK) (Dinu et al., 2011; Zhang et al., 2009) (Figure 10.4).

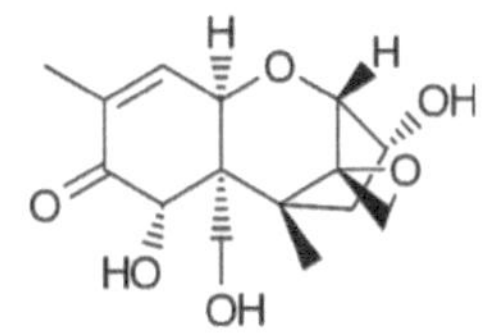

FIGURE 10.4 Chemical structure of deoxynivalenol (DON).

10.3 CONVENTIONAL DIAGNOSTIC TECHNIQUES FOR MYCOTOXIN DETECTION

In the past, several detection methods, including high-performance liquid chromatography (HPLC), liquid chromatography–mass spectrometer (LC–MS), enzyme–linked immunosorbent assay (ELISA), gas chromatography (GC), and gas chromatography–mass spectrometer (GC–MS) have been developed to diagnose mycotoxins. The following is a brief discussion of a few well-known methods.

10.3.1 Enzyme-Linked Immunosorbent Assay (ELISA)

An established bioanalytical instrument used as a biochemistry assay is the ELISA. The basis for ELISA is the concept of antibody–antigen interactions, in which immune responses occur between antibodies (cell-produced substances) and antigens. The quantitative and qualitative examination of antigens, proteins, hormones, nucleic acids, herbicides, other secondary metabolites, etc., is made possible by this very sensitive and selective technique. Enzyme immunoassay, which labels the molecules with the enzymes horseradish peroxidase (HRP) and galactosidase is used to measure the molecules (Swartzman et al., 1999).

10.3.2 High Performance Liquid Chromatography (HPLC)

HPLC is a traditional method that has been extensively employed in terms of selectivity, sensitivity, and time analysis. To separate complex mixtures, HPLC is frequently used. It also has the capacity to identify compounds at very low concentrations. Today, HPLC is frequently used for sample detection, separation, and quantification. It is also utilized in biochemistry to identify, separate, and quantify active chemicals (Tabari et al., 2011).

10.3.3 Liquid Chromatography–Mass Spectrometry (LC–MS)

Liquid chromatography (LC) and mass spectrometry (MS) are two traditional analytical techniques that are combined. LC is ineffective for identifying completely unknown, known substances, or complicated mixtures; nevertheless, a combination with MS can do so. This quantitative and qualitative analysis method is called LC–MS. It is widely employed in forensic, environmental, food, agricultural, pharmaceutical, and industrial applications. Additionally, it is employed in quality assurance testing and medicinal research (Tabari et al., 2011).

10.3.4 Limitations of Conventional Detection Techniques

When it comes to the detection of different analytes, such as foodborne pathogens and their eventual clinical applications, conventional approaches are typically used. These have some limitations because of their properties even though they are sensitive and specific to their analytes. These configurations need to be operated by a qualified technician and are exceedingly tedious, expensive, time consuming, and intricate. To identify different analytes, a number of methods, including

ELISA, HPLC, LC–MS, and others, are frequently utilized. However, ELISA has certain drawbacks: time consuming, requires expensive lab equipment and a large number of samples, which can occasionally result in normal assay errors. Similarly, HPLC and LC–MS are quite expensive, need a lot of samples for analysis, and are equally challenging to work with. These techniques require lengthier analysis durations, perhaps weeks. Therefore, from the perspective of detection, quick responsive procedures are needed.

10.3.5 Alternative Techniques

Sensing platforms solve the problems with traditional detection methods and offer a quick and effective solution to analyze various analytes. It is relatively inexpensive, portable, quick, only needs a small number of samples, simple to use, etc., A range of biosensors, including electrochemical, optical, thermal, piezoelectric, and others, are included in the sensing platform. The sections that follow address these methods.

10.4 BIOSENSORS

An analytical equipment called a biosensor converts biological signals from living things into electrical impulses. It is a sophisticated biological analysis tool that transforms biological signals into electrical impulses (or another signal which depends on the transducer). The first enzyme-based biosensor was created by the fathers of the biosensor field, Clark and Lyons, in 1962 for the purpose of detecting glucose in diabetic patients (Clark Jr & Lyons, 1962). The two main components playing a significant role in the creation and operation of the biosensor are transducers and bioreceptors. Bioreceptors are biological components that recognize their complementary analyte as an enzyme substrate; an antigen; complementary sequences; and other biological components like antibodies, nucleic acids, organelles, and microbial cells.

Transducers, on the other hand, change the signal that results from the biological interaction of the bioreceptor and the analyte. The transduction of a biosensor can be electrochemical, optical, chemical, thermal, mechanical, acoustic, or piezoelectric, for example. Biosensors are classified according to their biorecognition components.

Below is a definition of a biosensor:

1. A biosensor is a self-contained integrated device that can provide precise quantitative or semi-quantitative analytical information using a biological recognition element (biochemical receptor) that is in direct spatial contact with a transducer element, according to the definition given by International Union of Pure and Applied Chemistry (IUPAC). The difference between a biosensor and a bioanalytical system, which calls for additional processing steps like reagent addition, should be made very obvious. In addition, a biosensor should be distinguished from a bio-probe that can't continually monitor the analyte concentration or is disposable after a single measurement, or both (Thévenot et al., 2001).
2. The term that is most frequently used is "an analytical device that incorporates a biologically active element with an appropriate physical transducer to provide a quantifiable signal proportional to the concentration of chemical species in any type of sample" (Belkin, 2003; Eggins, 2002).

Typically, biosensors have several basic parts including bioreceptors, transducers, signal processors, and displays.

Biosensors are divided into a few categories based on the biological signaling and transduction mechanisms, such as electrochemical, optical, mass, and thermal, among others (Figure 10.5).

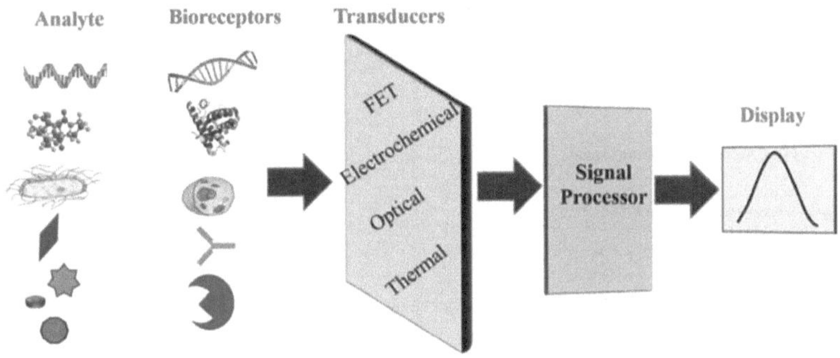

FIGURE 10.5 Schematic picture of a biosensor.

10.4.1 Electrochemical Biosensors

A quick analytical instrument for chemical, clinical, pharmacological, and environmental studies is the electrochemical biosensor (MacDiarmid, 2001; Arora et al., 2023). Electrochemical biosensors allow for extremely high sensitivity, specificity, selectivity, and speed of detection in the investigation of species. This method has good sensitivity for measuring femtoliter concentrations. Optical, pyroelectric, acoustic wave, and other nano-biosensors are more expensive and less cost-effective than electrochemical biosensors. This type of biosensor operates on the idea that an immobilized bioreceptor interacts with the analyte to provide a signal that changes the electrolyte solution's electrically detectable characteristics, such as electrical current, voltage, impedance, and capacitance. Electrochemical biosensors are ideal and helpful in miniaturization/microfabrication technologies because of their highly sensitive quality. For analyte measurement, this method has both qualitative and quantitative significance. Three primary subclasses of electrochemical biosensors are amperometry, potentiometry, and conductometry, also known as impedance spectroscopy.

10.4.2 Amperometry Biosensor

The basis of the amperometry biosensor is the detection of the oxidation and reduction currents produced by electroactive biological components. By using a constant potential and supplying precise quantitative analytical data, the current measuring procedure operates. A three-electrode system called an amperometry biosensor consists of a working electrode [such as an indium tin oxide (ITO) electrode made of noble metals], a reference electrode (such as an Ag/AgCl electrode), and an additional or counter electrode (Pt electrode). Amperometry is typically used to describe current measurements at a constant potential. Voltammetry is the term used for some current measurements at controlled fluctuations in potential (Belluzo et al., 2008).

Among electrochemical biosensors, amperometry has had greater success and inspired the development of commercial products based on the electron transfer mechanism. The majority of the signals produced by these biosensors are a result of an electron transfer between the electrodes and the biological interaction. In amperometry, the analyte typically undergoes a redox reaction in an electrochemical cell when voltage is applied between two electrodes. Due to the presence of biomolecules at the electrode surface, the oxidation state of the analyte changes, and the measured signal is proportional to the electroactive species. The Clark oxygen electrode is used in the most popular amperometric biosensor (Figure 10.6).

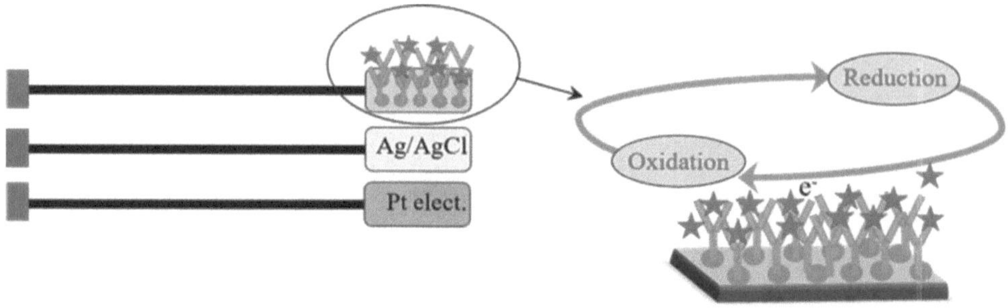

FIGURE 10.6 Schematic picture of amperometry biosensor.

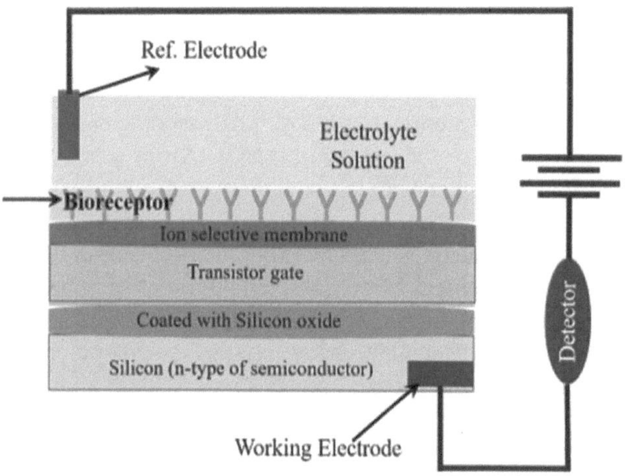

FIGURE 10.7 Schematic picture of potentiometric biosensor.

10.4.3 Potentiometric Biosensor

Ion-selective electrode (ISE) and ion-selective field effect transistor are related to the potentiometric biosensor (ISFET) (Wang, 2008; Ding & Qin, 2013). These biosensors are made by coupling biological receptors like DNA, aptamers, enzymes, antibodies, and peptides with a transducer that detects variations in proton concentrations. The recorded analytical signal is then correlated logarithmically with the analyte concentration. A potentiometric biosensor has two electrode systems: an Ag/AgCl reference electrode and a working electrode that serves as a matrix for encasing the biorecognition components or the outer surface for immobilizing biomolecules. The construction of a potentiometric biosensor uses the glass pH electrode as a simple but important transducer (Purvis et al., 2003).

In an ion-sensitive field effect transistor (ISFET), a selective membrane is used to cover the transistor gate surface. This membrane may be formed of materials such as silicon nitride (Si_3N_4), alumina (Al_2O_3), tantalum oxide (Ta_2O_5), zirconium oxide (ZrO_2), etc., The link between potential and concentration for potentiometric measurement is shown by the Nernst equation. Following is Nernst equation 120's relationship:

$$E_{\text{cell}} = E^0_{\text{cell}} - \frac{RT}{nF} \ln Q$$

Where n is the number of charges on the electrode, T denotes absolute temperature, F denotes the Faraday constant, and $\ln Q$ denotes the logarithmic analyte concentration. E_{cell} is the cell potential at zero current, E^0_{cell} denotes a constant potential contribution to the cell, R represents the universal gas constant (Figure 10.7).

10.4.4 Impedimetric Biosensor

A notable method based on the examination of interfacial characteristics connected to biorecognition activities taking place at electrode surfaces is electrochemical impedance spectroscopy (EIS). In the presence and absence of redox species, respectively, impedance is expressed in terms of Faradic and non-Faradaic states. The change in faradic current (interfacial electron transfer resistance) caused by biorecognition events is measured by a faradic biosensor. Due to biomolecular contact and electrostatic repulsion between the free charges of the target molecules and electroactive species in the electrolyte solution, the faradic current creates steric hindrance (Grieshaber et al., 2008).

A sensitive method for detecting antigen-antibody interaction is the impedimetric biosensor. Antibodies are typically immobilized on the electrode surface or semiconductor chips in impedimetric biosensors. Antigen-antibody binding causes the creation of immunocomplexes, which alters the kinetics of electron transfer. In comparison to unbound antigens with antibodies, antigen-antibody interaction increases the steric hindrance of the redox probe, and this hindrance raises the electron transfer resistance and lowers capacitance as well. The importance of this biosensor in producing reliable data cannot be overstated (Bahadır & Sezgintürk, 2016).

10.4.5 Characteristics of Biosensors

10.4.5.1 Selectivity

One of the key characteristics of biosensors that defines their effectiveness with regard to the targeted analyte is selectivity. It is based on the detection of a particular analyte in the presence of other analytes and solution pollutants. Selectivity is frequently demonstrated via interactions between bioreceptors and analytes, such as antibody–antigen, enzyme–ligand, and DNA–ligand interactions. A well-known example is selectivity toward antibody–antigen interaction (Nikhil et al., 2016).

10.4.5.2 Stability

During biosensing measurements, stability makes biosensors susceptible to environmental disruption. These alterations can impair the biosensor's precision and accuracy as well as cause errors in concentration measurements. Longer incubation times are necessary for biosensing monitoring, which is an essential component for stability. One of the key stability elements that affects the electrostatic bonding and covalent linkage with the analyte is the affinity of the bioreceptors. The stability of biosensors over time is also impacted by the deterioration of bioreceptors (Nikhil et al., 2016).

10.4.5.3 Sensitivity

The limit of detection (LOD) or sensitivity is expressed by the biosensor's ability to detect the smallest amount of analyte. Following applications for medical and environmental analysis, biosensors are needed to detect very low concentrations of analyte in the range of ng mL^{-1} to fg mL^{-1}, which shows the high efficiency of the biosensor. Sensitivity is a crucial characteristic of biosensors (Nikhil et al., 2016).

10.4.5.4 Reproducibility

The ability to produce the same results or signals for repeated trials is a crucial property of biosensors. Accuracy and precision of the transducer in a biosensor serve as a measure of reproducibility. While accuracy refers to the sensor's capacity to offer a mean value close to the genuine value during repeated measurements of the sample, precision is a crucial characteristic of sensors that produce results that are similar (Nikhil et al., 2016).

10.4.5.5 Linearity

The mathematical relationship $y=mx+c$, where c is the analyte concentration, y displays the output signal, and m is the sensitivity of the biosensor, indicates linearity and demonstrates the accuracy of measured response. The range and analyte concentrations have an impact on the linearity of biosensors. Additionally connected to linearity is the biosensor's resolution. The response of the biosensor, which is regarded as the biosensor's resolution, is influenced by even small changes in analyte concentration (Nikhil et al., 2016).

10.5 CONCLUSION

The existence of mycotoxins in food items has been found to have a harmful effect on animal and human health throughout history. In the past, the ability to dictate the poisoning of mycotoxins was very limited. However, nowadays, with our advanced knowledge and capabilities, we have the means to select and employ highly effective, scientifically validated methods to counter the harmful effects of mold infestations. To mitigate the agricultural, economic, and health risks posed by the spreading of mycotoxins, it will be imperative to develop solutions for early detection and prevention of mycotoxin contamination.

The emergence of biosensors has introduced innovative techniques for detecting analytes in the biomedical field as well as a broad range of other fields including point-of-care diagnosis of disease progression, food control, environmental monitoring, biomedical research, etc. Biosensor devices are based on the interactions of biorecognition elements with respect to their specific analytes that requires the immobilization of biomolecules on solid surfaces. Moreover, biosensors contain several characteristics compared to conventional techniques such as sensitive to their analytes, require very little amount of sample, robustness, specificity, quick response, portability, etc. These properties of biosensors indicate their efficiency toward their analytes.

REFERENCES

Abarca, M. L., Bragulat, M. R., Castella, G., & Cabanes, F. J. (1994). Ochratoxin A production by strains of Aspergillus niger var. niger. *Applied and Environmental Microbiology*, 60(7), 2650–2652.

Abbas, H. K., Duke, S. O., Shier, W. T., & Duke, M. V. (2002). Inhibition of ceramide synthesis in plants by phytotoxins. In: Upadhyay, R.K., ed. *Advances in Microbial Toxin Research and Its Biotechnological Exploitation*, Berlin/Heidelberg, Germany: Springer, 211–229.

Abbax, H. K., Gelderblom, W. C. A., Cawood, M. E., & Shier, W. T. (1993). Biological activities of fumonisins, mycotoxins from Fusarium moniliforme, in jimsonweed (Datura stramonium L.) and mammalian cell cultures. *Toxicon*, 31(3), 345–353.

Abdel-Wahhab, M. A., Omara, E. A., Abdel-Galil, M. M., Hassan, N. S., Nada, S. A., Saeed, A., & ElSayed, M. M. (2007). Zizyphus spina-christi extract protects against aflatoxin B1-intitiated hepatic carcinogenicity. *African Journal of Traditional, Complementary, and Alternative Medicines*, 4(3), 248.

Abramson, D., Usleber, E., & Märtlbauer, E. (1995). An indirect enzyme immunoassay for the mycotoxin citrinin. *Applied and Environmental Microbiology*, 61(5), 2007–2009.

Anwar, N., Vagin, M., Laffir, F., Armstrong, G., Dickinson, C., & McCormac, T. (2011). Transition metal ion-substituted polyoxometalates entrapped into polypyrrole as electrochemical sensor for hydrogen peroxide. *Analyst*, 137, 624–630.

Arbillaga, L., Azqueta, A., van Delft, J. H. M., & de Cerain, A. L. (2007). In vitro gene expression data supporting a DNA non-reactive genotoxic mechanism for ochratoxin A. *Toxicology and Applied Pharmacology*, 220(2), 216–224.

Arora, S., Nagpal, R., Gusain, M., Singh, B., Pan, Y., Yadav, D., Ahmed, I., Kumar, V., & Parshad, B. (2023). Organic–inorganic porphyrinoid frameworks for biomolecule sensing. *ACS Sensors*, 8(2), 443–464.

Authority, E. F. S. (2013). Deoxynivalenol in food and feed: Occurrence and exposure. *EFSA Journal*, 11(10), 3379.

Azziz-Baumgartner, E., Lindblade, K., Gieseker, K., Rogers, H. S., Kieszak, S., Njapau, H., Schleicher, R., McCoy, L. F., Misore, A., & DeCock, K. (2005). Case–control study of an acute aflatoxicosis outbreak, Kenya, 2004. *Environmental Health Perspectives*, 113(12), 1779–1783.

Bahadır, E. B., & Sezgintürk, M. K. (2016). A review on impedimetric biosensors. *Artificial Cells, Nanomedicine, and Biotechnology*, 44(1), 248–262.

Bakos, K., Kovács, R., Staszny, Á., Sipos, D. K., Urbányi, B., Müller, F., Csenki, Z., & Kovács, B. (2013). Developmental toxicity and estrogenic potency of zearalenone in zebrafish (Danio rerio). *Aquatic Toxicology*, 136, 13–21.

Bayman, P., Baker, J. L., Doster, M. A., Michailides, T. J., & Mahoney, N. E. (2002). Ochratoxin production by the Aspergillus ochraceus group and Aspergillus alliaceus. *Applied and Environmental Microbiology*, 68(5), 2326–2329.

Beasley, V. R. (2017). *Trichothecene Mycotoxicosis Pathophysiologic Effects (1989): Volume II*. CRC Press, Boca Raton, FL.

Belkin, S. (2003). Microbial whole-cell sensing systems of environmental pollutants. *Current Opinion in Microbiology*, 6(3), 206–212.

Belluzo, M. S., Ribone, M. É., & Lagier, C. M. (2008). Assembling amperometric biosensors for clinical diagnostics. *Sensors*, 8(3), 1366–1399.

Bennett, J., & Klich, M. (2003). Mycotoxins. *Clinical Microbiology Reviews*, 16, 497–516.

Berthiller, F., Crews, C., Dall'Asta, C., Saeger, S. De, Haesaert, G., Karlovsky, P., Oswald, I. P., Seefelder, W., Speijers, G., & Stroka, J. (2013). Masked mycotoxins: A review. *Molecular Nutrition & Food Research*, 57(1), 165–186.

Bezuidenhout, S. C., Gelderblom, W. C. A., Gorst-Allman, C. P., Horak, R. M., Marasas, W. F. O., Spiteller, G., & Vleggaar, R. (1988). Structure elucidation of the fumonisins, mycotoxins from Fusarium moniliforme. *Journal of the Chemical Society, Chemical Communications*, 11, 743–745.

Blount, W. P. (1961). Turkey "X" disease. *Turkeys*, 9(2), 52–55.

Bose, R., Verheij, M., Haimovitz-Friedman, A., Scotto, K., Fuks, Z., & Kolesnick, R. (1995). Ceramide synthase mediates daunorubicin-induced apoptosis: An alternative mechanism for generating death signals. *Cell*, 82(3), 405–414.

Braun, M. S., & Wink, M. (2018). Exposure, occurrence, and chemistry of fumonisins and their cryptic derivatives. *Comprehensive Reviews in Food Science and Food Safety*, 17(3), 769–791.

Bucci, T. J., Hansen, D. K., & Laborde, J. B. (1996). Leukoencephalomalacia and hemorrhage in the brain of rabbits gavaged with mycotoxin fumonisin B1. *Natural Toxins*, 4(1), 51–52.

Castegnaro, M., Bartsch, H., & Chernozemsky, I. (1987). Endemic nephropathy and urinary tract tumors in the balkans. *Cancer Research*, 47(13), 3608–3609.

Cavin, C., Delatour, T., Marin-Kuan, M., Fenaille, F., Holzhäuser, D., Guignard, G., Bezençon, C., Piguet, D., Parisod, V., & Richoz-Payot, J. (2009). Ochratoxin a–mediated DNA and protein damage: Roles of nitrosative and oxidative stresses. *Toxicological Sciences*, 110(1), 84–94.

Cendoya, E., Chiotta, M. L., Zachetti, V., Chulze, S. N., & Ramirez, M. L. (2018). Fumonisins and fumonisin-producing Fusarium occurrence in wheat and wheat by products: A review. *Journal of Cereal Science*, 80, 158–166.

Chen, W., Li, C., Zhang, B., Zhou, Z., Shen, Y., Liao, X., Yang, J., Wang, Y., Li, X., & Li, Y. (2018). Advances in biodetoxification of ochratoxin AA review of the past five decades. *Frontiers in Microbiology*, 9, 1386.

Ciegler, A. (1972). Ochratoxin synthesis by Penicillium species. *Naturwissenschaften*, 59(8), 365–366.

Cirlini, M., Generotti, S., Dall'Erta, A., Lancioni, P., Ferrazzano, G., Massi, A., Galaverna, G., & Dall'Asta, C. (2013). Durum wheat (Triticum Durum Desf.) lines show different abilities to form masked mycotoxins under greenhouse conditions. *Toxins*, 6(1), 81–95.

Clark Jr, L. C., & Lyons, C. (1962). Electrode systems for continuous monitoring in cardiovascular surgery. *Annals of the New York Academy of Sciences*, 102(1), 29–45.

Creppy, E. E., Baudrimont, I., Belmadani, A., & Betbeder, A. M. (1996). Aspartame as a preventive agent of chronic toxic effects of ochratoxin A in experimental animals. *Journal of Toxicology: Toxin Reviews*, 15(3), 207–221.

Dai, Q., Zhao, J., Qi, X., Xu, W., He, X., Guo, M., Dweep, H., Cheng, W.-H., Luo, Y., & Xia, K. (2014). MicroRNA profiling of rats with ochratoxin A nephrotoxicity. *BMC Genomics*, 15(1), 1–14.

Dall'Asta, C., & Battilani, P. (2016). Fumonisins and their modified forms, a matter of concern in future scenario? *World Mycotoxin Journal*, 9(5), 727–739.

Damiani, T., Righetti, L., Suman, M., Galaverna, G., & Dall'Asta, C. (2019). Analytical issue related to fumonisins: A matter of sample comminution? *Food Control*, 95, 1–5.

Ding, J., & Qin, W. (2013). Potentiometric sensing of nuclease activities and oxidative damage of single-stranded DNA using a polycation-sensitive membrane electrode. *Biosensors and Bioelectronics*, 47, 559–565.

Dinu, D., Bodea, G. O., Ceapa, C. D., Munteanu, M. C., Roming, F. I., Serban, A. I., Hermenean, A., Costache, M., Zarnescu, O., & Dinischiotu, A. (2011). Adapted response of the antioxidant defense system to oxidative stress induced by deoxynivalenol in Hek-293 cells. *Toxicon*, 57(7–8), 1023–1032.

Dixon, D. M., McNeil, M. M., Cohen, M. L., Gellin, B. G., & La Montagne, J. R. (1996). Fungal infections: A growing threat. *Public Health Reports*, 111(3), 226.

Dohnal, V., Wu, Q., & Kuča, K. (2014). Metabolism of aflatoxins: Key enzymes and interindividual as well as interspecies differences. *Archives of Toxicology*, 88(9), 1635–1644.

Duarte, S. C., Pena, A., & Lino, C. M. (2010). A review on ochratoxin A occurrence and effects of processing of cereal and cereal derived food products. *Food Microbiology*, 27(2), 187–198.
Eaton, D. L., & Groopman, J. D. (2013). *The Toxicology of Aflatoxins: Human Health, Veterinary, and Agricultural Significance*. Elsevier, Amsterdam, The Netherlands.
Eaton, D. L., & Ramsdell, H. S. (1992). Species and diet related differences in aflatoxin biotransformation. *Handbook of Applied Mycology*, 5, 157–182.
Eggins, B. R. (2002). *Chemical Sensors and Biosensors* (Vol. 2). John Wiley & Sons, Hoboken, NJ.
El-Nezami, H., Kankaanpää, P., Salminen, S., & Ahokas, J. (1998). Physicochemical alterations enhance the ability of dairy strains of lactic acid bacteria to remove aflatoxin from contaminated media. *Journal of Food Protection*, 61(4), 466–468.
Etzel, R. A. (2002). Mycotoxins. *Journal of the American Medical Association*, *287*(4), 425–427.
Fallah, A. A., Rahnama, M., Jafari, T., & Saei-Dehkordi, S. S. (2011). Seasonal variation of aflatoxin M1 contamination in industrial and traditional Iranian dairy products. *Food Control*, 22(10), 1653–1656.
FAO. (2001). Safety Evaluation of Certain Mycotoxins in Food. Prepared by the Fifty-sixth meeting of the Joint FAO/WHO Expert Committee on Food Additives (JECFA). FAO Food and Nutrition Paper No. 74. Rome, Italy.
Fink-Grernmels, J. (1999). Mycotoxins: Their implications for human and animal health. *Veterinary Quarterly*, 21(4), 115–120.
Food and Drug Administration. (2012). *Bad Bug Book: Handbook of Foodborne Pathogenic Microorganisms and Natural Toxins*. Center for Food Safety and Applied Nutrition, Washington, DC.
Foroud, N. A., & Eudes, F. (2009). Trichothecenes in cereal grains. *International Journal of Molecular Sciences*, 10(1), 147–173.
Freire, L., Passamani, F. R. F., Thomas, A. B., Nassur, R. de C. M. R., Silva, L. M., Paschoal, F. N., Pereira, G. E., Prado, G., & Batista, L. R. (2017). Influence of physical and chemical characteristics of wine grapes on the incidence of Penicillium and Aspergillus fungi in grapes and ochratoxin A in wines. *International Journal of Food Microbiology*, 241, 181–190.
Frisvad, J. C., Smedsgaard, J., Samson, R. A., Larsen, T. O., & Thrane, U. (2007). Fumonisin B2 production by Aspergillus niger. *Journal of Agricultural and Food Chemistry*, 55(23), 9727–9732.
Gajęcka, M., Stopa, E., Tarasiuk, M., Zielonka, Ł., & Gajęcki, M. (2013). The expression of type-1 and type-2 nitric oxide synthase in selected tissues of the gastrointestinal tract during mixed mycotoxicosis. *Toxins*, 5(11), 2281–2292.
Gallo, A., Ferrara, M., & Perrone, G. (2017). Recent advances on the molecular aspects of ochratoxin A biosynthesis. *Current Opinion in Food Science*, 17, 49–56.
Gallo, A., Solfrizzo, M., Epifani, F., Panzarini, G., & Perrone, G. (2016). Effect of temperature and water activity on gene expression and aflatoxin biosynthesis in Aspergillus flavus on almond medium. *International Journal of Food Microbiology*, 217, 162–169.
Gärtner, B. H., Munich, M., Kleijer, G., & Mascher, F. (2008). Characterisation of kernel resistance against Fusarium infection in spring wheat by baking quality and mycotoxin assessments. *European Journal of Plant Pathology*, 120, 61–68.
Gelderblom, W. C., Abel, S., Smuts, C. M., Marnewick, J., Marasas, W. F., Lemmer, E. R., & Ramljak, D. (2001). Fumonisin-induced hepatocarcinogenesis: Mechanisms related to cancer initiation and promotion. *Environmental Health Perspectives*, 109(suppl 2), 291–300.
Generotti, S., Cirlini, M., Malachova, A., Sulyok, M., Berthiller, F., Dall'Asta, C., & Suman, M. (2015). Deoxynivalenol & deoxynivalenol-3-glucoside mitigation through bakery production strategies: Effective experimental design within industrial rusk-making technology. *Toxins*, 7(8), 2773–2790.
Gillman, I. G., Clark, T. N., & Manderville, R. A. (1999). Oxidation of ochratoxin a by an Fe– porphyrin system: Model for enzymatic activation and DNA cleavage. *Chemical Research in Toxicology*, 12(11), 1066–1076.
Giray, B., Girgin, G., Engin, A. B., Aydın, S., & Sahin, G. (2007). Aflatoxin levels in wheat samples consumed in some regions of Turkey. *Food Control*, 18(1), 23–29.
Goto, T., Wicklow, D. T., & Ito, Y. (1996). Aflatoxin and cyclopiazonic acid production by a sclerotium-producing Aspergillus tamarii strain. *Applied and Environmental Microbiology*, 62(11), 4036–4038.
Grieshaber, D., MacKenzie, R., Vörös, J., & Reimhult, E. (2008). Electrochemical biosensors-sensor principles and architectures. *Sensors*, 8(3), 1400–1458.
Haschek, W. M., Rousseaux, C. G., Wallig, M. A., Bolon, B., & Ochoa, R. (2013). *Haschek and Rousseaux's Handbook of Toxicologic Pathology*. Academic Press, New York.
Heenan, C. N., Shaw, K. J., & Pitt, J. I. (1998). Ochratoxin A production by Aspergillus carbonarius and A. niger isolates and detection using coconut cream agar. *Journal of Food Mycology*, 1, 67–72.

Henry, S. H., Bosch, F. X., Troxell, T. C., & Bolger, P. M. (1999). Reducing liver cancer – global control of aflatoxin. *Science*, 286(5449), 2453–2454.

Hope, J. H., & Hope, B. E. (2012). A review of the diagnosis and treatment of Ochratoxin A inhalational exposure associated with human illness and kidney disease including focal segmental glomerulosclerosis. *Journal of Environmental and Public Health*, 2012, 835059.

Hope, R., Aldred, D., & Magan, N. (2005). Comparison of environmental profiles for growth and deoxynivalenol production by Fusarium culmorum and F. graminearum on wheat grain. *Letters in Applied Microbiology*, 40(4), 295–300.

Hussain, I., & Anwar, J. (2008). A study on contamination of aflatoxin M1 in raw milk in the Punjab province of Pakistan. *Food Control*, 19(4), 393–395.

Hussain, S. S., & Bloom, S. R. (2013). The regulation of food intake by the gut-brain axis: Implications for obesity. *International Journal of Obesity*, 37(5), 625–633.

IARC Working Group on the Evaluation of Carcinogenic Risks to Humans. (2002). *Some Traditional Herbal Medicines, some Mycotoxins, Naphthalene and Styrene* (Vol. 82). World Health Organization, Geneva, Switzerland.

Inan, F., Pala, M., & Doymaz, I. (2007). Use of ozone in detoxification of aflatoxin B1 in red pepper. *Journal of Stored Products Research*, 43(4), 425–429.

International Agency for Research on Cancer. (1982). Evaluation of the carcinogenic risk of chemicals to humans. *Chemicals, Industrial Processes and Industries Associated with Cancer in Humans*, 1982(Issue Suppl. 4), 1.

Jaimez, J., Fente, C. A., Vazquez, B. I., Franco, C. M., Cepeda, A., Mahuzier, G., & Prognon, P. (2000). Application of the assay of aflatoxins by liquid chromatography with fluorescence detection in food analysis. *Journal of Chromatography A*, 882(1–2), 1–10.

Jelinek, C. F., Pohland, A. E., & Wood, G. E. (1989). Worldwide occurrence of mycotoxins in foods and feeds—an update. *Journal of the Association of Official Analytical Chemists*, 72(2), 223–230.

Joint FAO. (1999). Third Joint FAO/WHO/UNEP International Conference on Mycotoxins [computer file]. Tunis, Tunisia.

Kelly, J. D., Eaton, D. L., Guengerich, F. P., & Coulombe Jr, R. A. (1997). Aflatoxin B1 activation in human lung. *Toxicology and Applied Pharmacology*, 144(1), 88–95.

Khaneghah, A. M., Martins, L. M., von Hertwig, A. M., Bertoldo, R., & Sant'Ana, A. S. (2018). Deoxynivalenol and its masked forms: Characteristics, incidence, control and fate during wheat and wheat based products processing-A review. *Trends in Food Science & Technology*, 71, 13–24.

Klich, M. A., Mullaney, E. J., Daly, C. B., & Cary, J. W. (2000). Molecular and physiological aspects of aflatoxin and sterigmatocystin biosynthesis by Aspergillus tamarii and A. ochraceoroseus. *Applied Microbiology and Biotechnology*, 53(5), 605–609.

Kőszegi, T., & Poór, M. (2016). Ochratoxin A: Molecular interactions, mechanisms of toxicity and prevention at the molecular level. *Toxins*, 8(4), 111.

Krishnamachari, K., Nagarajan, V., Bhat, R., & Tilak, T. B. G. (1975). Hepatitis due to aflatoxicosis: An outbreak in western India. *The Lancet*, 305(7915), 1061–1063.

Kumar, P., Mahato, D. K., Kamle, M., Mohanta, T. K., & Kang, S. G. (2017). Aflatoxins: A global concern for food safety, human health and their management. *Frontiers in Microbiology*, 7, 2170.

Lim, C. W., Yoshinari, T., Layne, J., & Chan, S. H. (2015). Multi-mycotoxin screening reveals separate occurrence of aflatoxins and ochratoxin A in Asian rice. *Journal of Agricultural and Food Chemistry*, 63(12), 3104–3113.

MacDiarmid, A. G. (2001). "Synthetic metals": A novel role for organic polymers (Nobel lecture). *Angewandte Chemie International Edition*, 40(14), 2581–2590.

Malbrán, I., Mourelos, C. A., Girotti, J. R., Balatti, P. A., & Lori, G. A. (2014). Toxigenic capacity and trichothecene production by Fusarium graminearum isolates from Argentina and their relationship with aggressiveness and fungal expansion in the wheat spike. *Phytopathology*, 104(4), 357–364.

Mally, A., Zepnik, H., Wanek, P., Eder, E., Dingley, K., Ihmels, H., Völkel, W., & Dekant, W. (2004). Ochratoxin A: Lack of formation of covalent DNA adducts. *Chemical Research in Toxicology*, 17(2), 234–242.

Marasas, W. F. (1996). *Fumonisins in Food*. Springer, Berlin/Heidelberg, Germany.

Marasas, W. F. O., Kellerman, T. S., Gelderblom, W. C. A., Thiel, P. G., Van der Lugt, J. J., & Coetzer, J. A. W. (1988). Leukoencephalomalacia in a horse induced by fumonisin B_1 isolated from Fusarium moniliforme. *The Onderstepoort Journal of Veterinary Research*, 55, 197–203.

Marasas, W. F. O., Miller, J. D., Riley, R. T., & Visconti, A. (2001). *Fumonisins-Occurrence, Toxicology, Metabolism and Risk Assessment*. Fusarium. Paul E. Nelson Memorial Symposium. APS PRESS. The American Phytopathological Society, St. Paul, Minnesota, 332–359.

Marin, D. E., & Taranu, I. (2015). Ochratoxin A and its effects on immunity. *Toxin Reviews*, 34(1), 11–20.
Mitchell, D., Parra, R., Aldred, D., & Magan, N. (2004). Water and temperature relations of growth and ochratoxin A production by Aspergillus carbonarius strains from grapes in Europe and Israel. *Journal of Applied Microbiology*, 97(2), 439–445.
Moss, M. O. (1998). Recent studies of mycotoxins. *Symposium Series (Society for Applied Microbiology)*, 27, 62S–76S.
Mukherjee, K., & Lakshminarasimham, A. V. (1995). Aflatoxin contamination of sorghum seeds during storage under controlled conditions. *Zentralblatt Für Bakteriologie*, 282(3), 237–243.
Nagy, C. M., Fejer, S. N., Berek, L., Molnar, J., & Viskolcz, B. (2005). Hydrogen bondings in deoxynivalenol (DON) conformations—a density functional study. *Journal of Molecular Structure: THEOCHEM*, 726(1–3), 55–59.
Narvaez, A., Izzo, L., Rodriguez-Carrasco, Y., & Ritieni, A. (2021). Citrinin dietary exposure assessment approach through human biomonitoring high-resolution mass spectrometry-based data. *Journal of Agricultural and Food Chemistry*, 69(22), 6330–6338.
Ngindu, A., Kenya, P., Ocheng, D., Omondi, T., Ngare, W., Gatei, D., Johnson, B., Ngira, J., Nandwa, H., & Jansen, A. (1982). Outbreak of acute hepatitis caused by aflatoxin poisoning in Kenya. *The Lancet*, 319(8285), 1346–1348.
Nikhil, B., Pawan, J., Nello, F., & Pedro, E. (2016). Introduction to biosensors. *Essays Biochemistry*, 60(1), 1–8.
Northolt, M. D., Van Egmond, H. P., & Paulsch, W. E. (1979). Ochratoxin A production by some fungal species in relation to water activity and temperature. *Journal of Food Protection*, 42(6), 485–490.
Payne, G. A., & Brown, M. P. (1998). Genetics and physiology of aflatoxin biosynthesis. *Annual Review of Phytopathology*, 36(1), 329–362.
Peterson, S. W., Ito, Y., Horn, B. W., & Goto, T. (2001). Aspergillus bombycis, a new aflatoxigenic species and genetic variation in its sibling species, A. nomius. *Mycologia*, 93(4), 689–703.
Pfohl-Leszkowicz, A., & Manderville, R. A. (2007). Ochratoxin A: An overview on toxicity and carcinogenicity in animals and humans. *Molecular Nutrition & Food Research*, 51(1), 61–99.
Pfohl-Leszkowicz, A., & Manderville, R. A. (2012). An update on direct genotoxicity as a molecular mechanism of ochratoxin a carcinogenicity. *Chemical Research in Toxicology*, 25(2), 252–262.
Pfohl-Leszkowicz, A., Petkova-Bocharova, T., Chernozemsky, I. N., & Castegnaro, M. (2002). Balkan endemic nephropathy and associated urinary tract tumours: A review on aetiological causes and the potential role of mycotoxins. *Food Additives & Contaminants*, 19(3), 282–302.
Pietsch, C., Michel, C., Kersten, S., Valenta, H., Dänicke, S., Schulz, C., Kloas, W., & Burkhardt-Holm, P. (2014). In vivo effects of deoxynivalenol (DON) on innate immune responses of carp (Cyprinus carpio L.). *Food and Chemical Toxicology*, 68, 44–52.
Pitt, J. I. (2000). Toxigenic fungi and mycotoxins. *British Medical Bulletin*, 56(1), 184–192.
Pitt, J. I., & Hocking, A. D. (2009). *Fungi and Food Spoilage* (Vol. 519). Springer, Berlin.
Pohland, A. E., Nesheim, S., & Friedman, L. (1992). Ochratoxin A: A review (technical report). *Pure and Applied Chemistry*, 64(7), 1029–1046.
Program, N. T. (2009). Chemical Information Review Document for Deoxynivalenol [CAS No. 51481-10-8]. https://www.semanticscholar.org/paper/Chemical-Information-Review-Document-for-No/84fbdb5ae44db54619cd724428d941591d07e1f8. Last Accessed on March, 16, 2014.
Purvis, D., Leonardova, O., Farmakovsky, D., & Cherkasov, V. (2003). An ultrasensitive and stable potentiometric immunosensor. *Biosensors and Bioelectronics*, 18(11), 1385–1390.
Quintela, S., Villarán, M. C., de Armentia, I. L., & Elejalde, E. (2013). Ochratoxin A removal in wine: A review. *Food Control*, 30(2), 439–445.
Raj, H. G., Prasanna, H. R., Magee, P. N., & Lotlikar, P. D. (1986). Effect of purified rat and hamster hepatic glutathione S-transferases on the microsome mediated binding of aflatoxin B1 to DNA. *Cancer Letters*, 33(1), 1–9.
Ramos, A. J., Labernia, N., Marın, S., Sanchis, V., & Magan, N. (1998). Effect of water activity and temperature on growth and ochratoxin production by three strains of Aspergillus ochraceus on a barley extract medium and on barley grains. *International Journal of Food Microbiology*, 44(1–2), 133–140.
Reddy, P., Hemsworth, J., Guthridge, K. M., Vinh, A., Vassiliadis, S., Ezernieks, V., Spangenberg, G. C., & Rochfort, S. J. (2020). Ergot alkaloid mycotoxins: Physiological effects, metabolism and distribution of the residual toxin in mice. *Scientific Reports*, 10(1), 9714.
Richard, J. L. (2007). Some major mycotoxins and their mycotoxicoses—An overview. *International Journal of Food Microbiology*, 119(1–2), 3–10.

Riley, R. T., Voss, K. A., Norred, W. P., Sharma, R. P., Wang, E., & Merrill, A. H. (1998). Fumonisins: Mechanism of mycotoxicity. *Revue de Medecine Veterinaire (France)*, 149, 617–626.

Rodricks, J. V, Hesseltine, C. W., & Mehlman, M. A. (1977). *Mycotoxins in Human and Animal Health.* Pathotox Publishers, Incorporated, Park Forest South, IL.

Rotter, B. A. (1996). Invited review: Toxicology of deoxynivalenol (vomitoxin). *Journal of Toxicology and Environmental Health Part A*, 48(1), 1–34.

Russo, A., La Fauci, L., Açquaviva, R., Campisi, A., Raciti, G., Scifo, C., Renis, M., Galvano, G., Vanella, A., & Galvano, F. (2005). Ochratoxin A-induced DNA damage in human fibroblast: Protective effect of cyanidin 3-O-β-D-glucoside. *The Journal of Nutritional Biochemistry*, 16(1), 31–37.

Rychlik, M., Humpf, H.-U., Marko, D., Dänicke, S., Mally, A., Berthiller, F., Klaffke, H., & Lorenz, N. (2014). Proposal of a comprehensive definition of modified and other forms of mycotoxins including "masked" mycotoxins. *Mycotoxin Research*, 30, 197–205.

Sanchis, V., & Magan, N. (2004). Environmental conditions affecting mycotoxins. In: Magan, N., & Olsen, M., eds. *Mycotoxins in Food: Detection and Control*, Sawston: Woodhead Publishing, 174–189.

Sarma, P. U. (2010). *Fascinating Potential of Aspergilli.* Springer, Berlin.

Sarma, U. P., Bhetaria, P. J., Devi, P., & Varma, A. (2017). Aflatoxins: Implications on health. *Indian Journal of Clinical Biochemistry*, 32(2), 124–133.

Schmidt-Heydt, M., Abdel-Hadi, A., Magan, N., & Geisen, R. (2009). Complex regulation of the aflatoxin biosynthesis gene cluster of Aspergillus flavus in relation to various combinations of water activity and temperature. *International Journal of Food Microbiology*, 135(3), 231–237.

Schmidt-Heydt, M., Rüfer, C. E., Abdel-Hadi, A., Magan, N., & Geisen, R. (2010). The production of aflatoxin B1 or G1 by Aspergillus parasiticus at various combinations of temperature and water activity is related to the ratio of afl S to afl R expression. *Mycotoxin Research*, 26(4), 241–246.

Severns, D. E., Clements, M. J., Lambert, R. J., & White, D. G. (2003). Comparison of Aspergillus ear rot and aflatoxin contamination in grain of high-oil and normal-oil corn hybrids. *Journal of Food Protection*, 66(4), 637–643.

Singh, A. K., Dhiman, T. K., Lakshmi, G., & Solanki, P. R. (2021). Dimanganese trioxide (Mn_2O_3) based label-free electrochemical biosensor for detection of Aflatoxin-B1. *Bioelectrochemistry*, 137, 107684.

Smoke, T., & Smoking, I. (2004). *IARC Monographs on the Evaluation of Carcinogenic Risks to Humans.* IARC, Lyon, 1–1452.

Sobrova, P., Adam, V., Vasatkova, A., Beklova, M., Zeman, L., & Kizek, R. (2010). Deoxynivalenol and its toxicity. *Interdisciplinary Toxicology*, 3(3), 94–99.

Soriano, J. M., Gonzalez, L., & Catala, A. I. (2005). Mechanism of action of sphingolipids and their metabolites in the toxicity of fumonisin B1. *Progress in Lipid Research*, 44(6), 345–356.

Suárez-Bonnet, E., Carvajal, M., Méndez-Ramírez, I., Castillo-Urueta, P., Cortés-Eslava, J., Gómez-Arroyo, S., & Melero-Vara, J. M. (2013). Aflatoxin (B1, B2, G1, and G2) contamination in rice of Mexico and Spain, from local sources or imported. *Journal of Food Science*, 78(11), T1822–T1829.

Sun, X. D., Su, P., & Shan, H. (2017). Mycotoxin contamination of maize in China. *Comprehensive Reviews in Food Science and Food Safety*, 16(5), 835–849.

Swartzman, E. E., Miraglia, S. J., Mellentin-Michelotti, J., Evangelista, L., & Yuan, P.-M. (1999). A homogeneous and multiplexed immunoassay for high-throughput screening using fluorometric microvolume assay technology. *Analytical Biochemistry*, 271(2), 143–151.

Tabari, M., Karim, G., Ghavami, M., & Chamani, M. (2011). Method validation for aflatoxin M1 determination in yoghurt using immunoaffinity column clean-up prior to high-performance liquid chromatography. *Toxicology and Industrial Health*, 27(7), 629–635.

Tejada-Simon, M. V., Marovatsanga, L. T., & Pestka, J. J. (1995). Comparative detection of fumonisin by HPLC, ELISA, and immunocytochemical localization in Fusarium cultures. *Journal of Food Protection*, 58(6), 666–672.

Thévenot, D. R., Toth, K., Durst, R. A., & Wilson, G. S. (2001). Electrochemical biosensors: Recommended definitions and classification. *Biosensors and Bioelectronics*, 16(1–2), 121–131.

Tola, M., & Kebede, B. (2016). Occurrence, importance and control of mycotoxins: A review. *Cogent Food & Agriculture*, 2(1), 1191103.

Tozlovanu, M., Faucet-Marquis, V., Pfohl-Leszkowicz, A., & Manderville, R. A. (2006). Genotoxicity of the hydroquinone metabolite of ochratoxin A: Structure-activity relationships for covalent DNA adduction. *Chemical Research in Toxicology*, 19(9), 1241–1247.

Van der Merwe, K. J., Steyn, P. S., Fourie, L., Scott, D. B., & Theron, J. J. (1965). Ochratoxin A, a toxic metabolite produced by Aspergillus ochraceus Wilh. *Nature*, 205(4976), 1112–1113.

Wang, J. (2008). Electrochemical glucose biosensors. *Chemical Reviews*, 108(2), 814–825.

Wilson, T. M., Ross, P. F., Owens, D. L., Rice, L. G., Green, S. A., Jenkins, S. J., & Nelson, H. A. (1992). Experimental reproduction of ELEM. *Mycopathologia*, 117(1), 115–120.

Yazar, S., & Omurtag, G. Z. (2008). Fumonisins, trichothecenes and zearalenone in cereals. *International Journal of Molecular Sciences*, 9(11), 2062–2090.

Zhang, M. (1997). Tumor necrosis factor. In: Thomson, A.W., ed. *The Cytokine Handbook*, San Diego, CA: Academic Press, 517–548.

Zhang, X., Jiang, L., Geng, C., Cao, J., & Zhong, L. (2009). The role of oxidative stress in deoxynivalenol-induced DNA damage in HepG2 cells. *Toxicon*, 54(4), 513–518.

11 Reviewing the Prospects of Machine Learning for High-Performance Electrochemical Sensors

Vipin Shukla, Paawan Sharma, and Manish Kumar

11.1 INTRODUCTION

The extraordinary development of machine learning (ML) techniques is making an indelible impact on a variety of scientific fields, and the landscape of electrochemical sensing is not an exception to this rule [1]. Electrochemical sensors, renowned for their capacity to detect and quantify analytes across various applications, including environmental monitoring, medical diagnostics, industrial process control, and more, are currently undergoing a paradigm shift due to the incorporation of ML techniques [2]. This shift is being brought about by the fact that electrochemical sensors can now use ML methodologies [3]. regression, classification, and clustering stand out as transformational techniques that can catapult electrochemical sensors into a new era of high-performance capabilities and multifarious applications [4–7]. Among them, these three stand out as the most prominent transformative technologies.

It turns out that regression, which is a fundamental part of ML, is a crucial method that may be utilized to deal with the complexities of electrochemical sensing [8]. Regression analysis aims to establish a link between the variables that serve as input and a continuous target variable. In the context of electrochemical sensors, this translates into the prediction of analyte concentrations based on the sensor's response. These kinds of predictive skills have far-reaching ramifications and provide a way to calibrate sensors to work at their very best [9]. Calibration allows sensors to reliably translate readings into meaningful concentration values by determining the connection between raw sensor outputs and actual analyte concentrations. This is accomplished by comparing raw sensor outputs to actual analyte concentrations. Notably, ML regression models also enable real-time monitoring, a process in which analyte concentrations are rapidly anticipated based on sensor outputs [10]. This paves the way for quick interventions and informed decision-making in applications that require speedy replies.

Within the realm of electrochemical sensing, the applications of ML regression are diverse and constantly developing, making them one of the field's most promising areas. Take, for example, the field of diabetes management, where the concept of continuous glucose monitoring comes into play as being of the utmost importance in maintaining stable glycemic control [11]. The data from electrochemical sensors may be used to feed ML regression models, which can then be used to predict glucose levels. This gives patients and healthcare practitioners real-time insights into glucose swings' dynamics. In a similar vein, the application of ML regression techniques in the field of environmental monitoring has proven to be quite helpful in anticipating pollutant concentrations. This, in turn, has provided crucial information for developing pollution management measures. Furthermore, the predictive power of regression-driven ML approaches finds resonance in scenarios involving corrosion rate prediction, material characterization, and battery health assessment, collectively highlighting the versatility of this technique within the electrochemical domain [12,13].

DOI: 10.1201/9781032654874-11

However, the application of ML in electrochemical sensing goes beyond simple prediction and delves into classification [14]. Classification is an essential step in organizing data into discrete categories. When it comes to distinguishing between the numerous classes contained in electrochemical data, classification algorithms that are infused with ML skills are extremely helpful. This enables tasks such as the identification of analytes, the detection of anomalies, and the control of quality. The complexities that frequently permeate electrochemical data, characterized by noise and interference, require reliable classification strategies [15]. This difficulty is met head on by ML classification algorithms, which use advanced learning skills to decipher deep patterns and correlations hidden within the data. When it comes to medical diagnostics, classification models have the potential to play a pivotal role in distinguishing between healthy and diseased states based on electrochemical biomarker profiles [16]. In addition, these models contribute to detecting faults in sensors, which ensure the dependability of data and the sustained accuracy of sensors in real-world applications. As a result, incorporating ML classification algorithms into electrochemical sensing enhances the efficacy and usefulness of sensors across various use scenarios [17,18].

Clustering brings a new facet to the electrochemical sensing environment, in contrast to the more task-oriented approaches of regression and classification. Clustering is an unsupervised ML technique that involves grouping related data points based on inherent qualities [19]. This technique clears the way for finding patterns and studying data, which, in turn, facilitates a better understanding of complex data. In the context of electrochemical sensors, clustering algorithms play a crucial role in extracting characteristic clusters within sensor responses [20]. This allows for insights into uncharacterized analytes and unanticipated anomalies to be gained. Clustering comes into play when classifying electrochemical signals, where its relevance becomes apparent. This method makes it possible to identify analytes with a high degree of precision, which results in greater accuracy in the quantification of analytes. In addition, Clustering has the potential to improve sensor calibration by classifying responses that are similar to one another, which would increase the precision of calibration curves [21]. This combination of clustering and electrochemical sensing is positioned to untangle nuanced patterns inside complicated datasets, ultimately boosting sensor performance and expanding data comprehension.

Combining ML techniques with electrochemical sensing raises performance standards and opens up new avenues for innovation and research. The convergence of ML with electrochemical sensing has resulted in the developing of "smart" sensors, which display adaptive learning characteristics and constantly improve their performance by assimilating new data inputs. These sensors are also known as "smart" gadgets. In addition, they are combining ML with sensor arrays, which ushers in a new era of more excellent selectivity by leveraging the collective power of several sensors to deliver expanded analytical capabilities. This can be done by capitalizing on the strength of multiple sensors working together. Figure 11.1 depicts the different categories of ML models and their potential applications in electrochemical sensing.

ML can be applied to several parameters such as current-voltage characteristics, sensor response kinetics, signal-to-noise ratio enhancement, analyte concentration estimation, sensor drift correction, etc. The convergence of ML methods, particularly regression, classification, and clustering, with high-performance electrochemical sensors opens up possibilities. This is because ML approaches can more accurately predict the results of experiments. These ML algorithms offer a significant advance in sensor capabilities, expanding their applicability to a broader range of domains due to their ability to reveal previously hidden patterns, make accurate predictions of analyte concentrations, and identify intricate correlations buried within electrochemical data. The harmonious integration of ML algorithms with electrochemical sensing is positioned to alter domains spanning healthcare diagnostics, environmental preservation, and beyond. As ML algorithms continue to improve, their integration with electrochemical sensing is poised to become more harmonious. In electrochemical sensing, the groundwork has been laid for a brand new era that will usher in unprecedented precision, sensitivity, and game-changing innovation.

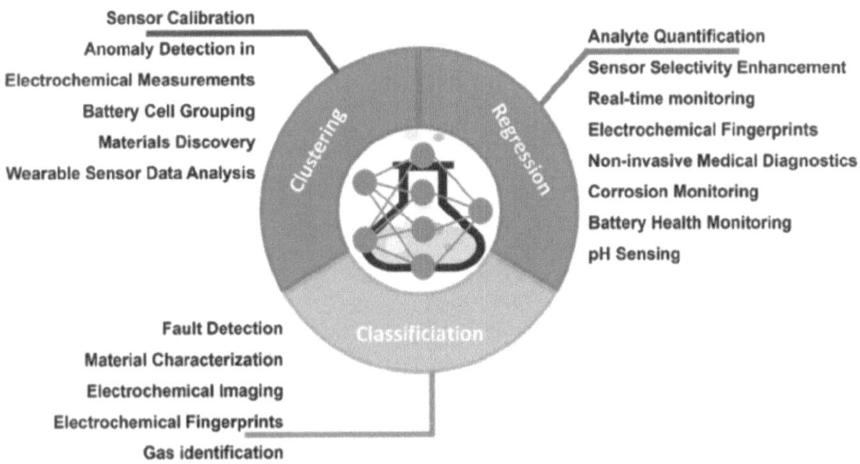

FIGURE 11.1 ML applications in several fields of electrochemical sensing.

11.2 ML MODEL BUILDING PROCESS FOR ELECTROCHEMICAL SENSORS

The development of ML models for electrochemical sensors is a multi-step process that requires the completion of various essential phases. This method involves a variety of steps and components. Using ML to improve the capabilities of electrochemical sensors requires many steps, each of which plays an essential part in the overall process. These steps include performing tests and assessing the performance of models. Let's go deeper into the specifics of each stage, paying particular attention to data preparation, data preprocessing, supervised learning (including regression and classification), and unsupervised learning processes. Figure 11.2 gives an overview of the steps involved in the ML model-building process.

11.2.1 Experimentation

Experimentation is essential to any ML project focusing on electrochemical sensors since it provides the data necessary to validate hypotheses and guide future work. In these investigations, sensors are subjected to carefully controlled environments in which specific analytes are present, and data are collected regarding the reactions of the sensors. These data will be the foundation for future model building and validation efforts.

11.2.2 Preparing the Data

Once the data from the experiment have been gathered, the next stage, which is the data preparation stage, may begin. In this stage, the data are organized in a structured manner appropriate for ML model training. It is necessary to identify the pertinent characteristics and qualities. This may contain, but is not limited to, the raw sensor outputs, analyte concentrations, experimental conditions (such as temperature and pH), and perhaps the electrode type used. Supervised and unsupervised learning will benefit significantly from access to this information.

11.2.2.1 Data Preprocessing

The data preprocessing stage is a crucial step that guarantees the data for the model training are of high quality and can be used effectively [22]. In the field of electrochemical sensing, this phase is comprised of numerous crucial components, including the following:

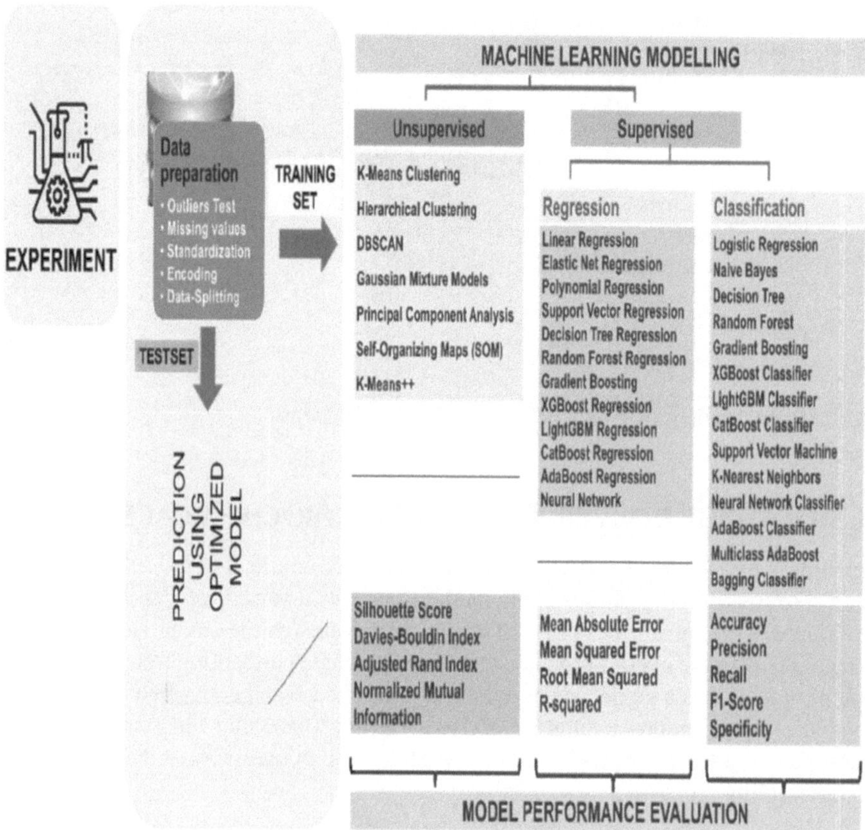

FIGURE 11.2 Overview of ML model-building process for electrochemical sensing.

11.2.2.2 Managing Missing Values

The presence of noise or experimental conditions might make electrochemical data susceptible to missing values, which must be managed [23]. Filling up these gaps is absolutely necessary for achieving proper model training. Depending on the data's composition, one may use several different approaches, such as mean imputation or interpolation.

11.2.2.3 Detecting and Removing Outliers

It is important because outliers can skew the learning process of ML models. Extreme sensor responses might result in erroneous predictions or classifications, making it necessary to identify and handle any outliers that may occur [24]. Methods such as the Z-score and the interquartile range (IQR) can assist in detecting and managing outliers.

11.2.2.4 Normalization and Scaling

Data from electrochemical sensors may be provided in a variety of scales, which has the potential to have an effect on the efficiency of certain algorithms [25]. Normalization and scaling guarantee that all characteristics are brought to a similar scale, preventing some traits from taking precedence over others throughout the learning process. Scaling by min-max or normalizing by z-score are two methods that are often used.

11.2.2.5 Engineering Features

The process of engineering features entails selecting, altering, or developing new features that are pertinent and instructive for the activity that is currently being performed [26]. This may require determining features that capture certain aspects of the sensor response in the case of electrochemical sensing.

11.2.3 Regression Algorithms for Electrochemical Sensing

The creation of ML regression models in the field of electrochemical sensing is a potent method for predicting continuous variables from sensor responses, such as analyte concentrations. The next part outlines common ML regression techniques used in electrochemical sensing, along with descriptions of how these algorithms function and sources pertinent to the topic.

11.2.3.1 Linear Regression (LR)

One of the most fundamental algorithms in ML is known as LR. It does this by establishing a linear link between the entered characteristics and the variable being targeted. In electrochemical sensing, it may be utilized to make predictions on the concentrations of analytes based on the sensors' responses [27]. The goal of the method is to reduce, as much as possible, the total squared sum of the disparities between the projected and actual values. LR is a method that, despite its apparent lack of complexity, has shown to be helpful in accurately capturing linear connections in data [28].

11.2.3.2 Support Vector Regression (SVR)

It is a regression-based variation of support vector machines (SVM). The data are mapped using SVR into a space with a higher dimension to locate a hyperplane that best matches the data while simultaneously reducing deviations (epsilon-insensitive loss) [29]. When used in electrochemical sensing, SVR can capture the complicated interactions between sensor responses and analyte concentrations. It is beneficial when dealing with non-linear interactions because of its success.

The physical knowledge of the sensor operating principles and a state-of-the-art ML approach called SVM are used to develop a gray-box machine-learning-based model of an electrochemical O_2–NO_x sensor [30]. Figure 11.3 shows the sensor and the SVR strategy. This model is based on ML. The model is utilized to make predictions regarding the response of the sensor under a wide variety of sensor operating settings and in the presence of varying levels of NO_x and ammonia. The production sensor is initially installed on a spark ignition's exhaust system; then on a diesel engine's exhaust system; and, finally, on a fully controlled sensor test rig. This is done to produce a thorough

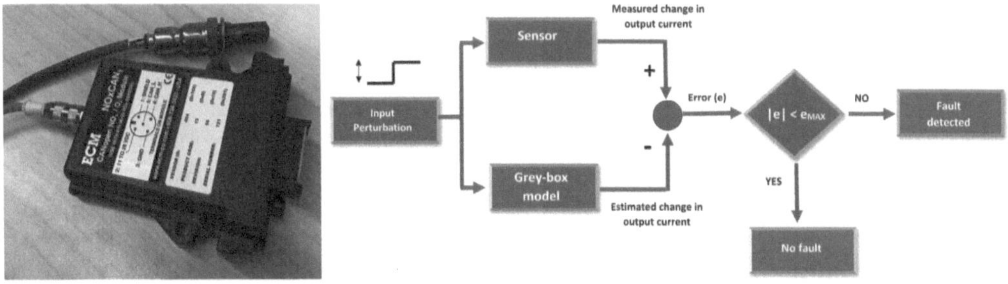

FIGURE 11.3 (a) The engine control and monitoring (ECM) production NO_x sensor and the corresponding control module; (b) An example for sensor diagnostics using the developed gray-box model: an input-perturbation-based sensor diagnostics strategy.

Source: Reprinted from Ref. [30] with permission from Elsevier.

training and test dataset. The sensor itself is not changed; relatively, the sensor's operating temperature, the potentials of all of the sensor cells, and the pumping current of the O_2 sensing cell are the model inputs that may be adjusted.

The pumping current of the NO_x-detecting cell is regarded to be the model output. A low-order model (LOM) with 9 features and a high-order model (HOM) with 45 features are constructed using linear and Gaussian kernels, respectively. After that, an input-output trend analysis is carried out so that the performance of the model and its generalizability may be confirmed. Both the LOM with a Gaussian kernel and the HOM with a linear kernel have demonstrated the highest level of accuracy and the best ability to forecast the response trend.

11.2.3.3 Random Forest Regression

Random forest regression is an ensemble learning technique that, in order to prevent overfitting, generates numerous decision trees and then takes the average of those trees' predictions [31]. It is resilient and can capture non-linearities properly. Random forest regression is capable of handling noisy data as well as interactions between sensor responses and analyte concentrations when applied to the field of electrochemical sensing [32].

11.2.3.4 Gradient Boosting Regression

It is another ensemble approach that sequentially creates decision trees, with each subsequent tree fixing the mistakes made by the one before it [33]. Both the ability to capture complicated relationships and the ability to make accurate predictions are strengthened by its use. Gradient boosting regression can handle high-dimensional data and produce excellent predictions when used for electrochemical sensing [34].

11.2.3.5 Neural Networks (NN)

NN are made up of linked nodes, or neurons, that process information [35]. In tasks involving regression, an NN can identify detailed patterns between the sensors' responses and the concentrations of analytes [21]. It has the capability of approximating relationships that are substantially non-linear. NN versions like convolutional neural networks (CNNs) and recurrent neural networks (RNNs) may also be used to analyze particular kinds of electrochemical data. These networks are referred to by their respective acronyms.

Iwata et al. [36] used an ML method and neural-network-based regression to carry out a gas analysis employing integrated gas sensors of the TiO_2 nanotube (NT) type. A TiO_2-NT-integrated gas sensor with several sensing elements was created, each of which had a unique response characteristic. Then, the output signals of each sensing element were monitored when it was exposed to a gas mixture, the primary components of which were nitrogen gas and oxygen gas with a tiny quantity of carbon monoxide gas. To make an educated guess as to the CO and O_2 concentrations to which the TiO_2-NT gas sensors were most sensitive, the ML approach was applied to analyze the output signals from the sensor components. Data on the output of the sensors were gathered for seven distinct sets of mixed gas concentrations, each of which had a different concentration of its component gas. Figure 11.4 shows the pictorial summary of the work done in Ref. [36]. For ML training, four or five of the seven datasets were utilized, and for the other three or two datasets, the concentrations of CO and O_2 were forecasted. As a result, the authors validated that increasing the number of sensor components improved the accuracy of the gas concentration forecast. It was possible to forecast the gas concentration with an accuracy of less than 0.001% when the output signals from ten sensor elements were employed. This was the case for a concentration of 0.02% carbon monoxide. This level of precision was enough for use in real situations.

In addition to the ML algorithms mentioned above regression algorithms, several other algorithms exist, such as kernel ridge regression (KRR), lasso regression, elastic net regression, and Bayesian ridge regression. Kernel ridge regression combines ridge regression with the kernel technique, which implicitly transfers data into a higher-dimensional space. This results in ridge regression being

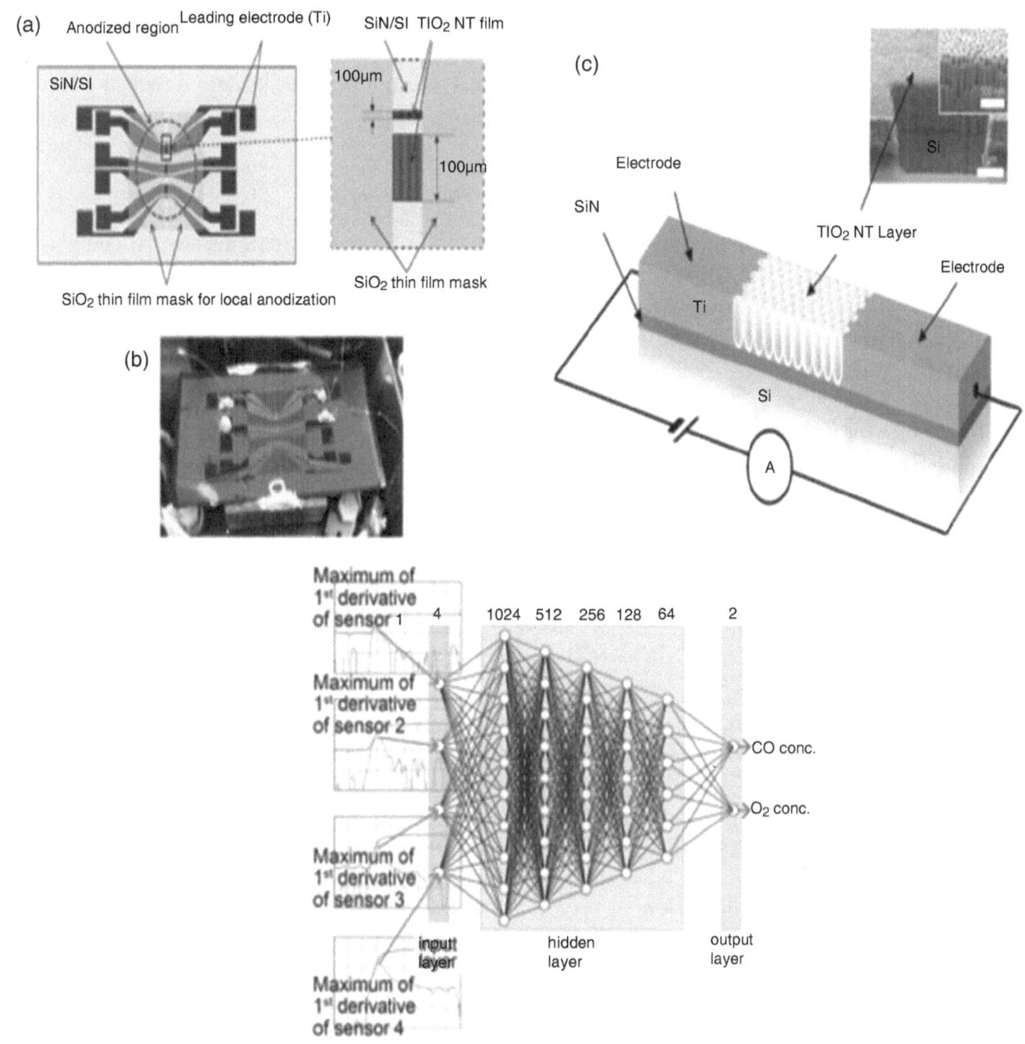

FIGURE 11.4 (a) Schematic (b) photograph of sensor elements fabricated on Si substrate (Si wafer). Six sensor elements with different line widths were arranged on the silicon substrate (c) Schematic of the fabricated gas sensor. The gas detection medium is a film of regularly arranged titanium dioxide (TiO_2) NTs with a diameter of approximately 100 nm. (d) Neural network model used in this study. The neural network is composed of an input layer, a hidden layer of five full-connection layers, and an output layer. In the hidden layer, the five full-connection layers have $n = 1024$, 512, 256, 128, and 64 nodes.

Source: Reprinted from Ref. [36] with permission from Elsevier.

renamed kernel ridge regression [37]. Because of this, KRR is well suited for dealing with non-linear connections. When used in electrochemical sensing, KRR can capture the intricate interactions between sensor responses and analyte concentrations. Lasso regression is a version of linear regression that adds a penalty term to the loss function [38]. This encourages the model to choose a subset of valuable characteristics while setting others to zero. In electrochemical sensing, lasso regression is a technique that helps with feature selection by determining the sensor responses that are the most informative. Elastic net regression combines lasso regression and ridge regression, incorporating both L1 (Lasso) and L2 (Ridge) penalties. This can potentially be useful in situations with many features and collinearity. Elastic net can strike a compromise between selecting features and managing multicollinearity in electrochemical sensing [39].

Further, Bayesian ridge regression is a type of regression that integrates a probabilistic framework into the analysis [40]. It is not susceptible to overfitting and can deal with noisy data. The Bayesian ridge regression technique can offer uncertainty estimates in addition to forecasts in electrochemical sensing, which can help with decision-making.

11.2.4 Classification Algorithms for Electrochemical Sensing

In electrochemical sensing, using ML classification models is a transformational way to categorize data into distinct groups. This helps with tasks like analyte identification and anomaly detection. In the next part, you will find an overview of some of the most common ML classification algorithms utilized extensively in electrochemical sensing, as well as explanations of their underlying mechanics and pertinent citations.

11.2.4.1 Logistic Regression

It is a fundamental classification technique that calculates the likelihood of an instance belonging to a specific class. Logistic regression is sometimes referred to as logistic modeling. The logistic function creates a model of the relationship between the features fed into the system and the likelihood of belonging to a particular category [41]. Logistic regression is a technique that may be applied in the context of electrochemical sensing for activities such as the binary categorization of analytes or the identification of anomalies [42].

11.2.4.2 Decision Trees

Decision trees establish a tree-like structure of decisions to categorize instances and divide the data into subsets depending on the feature values [43]. These values are used to partition the data. Each node in the tree indicates a choice based on a feature, ultimately leading to class assignments. Because decision trees can capture non-linear correlations in the data collected by electrochemical sensors, they are appropriate for various applications [44,45].

11.2.4.3 Random Forest Classifier

The random forest classifier is an ensemble method combining numerous decision trees to provide a more reliable and accurate classification [46,47]. Each tree in the forest is educated using a distinctive slice of the total data. Random Forests can provide reliable classification outputs despite noise and unpredictability in the data when used in electrochemical sensing [48].

11.2.4.4 Support Vector Machines (SVM)

An SVM is a robust algorithm that locates a hyperplane to distinguish instances of distinct classes with the highest possible margin [49,50]. The SVM can handle complicated classification tasks and excels in high-dimensional areas because of its effectiveness. In electrochemical sensing, SVMs can differentiate between analytes based on sensor responses. Also, SVM is able to convert the data into a higher-dimensional space when the kernel trick is used. This allows it to capture non-linear correlations between sensor responses and classes. This is especially helpful in electrochemical sensing situations with complicated connection dynamics [51].

11.2.4.5 k-Nearest Neighbors (k-NN)

k-NN is a classification method that sorts instances into categories by giving them the most prevalent class among their k-NN in the training data [52]. It's a straightforward method that does the job well for categorization projects. In electrochemical sensing, k-NNmay be utilized for activities such as analyte identification by comparing the results to previously established sensor response patterns.

11.2.4.6 Naive Bayes

Naive Bayes is a statistical method derived from Bayes' theorem and makes the "naive" assumption that features are independent [53]. It makes effective use of high-dimensional data and is efficient in its operation. When it comes to electrochemical sensing, the Naive Bayes algorithm may be utilized to classify analytes based on the response patterns of the sensors.

11.2.4.7 Gradient Boosting Classifier

The Gradient boosting classifier constructs an ensemble of decision trees sequentially, with each tree fixing the mistakes of the ones that came before it in the construction process [54]. It helps prevent overfitting and is potent when capturing complicated correlations in data. Gradient boosting is a method that can manage numerous sensor responses in electrochemical sensing, allowing for more accurate categorization. When applied to a drifting chemical sensor array, Yan et al. [55] provide a unique framework that improves gas detection performance. This system has been given the moniker domain-adaptation-based active ensemble learning (DAEL). The proposed framework uses samples from the source domain in conjunction with a selected number of guiding samples taken from the target domain to build a reliable ensemble classifier. These methods train ensemble classifier models for domain knowledge transfer by utilizing various learning strategies for samples from the source domain and guidance samples from the target domain, respectively. These algorithms are referred to as cross-domain-adaptation-based active ensemble learning (DAEL-C) and discriminative-domain-adaptation-based active ensemble learning (DAEL-D), and they are both proposed within this framework. The presented DAEL methodology demonstrates a more successful drift compensation mechanism than some of the comparable approaches. This was demonstrated through experimental validation that was carried out on a drifted sensor dataset. Figure 11.5 summarizes the work done by [55].

11.2.4.8 Convolutional Neural Networks (CNNs)

Classification is another use for neural networks, which may be accomplished using the NN classifier. MLPs, which stand for multi-layer perceptrons, are a prevalent type of architecture [56]. CNNs perform exceptionally well in image-based classification tasks. Regarding electrochemical sensing, NN classifiers can recognize subtle patterns in sensor responses, which is necessary for precise analyte identification [57].

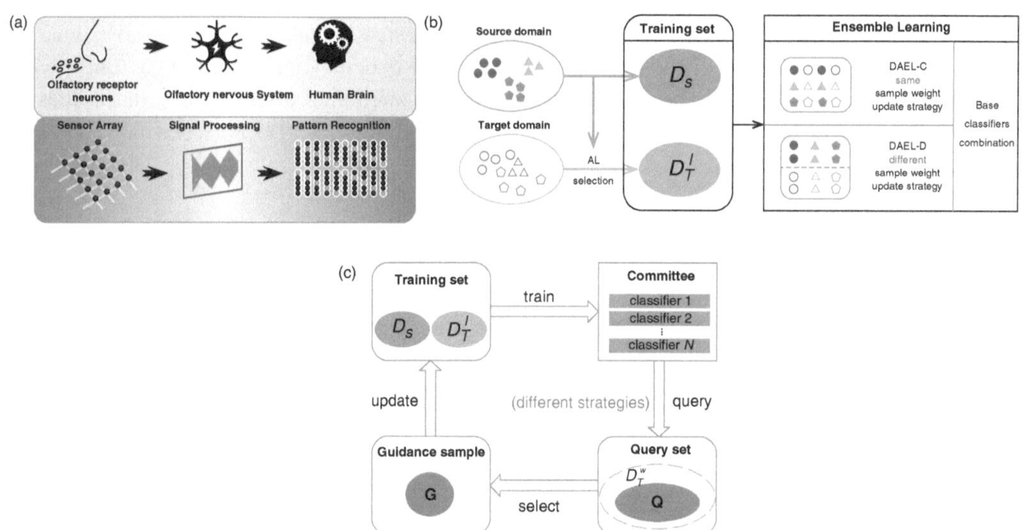

FIGURE 11.5 (a) Mapping of E-nose components to the human olfactory system, (b) the structure of DAEL. Triangles, circles, and pentagons represent samples belonging to different categories. (c) The process of guidance sample selection. G and Q represent the guidance samples and query set, respectively.

Source: Reprinted from Ref. [55] with permission from Elsevier.

11.2.5 Clustering Algorithms for Electrochemical Sensing

In electrochemical sensing, clustering is a ML solid approach used to discover similar patterns and group data points based on the data points' inherent qualities. This part discusses standard clustering methods and their applications in electrochemical sensing, along with necessary explanations and references for each topic.

11.2.5.1 Clustering Using K-Means

K-means is one of the clustering techniques that is used the most frequently. It does this by repeatedly assigning data points to the nearest cluster centroid and then recalculating centroids based on the assigned points [58]. This separates the data into "k" clusters. K-means may be used in electrochemical sensing to classify sensor responses into various groups [59]. This can help identify analytes or find anomalies in the data.

11.2.5.2 Hierarchical Clustering

Hierarchical clustering creates a hierarchy of clusters by either beginning with individual data points as clusters (or agglomerative clustering) or beginning with all of the data points in one cluster and recursively splitting them (divisive clustering) [60]. Hierarchical clustering is an example of a data analysis technique. This results in the production of a dendrogram that depicts the hierarchy. Hierarchical clustering has the potential to reveal features hidden within sensor response data when used for electrochemical sensing [61].

11.2.5.3 Density-Based Spatial Clustering of Applications with Noise (DBSCAN)

DBSCAN is an acronym for "Density-Based Spatial Clustering of Applications with Noise." DBSCAN is a density-based clustering technique that determines clusters to be regions of high data density that are separated by areas of low data density [62]. It can recognize groups of arbitrary shapes and label outliers as noise. In electrochemical sensing, the DBSCAN technique may be utilized to differentiate between different analyte classes by analyzing the sensor response densities of each.

11.2.5.4 The Gaussian Mixture Model (GMM)

GMM is based on the supposition that a combination of multiple different Gaussian distributions produces individual data points. It does so by modeling clusters using the Gaussian distributions and estimating the parameters of those distributions. GMM is beneficial when the genuine cluster boundaries are not well defined since it can capture complicated data distributions, and it does so effectively. GMM can detect underlying Gaussian distributions in sensor responses, which is helpful in electrochemical sensing [63].

11.3 PERFORMANCE EVALUATION OF ML MODELS FOR ELECTROCHEMICAL SENSING

When determining whether an ML model is reliable and successful, performance measures and parameters are the most essential tools to use. These metrics inform decisions about model selection, tuning, and deployment while providing insights into how well a model functions. The exact job, whether classification, regression, or clustering, will dictate which performance measures are the most appropriate. Let's investigate some of the popular performance measures and parameters used for ML models.

11.3.1 Classification Metrics

Classification metrics are used to evaluate the performance of models that classify data into discrete categories or classes. Classification metrics are also known as categorization metrics.

11.3.1.1 Accuracy

It can be defined as the proportion of successfully predicted occurrences. It gives a comprehensive picture of the model's performance, but it might be deceiving if the classes are not evenly distributed. The percentage of genuine positive predictions relative to the total number of positive forecasts measures accuracy [64]. It determines the degree to which optimistic forecasts are accurate.

11.3.1.2 Recall

The proportion of genuine positive predictions made out of the total number of real positive cases is referred to as the recall, sensitivity, or true positive rate. It evaluates the model's effectiveness in capturing good examples [65].

11.3.1.3 Specificity

Specificity, also known as the true negative rate, is the percentage of forecasts that turn out to be accurate when compared to the total number of real negative occurrences [66].

11.3.1.4 F1-Score

The F1-score is the harmonic mean of both accuracy and recall, offering a balance between the two aspects of the test [67].

11.3.1.5 Receiver Operating Characteristic - Area Under Curve (ROC-AUC)

ROC-AUC, which stands for Receiver Operating Characteristic - Area Under Curve, is a graphical depiction of the trade-off between the actual positive rate and the false positive rate. The model's capacity to differentiate between categories is measured by the amount of space beneath the ROC curve [68].

11.3.2 Metrics for Regression

Models that predict continuous values can have their performance evaluated using regression metrics. mean absolute error, often known as MAE, refers to the average absolute difference between the values that were anticipated and those that were actually observed [69]. The average squared deviation from the anticipated values to the actual values is referred to as the mean squared error (MSE) [69]. RMSE stands for "root mean squared error," and it is calculated by taking the square root of MSE. This gives a measurement that is in the same unit as the target variable.

R-squared, often known as the coefficient of determination, is statistics that measures the amount of the dependent variable's variation that can be predicted based on its relationship to the independent variables [70]. It can vary anywhere from 0 to 1, with higher numbers indicating a more suitable match.

11.3.3 Clustering Metrics

It is used to assess the quality of the clusters that are produced by unsupervised learning algorithms. The silhouette score is a metric that evaluates how similar data points are to one another within the same cluster in comparison to how different they are from data points in other clusters [71]. The values range from –1 to 1, with higher values suggesting more well-defined clusters. The Davies–Bouldin index is a measurement that determines the average degree of similarity between each cluster and the cluster that is most like it. Lower numbers represent a more effective grouping.

Cross-validation is a method that evaluates the performance of a model with data that has not been seen before. The data are often segmented into "k" subsets before the model is trained and assessed "k" times. One example of a common approach is called k-fold cross-validation.

In general, performance measurements and parameters give valuable insights into the model's behavior and assist the iterative process of developing, fine-tuning, and optimizing the model.

When analyzing the performance of a model comprehensively, it is vital to identify metrics that match the particular objectives of the study and to consider a number of different indicators.

11.4 GUIDANCE FOR THE SCIENTIFIC COMMUNITY

The convergence of ML and electrochemical sensing presents the scientific community with a wealth of opportunities to advance the field and drive innovation. However, navigating this synergy requires thoughtful consideration, strategic planning, and a clear understanding of the challenges and possibilities. This comprehensive description outlines key guidance for the scientific community to effectively employ ML in electrochemical sensing, fostering impactful research and meaningful applications.

Before diving into the integration of ML, it's crucial to grasp the foundational principles of this technology. Familiarize yourself with different ML algorithms, their strengths, weaknesses, and the types of problems they are best suited to address. Resources such as [72]. "Pattern Recognition and Machine Learning" provide insights into the theory behind ML techniques, helping researchers make informed choices. Establishing clear objectives is paramount. Determine the specific problems or challenges you intend to address using ML in electrochemical sensing. Whether it's enhancing sensor accuracy, optimizing parameter settings, or predicting analyte concentrations, well-defined objectives will guide your approach and shape your data collection strategy [73]. ML thrives on data. Collect, preprocess, and curate high-quality datasets encompassing diverse scenarios, analytes, and conditions relevant to your objectives. Quality data are essential for training accurate and reliable ML models. Select ML algorithms that align with your objectives and data characteristics. For example, NN might excel in pattern recognition tasks, while support vector machines might be better suited for classification [74]. Study the strengths and limitations of each algorithm and opt for those that resonate with your research goals. Feature selection and engineering play a pivotal role in model performance. Identify relevant features that directly influence sensor responses and analyte concentrations.

Finally, thoroughly evaluate and validate your ML models. Implement appropriate cross-validation techniques to ensure the robustness of your models against unseen data. Consider metrics relevant to your research objectives, such as accuracy, precision, recall, or F1-score. The intersection of electrochemistry and ML often requires multidisciplinary collaboration. Engage with experts from both domains to ensure a holistic and impactful approach. Cross-disciplinary collaboration can uncover novel insights and foster innovation. ML is a dynamic field. Stay updated with the latest advancements, techniques, and best practices. Continually refine your models and approaches based on new knowledge and emerging trends to ensure your research remains relevant and impactful. ML applications raise ethical concerns, such as bias, privacy, and data security. Be mindful of these considerations and adhere to ethical guidelines and regulations when collecting, storing, and using data. Guided by these principles, the scientific community can leverage ML to transform electrochemical sensing into a realm of precision, adaptability, and innovation [75]. By blending domain expertise with ML prowess, researchers can forge a path toward data-driven insights, pushing the boundaries of what is achievable in sensor design, performance optimization, and real-world applications.

11.5 CONCLUSION

In summary, the rapid advancement of artificial intelligence has brought about a hopeful period for electrochemical sensing, whereby ML has emerged as a crucial instrument for enhancing the capabilities of sensors. The present chapter extensively examined the intrinsic capabilities of ML methodologies, specifically emphasizing their capacity to propel the progress of high-performance electrochemical sensors in many fields of application. The incorporation of ML methodologies has the potential to significantly enhance the design and operational capabilities of electrochemical sensors, leading to a notable surge of interest among the scientific community.

The intrinsic flexibility of ML algorithms allows them to effectively handle complex electrochemical data, facilitating the extraction of crucial characteristics, the discovery of detailed patterns, and the efficient development of data-driven models. This work extensively investigates the significant contribution of ML in enhancing the performance of sensors, leading to improved accuracy, sensitivity, and dependability. The study comprehensively analyzes the advantages of advanced predictive modeling techniques in essential fields such as environmental monitoring, non-invasive healthcare, and biological diagnostics.

Moreover, besides its pragmatic usefulness, this chapter is a helpful tool for understanding current scientific and technical trends. This publication provides concise and informative instructions to the scientific community on efficiently using ML techniques to advance the development of electrochemical sensors. Integrating ML with electrochemical sensing enhances the capabilities of sensors and facilitates progress in the multidisciplinary domain by expanding knowledge and practical applications.

REFERENCES

1. V. Shukla, M. Bandyopadhyay, V. Pandya, and A. Pandey, "Prediction of axial variation of plasma potential in helicon plasma source using linear regression techniques," *Int. J. Math. Eng. Manag. Sci.*, vol. 5, no. 6, pp. 1–16, 2020.
2. C. Zhu, G. Yang, H. Li, D. Du, and Y. Lin, "Electrochemical sensors and biosensors based on nanomaterials and nanostructures," *Anal. Chem.*, vol. 87, no. 1, pp. 230–249, 2015.
3. A. T. Lawal, "Recent developments in electrochemical sensors based on graphene for bioanalytical applications," *Sens. Bio-Sensing Res.*, vol. 41, p. 100571, 2023.
4. D. R. Thévenot, K. Toth, R. A. Durst, and G. S. Wilson, "Electrochemical biosensors: Recommended definitions and classification," *Biosens. Bioelectron.*, vol. 16, no. 1, pp. 121–131, 2001.
5. I. G. Munteanu and C. Apetrei, "Classification and antioxidant activity evaluation of edible oils by using nanomaterial-based electrochemical sensors," *Int. J. Mol. Sci.*, vol. 24, no. 3, p. 3010, 2023.
6. V. Shukla, M. Bandyopadhyay, V. Pandya, and A. Pandey, "Prediction of axial variation of plasma potential in helicon plasma source using linear regressiontechniques," *Int. J. Math. Eng. Manag. Sci.*, vol. 5, no. 6, pp. 1284–1299, 2020.
7. V. Shukla and M. Bandyopadhyay, "Optimization of input parameters of ANN–driven plasma source through nature-inspired evolutionary algorithms," *Intell. Syst. Appl.*, vol. 18, p. 200200, 2023.
8. H. Chen, E. Kätelhön, and R. G. Compton, "Machine learning in fundamental electrochemistry: Recent advances and future opportunities," *Curr. Opin. Electrochem.*, vol. 38, p. 101214, 2023.
9. S. De Vito, G. Di Francia, E. Esposito, S. Ferlito, F. Formisano, and E. Massera, "Adaptive machine learning strategies for network calibration of IoT smart air quality monitoring devices," *Pattern Recognit. Lett.*, vol. 136, pp. 264–271, 2020.
10. P. P. Mondal et al., "Review on machine learning-based bioprocess optimization, monitoring, and control systems," *Bioresour. Technol.*, vol. 370, p. 128523, 2023.
11. A. N. Rao, M. N. Avula, and D. W. Grainger, "*3.34 Biomaterials Challenges in Continuous Glucose Monitors In Vivo*," Oxford: Elsevier, 2017, pp. 755–770.
12. S. M. Moosavi, K. M. Jablonka, and B. Smit, "The role of machine learning in the understanding and design of materials," *J. Am. Chem. Soc.*, vol. 142, no. 48, pp. 20273–20287, 2020.
13. G. Han, J. Yan, Z. Guo, D. Greenwood, J. Marco, and Y. Yu, "A review on various optical fibre sensing methods for batteries," *Renew. Sustain. Energy Rev.*, vol. 150, p. 111514, 2021.
14. I. Kavakiotis, O. Tsave, A. Salifoglou, N. Maglaveras, I. Vlahavas, and I. Chouvarda, "Machine learning and data mining methods in diabetes research," *Comput. Struct. Biotechnol. J.*, vol. 15, pp. 104–116, 2017.
15. S. Mourdikoudis, R. M. Pallares, and N. T. K. Thanh, "Characterization techniques for nanoparticles: Comparison and complementarity upon studying nanoparticle properties," *Nanoscale*, vol. 10, no. 27, pp. 12871–12934, 2018.
16. M. M. Shanbhag, G. Manasa, R. J. Mascarenhas, K. Mondal, and N. P. Shetti, "Fundamentals of bio-electrochemical sensing," *Chem. Eng. J. Adv.*, vol. 16, p. 100516, 2023.
17. J. Baranwal, B. Barse, G. Gatto, G. Broncova, and A. Kumar, "Electrochemical sensors and their applications: A review," *Chemosensors*, vol. 10, no. 9, p. 363, 2022.

18. K. E. Schackart and J.-Y. Yoon, "Machine learning enhances the performance of bioreceptor-free biosensors," *Sensors*, vol. 21, no. 16, p. 5519, 2021.
19. P. Wittek, Quantum Machine Learning, "Chapter 5 – Unsupervised learning," Boston, MA: Academic Press, 2014, pp. 57–62.
20. M. Cuartero, "Electrochemical sensors for in-situ measurement of ions in seawater," *Sensors Actuators B Chem.*, vol. 334, p. 129635, 2021.
21. N. U. Okafor, Y. Alghorani, and D. T. Delaney, "Improving data quality of low-cost IoT sensors in environmental monitoring networks using data fusion and machine learning approach," *ICT Express*, vol. 6, no. 3, pp. 220–228, 2020.
22. V. Shukla, V. Pandya, M. Bandyopadhyay, and A. Pandey, "Input parameter optimization with simulated annealing algorithm for predictive HELEN-I ion source," in *Congress on Intelligent Systems CIS 2020*, Springer, Singapore, 2020, pp. 1–10.
23. S. K. Kwak and J. H. Kim, "Statistical data preparation: Management of missing values and outliers.," *Korean J. Anesthesiol.*, vol. 70, no. 4, pp. 407–411, 2017.
24. J. Han, M. Kamber, and J. Pei, "Chapter 12 – Outlier detection," in J. Han, M. Kamber, and J. Pei (Eds). *The Morgan Kaufmann Series in Data Management Systems.* Boston, MA: Morgan Kaufmann, 2012, pp. 543–584.
25. Q. He et al., "Research on the construction of portable electrochemical sensors for environmental compounds quality monitoring," *Mater. Today Adv.*, vol. 17, p. 100340, 2023.
26. Y. Liu, T. Zhao, W. Ju, and S. Shi, "Materials discovery and design using machine learning," *J. Mater.*, vol. 3, no. 3, pp. 159–177, 2017.
27. X. Lu et al., "An emerging machine learning strategy for electrochemical sensor and supercapacitor using carbonized metal–organic framework," *J. Electroanal. Chem.*, vol. 920, p. 116634, 2022.
28. J. Lee et al., "Improved predictive models for plasma glucose estimation from multi-linear regression analysis of exhaled volatile organic compounds," *J. Appl. Physiol.*, vol. 107, no. 1, pp. 155–160, 2009.
29. Y.-J. Lee, W. F. Hsieh, and C. M. Huang, "Epsilon-SSVR: A smooth support vector machine for epsilon-insensitive regression," *IEEE Trans. Knowl. Data Eng.*, vol. 17, pp. 678–685, 2005.
30. M. Aliramezani, A. Norouzi, and C. R. Koch, "A grey-box machine learning based model of an electrochemical gas sensor," *Sensors Actuators, B Chem.*, vol. 321, p. 128414, 2020.
31. V. Shukla, V. Pandya, M. Bandyopadhyay, and A. Pandey, "Plasma density prediction for helicon negative hydrogen plasma source using decision tree and random forest algorithm," in *Congress on Intelligent Systems: Proceedings of CIS 2020, Volume 2*, Springer, Singapore, 2021, pp. 357–368.
32. A. Pennacchio et al., "A machine learning-enhanced biosensor for mercury detection based on an hydrophobin chimera," *Biosens. Bioelectron.*, vol. 196, p. 113696, 2022.
33. V. Kotu and B. Deshpande, "Chapter 2 - Data mining process," in V. Kotu and D. M. Deshpande (Eds). *The Morgan Kaufmann Series in Data Management Systems.* Boston, MA: Morgan Kaufmann, 2015, pp. 17–36.
34. B. Farhadi et al., "Machine learning for fast development of advanced energy materials," *Next Mater.*, vol. 1, no. 3, p. 100025, 2023.
35. K. Ramesh, S. Desai, D. Jariwala, and V. Shukla, "AI modelled clutch operation for automobiles," in *2022 IEEE World Conference on Applied Intelligence and Computing (AIC)*, Sonbhadra, India, 2022, pp. 487–491.
36. K. Iwata et al., "Application of neural network based regression model to gas concentration analysis of TiO_2 nanotube-type gas sensors," *Sensors Actuators B Chem.*, vol. 361, p. 131732, 2022.
37. S. Theodoridis, "Chapter 11 - Learning in reproducing kernel hilbert spaces," in S. Theodoridis (Ed). *Machine Learning.* San Diego, CA: Academic Press, 2020, pp. 531–594.
38. W. K. Härdle and D. D. Prastyo, "Chapter 7 - Embedded predictor selection for default risk calculation: A southeast asian industry study," in G. N. Gregoriou and A. F. Chuen (Eds). *Handbook of Asian Finance: Financial Markets and Sovereign Wealth Funds.* San Diego, CA: Academic Press, 2014, pp. 131–148.
39. N. N. Narisetty, "Chapter 4 - Bayesian model selection for high-dimensional data," in A. S. R. Srinivasa Rao and C. R. Rao (Eds). *Principles and Methods for Data Science*, vol. 43. Amsterdam, The Netherlands: Elsevier, 2020, pp. 207–248.
40. K. A. Venkatesh, D. Mishra, and T. Manimozhi, "Chapter 9 - Model selection and regularization," in T. Goswami and G. R. Sinha (Eds). *Statistical Modeling in Machine Learning.* San Diego, CA: Academic Press, 2023, pp. 159–178.
41. A. Singh, N. Thakur, and A. Sharma, "A review of supervised machine learning algorithms," in *2016 3rd International Conference on Computing for Sustainable Global Development (INDIACom)*, New Delhi, India, 2016, pp. 1310–1315.

42. Y. Zhao et al., "AI powered electrochemical multi-component detection of insulin and glucose in serum," *Biosens. Bioelectron.*, vol. 186, p. 113291, 2021.
43. W.-Y. Loh, "Classification and regression tree methods," *Wiley StatsRef: Statistics Reference Online*. 29-Sep-2014.
44. V. Kammarchedu, D. Butler, and A. Ebrahimi, "A machine learning-based multimodal electrochemical analytical device based on eMoSx," *Anal. Chim. Acta*, vol. 1232, p. 340447, 2022. https://doi.org/10.1149/osf.io/67h28.
45. S. Kumar, R. B. Grover, P. K. Vijayan, and U. Kannan, "Numerical investigation on the effect of shrouds around an immersed isolation condenser on the thermal stratification in large pools," *Ann. Nucl. Energy*, vol. 110, pp. 109–125, 2017.
46. L. F. M. Mota et al., "Evaluating the performance of machine learning methods and variable selection methods for predicting difficult-to-measure traits in Holstein dairy cattle using milk infrared spectral data," *J. Dairy Sci.*, vol. 104, no. 7, pp. 8107–8121, 2021.
47. S. Kumar, R. B. Grover, H. Yadav, P. K. Vijayan, U. Kannan, and A. Agrawal, "Experimental and numerical investigation on suppression of thermal stratification in a water-pool: PIV measurements and CFD simulations," *Appl. Therm. Eng.*, vol. 138, pp. 686–704, 2018.
48. H. Kim, A. L. Gebreselassie, S. Dan, and D. Shin, "Random forest classifier for real-time chemical leak source tracking using fence-monitoring sensors," *Korean J. Chem. Eng.*, vol. 35, no. 6, pp. 1231–1239, 2018.
49. A. Tuor, S. Kaplan, B. Hutchinson, N. Nichols, and S. Robinson, "Deep learning for unsupervised insider threat detection in structured cybersecurity data streams," *Tech. Rep.*, vol. WS-17-01, pp. 224–234, 2017.
50. S. Kumar and M. D. King, "Numerical investigation on indoor environment decontamination after sneezing," *Environ. Res.*, vol. 213, p. 113665, 2022.
51. C. Wang et al., "Machine learning-assisted cell-imprinted electrochemical impedance sensor for qualitative and quantitative analysis of three bacteria," *Sensors Actuators B Chem.*, vol. 384, p. 133672, 2023.
52. S. Shahub, S. Upasham, A. Ganguly, and S. Prasad, "Machine learning guided electrochemical sensor for passive sweat cortisol detection," *Sens. Bio-Sensing Res.*, vol. 38, p. 100527, 2022.
53. S. Nam and J. Hur, "Probabilistic forecasting model of solar power outputs based on the Na ve bayes classifier and kriging models," *Energies*, vol. 11, no. 11, p. 2982, 2018.
54. Y. Zhang and A. Haghani, "A gradient boosting method to improve travel time prediction," *Transp. Res. Part C Emerg. Technol.*, vol. 58, pp. 308–324, 2015.
55. J. Yan, R. Sun, T. Liu, and S. Duan, "Domain-adaptation-based active ensemble learning for improving chemical sensor array performance," *Sensors Actuators A Phys.*, vol. 357, p. 114411, 2023.
56. J. Raitoharju, "Chapter 3 – Convolutional neural networks," in A. Iosifidis and C. Tefas (Eds). *Deep Learning for Robot Perception and Cognition*. San Diego, CA: Academic Press, 2022, pp. 35–69.
57. A. Kale, Y. J. Chen, and M. Sundaram, "Predicting the sparks occurrence in electrochemical discharge machining by machine learning using convolutional neural networks," *Procedia CIRP*, vol. 113, pp. 427–431, 2022.
58. T. Sterling, M. Anderson, and M. Brodowicz, "Chapter 19 – MapReduce," in T. Sterling, M. Anderson, and M. B. T. Brodowicz (Eds). *The Morgan Kaufmann Series in Data Management Systems*. Boston: Morgan Kaufmann, 2018, pp. 579–589.
59. O. Capron and L. D. Couto, "An efficient methodology combining K-means machine learning and electrochemical modelling for the determination of ionic diffusivity and kinetic properties in battery electrodes," *Materials (Basel)*, vol. 16, no. 14, p. 5146, 2023.
60. O. Akman, T. Comar, D. Hrozencik, and J. Gonzales, "Chapter 11 – Data clustering and self-organizing maps in biology," in R. Robeva and C. C. B. Macauley (Eds). *MSE/Mathematics in Science and Engineering*. San Diego, CA: Academic Press, 2019, pp. 351–374.
61. A. Smoliński, B. Walczak, and J. W. Einax, "Hierarchical clustering extended with visual complements of environmental data set," *Chemom. Intell. Lab. Syst.*, vol. 64, no. 1, pp. 45–54, 2002.
62. B. Nidhi and K. A. Patel, "An efficient and scalable density-based clustering algorithm for normalize data," *Procedia Comput. Sci.*, vol. 92, pp. 136–141, 2016.
63. A. A. Mas'ud, A. Sundaram, J. A. Ardila-Rey, R. Schurch, F. Muhammad-Sukki, and N. A. Bani, "Application of the gaussian mixture model to classify stages of electrical tree growth in epoxy resin," *Sensors*, vol. 21, no. 7, p. 2562, 2021.
64. N. Bu, T. Tsuji, and O. Fukuda, "Chapter 3 - EMG-Controlled human-robot interfaces: A hybrid motion and task modeling approach," in J. Ueda and I. R. Kurita (Eds). *Human Modelling for Bio-Inspired Robotics*. San Diego, CA: Academic Press, 2017, pp. 75–109.

65. D. Sengupta, V. K. Sharma, S. K. Hota, R. B. Srivastava, and P. K. Naik, "Chapter 10 – An ensemble approach for evaluating the cognitive performance of human population at high altitude," in P. Kumar, Y. Kumar, and M. A. Tawhid (Eds). *Intelligent Data-Centric Systems.* San Diego, CA: Academic Press, 2021, pp. 165–178.
66. M. Barua et al., "Sub-clinical mastitis prevalent in dairy cows in Chittagong district of Bangladesh: Detection by different screening tests," *Vet. World*, vol. 7, no. 7, pp. 483–488, 2014.
67. A. J. Hall, N. Pitropakis, W. J. Buchanan, and N. Moradpoor, "Predicting malicious insider threat scenarios using organizational data and a heterogeneous stack-classifier," *2018 IEEE International Conference on Big Data (Big Data)*, Seattle, WA, USA, pp. 5034–5039, 2019.
68. J. N. Mandrekar, "Receiver operating characteristic curve in diagnostic test assessment," *J. Thorac. Oncol.*, vol. 5, no. 9, pp. 1315–1316, 2010.
69. A. Botchkarev, "Evaluating performance of regression machine learning models using multiple error metrics in azure machine learning studio," *SSRN Electron. J.*, https://ssrn.com/abstract=3177507, 2018.
70. V. Shukla, V. Pandya, M. Bandyopadhyay, and A. Pandey, "Plasma density prediction for helicon negative hydrogen plasma source using decision tree and random forest algorithm," in H. Sharma, et al. (Eds). *Congress on Intelligent Systems*, Singapore: Springer, 2021, pp. 357–368.
71. M. R. M. Talabis, R. McPherson, I. Miyamoto, J. L. Martin, and D. Kaye, "Chapter 1 – Analytics defined," in M. R. M. Talabis, R. McPherson, I. Miyamoto, J. L. Martin, and D. Kaye (Eds). *Information Security Analytics.* Boston: Syngress, 2015, pp. 1–12.
72. C. M. Bishop, *Pattern Recognition and Machine Learning.* Singapore: Springer Science+Business Media, 2006.
73. W. Y. Tsai, Y. Chi, and C. M. Chen, "Interval estimation of binomial proportion in clinical trials with a two-stage design," *Stat. Med.*, vol. 27, no. 1, pp. 15–35, 2008.
74. K. P. Singh, N. Basant, and S. Gupta, "Support vector machines in water quality management," *Anal. Chim. Acta*, vol. 703, no. 2, pp. 152–162, 2011.
75. M. T. Ribeiro, S. Singh, and C. Guestrin, "'Why Should I Trust You?' Explaining the Predictions of Any Classifier," *NAACL-HLT 2016-2016 Proceedings of the 2016 Conference of the North American Chapter of the Association for Computational Linguistics: Human Language Technologies*, San Diego, CA, pp. 97–101, 2016.

12 Nanomaterials for Electrochemical Sensing of Explosive Substances

Rama Gaur, Rachna Sharma, and Nandini Mukherjee

12.1 INTRODUCTION

The detection of explosive substances is a crucial issue in the fields of national security, homeland defense, and counter-terrorism. The threat of explosives has become a significant concern due to their widespread use in terrorist attacks and criminal activities. Several analytical techniques have been developed to detect and identify explosive substances in different environments. However, the development of highly sensitive and selective sensors for the detection of explosives remains a significant challenge.

According to available data from Action on Armed Violence (AOAV), there have been nearly 125 instances of explosive violence globally, resulting in a reported total of 563 casualties. Out of these casualties, 309, constituting 55%, were civilians. The data is sourced from incident-specific English-language reports between March 24 and March 30, 2023 [1]. To add to the gravity of the situation, there has been an 83% increase in global explosive violence harm to civilians from 2021 to 2022, almost 20,793 global civilian casualties from explosive weapons in 2022 [2]. To tackle the prevailing issues of the use of explosives, it is imperative and need of the hour to develop systems with high sensitivity for the quick detection of the explosives and even their precursor molecules.

Nanotechnology has emerged as a promising field for the development of highly sensitive and selective sensors. The utilization of nanomaterials in the sensing area has led to the development of improved sensing technologies, particularly in electrochemical sensing, which is known for its high sensitivity, selectivity, and low detection limits. The unique properties of nanomaterials have enabled the development of electrochemical sensors for the detection of explosive substances. The present chapter discusses explosives, their classification, and methods for sensing explosive substances with a special emphasis on the electrochemical mode of sensing, and the role of various nanomaterials in the sensing of explosives.

12.1.1 Explosives Substances

Explosives refer to chemical compounds that initiate explosive reactions, resulting in a release of high-energy explosion. When these substances undergo a reaction, they produce a substantial volume of gas that rapidly expands over a short period. This process is commonly termed "Chemical Detonation," denoting an explosion caused by a series of chemical reactions. An explosion involves the transformation of chemical energy stored in the substance into a shock wave that rapidly travels, causing destruction in the surrounding environment. As an example, the rapid and exothermic decomposition of trinitrotoluene generates significant quantities of gases like hydrogen (H_2), water vapor (H_2O), nitrogen (N_2), carbon monoxide (CO), and carbon (C) [3].

The speed of a shock wave is called as velocity of detonation. The relation between the velocity of detonation (v_d) and velocity of sound (v_s) decides the nature of the explosion. For the condition (1) $v_d > v_s$, detonation occurs, and (2) $v_d < v_s$, deflagration occurs. Figure 12.1 shows the schematic

DOI: 10.1201/9781032654874-12

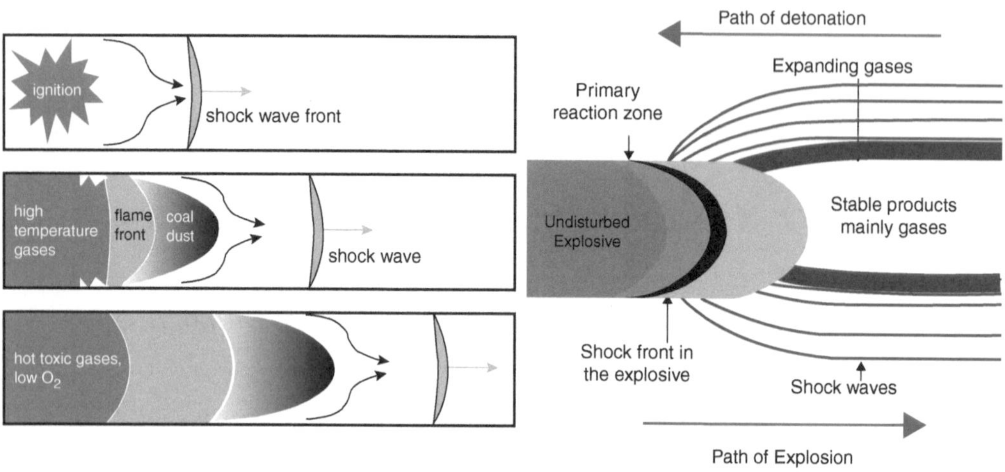

FIGURE 12.1 Schematic representation of the explosion and generation of an expansive shock wave [4].

representation of the mechanism involved in the explosive reaction and the propagation of the shock wave. A chemical explosion involves a series of ultrafast chemical reactions classified as unimolecular decomposition reactions and multimolecular reduction and oxidation reactions. Unimolecular reactions involve decomposition of a single component, while multimolecular reactions require two or more reactants. One major requirement is that the reaction should be ultrafast to generate a supersonic shock wave.

12.1.2 Classification of Explosives Substances

Looking from the chemical composition perspective, explosives can be classified as pure individual substances or combinations of two or more substances. The pristine explosive chemicals can be further categorized as inorganic and organic substances. The detailed classification of the explosive based on the chemical composition is depicted in Figure 12.2. Typical representatives from different chemical classes are shown in Figure 12.3.

12.1.3 Methods for the Sensing of Explosive Substances

As discussed in the above section, due to the increased severity of accidents due to explosive reactions, it is very important to detect the explosive substances and their precursor molecules effectively and vigilantly even in trace amounts to have control of the situation. Various techniques have been reported for the effective detection and sensing of explosive substances. The detection approaches have been categorized into three broad categories, namely, spectroscopic sensors, olfactory sensors, and other sensor techniques that comprise different analytical techniques. Figure 12.4 shows the different approaches reported for the detection of the explosives. Table 12.1 lists the different techniques employed for the application of explosive detection.

The reported techniques have certain advantages and disadvantages associated with them. The details about some of the characteristics and advanced detection techniques used for sensing explosive substances such as vapor sensing, fluorescence sensing, thermodynamic sensing, biosensors, and laser-induced immunofluorometric biosensors are discussed briefly as follows.

12.1.3.1 Vapor Sensing of Explosive Materials

Sensing of explosives involves two types of analyte: (1) solid explosive substances which are generally scanned using various techniques and (2) vapor of the explosives or their byproducts. Among the two methods former fails to detect the traces of the explosives while the latter is highly sensitive

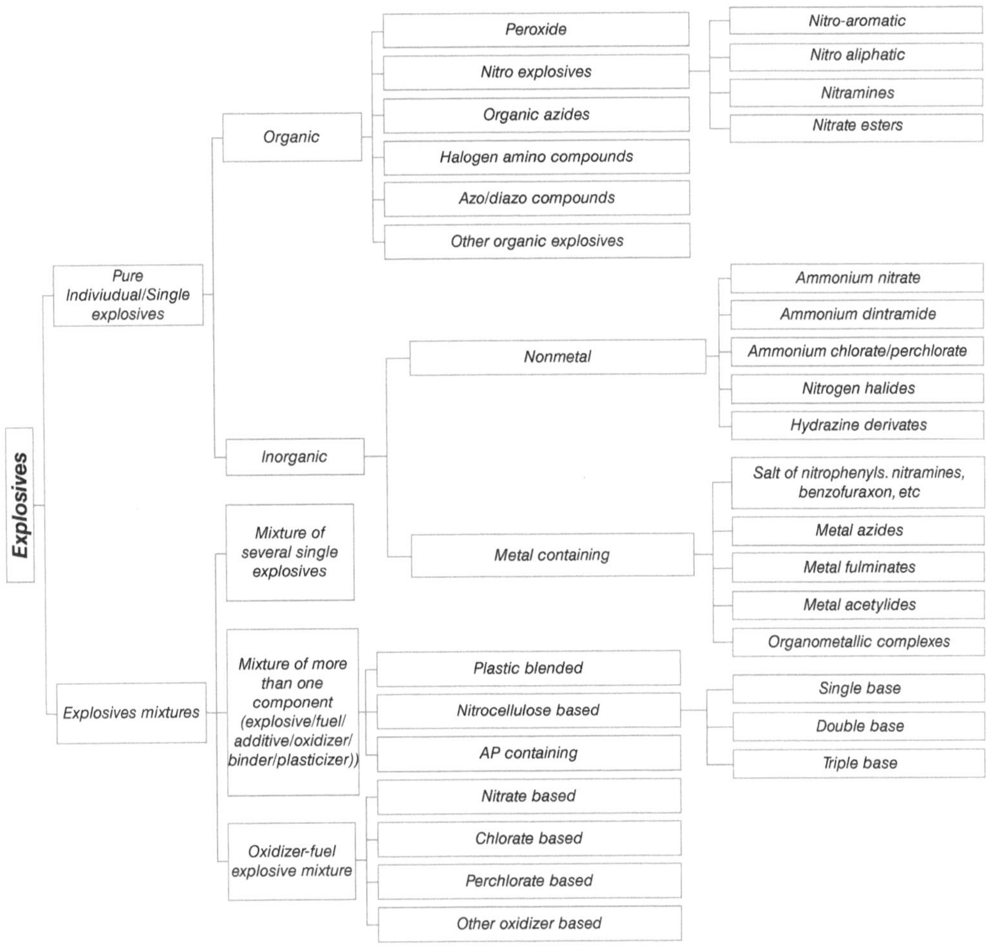

FIGURE 12.2 Classification of explosives based on their chemical composition.

to explosives present in trace amounts. Vapor sensing involves direct and indirect sensing of the vapors emitted by bulk and explosive residue, respectively. This helps in the sensing of concealed explosive substances. Vapor sensing is considered a future avenue for existing animal olfactory sensing. The use of trained animals such as sniffer dogs, rats, and honey bees [4] involves the detection of explosives by the high sensitivity toward the odor i.e., vapors of the explosives. The lowest detection limit for canine olfaction has been reported as 1.14 ppt and is generally considered a reference for the comparison of other reported methods or novel methods being developed.

For the development of novel sensors, it is important to understand the animal olfaction. The explosive vapors from the source enter the nose and are dissolved in the mucus of the nasal cavity. The dissolved vapors interact with receptors present in the nasal cavity and a signal is created. The signal travels via the olfactory nerve and is transmitted to the brain where it is interpreted for further action. Similarly, the use of gas chromatography and allied techniques for vapor sensing comes in handy for the detection of explosives. Figure 12.5 shows the comparison of the principles of operation of the canine sense of smell with bioelectronic analog sensing mechanism.

In materials science, gold nanoparticles (NPs) are used as a standard and are considered equivalent to a trained canine olfaction. Efforts are being made toward the development of novel materials with high sensitivity and low detection limits.

One of the major challenges faced in the area of vapor sensing is that most explosive materials have low vapor pressure, and hence, they need to be pre-treated before detection which is not

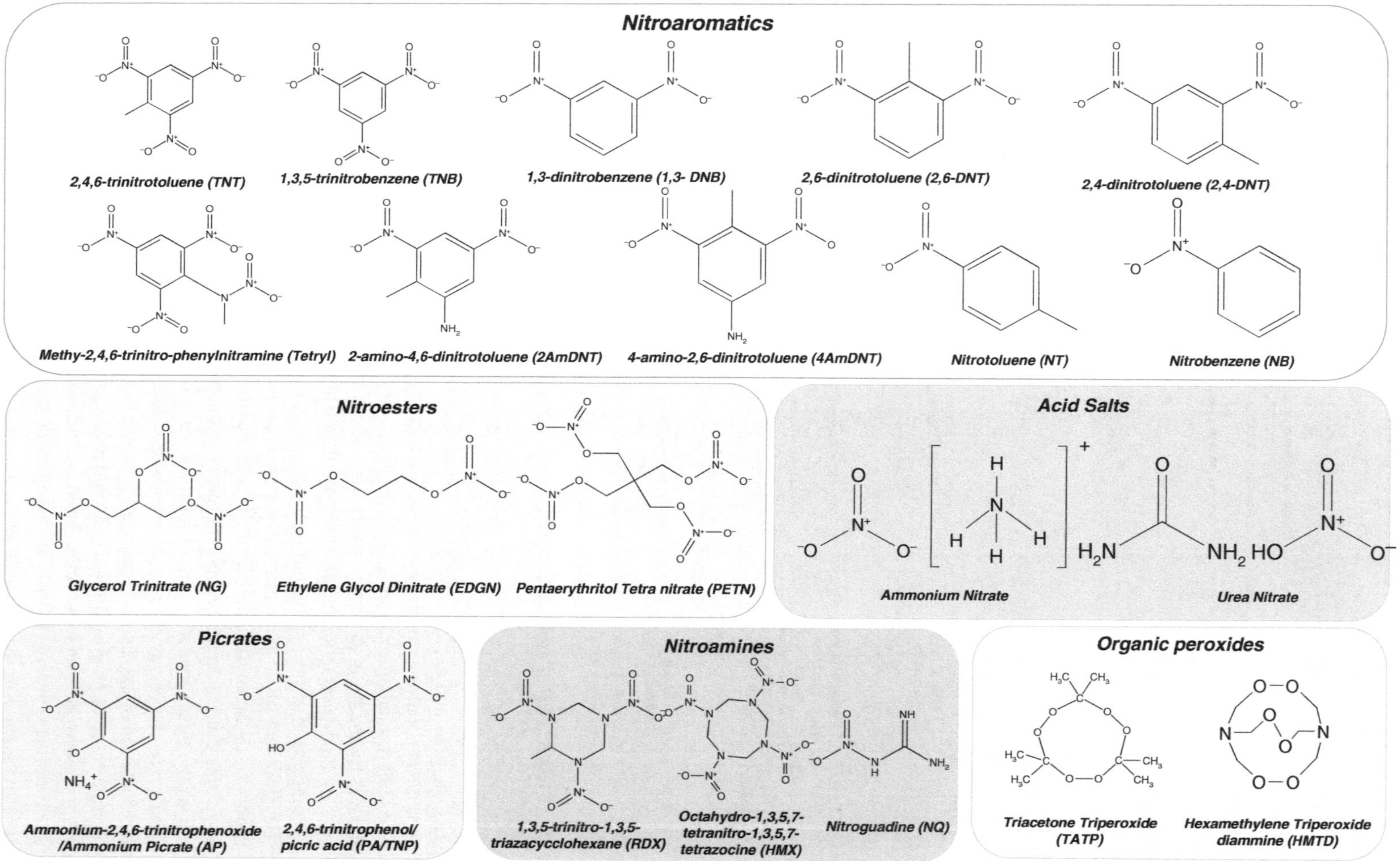

FIGURE 12.3 Representative explosives from different Chemical Groups.

Source: Structures have been drawn by ChemDraw [5].

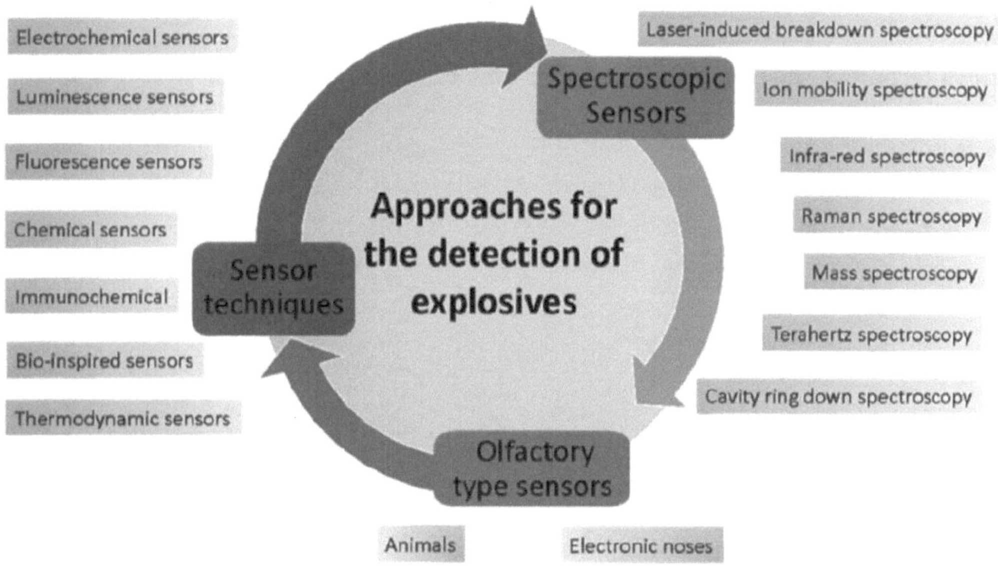

FIGURE 12.4 Approaches for the detection of explosives.

practical. Researchers are working in the direction of developing and establishing highly sensitive methods for the detection of explosives in trace amounts. This will not only help in improved sensitivity but reduced dependence on animal olfaction. The conventional methods involve separation techniques such as mass spectrometry, ion mobility spectrometry, and gas chromatography. The novel methods involve the use of nanomaterials as a sensing component in the electrochemical devices which will be discussed in detail in Section 12.3: Nanomaterial-Based Electrochemical Sensors for Explosive Detection later in this chapter. The majority of ongoing and future research focuses either on creating new techniques or on enhancing conventional sensing techniques to boost sensitivity and speed up detection times. Particularly interesting are electronic nasal approaches combined with small-scale, easily functionalizable, and processed materials.

12.1.3.2 Thermodynamic Sensor

The detection of peroxide and plastic-based explosives with low volatility is still a challenge for researchers and task forces. The majority of the improvised explosive devices (IEDs) used by the terrorists for the attack are comprised of nitrogen-based or peroxide-based chemicals, where these chemicals are used as either detonators or initiators, or both. The majority of the sensors and detectors are designed for nitrogen-based explosives, and hence, the detection of peroxide-based explosives such as triacetone triperoxide (TATP), and nitramines like research department explosive (RDX), high melting explosive (HMX) with low vapor pressure remains a big challenge [7]. In this direction, Ricci & Gregory have demonstrated a robust free-standing thin-film thermodynamic sensor for the trace detection of explosives with low vapor pressure. They have exploited the interdiffusion characteristics in the Cu/Pd system and the highly anisotropic thermal conductivity of ultrathin yttria-stabilized zirconia (YSZ) substrates, a manufacturable fabrication sequence for the development of a thermodynamic sensor for the detection of a series of nitramines, nitrates, and peroxides based at the parts per trillion (ppt) level or lower. As per the reports, Gregory and team reported the detection of plastic explosives such as RDX and HMX at concentrations of 0.5 ppt and 0.02 parts per quadrillion (ppq) with remarkable stability and reproducibility up to hundreds of cycles. Figure 12.6 shows the schematic representation of the thermodynamic sensor and responses of the free-standing sensor to a variety of peroxide-based explosives as well as nitramines and nitrites at known concentrations in comparison to an 8-μm YSZ sensor [7].

TABLE 12.1
Various Techniques Employed for the Application of Explosive Detection

Technique	Details	Explosive	Limit of Detection (LOD)	References
Ion mobility spectroscopy	Positive corona discharge technique	PETN	1 ng	9
		RDX	10 ng	
		TNT	40 ng	
		NTO	1000 ng	
		HMX	1000 ng	
	Electron spray ionization/ion mobility spectrometry (ESI/IMS)	TNT, DNT, 2-ADNT, 4-NT, TNB, RDX, HMX, EGDN, and NG	15–190 μg L^{-1}	10
	Negative corona discharge ion mobility spectrometry	PETN, TNT, and RDX	8×10^{-11}, 7×10^{-11} and 3×10^{-10} g	11
Mass spectroscopy	Atmospheric pressure chemical ionization (APCI) and desorption atmospheric pressure chemical ionization	TNB, TNT, 2,4-DNT, m-DNB, p-DNB		12
	Single photon laser ionization time-of-flight mass spectrometry	NB, 1,3-DNB o-NT, 2,4-DNT, TNT,	17–24 ppb 40 ppb	13
	Desorption electrospray ionization with a portable mass spectrometer	RDX	<1 ng	14
	Direct Analysis in Real Time (DART) and Desorption Electrospray Ionization (DESI) ionization on a fieldable mass spectrometer	RDX and PETN	1 ng and 250 pg	15
Terahertz spectroscopy	THz-TDS coupled with diffuse reflectance spectroscopy	RDX	–	16
	THz-TDS coupled with absorption spectroscopy	TNT, HMX, RDX, and PETN		17
Infra-red spectroscopy	Fiber-optic-coupled reflection/absorption infrared spectroscopy (RAIRS)	TNT and Tetryl	160 ng cm^{-2}	18
		PETN and DNT	2220 ng cm^{-2}	
		HMX	2400 ng cm^{-2}	
	Attenuated total reflection (ATR)-Fourier transform infrared (FTIR) spectro-microscopy	TNT, 1,3,5- trinitrobenzene (TNB), ammonium nitrate	–	19
Raman spectroscopy	Standoff infrared and Raman spectroscopy (SIRS and SRS) detection systems	DNT and TNT	2 mg;	20
		RDX	3 mg	
		C4 and TATP	10 mg	
	Confocal Raman microscopy	PETN, AN, and TNT	–	21
	Raman spectroscopy	PETN, TNT, and ammonium nitrate as well as the explosives precursors hexamethylenetetramine (HMTA)	–	22
Cavity Ring-Down Spectroscopy (CRDS)	Ultraviolet cavity ringdown spectroscopy	2,4-DNT and 2,6-DNT	–	23
	Near-infrared cavity, ringdown spectroscopy	2,4-DNT, 2,6-DNT and TNT	10–100 ng	24

(Continued)

TABLE 12.1 (*Continued*)
Various Techniques Employed for the Application of Explosive Detection

Technique	Details	Explosive	Limit of Detection (LOD)	References
Chemical Sensors	Copolypeptide-doped polyaniline nanofibers for electrochemical detection of ultra-trace trinitrotoluene adsorptive stripping voltammetry	TNT	0.5–10 μM	25
	MWNTs–PANI composite	TNT	–	26
Electrochemical sensors	Carbon-fiber electrodes by voltametric determination	Nitroaromatic explosives	30 μg L^{-1}	27
	Integrated explosive preconcentrator and electrochemical detection system	TNT	–	28
	{MSU/PDDA}n LBL [layer by layer structure created by Mesoporous Silica nanoparticles and poly(diallyldimethylammonium chloride)] assembled modified sensor	TNT	340 ng L^{-1}	29
	Amino-functionalized mesoporous silica microspheres	TNT	114 ng L^{-1}	30
	Prussian-blue "artificial-peroxidase" modified electrode	Liquid peroxide explosives	–	31
Immunochemical	Multiplexed immunoassay	TNT	1.0 μg L^{-1}	32
	Monoclonal and one polyclonal commercial antibodies	PETN.		33
	Sensitive surface plasmon resonance immunosensor for novel oligo (ethylene glycol)-based sensor surface	PBS	20 ng L^{-1}	34
Luminescence sensors	Metallole-containing polymer	TNT	5 ng	35
	Silver nitrate	TATP	7.4 μg L^{-1}	36
	DAN forms a blue luminescent complex	2,3-diaminonaphthalene (DAN)		37
	Polymers and copolymers	Nitroaromatic, nitro-amine, and nitro-ester explosive		38
	Sila fluorene-fluorene copolymer	Nitro-based explosives	1 pg cm^{-1}	38
	Pyrene-functionalized fluorescent film with a benzene ring in spacer	Picric acid (PA)	2.3 μg L^{-1}	39
	Oleic-acid-covered cadmium selenide (CdSe) quantum dots	TNT	340 μg L^{-1}	40
	Fluorescent detector	TNT and DNT	10^{-6} to 10^{-1} mol L^{-1}	41

Gregory & co-workers have developed a metal-oxide-based thermodynamics-based gas sensor for the detection of TATP; IEDs, commonly used by terrorists; and its decomposition products [8]. The development of thermodynamic sensors with low-power characteristics will help in continuous monitoring and detection of the explosives in real time.

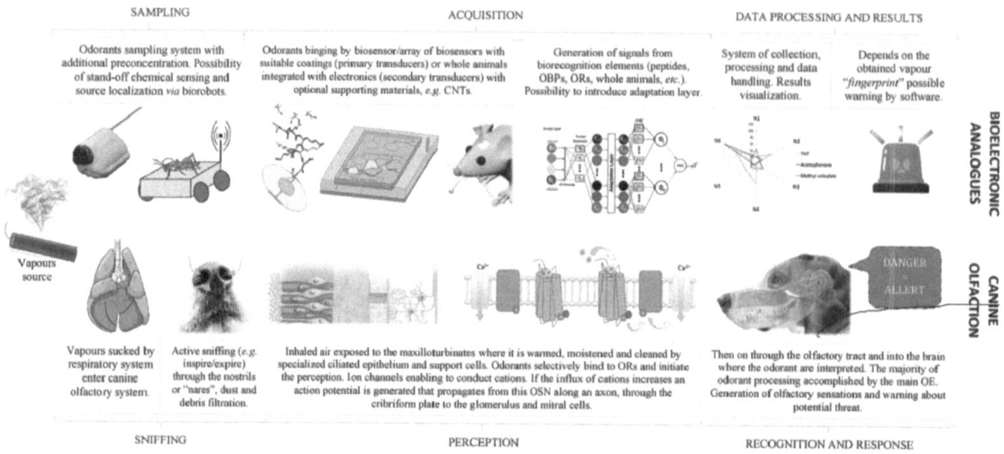

FIGURE 12.5 Comparison of the principles of operation of canine sense of smell with bioelectronic analogs sensing mechanism.

Source: Reproduced with permission from Ref. [6] © 2021 The Author(s). Published by Elsevier B.V.

12.1.3.3 Bio-Inspired Approaches

Animals continue to be the gold standard in the analysis of explosives because of their special aptitude for analyzing complicated gas compounds. The development of devices that closely imitate the activity and operational parameters of biological chemosensory systems, including those of both vertebrates and invertebrates, was sparked by the unusual powers of these systems. Numerous research institutions are looking into electronic equivalents, which brings their successful commercialization closer. When analyzing gas chemicals, notably explosive and dangerous ones, they are thought to replace or supplement animals. By employing techniques drawn from those employed by biological systems, it is possible to get around the limits of conventional gas sensors. In addition to the high selectivity and sensitivity needed for explosives analysis, emulating biological systems helps solve other issues, namely those related to efficient sampling and odor localization.

Wasilewski et al have reviewed the biomimetic devices, which directly imitate the sense of smell and serve as design inspiration for explosives detection equipment. It was assessed whether biosensors and bioelectronic noses (B-ENs) had the potential to emulate the remarkably precise and adaptable "biological noses." A review of approaches drawn from biological olfactory systems should make it easier to approach the issue of designing artificial instruments and to establish an approach for using these systems to analyze explosives [6]. Figure 12.7 demonstrates the scheme of sensitivity levels of humans' and dogs' noses with detection levels and vapor concentrations of different chemicals [6].

The sensors used in the arrays' functionalized receptor materials allow for improved selectivity with regard to substances with identical characteristics or chemical groups. The signal, which resembles a fingerprint, can be recorded and interpreted by using a well-chosen (bio) sensor array. It is possible to use a variety of sensors with various functionalities in the array to give the appropriate selectivity, providing various responses to various materials. It uses a method akin to the mammalian sense of smell, which uses a large number of non-selective receptors whose broad response is perceived by the brain as a particular odor that is sufficient for olfactory sensation.

12.2 ELECTROCHEMICAL METHODS FOR SENSING

The detection of explosive substances is challenging due to several factors, including lack of selectivity, sensitivity, and the complexity of the sample. Explosives are often present in complicated matrices, which are affected by various environmental factors, including temperature, humidity,

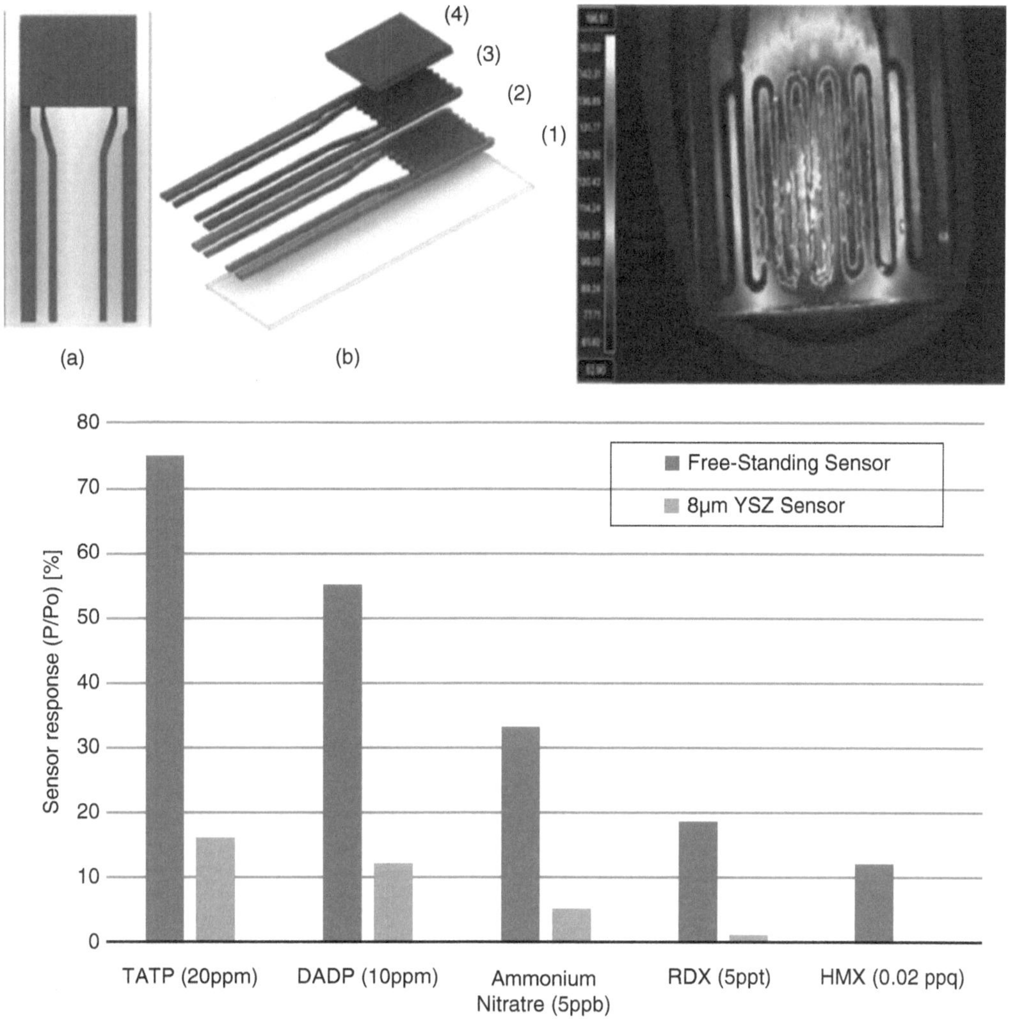

FIGURE 12.6 Schematic of a thermodynamic sensor (a and b) showing the various layers comprising the sensor: YSZ substrate (1), copper adhesion layer (2), palladium microheater (3), metal oxide catalyst (4) and a high-resolution IR image of a thin film microheater fabricated onto a 20-μm thick YSZ substrate responding to an explosive analyte (c). Responses of an 8-μm YSZ sensor and free-standing sensor to a variety of peroxide-based explosives as well as nitramines and nitrites at known concentrations.

Source: Reproduced with permission from Ref. [7], Copyright © 2021, The Author(s).

and pH. The detection of explosives requires sensors that can accurately detect trace amounts of the explosive in these complex matrices. Another significant challenge in explosive detection is the differentiation between various explosive substances. Often, the structural similarities between different explosives make it challenging to differentiate between them. Hence, the development of highly selective sensors that can differentiate between structurally similar explosive compounds is essential. Electrochemical detection is superior to other techniques in a number of ways, including its simplicity and suitability for use in the construction of low-cost, portable detectors. The target chemicals, such as nitroaromatic compounds (NACs), will selectively adsorb onto and enrich the chemically changed surface, enabling ultra-trace detection. Electrodes employed in an electrochemical test can be chemically modified with designable molecules [42].

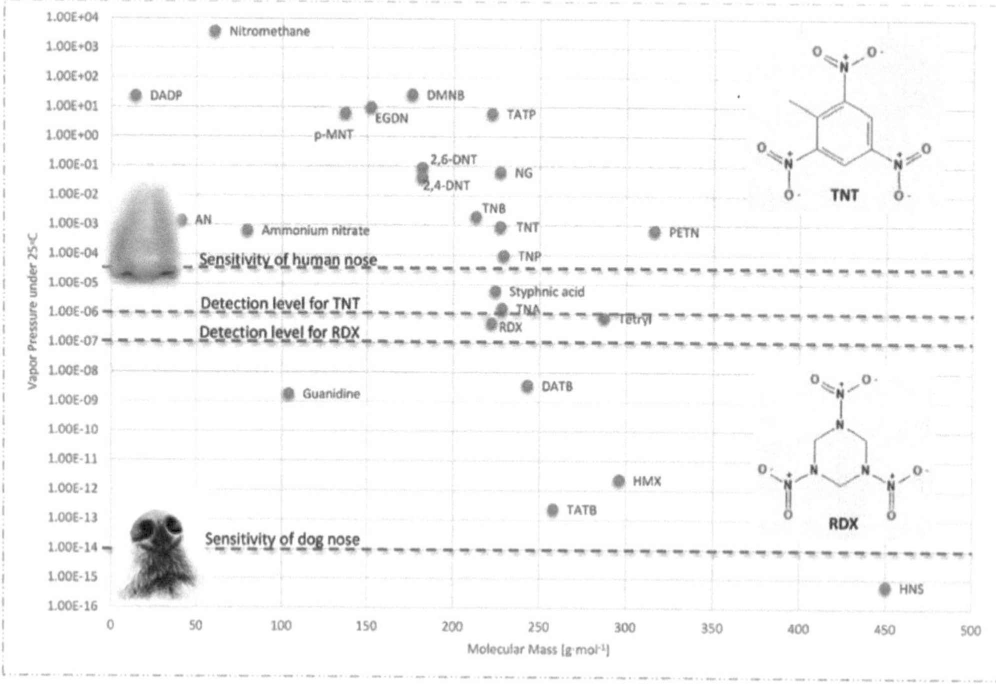

FIGURE 12.7 Scheme of sensitivity levels of humans' and dogs' noses with detection levels and vapor concentrations of different chemicals with structural formulas of 2,4,6-Trinitrotoluene (TNT) and 1,3,5-Trinitro-1,3,5-triazinane (RDX). Styphnic acid – 2,4,6-trinitrobenzene-1,3-diol; 2,3-Dimethyl-2,3-dinitrobutane (DMNB); 2,20,4,40,6,60-Hexanitrophenylethylene (HNS); 4-Nitrotoluene (4-NT), 2,4-Dinitrotoluene (2,4-DNT); 2,6-Dinitrotoluene (2,6-DNT); 1,3,5-Trinitrobenzene (TNB); 2,4,6-Trinitrophenol (TNP); 2,4,6-Triamino-1,3,5-trinitrobenzene (TATB); 3,3,6,6-Tetramethyl-1,2,4,5-tetraoxane (DADP); 3,3,6,6,9,9-Hexamethyl-1,2,4,5,7,8-hexaoxacyclononane (TATP); 1,2-Ethanediol dinitrate (EGDN); Propane-1,2,3-triyl trinitrate (PETN); 2,2-Bis[(nitrooxy)methyl]propane-1,3-diyldinitrate; N-Methyl-N-(2,4,6-trini- trophenyl)nitramide (Tetryl); 2,4,6-Trinitrobenzene-1,3-diamine (DATB); 1,3,5,7-Tetranitro-1,3,5,7-tetrazocane (HMX); Acetonitrile (AN); 2,4,6-Trinitroaniline (TNA); 1-Nitroguanidine (Guanidine).

Source: Reproduced with permission from Ref. [6], © 2021 The Author(s). Published by Elsevier B.V.

12.3 NANOMATERIAL-BASED ELECTROCHEMICAL SENSORS FOR EXPLOSIVE DETECTION

Nanomaterials are characterized by their high surface-area-to-volume ratio, which enables them to interact more efficiently with the target analyte, resulting in greater sensitivity. The use of nanomaterials as electrochemical sensing materials offers a range of benefits, including improved selectivity, sensitivity, and stability. There are several types of nanomaterials that have been explored for electrochemical sensing applications, including carbon-based nanomaterials, metal NPs, metal oxide NPs, and quantum dots. Carbon-based nanomaterials such as graphene and carbon nanotubes (CNTs) have been widely studied for their exceptional properties, including high conductivity, surface area, and stability. Metal NPs such as gold (Au) and silver (Ag) have been used in the development of electrochemical biosensors due to their unique optical and electrical properties. Metal oxide NPs, including zinc oxide (ZnO), titanium dioxide (TiO_2), and iron oxide (Fe_3O_4), have found applications in sensing due to their high surface area, stability, and catalytic activity. Quantum dots have unique electronic properties, including size-dependent and wavelength-tunable fluorescence and have been used as sensing materials for their high sensitivity and specificity.

The use of nanomaterials in electrochemical sensing has enabled the development of highly sensitive and selective sensors for the detection of explosive substances. The incorporation of nanomaterials in electrochemical sensors has provided a large surface area for interaction with explosive compounds, enhancing sensitivity and selectivity. In addition, other nanomaterials such as graphene, quantum dots, and metal NPs have been used to develop sensors for the detection of explosive compounds. The unique properties of these materials, including high sensitivity, selectivity, and stability, offer new possibilities for the development of electrochemical sensors with improved performance. Various nanomaterials have been explored for the development of electrochemical sensors for the detection of explosive substances. The following section discusses in detail various classes of nanomaterials explored for their potential as highly sensitive electrochemical sensors for explosives.

12.3.1 Graphene-Based Explosive Detection

Graphene-based materials have high conductivity, good biocompatibility, and high surface area, making them ideal for electrochemical sensing. Graphene and its derivatives due to their special characteristics are considered potential candidates for sensing applications. The chemical and structural modifications in graphene result in enhanced performance in terms of sensitivity and selectivity. As per the available reports, Jie Zhang & Eric P. Fahrenthold have developed ab initio models of five different graphene nanoribbon sensor configurations and have investigated their chemiresistive response toward explosives [HMX, RDX, & 2,4,6-trinitrotoluene (TNT)] [43]. Graphene nanoribbons and films have been used for the detection of explosives and simulants such as, TNT, NO_2, and 2,4-dinitrotoluene (DNT) [44–46].

Rao and co-workers have developed graphene-based strain sensors for the TNT detection down to ppb levels. The research team exploited the piezoelectric property of graphene and substituted silicon-based cantilevers with graphene nanoplatelets [45]. In the same direction, Castro et al. reported the use of graphite sheets (graphite papers) as a new platform for the collection and sensing of explosive residues. This material offered a lightweight, highly conductive, flexible platform that can be cut in several ways, enabling the collection of explosive residues at the place of interest, without any further sample preparation steps. Graphite sheet is a novel material for the collection and electrochemical sensing of explosive residues [47]. Dettlaff et al. have reported the use of boron-doped diamond/graphene nano-wall electrodes for the detection of explosives in real liquid effluents like wastewater and landfill leachates [48]. Figure 12.8 shows the schematic representation of the electrochemical determination of nitroaromatic explosives at boron-doped diamond/graphene nano-wall electrodes: TNT and 2,4,6-trinitroanisole (TNA) in liquid effluents.

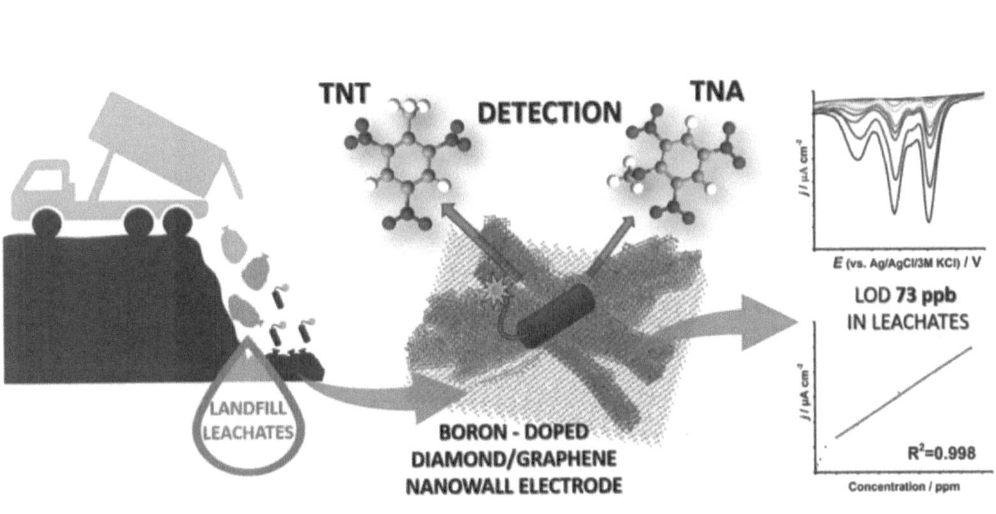

FIGURE 12.8 Schematic representation of the electrochemical determination of nitroaromatic explosives at boron-doped diamond/graphene nano-wall electrodes: TNT and TNA in liquid effluents.

12.3.2 CNTs for Explosive Detection

CNTs are considered an important form of carbon structures as an intermediate form of fullerenes and graphite. CNTs are well known for their characteristics and unique properties as compared to other forms of carbon. Owing to the enormous surface area, CNTs are potential candidates for application in sensors with high sensitivity [49]. The CNTs have special surface-related properties and their most of the weight is concentrated on the surface only. It allows the tunable and tailored properties of CNTs by simple surface modification and functionalization. Several surface modifications have been reported for the enhancement in the sensitivity and selectivity of the CNTs'. Zhang et al. have demonstrated the use of oligomer-coated carbon nanotube chemiresistive sensors for selective detection of nitroaromatic explosives like nitrotoluene, trinitrotoluene, and dinitrotoluene (NT, TNT, and DNT) up to 95 ppb [50]. Similarly, Ferrrari et al. have reported the detection of nitroaromatic explosives in air by amino-functionalized carbon nanotubes. Electrochemical detection of TNT by differential pulse adsorptive stripping voltammetry at carbon paste electrode modified by 1-butyl-3-methylimidazolium tetrafluoroborate with a limit of detection (LOD) as 88.6 ppb has been demonstrated by Vu et al. [51].CNTs have been utilized in the detection of explosives, such as nitroaromatics, by modifying their surface with electron-withdrawing groups to enhance selectivity. The sensitivity of CNT-based sensors can be improved further by incorporating metal NPs on their surface, which provides additional catalytic activity. In the same direction, Shafi et al. have reported the simultaneous detection of two nitroaromatic compounds by g-C_3N_4-decorated $BiVO_4/Ag_2CO_3$ hierarchical nanocomposite [52] with a low detection limit of 0.012 μmol L^{-1} and 0.028 μmol L^{-1}. Figure 12.9 shows the simultaneous detection of two nitroaromatic compounds by g-C_3N_4-decorated $BiVO_4/Ag_2CO_3$ hierarchical nanocomposite.

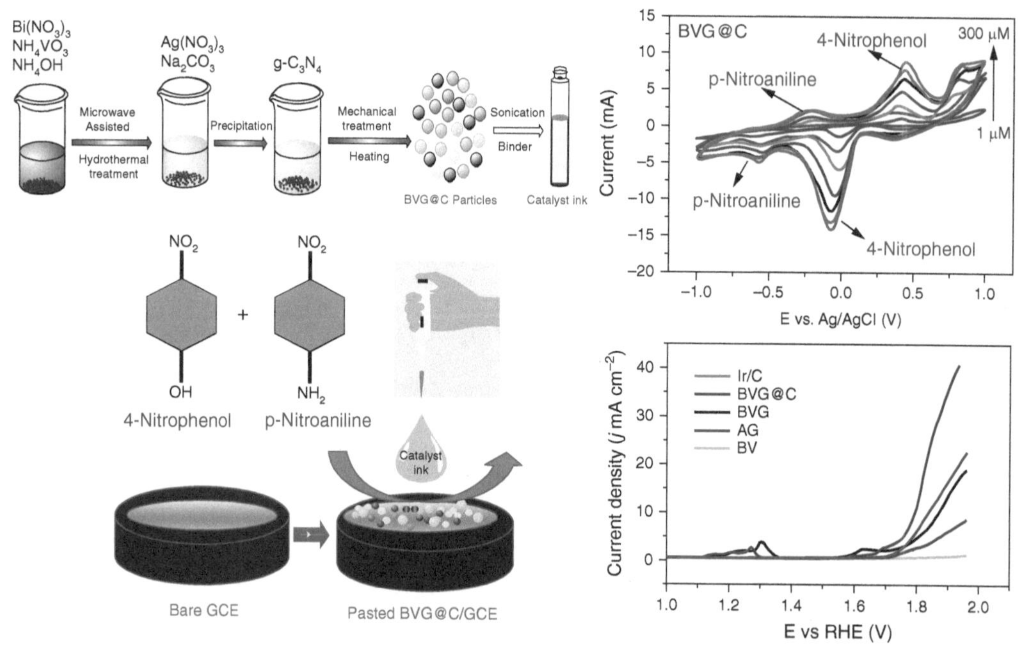

FIGURE 12.9 Simultaneous detection of two nitroaromatic compounds by g-C_3N_4-decorated $BiVO_4/Ag_2CO_3$ hierarchical nanocomposite.

Source: Reproduced with permission from Ref. [52], © 2022 Published by Elsevier B.V.

12.3.3 NPs and Nano Porous Materials for Explosive Sensors

Metal oxide NPs such as TiO_2 and ZnO have been used to fabricate sensors for the detection of nitroaromatics and other explosive compounds. The incorporation of metal oxide NPs enhances the sensitivity of the electrochemical sensor, allowing for the detection of low concentrations of explosive compounds. Table 12.2 summaries the research papers on the use of nanomaterials as an electrochemical sensor.

12.3.4 Covalent Organic Frameworks (COF) in Electrochemical Sensors

Covalent organic frameworks (COFs), a new kind of non-metallic material, are made of organic structural units linked together by covalent bonds (–C–C, –C=C, –C=N, –B-O, etc.) [53]. They have a periodic structure and crystallinity. Building planes with various symmetries and molecular sizes connects two-dimensional (2D) COFs, resulting in structures of various sizes and shapes. In order to achieve excellent conductivity, 2D COFs typically have a layered structure resembling graphene and comprise of conjugated electron-rich or electron-deficient units [53]. In various environments and situations, COF, a novel class of porous crystalline polymers, has been shown to exhibit greater thermodynamic stability. COF has become a heterogeneous functionalized material for a variety of applications because of its predesignable, functionalizable porosity architectures and highly ordered nature. The COF can therefore be an excellent support for loading metal NPs. In terms of catalysis and energy storage, the produced composite exhibits outstanding stability and strong catalytic activity. In order to increase the sensitive activity for analytes, the noble-metal-NP-modified COF may therefore function as a promising material in the sensing sector. A successful electrochemical sensing platform for the detection of hazardous and explosive sodium picrate (SP) and analog picric acid (PA) based on 2D heterogeneous functionalized material has been developed by Tan et al. and as shown in Figure 12.10 [54].

12.3.5 Ionic Liquid Gel-Polymer Electrolyte Film

Room-temperature ionic liquids (RTILs) have been researched as replacement solvents/electrolytes in electrochemical reactions and electrochemical sensors. RTILs are salts made entirely of ions (usually a bulky organic cation and an inorganic anion). Silvester and co-workers have demonstrated a novel electrochemical technique to identify and measure the explosive 2,4,6-trinitrotoluene (TNT) in aqueous solutions as shown in Figure 12.11. For the same TNT samples were placed directly into disposable thin-film electrodes that had been modified with a droplet of a gel-polymer electrolyte (GPE). Trihexyltetradecylphosphonium bis(trifluoromethyl sulfonyl)imide ([P14,6,6,6][NTf2]), a hydrophobic RTIL, and the polymer poly(hexyl methacrylate were both present in the GPE [55].

Silvester and co-workers have studied the reduction mechanism of DNT in eight RTILs using cyclic voltammetry (CV), square wave voltammetry (SWV), chronoamperometry, and digital simulation [56]. Silvester and co-workers also investigate the influence of oxygen, moisture, and other nitroaromatics on the sensing response during the electrochemical detection of explosive compounds in an ionic liquid in mixed environments [57].

Studies have been done on CV and SWV for TNT and DNT reduction in a variety of mixed situations. It has been demonstrated that oxygen greatly affects the current for DNT but not the analyte current for TNT (first peak). As a result, before accurately quantifying DNT in the environment, samples would need to be degassed or flushed with an electrochemically inert gas (such as N_2 or Ar). Both TNT and DNT sensing signals were found to be significantly impacted by water, necessitating either drying the sample before measurement or calibrating the current signals in various humidity settings. Additionally, the behavior of TNT and DNT was compared to that of other di- and trinitro aromatic compounds, indicating the potential [57].

TABLE 12.2
Summary of Research Papers on the Use of Nanomaterials as Electrochemical Sensors for the Detection of Explosives

S. No.	Sensing Element/Nanomaterial	Analyte	LOD	References
1.	MCM-41	NACs TNT	1.8 nM	[42]
1.	SiO_2	TNB	2.8 nM	
1.		DNT	6.8 nM	
1.		1,3-DNB	7.1 nM	
2.	Carbon Paste Electrode Modified by 1-Butyl-3-methylimidazolium tetrafluoroborate by differential pulse adsorptive stripping voltammetry	TNT	88.6 ppb	[51]
3.	Graphite sheet (graphite papers) by square-wave voltammetry (SWV)	TNT	0.06 μmol dm^{-3}	[47]
4.	Graphene oxide	DNT	DFT	[59]
5.	Ionic liquid gel-polymer electrolyte film (bis(trifluoromethylsulfonyl)imide ([P14,6,6,6] [NTf2]) and the polymer poly(hexyl methacrylate) (PHMA)		0.37 μg mL^{-1}	[55]
6.	VO_2 film	TNT	1 mg L^{-1}	[60]
7.	GCE/P (o-PDA-co-ANI)-Au	TNT	2.1 mg L^{-1}	[61]
8.	GCE/P(Cz-co-ANI)-Au	TNT	25 mg L^{-1}	[62]
9.	Pd NC	TNT	0.01 ppm	[63]
10.	PtPd concave nanocubes anchored on graphene nanoribbons (PtPd-rGONRs)	TNT	0.8 ppb	[64]
11.	WP6-N-CQD	TNT	0.95 mM	[65]
12.	N-rich carbon dots (CDs)	TNT	1nM	[66]
13.	Pt nanoparticle (NP) ensemble-on-graphene hybrid nanosheet (PNEGHNs/GCE)	TNT	3 ppm	[67]
14.	Ap/AgNPs/thiol-GQD/GCE	TNT	3.33×10^{-7} nM	[68]
15.	AgNPs/CS-G	TNT	0.45 ppm	[69]
16.	rGO/MWCNT	TNT	0.019 mM	[70]
17.	Ordered mesoporous carbon NG	TNT	0.2 ppb	[71]
18.	BW-NS-rGONRs	TNT	0.1 ppb	[72]

(Continued)

TABLE 12.2 (*Continued*)
Summary of Research Papers on the Use of Nanomaterials as Electrochemical Sensors for the Detection of Explosives

S. No.	Sensing Element/Nanomaterial	Analyte	LOD	References
19.	BASMM/ITO 10–1000 ppb	TNT	4.97 ppb	[73]
20.	Mesoporous SiO_2 NG	TNT	7.2 nM	[42]
21.	MCM-41 NG	TNT	1.8 nM	[42]
22.	SiO_2@mSiO_2@NH_2/GCE	TNT	0.5 nM	[61]
23.	MIP-CP	TNT	1.5 nM	[74]
24.	MIMs-AuNPs-GCE	TNT	1.3 nM	[75]
25.	Molecularly imprinted polymers (MIP)-based polymer/AuNPs	TNT	0.044 fM	[76]
26.	N-GQDs	TNT	0.2 ppb	[77]
27.	Disposable screen-printed electrodes	TNT	200 (ppb)	[78]
28.	Microfabricated capillary electrophoresis chips	TNT	24–36 μg L^{-1}	[79]
29.	Gel polymer electrolyte (hydrophobic polymer and RTIL mixture),1-ethyl-3-methylimidazolium bis(trifluoromethylsulfonyl)imide ([C2mim][NTf2]), 1-butyl-3-methylimidazolium bis(trifluoromethylsulfonyl)imide ([C4mim][NTf2]), N-butylN-methylpyrrolidinium bis(tri-fluoromethylsulfonyl)imide ([C4mpyrr][NTf2]), and 1-butyl-3-methylimidazolium tetrafluoroborate ([C4mim][BF4])	TNT		[56]
30.	Graphite sheet (sensing on from bank notes)	TNT	0.06 μmol dm^{-3}	[57]
31.	RTILs [1-butyl-3-methylimidazolium hexafluoro- phosphate (BMIM-PF6)]	TNT, DNT, PA	1 ppb	[58]
32.	"Forensic Finger" 1-Ethyl-3-methylimidazolium tetrafluoroborate (C2mimBF4),	GSR & DNT		[80]
33.	Boron-doped diamond/graphene nano-wall electrodes	TNT & TNA	3 ppb 270 ppb	[48]
34.	Si-doped graphene nanosheets	NB, 2-nitrotoluene (2-NT), 4-nitrotoluene (4-NT), DNT, TNT	–	[81]
35.	g-C_3N_4-decorated $BiVO_4$/Ag_2CO_3 hierarchical nanocomposite	4-NP & PNA	0.012 μmol L^{-1} & 0.028 μmol L^{-1}	[52]
36.	Exfoliated graphene	DNT & TNT	–	[82]

(*Continued*)

TABLE 12.2 (*Continued*)
Summary of Research Papers on the Use of Nanomaterials as Electrochemical Sensors for the Detection of Explosives

S. No.	Sensing Element/Nanomaterial	Analyte	LOD	References
37.	Dihydroxylatopillar [6] arene-modified gold nanoparticles and cationic Pillar [6] arene-functionalized covalent organic framework, 2HP6@Au@CP6@COF	SP	0.0017 μM	[54]
38.	Kadsura-shaped fluorescent COFs	TNP		[83]
39.	Pythz-COF and Pyaph-COF	TNP	7.9×10^4 M^{-1}	[84]
40.	COFs with diverse hydrazone moieties	TCCA	5.38 nmol L^{-1}	[84]
41.	Graphene-based	TNT in seawater	1 μg mL^{-1}	[45]
42.	Graphene	NO_2, NH_3, and 2,4-dinitrotoluene	28 ppb	[44]
43.	Uniform and rich-wrinkled electrophoretic deposited graphene film: a robust electrochemical platform for TNT sensing	TNT	0.2 ppb	[46]
44.	Oligomer-Coated carbon nanotube	NT, TNT, and DNT	95 ppb	[50]
45.	Amino-functionalized multi-walled carbon nanotubes (MWCNTs)	nitroaromatic explosives	Few ppb (T=28°C)	[85]
46.	ZnO/(rGO)	DNT	1 ppb	[86]
47.	TiO_2 nanosheet, a chemiresistive gas sensor	DNT	9 ppb	[87]
48.	Sulfonated-dye-doped-conducting polypyrrole chemiresistor	TNT	200 ppt	[88]
49.	ZnO nanoparticles (NPs)	DNT	9 ppb	[89]
50.	g-C_3N_4-$BiVO_4$/Ag_2CO_3	Nitroaniline p-nitrophenol	0.012 μmol L^{-1} 0.028 μmol L^{-1}	[52]

TNT, 2,4,6-trinitrotoluene; TCAA, trichloroacetic acid; TNP, 2,4,6-trinitrophenol; DNT, 2,4-dinitrotoluene; 4-NP, 4-Nitrophenol; PNA, 4-Nitroaniline; NB, nitrobenzene, 2-NT, 2-nitrotoluene, 4-NT, 4-nitrotoluene, TNA, 2,4,6-trinitroanisole; GSR, gunshot residue; SP, Sodium picrate; 1,3-DNB, 1,3-dinitrobenzene; TNB, 1,3,5- trinitrobenzene; NAC, nitroaromatic compounds; PA, picric acid.

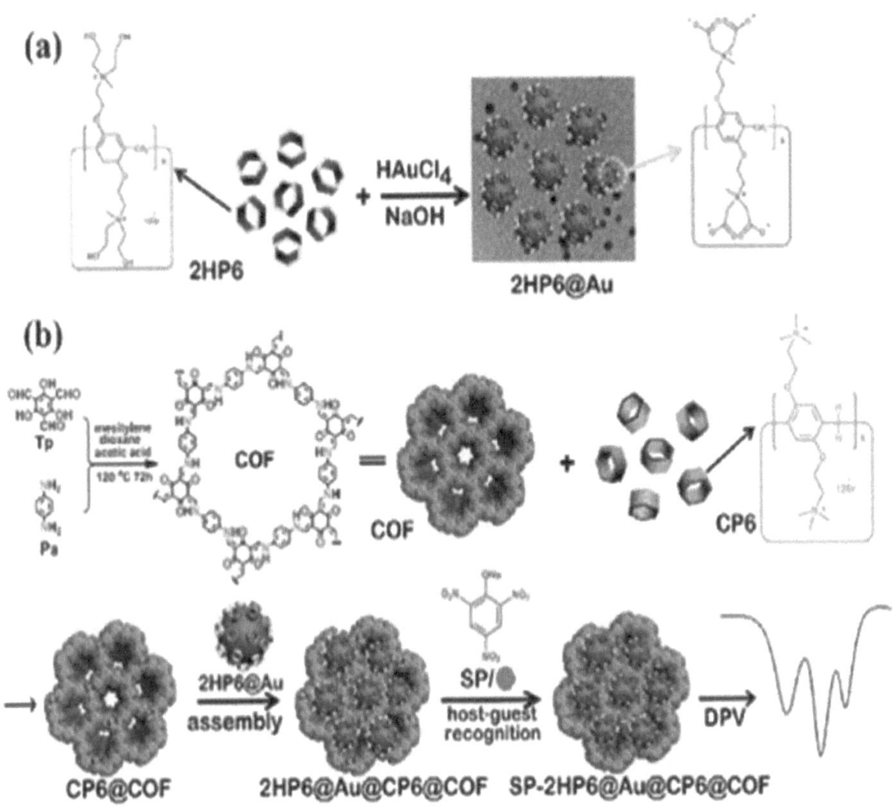

FIGURE 12.10 (a) Preparation of 2HP6 reduced and stabilized Au NPs, (b) the sensing process based on 2HP6@Au@CP6@COF composite toward SP.

Source: Reproduced with permission from Ref. [54], Copyright 2020, Elsevier Ltd.

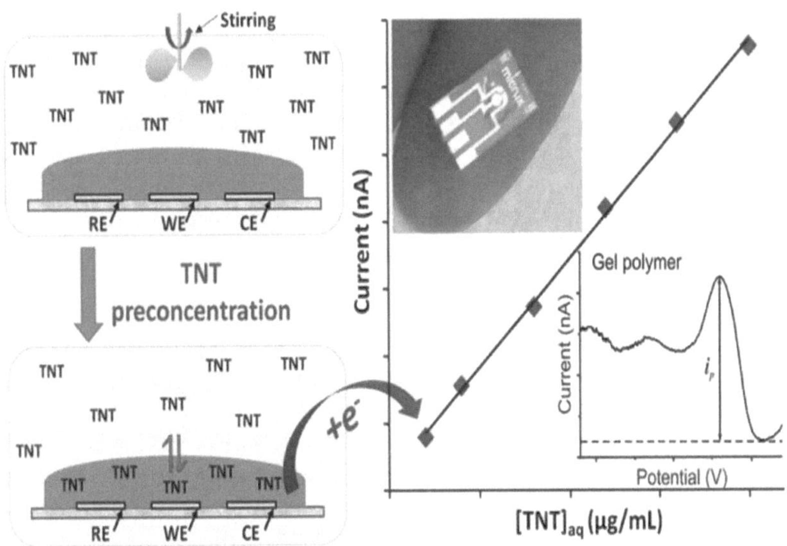

FIGURE 12.11 Schematic representation of detection of 2,4,6-trinitrotoluene using a miniaturized, disposable electrochemical sensor with an ionic liquid gel-polymer electrolyte film.

Source: Reproduced with permission from Ref. [55], Copyright © 2017, American Chemical Society.

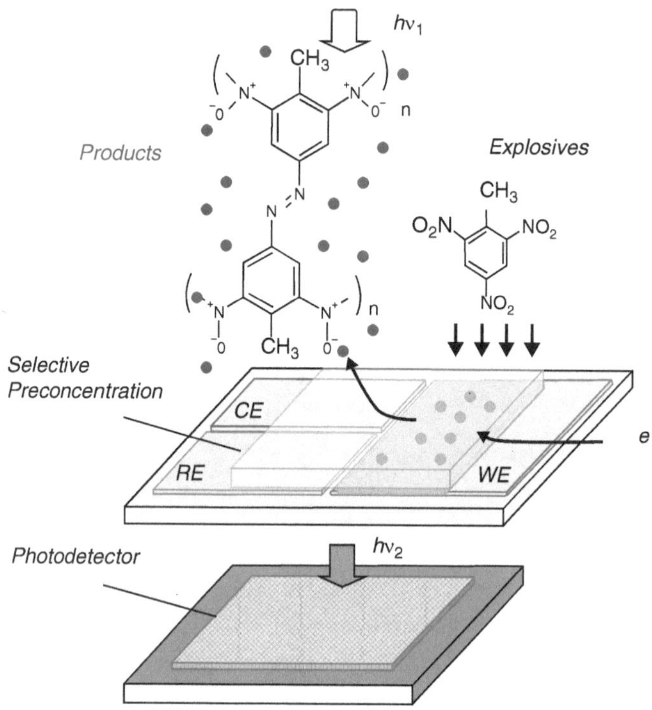

FIGURE 12.12 Schematic representation of the hybrid electrode with an integrated electrochemical and colorimetric approach for the detection of explosives in an ionic liquid.

Source: Reproduced with permission from Ref. [58], Copyright © 2009, American Chemical Society.

12.3.6 Hybrid Sensors

Combining two or more sensing techniques leads to enhancement in the selectivity, sensitivity, and effective sensing ability of the electrodes. The integration of electrochemical and colorimetric approaches toward the detection of explosives has been demonstrated by Forzani et al. The developed hybrid sensor exploited the unique chemical properties of the explosives in the ionic liquids. In electrochemical sensing mode the change in current during the reaction acts a signal. In this study, Forzani et al. exploited the electrochemical reactions to bring out optical changes in the reaction products. Figure 12.12 demonstrates a highly selective, sensitive, and low-cost hybrid sensing platform is developed based on extraordinary properties of explosives in an ionic liquid and an integrated electrochemical and colorimetric approach [58].

12.4 REAL-WORLD APPLICATIONS

The development of nanomaterial-based electrochemical sensors for explosive detection has significant real-world applications in different fields. Homeland security and defense are major applications of explosive detection, where sensors can be used to detect explosives in various environments, including airports, government buildings, and public transportation. The use of electrochemical sensors that can detect explosives at low concentrations can improve security and safety measures in these critical areas. Explosive detection is also important in the military, where sensors can be used to detect IEDs in combat zones. The use of nanomaterial-based electrochemical sensors in this field can potentially save lives by enabling soldiers to detect explosives from a distance and avoid harm.

Choi and Ahn (2023) demonstrated the use of gas-phase scanning electron microscopy for the detection of NAC via real-time monitoring and 3D mapping approach. Figure 12.13 depicts the

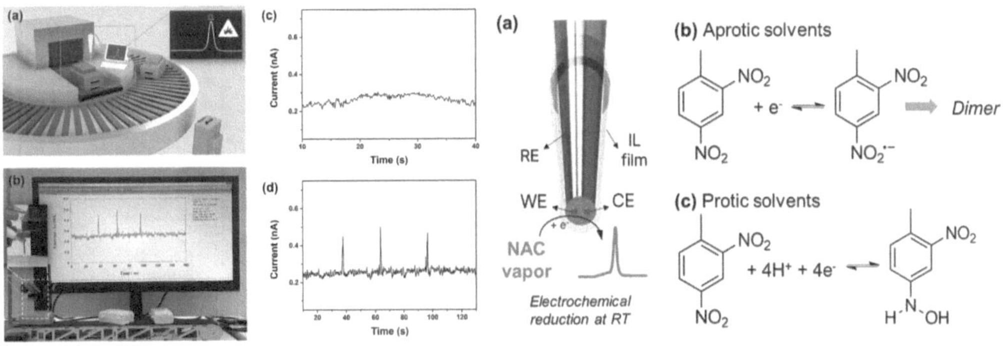

FIGURE 12.13 (a) Schematic representation of the electrochemical cell for the detection of NAC at the airport and the electrochemical response of the empty bag and bag with explosives. (b) Ionic liquid nanoprobe and the reduction mechanism of DNT in aprotic and protic solvent.

Source: Reproduced with permission from Ref. [90], Copyright © 2023, American Chemical Society.

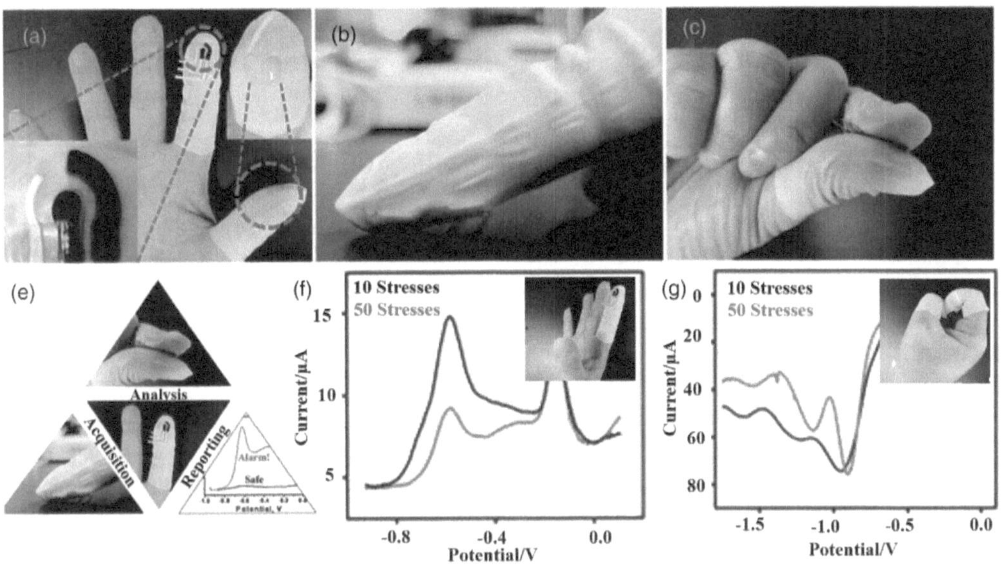

FIGURE 12.14 Schematic representation of the mode of operation of the forensic finger by the collection of the sample (by swipe) from the surface and completing the circuit by joining the fingers and the effect of the mechanical stress on the response.

Source: Reproduced with permission from Ref. [80].

schematic representation of NAC sensors for real-time detection of baggage at the airport. The developed nanoprobe was proven to have a low LOD, quick response, direct assessment without the need for pretreatment along with the real-time mapping of the concentration gradient [90].

Bandodkar et al., developed a solid-state wearable fingertip sensor name as forensic finger for combined sampling and detection of gunshot remains (GSR) and explosives by rapid on-site voltametric screening [80]. According to a thorough investigation, the fingertip detection device can quickly and accurately detect GSR and nitroaromatic chemical residues without degrading its appealing behavior even when subjected to repetitive mechanical stress. This innovative combined sampling/detection fingertip technique has a lot of potential as a quick, inexpensive, and practical method for forensic examinations at crime scenes. Figure 12.14 shows the mode of operation of the

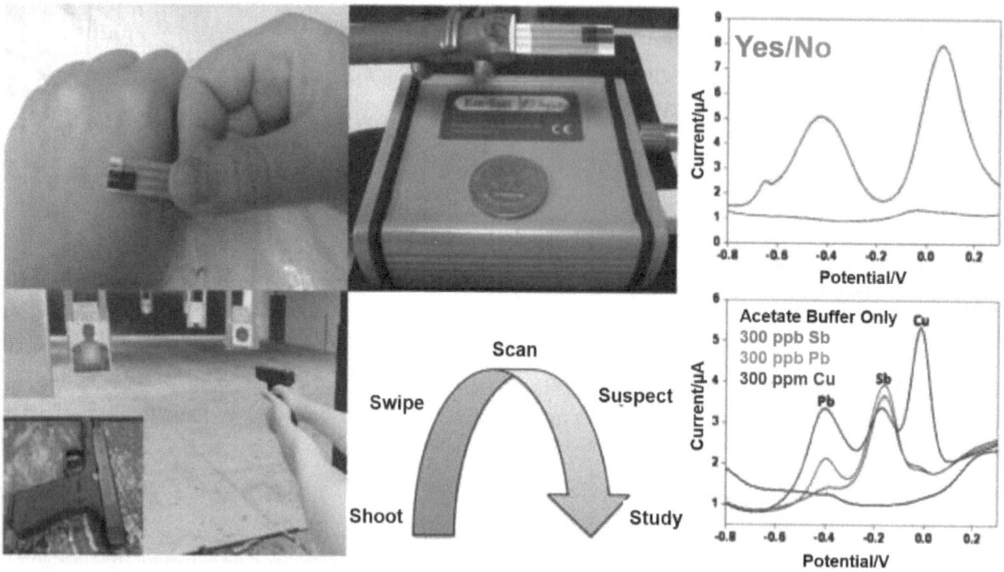

FIGURE 12.15 The digital images for the abrasive stripping voltammetry-based "Swipe-and-Scan" sequence for GSR: Shoot, swipe, scan, signatures (for discharge and non-discharge of a firearm), and research (using pre-made solutions).

Source: Reproduced with permission from Ref. [91], Copyright © 2012 Elsevier B.V.

forensic finger by collecting sample (by swipe) from the surface and completing the circuit by joining the fingers and the effect of the mechanical stress on the response [80].

An innovative integrated system of sampling and analysis of gunshot residue from the hands of persons discharging a handgun was demonstrated by Wang and colleagues. The team used the abrasive stripping voltammetry (AbrSV) procedure, which relies on an initial mechanical transfer of minute amounts of the immobilized GSR from the suspect's hand directly to the working electrode surface of the sensor strip, obviating the need for additional processing stages [91]. Figure 12.15 depicts the images for the abrasive stripping voltammetry-based "Swipe-and-Scan" sequence for GSR: shoot, swipe, scan, signatures (for discharge and non-discharge of a firearm), and research (using pre-made solutions).

12.5 CONCLUSION

Nanomaterial-based electrochemical sensors have shown significant promise in the detection of explosive substances. The incorporation of nanomaterials in electrochemical sensors has led to considerable improvements in sensitivity, selectivity, and stability, providing the potential for lower detection limits and the ability to differentiate between structurally similar compounds. While explosive detection remains a significant challenge, nanostructured electrochemical sensors offer new possibilities for the development of highly sensitive and selective sensors. The real-world applications of these sensors show the potential to improve safety and security measures in different fields, including homeland defense, military, and food safety.

REFERENCES

1. https://aoav.org.uk/2023/explosive-violence-monitor-31-march/
2. https://aoav.org.uk/
3. Lefferts, M. J., and M. R. Castell. "Vapor sensing of explosive materials." *Analytical Methods* 7, no. 21 (2015): 9005–9017.

4. https://commons.wikimedia.org/wiki/File:Coal_dust_explosion_schematic.svg
5. Vaishnav, S. K., H. S. Bhawara, and R. Mishra. "Overview of nano enabled sensor for analysis of explosive substances." *International Journal of Science and Research Archive* 7, no. 2 (2022): 487–500.
6. Wasilewski, T., J. Gębicki, and W. Kamysz. "Bio-inspired approaches for explosives detection." *TrAC Trends in Analytical Chemistry* 142 (2021): 116330.
7. Ricci, P. P., and O. J. Gregory. "Free-standing, thin-film sensors for the trace detection of explosives." *Scientific Reports* 11, no. 1 (2021): 6623.
8. Amani, M., Y. Chu, K. L. Waterman, C. M. Hurley, M. J. Platek, and O. J. Gregory. "Detection of triacetone triperoxide (TATP) using a thermodynamic based gas sensor." *Sensors and Actuators B: Chemical* 162, no. 1 (2012): 7–13.
9. Tabrizchi, M., and V. ILbeigi. "Detection of explosives by positive corona discharge ion mobility spectrometry." *Journal of Hazardous Materials* 176, no. 1–3 (2010): 692–696.
10. Asbury, G. R., J. Klasmeier, and H. H. Hill Jr. "Analysis of explosives using electrospray ionization/ion mobility spectrometry (ESI/IMS)." *Talanta* 50, no. 6 (2000): 1291–1298.
11. Khayamian, T., M. Tabrizchi, and M. T. Jafari. "Analysis of 2, 4, 6-trinitrotoluene, pentaerythritol tetranitrate and cyclo-1, 3, 5-trimethylene-2, 4, 6-trinitramine using negative corona discharge ion mobility spectrometry." *Talanta* 59, no. 2 (2003): 327–333.
12. Song, Y., and R. G. Cooks. "Atmospheric pressure ion/molecule reactions for the selective detection of nitroaromatic explosives using acetonitrile and air as reagents." *Rapid Communications in Mass Spectrometry: An International Journal Devoted to the Rapid Dissemination of Up-to-the-Minute Research in Mass Spectrometry* 20, no. 20 (2006): 3130–3138.
13. Mullen, C., A. Irwin, B. V. Pond, D. L. Huestis, M. J. Coggiola, and H. Oser. "Detection of explosives and explosives-related compounds by single photon laser ionization time-of-flight mass spectrometry." *Analytical Chemistry* 78, no. 11 (2006): 3807–3814.
14. Mulligan, C. C., N. Talaty, and R. G. Cooks. "Desorption electrospray ionization with a portable mass spectrometer: In situ analysis of ambient surfaces." *Chemical Communications* 16 (2006): 1709–1711.
15. Wells, J. M., M. J. Roth, A. D. Keil, J. W. Grossenbacher, D. R. Justes, G. E. Patterson, and D. J. Barket. "Implementation of DART and DESI ionization on a fieldable mass spectrometer." *Journal of the American Society for Mass Spectrometry* 19, no. 10 (2008): 1419–1424.
16. Liu, H.-B., Y. Chen, G. J. Bastiaans, and X. C. Zhang. "Detection and identification of explosive RDX by THz diffuse reflection spectroscopy." *Optics Express* 14, no. 1 (2006): 415–423.
17. Chen, J., Y. Chen, H. Zhao, G. J. Bastiaans, and X. C. Zhang. "Absorption coefficients of selected explosives and related compounds in the range of 0.1–2.8 THz." *Optics Express* 15, no. 19 (2007): 12060–12067.
18. Primera-Pedrozo, O. M., Y. M. Soto-Feliciano, L. C. Pacheco-Londono, and S. P. Hernández-Rivera. "Detection of high explosives using reflection absorption infrared spectroscopy with fiber coupled grazing angle probe/FTIR." *Sensing and Imaging: An International Journal* 10, no. 1–2 (2009): 1–13.
19. Mou, Y., and J. W. Rabalais. "Detection and identification of explosive particles in fingerprints using attenuated total reflection-Fourier transform infrared spectromicroscopy." *Journal of Forensic Sciences* 54, no. 4 (2009): 846–850.
20. Pacheco-Londoño, L. C., W. Ortiz-Rivera, O. M. Primera-Pedrozo, and S. P. Hernández-Rivera. "Vibrational spectroscopy standoff detection of explosives." *Analytical and Bioanalytical Chemistry* 395 (2009): 323–335.
21. Ali, E. M. A., H. G. M. Edwards, and I. J. Scowen. "In-situ detection of single particles of explosive on clothing with confocal Raman microscopy." *Talanta* 78, no. 3 (2009): 1201–1203.
22. Ali, E. M. A., H. G. M. Edwards, and I. J. Scowen. "Raman spectroscopy and security applications: The detection of explosives and precursors on clothing." *Journal of Raman Spectroscopy: An International Journal for Original Work in all Aspects of Raman Spectroscopy, Including Higher Order Processes, and also Brillouin and Rayleigh Scattering* 40, no. 12 (2009): 2009–2014.
23. Ramos, C., and P. J. Dagdigian. "Detection of vapors of explosives and explosive-related compounds by ultraviolet cavity ringdown spectroscopy." *Applied Optics* 46, no. 4 (2007): 620–627.
24. Snels, M., T. Venezia, and L. Belfiore. "Detection and identification of TNT, 2, 4–DNT and 2, 6–DNT by near-infrared cavity ringdown spectroscopy." *Chemical Physics Letters* 489, no. 1–3 (2010): 134–140.
25. Wang, F., W. Wang, B. Liu, Z. Wang, and Z. Zhang. "Copolypeptide-doped polyaniline nanofibers for electrochemical detection of ultratrace trinitrotoluene." *Talanta* 79, no. 2 (2009): 376–382.
26. Yuan, C.-L., C.-P. Chang, Y.-S. Hong, and Y. Sung. "Fabrication of MWNTs–PANI composite—A chemiresistive sensor material for the detection of explosive gases." *Materials Science-Poland* 27, no. 2 (2009): 509–520.

27. Agüí, L., D. Vega-Montenegro, P. Yáñez-Sedeño, and J. M. Pingarrón. "Rapid voltammetric determination of nitroaromatic explosives at electrochemically activated carbon-fiber electrodes." *Analytical and Bioanalytical Chemistry* 382 (2005): 381–387.
28. Cizek, K., C. Prior, C. Thammakhet, M. Galik, K. Linker, R. Tsui, A. Cagan, J. Wake, J. La Belle, and J. Wang. "Integrated explosive preconcentrator and electrochemical detection system for 2, 4, 6-trinitrotoluene (TNT) vapor." *Analytica Chimica Acta* 661, no. 1 (2010): 117–121.
29. Shi, G., Y. Qu, Y. Zhai, Y. Liu, Z. Sun, J. Yang, and L. Jin. "{MSU/PDDA} n LBL assembled modified sensor for electrochemical detection of ultratrace explosive nitroaromatic compounds." *Electrochemistry Communications* 9, no. 7 (2007): 1719–1724.
30. Fu, X.-C., X. Chen, J. Wang, J.-H. Liu, and X.-J. Huang. "Amino functionalized mesoporous silica microspheres with perpendicularly aligned mesopore channels for electrochemical detection of trace 2, 4, 6-trinitrotoluene." *Electrochimica Acta* 56, no. 1 (2010): 102–107.
31. Lu, D., A. Cagan, R. A. A. Munoz, T. Tangkuaram, and J. Wang. "Highly sensitive electrochemical detection of trace liquid peroxide explosives at a Prussian-blue 'artificial-peroxidase'modified electrode." *Analyst* 131, no. 12 (2006): 1279–1281.
32. Anderson, G. P., S. C. Moreira, P. T. Charles, I. L. Medintz, E. R. Goldman, M. Zeinali, and C. R. Taitt. "TNT detection using multiplexed liquid array displacement immunoassays." *Analytical Chemistry* 78, no. 7 (2006): 2279–2285.
33. Anderson, G. P., M. Moore, P. T. Charles, and E. R. Goldman. "Bead-based fluid array detection of pentaerythritol tetranitrate: Comparison of monoclonal vs. llama polyclonal antibodies." *Analytical Letters* 43, no. 18 (2010): 2913–2922.
34. Nagatomo, K., T. Kawaguchi, N. Miura, K. Toko, and K. Matsumoto. "Development of a sensitive surface plasmon resonance immunosensor for detection of 2, 4-dinitrotoluene with a novel oligo (ethylene glycol)-based sensor surface." *Talanta* 79, no. 4 (2009): 1142–1148.
35. Toal, S. J., J. C. Sanchez, R. E. Dugan, and W. C. Trogler. "Visual detection of trace nitroaromatic explosive residue using photoluminescent metallole-containing polymers." *Journal of Forensic Sciences* 52, no. 1 (2007): 79–83.
36. Parajuli, S., and W. Miao. "Sensitive determination of hexamethylene triperoxide diamine explosives, using electrogenerated chemiluminescence enhanced by silver nitrate." *Analytical Chemistry* 81, no. 13 (2009): 5267–5272.
37. Sanchez, J. C., S. J. Toal, Z. Wang, R. E. Dugan, and W. C. Trogler. "Selective detection of trace nitroaromatic, nitramine, and nitrate ester explosive residues using a three-step fluorimetric sensing process: A tandem turn-off, turn-on sensor." *Journal of Forensic Sciences* 52, no. 6 (2007): 1308–1313.
38. Sanchez, J. C., A. G. DiPasquale, A. L. Rheingold, and W. C. Trogler. "Synthesis, luminescence properties, and explosives sensing with 1, 1-tetraphenylsilole-and 1, 1-silafluorene-vinylene polymers." *Chemistry of Materials* 19, no. 26 (2007): 6459–6470.
39. Du, H., G. He, T. Liu, L. Ding, and Y. Fang. "Preparation of pyrene-functionalized fluorescent film with a benzene ring in spacer and sensitive detection to picric acid in aqueous phase." *Journal of Photochemistry and Photobiology A: Chemistry* 217, no. 2–3 (2011): 356–362.
40. Shi, G. H., Z. B. Shang, Y. Wang, W. J. Jin, and T. C. Zhang. "Fluorescence quenching of CdSe quantum dots by nitroaromatic explosives and their relative compounds." *Spectrochimica Acta Part A: Molecular and Biomolecular Spectroscopy* 70, no. 2 (2008): 247–252.
41. Caron, T., M. Guillemot, P. Montméat, F. Veignal, F. Perraut, P. Prené, and F. Serein-Spirau. "Ultra trace detection of explosives in air: Development of a portable fluorescent detector." *Talanta* 81, no. 1–2 (2010): 543–548.
42. Zhang, H.-X., A.-M. Cao, J.-S. Hu, L.-J. Wan, and S.-T. Lee. "Electrochemical sensor for detecting ultratrace nitroaromatic compounds using mesoporous SiO_2-modified electrode." *Analytical Chemistry* 78, no. 6 (2006): 1967–1971.
43. Zhang, J., and E. P. Fahrenthold. "Graphene-based sensing of gas-phase explosives." *ACS Applied Nano Materials* 2, no. 3 (2019): 1445–1456.
44. Fowler, J. D., M. J. Allen, V. C. Tung, Y. Yang, R. B. Kaner, and B. H. Weiller. "Practical chemical sensors from chemically derived graphene." *ACS Nano* 3, no. 2 (2009): 301–306.
45. Goh, M. S., and M. Pumera. "Graphene-based electrochemical sensor for detection of 2, 4, 6-trinitrotoluene (TNT) in seawater: The comparison of single-, few-, and multilayer graphene nanoribbons and graphite microparticles." *Analytical and Bioanalytical Chemistry* 399 (2011): 127–131.
46. Tang, L., H. Feng, J. Cheng, and J. Li. "Uniform and rich-wrinkled electrophoretic deposited graphene film: A robust electrochemical platform for TNT sensing." *Chemical Communications* 46, no. 32 (2010): 5882–5884.

47. Castro, S. V. F., R. M. Cardoso, M. H. P. Santana, E. M. Richter, and R. A. A. Munoz. "Graphite sheet as a novel material for the collection and electrochemical sensing of explosive residues." *Talanta* 203 (2019): 106–111.
48. Dettlaff, A., P. Jakóbczyk, M. Ficek, B. Wilk, M. Szala, J. Wojtas, T. Ossowski, and R. Bogdanowicz. "Electrochemical determination of nitroaromatic explosives at boron-doped diamond/graphene nanowall electrodes: 2, 4, 6–trinitrotoluene and 2, 4, 6–trinitroanisole in liquid effluents." *Journal of Hazardous Materials* 387 (2020): 121672.
49. Zaporotskova, I. V., N. P. Boroznina, Y. N. Parkhomenko, and L. V. Kozhitov. "Carbon nanotubes: Sensor properties. A review." *Modern Electronic Materials* 2, no. 4 (2016): 95–105.
50. Zhang, Y., M. Xu, B. R. Bunes, N. Wu, D. E. Gross, J. S. Moore, and L. Zang. "Oligomer-coated carbon nanotube chemiresistive sensors for selective detection of nitroaromatic explosives." *ACS Applied Materials & Interfaces* 7, no. 14 (2015): 7471–7475.
51. Vu, H. T. T., H. T. V. Le, Y. T. H. Pham, H. Q. Le, and P. H. Pham. "Electrochemical detection of TNT by differential pulse adsorptive stripping voltammetry at carbon paste electrode modified by 1-butyl-3-methylimidazolium tetrafluoroborate." *Bulletin of the Korean Chemical Society* 37, no. 3 (2016): 378–385.
52. Shafi, A., S. Bano, L. Sharma, A. Halder, S. Sabir, and M. Z. Khan. "Exploring multifunctional behaviour of g-C_3N_4 decorated $BiVO_4/Ag_2CO_3$ hierarchical nanocomposite for simultaneous electrochemical detection of two nitroaromatic compounds and water splitting applications." *Talanta* 241 (2022): 123257.
53. Zhu, J., W. Wen, Z. Tian, X. Zhang, and S. Wang. "Covalent organic framework: A state-of-the-art review of electrochemical sensing applications." *Talanta* 260 (2023): 124613.
54. Tan, X., Y. Fan, S. Wang, Y. Wu, W. Shi, T. Huang, and G. Zhao. "Ultrasensitive and highly selective electrochemical sensing of sodium picrate by Dihydroxylatopillar [6] arene-Modified gold nanoparticles and cationic Pillar [6] arene functionalized covalent organic framework." *Electrochimica Acta* 335 (2020): 135706.
55. Yu, H. A., J. Lee, S. W. Lewis, and D. S. Silvester. "Detection of 2, 4, 6-trinitrotoluene using a miniaturized, disposable electrochemical sensor with an ionic liquid gel-polymer electrolyte film." *Analytical Chemistry* 89, no. 8 (2017): 4729–4736.
56. Lee, J., C. E. Hay, and D. S. Silvester. "Electrochemical reduction of 2, 4-dinitrotoluene in room temperature ionic liquids: A mechanistic investigation." *Australian Journal of Chemistry* 71, no. 10 (2018): 818–825.
57. Lee, J., and D. S. Silvester. "Electrochemical detection of explosive compounds in an ionic liquid in mixed environments: Influence of oxygen, moisture, and other nitroaromatics on the sensing response." *Australian Journal of Chemistry* 72, no. 2 (2018): 122–129.
58. Forzani, E. S., D. Lu, M. J. Leright, A. D. Aguilar, F. Tsow, R. A. Iglesias, Q. Zhang, J. Lu, J. Li, and N. Tao. "A hybrid electrochemical–colorimetric sensing platform for detection of explosives." *Journal of the American Chemical Society* 131, no. 4 (2009): 1390–1391.
59. Abdollahi, H., A. Kari, and F. Samaeifar. "Detection of 2, 4-dinitrotoluene by graphene oxide: First principles study." *Materials Research Express* 5, no. 5 (2018): 056404.
60. Casey, M. C., and D. E. Cliffel. "Surface adsorption and electrochemical reduction of 2, 4, 6-trinitrotoluene on vanadium dioxide." *Analytical Chemistry* 87, no. 1 (2015): 334–337.
61. Li, P., X. Li, and W. Chen. "Recent advances in electrochemical sensors for the detection of 2, 4, 6-trinitrotoluene." *Current Opinion in Electrochemistry* 17 (2019): 16–22.
62. Sağlam, Ş., A. Üzer, E. Erçağ, and R. Apak. "Electrochemical determination of TNT, DNT, RDX, and HMX with gold nanoparticles/poly (carbazole-aniline) film–modified glassy carbon sensor electrodes imprinted for molecular recognition of nitroaromatics and nitramines." *Analytical Chemistry* 90, no. 12 (2018): 7364–7370.
63. Soomro, R. A., O. P. Akyuz, H. Akin, R. Ozturk, and Z. H. Ibupoto. "Highly sensitive shape dependent electro-catalysis of TNT molecules using Pd and Pd–Pt alloy based nanostructures." *RSC Advances* 6, no. 51 (2016): 44955–44962.
64. Zhang, R., C.-L. Sun, Y.-J. Lu, and W. Chen. "Graphene nanoribbon-supported PtPd concave nanocubes for electrochemical detection of TNT with high sensitivity and selectivity." *Analytical Chemistry* 87, no. 24 (2015): 12262–12269.
65. Ran, X., Q. Qu, X. Qian, W. Xie, S. Li, L. Li, and L. Yang. "Water-soluble pillar [6] arene functionalized nitrogen-doped carbon quantum dots with excellent supramolecular recognition capability and superior electrochemical sensing performance towards TNT." *Sensors and Actuators B: Chemical* 257 (2018): 362–371.

66. Zhang, L., Y. Han, J. Zhu, Y. Zhai, and S. Dong. "Simple and sensitive fluorescent and electrochemical trinitrotoluene sensors based on aqueous carbon dots." *Analytical Chemistry* 87, no. 4 (2015): 2033–2036.
67. Guo, S., D. Wen, Y. Zhai, S. Dong, and E. Wang. "Platinum nanoparticle ensemble-on-graphene hybrid nanosheet: One-pot, rapid synthesis, and used as new electrode material for electrochemical sensing." *ACS Nano* 4, no. 7 (2010): 3959–3968.
68. Shahdost-Fard, F., and M. Roushani. "Designing an ultra-sensitive aptasensor based on an AgNPs/thiol-GQD nanocomposite for TNT detection at femtomolar levels using the electrochemical oxidation of Rutin as a redox probe." *Biosensors and Bioelectronics* 87 (2017): 724–731.
69. Lu, X., H. Qi, X. Zhang, Z. Xue, J. Jin, X. Zhou, and X. Liu. "Highly dispersive Ag nanoparticles on functionalized graphene for an excellent electrochemical sensor of nitroaromatic compounds." *Chemical Communications* 47, no. 46 (2011): 12494–12496.
70. Castro, S. V. F., M. N. T. Silva, T. F. Tormin, M. H. P. Santana, E. Nossol, E. M. Richter, and R. A. A. Munoz. "Highly-sensitive voltammetric detection of trinitrotoluene on reduced graphene oxide/carbon nanotube nanocomposite sensor." *Analytica Chimica Acta* 1035 (2018): 14–21.
71. Zang, J., C. X. Guo, F. Hu, L. Yu, and C. M. Li. "Electrochemical detection of ultratrace nitroaromatic explosives using ordered mesoporous carbon." *Analytica Chimica Acta* 683, no. 2 (2011): 187–191.
72. Zhang, R., C. Zhang, F. Zheng, X. Li, C.-L. Sun, and W. Chen. "Nitrogen and sulfur co-doped graphene nanoribbons: A novel metal-free catalyst for high performance electrochemical detection of 2, 4, 6–trinitrotoluene (TNT)." *Carbon* 126 (2018): 328–337.
73. Yan, F., Y. He, L. Ding, and B. Su. "Highly ordered binary assembly of silica mesochannels and surfactant micelles for extraction and electrochemical analysis of trace nitroaromatic explosives and pesticides." *Analytical Chemistry* 87, no. 8 (2015): 4436–4441.
74. Alizadeh, T., M. Zare, M. R. Ganjali, P. Norouzi, and B. Tavana. "A new molecularly imprinted polymer (MIP)-based electrochemical sensor for monitoring 2, 4, 6-trinitrotoluene (TNT) in natural waters and soil samples." *Biosensors and Bioelectronics* 25, no. 5 (2010): 1166–1172.
75. Nie, D., D. Jiang, D. Zhang, Y. Liang, Y. Xue, T. Zhou, L. Jin, and G. Shi. "Two-dimensional molecular imprinting approach for the electrochemical detection of trinitrotoluene." *Sensors and Actuators B: Chemical* 156, no. 1 (2011): 43–49.
76. Guo, Z. Z., A. Florea, C. Cristea, F. Bessueille, F. Vocanson, F. Goutaland, A. Zhang, R. Săndulescu, F. Lagarde, and N. Jaffrezic-Renault. "1, 3, 5-Trinitrotoluene detection by a molecularly imprinted polymer sensor based on electropolymerization of a microporous-metal-organic framework." *Sensors and Actuators B: Chemical* 207 (2015): 960–966.
77. Cai, Z., F. Li, P. Wu, L. Ji, H. Zhang, C. Cai, and D. F. Gervasio. "Synthesis of nitrogen-doped graphene quantum dots at low temperature for electrochemical sensing trinitrotoluene." *Analytical Chemistry* 87, no. 23 (2015): 11803–11811.
78. Wang, J., F. Lu, D. MacDonald, J. Lu, M. E. S. Ozsoz, and K. R. Rogers. "Screen-printed voltammetric sensor for TNT." *Talanta* 46, no. 6 (1998): 1405–1412.
79. Hilmi, A., and J. H. T. Luong. "Electrochemical detectors prepared by electroless deposition for microfabricated electrophoresis chips." *Analytical Chemistry* 72, no. 19 (2000): 4677–4682.
80. Bandodkar, A. J., A. M. O'Mahony, J. Ramírez, I. A. Samek, S. M. Anderson, J. R. Windmiller, and J. Wang. "Solid-state forensic finger sensor for integrated sampling and detection of gunshot residue and explosives: Towards 'Lab-on-a-finger'." *Analyst* 138, no. 18 (2013): 5288–5295.
81. Niu, F., Z.-W. Shao, L.-M. Tao, and Y. Ding. "Si-doped graphene nanosheets as a metal-free catalyst for electrochemical detection of nitroaromatic explosives." *Journal of Colloid and Interface Science* 594 (2021): 848–856.
82. Yew, Y. T., A. Ambrosi, and M. Pumera. "Nitroaromatic explosives detection using electrochemically exfoliated graphene." *Scientific Reports* 6, no. 1 (2016): 33276.
83. Liu, X., Y. Han, G. Wang, Y. Wang, J. Chen, Y. Shu, J.-H. Wang, and H. Qiu. "Kadsura-shaped covalent–organic framework nanostructures for the sensitive detection and removal of 2, 4, 6-trinitrophenol." *ACS Applied Nano Materials* 5, no. 5 (2022): 6422–6429.
84. Jiang, S., L. Meng, M.-X. Lv, F.-Y. Bai, W. Tian, and Y.-H. Xing. "Molecular insights into multi-channel detection of nitrophenol explosives and acids in covalent organic frameworks with diverse hydrazone moieties." *Microporous and Mesoporous Materials* 348 (2023): 112408.
85. Ferrari, C., G. Attolini, M. Bosi, C. Frigeri, P. Frigeri, E. Gombia, L. Lazzarini et al. "Detection of nitroaromatic explosives in air by amino-functionalized carbon nanotubes." *Nanomaterials* 12, no. 8 (2022): 1278.

86. Guo, L., Z. Yang, Y. Li, B. Zu, and X. Dou. "Sensitive, real-time and anti-interfering detection of nitro-explosive vapors realized by ZnO/rGO core/shell micro-Schottky junction." *Sensors and Actuators B: Chemical* 239 (2017): 286–294.
87. Li, Y., W. Zhou, B. Zu, and X. Dou. "Qualitative detection toward military and improvised explosive vapors by a facile TiO_2 nanosheet-based chemiresistive sensor array." *Frontiers in Chemistry* 8 (2020): 29.
88. Ghoorchian, A., and N. Alizadeh. "Chemiresistor gas sensor based on sulfonated dye-doped modified conducting polypyrrole film for high sensitive detection of 2, 4, 6-trinitrotoluene in air." *Sensors and Actuators B: Chemical* 255 (2018): 826–835.
89. Ge, Y., Z. Wei, Y. Li, J. Qu, B. Zu, and X. Dou. "Highly sensitive and rapid chemiresistive sensor towards trace nitro-explosive vapors based on oxygen vacancy-rich and defective crystallized In-doped ZnO." *Sensors and Actuators B: Chemical* 244 (2017): 983–991.
90. Choi, S., and H. S. Ahn. "Real-time monitoring and 3D mapping of trace vapors of explosive nitroaromatic compounds at room temperature by gas-phase scanning electrochemical microscopy." *ACS Sensors* 8, no. 4 (2023): 1750–1755.
91. O'Mahony, A. M., J. R. Windmiller, I. A. Samek, A. J. Bandodkar, and J. Wang. ""Swipe and Scan": Integration of sampling and analysis of gunshot metal residues at screen-printed electrodes." *Electrochemistry Communications* 23 (2012): 52–55.

13 Fabrication of MOF-Based Electrochemical Sensor

Meetkumar Y. Chaudhary, Dashrathbhai B. Kanzariya, and Tapan K. Pal

13.1 INTRODUCTION

MOFs are widely acknowledged as crystalline porous coordination polymeric materials, and the structural network is developed through the linking up among organic linkers and suitable metal ions [metal nodes or secondary building unity (SBU)].[1] MOFs offer magnificent structural features such as organic-inorganic crossed network, tunable pore size,[2] large surface area,[3,4] structural flexibility,[5,6,7] smooth chemical functionalization,[8] high chemical stability,[9] and the ability to show nice host–guest interaction.[10,11] These qualitative features of MOFs would be nicely tuned at the molecular level by judicial selection of the MOF forming central components, i.e., linker and metal ion, and by changing the reaction conditions.[12] These special features of MOFs are responsible to show a variety of potential applications like gas adsorption and separation,[13,14] CO_2 valorization,[15,16] energy storage,[17] drug delivery,[18] sensing,[19,20,21] catalysis,[22,23] separation,[24,25] etc. Thus, in the last two decades, MOFs are the center of attraction for material scientists, chemists, and industrial engineers.

Recently, MOFs have emerged as potential sensor, particularly electrochemical sensors and they offer excellent sensitivity and selectivity.[26,27,28] The electrochemical sensing process is low cost, quickly responsive, straightforward, and are largely accepted in comparison to other conventional techniques like ICP-MS, AES, HPLC, etc.[29,30] Scifinder literature survey revealed that the electrochemical sensing applications by MOF for the detection of diverse analytes are gradually increasing (SciFinder^n search terms: "Metal organic framework" by applying concept filter of "Electrochemical sensor") (Figure 13.1). Compared to the luminescence-based MOF sensing technique, the electrochemical sensing process is relatively less explored. The main issue in the latter process is the MOF should be electrically conductive and preferably aqua-phase stable.[31] The non-conducting nature of most of the MOFs prohibits charge mobility in the electrochemical method, and this is a significant concern to the researchers. Still, substantial numbers of conductive MOFs have been employed for the electrochemical sensing process. It has been noted that the following crucial parameters are very important to establish the MOF as an electrochemical sensor: (1) the easy coating of the MOF materials on the electrocatalytic electrodes; (2) the large cavity or high surface area of the MOFs allow the high concentration as well as facile mass mobility of the analytes through the MOF that may effectively enhance the electrical signal and hence recognition sensitivity increases i.e., reduces the detection limit; (3) the tunable pore size of the MOF materials offers the selective detection of the analytes; (4) single crystal structure determination allows the precise location of the atomic site (interaction site) that helps to institute the sensing mechanism, (5) the structure property correlation; and last but not least, (6) the functional modification of the MOF materials would control the selectivity and sensitivity of the sensing phenomena.

In this book chapter, we have documented MOF as an electrochemical sensor for the recognition of various analytes like biomolecules, pollutants, metal ions, etc. We have arranged this book chapter in the following order: First, the fundamental discussion on various synthetic techniques of MOFs are covered; then, the electrochemical sensing application by the synthesized MOFs, and finally, this chapter ends with the conclusion and future outlook.

 DOI: 10.1201/9781032654874-13

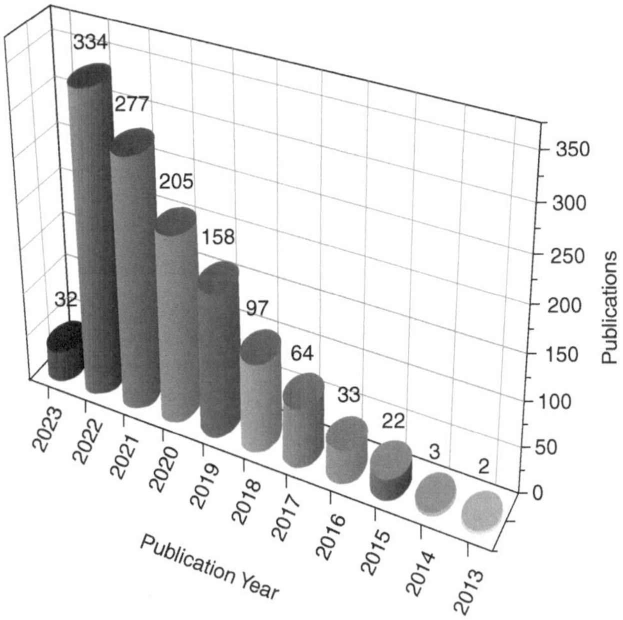

FIGURE 13.1 Sci-finder literature survey revealed the number of published literature in the year (2013–2023) based on the electrochemical sensing applications by MOF for the detection of diverse analytes.

13.2 DIVERSE METHODOLOGY FOR THE SYNTHESIS OF MOFs

13.2.1 Hydrothermal and Solvothermal Process

While talking about the synthesis of MOFs, the first term that usually comes to our mind is the hydro- or solvo-thermal techniques. The hydrothermal approach is widely used to produce various types of MOF materials. Hydrothermal synthesis occurs in a water medium under milder conditions. The reaction occurs inside the autoclave in a Teflon reactor under controlled temperature and autogenous pressure. Whereas the solvothermal method generally requires organic solvents instead of water for the synthesis of MOFs, commonly used solvents are diethyl formamide (DEF), methanol (MeOH), dimethylformamide (DMF), dimethyl sulfoxide (DMSO), butanol (Bu-OH), acetonitrile (ACN), acetone, etc.[32] Typically, the organic solvents used in the solvothermal process are highly expensive and some of them are very hazardous and toxic. To resolve this difficulty, the hydrothermal approach is generally preferred in which deionized water (green solvent and having good polarity) was used as a solvent to make it an eco-friendly and low-cost process. The characteristic features of water are vital in biology because many biochemical reactions are taking place only in aqueous solution. For example, the water-based synthesis of zeolitic imidazolate framework-90 (ZIF-90) with manageable particle size was reported where the different particle sizes can be obtained using water as a solvent. Most of the previously reported literature for the synthesis of ZIF-90 was involved in the utilization of organic solvents like DMF and methanol. The solvent DMF is responsible for reducing the surface area of MOF material. Due to its high boiling point and large size, it is very difficult to remove it from the pores of the material compared to other low boiling solvents such as methanol and ethanol. Indeed, it is possible to replace these large-sized solvent molecules with smaller ones through a solvent exchange procedure. Quite interestingly, the other two well-known MOFs, ZIF-8 and ZIF-67 can also be synthesized using water as a solvent. In the solvothermal approach, the reaction temperature goes up to 220°C, and it may take many hours or days for the complete formation of the crystal or crystalline MOF materials. It has been noticed that such crystallization process is greatly affected by a number of important parameters like concentration of both linker and metal ion, judicial selection of the solvents, temperature, etc.[33] All these factors directly affect the crystallization process and yield of the desired products. Though pros and cons are associated with both the processes, but sometimes, we used the more advantageous process to get the desired MOF network for practical application. In this case, the cons are balanced by the profit gained from the superior application.

13.2.2 Electrochemical Process

The electrochemical method of MOFs can be achieved in mild conditions. In most cases, this type of method does not require high temperature or high pressure, perhaps the reaction can be done at RT and atmospheric pressure. The electrochemical method is more economical compared to the solvothermal method.[34] Due to these benefits, this method gained consideration from scientists and researchers. Consequently, various types of MOFs have been synthesized using electrochemical syntheses, such as aluminum, copper, and zinc-based MOFs.[35,36,37] The well-known MOF, isoreticular metal-organic framework-3 (IRMOF-3), was also synthesized by using electrochemical processes in which metallic zinc was used as the metal source.[38] Although, synthesis of zirconium-based metal-organic frameworks (ZrMOFs) is challenging through electrochemical methods where zirconium is used as a metal source, there are few reports of ZrMOFs that are obtained by using this method. One of the well-known MOFs, UiO-66-NH_2, was derived where pure zirconium metal is taken in DMF as the source of Zr for the construction of MOFs. This process allows the large-scale synthesis of UiO-66-NH_2 with high porosity and unvarying morphology and is economically a better method.[39] Like in the case of hydrothermal/solvothermal technique, this process also depends on certain important parameters (vide supra). There are examples of some MOFs which are synthesized by the electrochemical method that showed better results in some applications compared to solvothermally synthesized MOFs. For example, UiO-66-NH_2, obtained electrochemically, demonstrated sensitive recognition of Fe^{3+} with limit of detection (LOD), which is much less than the LOD showed by solvothermally obtained UiO-66-NH_2.[39]

Furthermore, by using this process, the thin film of MOF can be constructed either directly or indirectly. With direct electrochemical techniques like reductive electrosynthesis and anodic dissolution, on the surface of the electrode, the desired MOFs are formed. Whereas, in the indirect method, the MOFs can be produced by processes where an electrochemical reaction is just one of the steps. There are various types of indirect electrochemical methods such as galvanic displacement synthesis, electrophoretic deposition, and self-templated synthesis.

13.2.3 Microwave-Assisted Process

The microwave-assisted synthetic technique is comparatively a new chemical technique that simplifies green syntheses of nano- and nanoporous materials. This synthetic technique is nowadays adopted by several researchers to synthesize various MOFs. Many novel and recognized MOFs structures are formed using microwave electromagnetic irradiation. In this process, microwave irradiation is used as a heat source for the construction of MOFs. This type of radiation lies between infrared and radio waves, with comparatively larger wavelengths of 1 mm to 1 m. Generally, conventional heating techniques involve electric heating, heating covers, sand baths, and oil baths. Several efforts have been put in the last decade to introduce alternative energy sources by seeing energy consumption, working productivity, and eco-friendliness. These synthesis processes typically require temperatures above 100°C, and reaction times are carefully adjusted to optimize the yield and purity of the target product. For MOF synthesis, crystallization time plays a very important role, but in this process, crystallization time is less compared to other processes of MOF synthesis.[40] These are core advantages of the microwave-assisted approach that is less time consuming for crystallization and has less reaction time. Due to microwave heating, the obtained MOF particles are smaller in size and have an extremely porous structure. The first MOF synthesis using a microwave-assisted approach was Cr-MIL-100.[41] Numerous MOFs have been successfully developed by using microwave-assisted techniques such as MOF-205, Mg-MOF, etc.

13.2.4 Sonochemical Process

The term "Sono" is associated with ultrasound electromagnetic radiation. In this process, a heat is generated from ultrasound waves which expedite the nucleation rates. In this process, maximum reactions occur at room temperature and the crystallization time is less. [42] In this process, ultrasound is generated with the help of an instrument called an ultrasonic transducer device which converts electrical or mechanical energy to ultrasound. Typically, the solution of organic ligand and metal salt is placed in the sonicator with a variable power output that ensues rapid heating which afforded the MOF.[43] Numerous MOFs were synthesized using the sonochemical process, for example, Zn-MOF, PCN-6, ZIF-8, etc. Interestingly, this process was utilized for the synthesis of MOF-5 with a crystal size of 5–20 nm.[44] The advantages of using this technique are less energy consumption, generation of zero toxic waste, and requirement of less toxic compounds.

13.2.5 Mechanochemical Process

The mechanochemical method for the synthesis of MOF is considered an eco-friendly and non-toxic technique because this method is solvent free, and sometimes, a very negligible quantity of solvent is required. It is an extremely effective and simple approach to synthesize MOFs. This process does not require a heating medium.[45] This technique contains three diverse methods such as ion- and liquid assisted grinding (ILAG), neat grinding (NG), and liquid assisted grinding (LAG). The ILAG synthetic process requires a catalytic liquid phase with a trace quantity of salt to quicken the crystallization of MOFs. The NG technique does not require any solvents, whereas LAG is fast and requires a very scanty liquid phase.[46,47] The first MOF was synthesized using this technique was Cu(isonicotinic acid)$_2$. In this process, isonicotinic acid and $Cu(OAc)_2{\cdot}H_2O$ were ground in the presence of a solvent, and the product obtained are extremely crystalline materials.[48] Other well-known MOFs like Cu-MOF, Hong Kong University of Science and Technology-1 (HKUST-1), etc., were also produced using this technique.[49]

13.3 GENERAL MECHANISM OF ELECTROCHEMICAL SENSORS

A chemical sensor is typically made up of two key functional components: a receptor and a transducer. Different types of receptors are available and can be activated or doped onto the surfaces of materials that interact with analytes in unique ways to produce sensing signals (Scheme 13.1).

The different types of electrochemical sensors are amperometric, potentiometric, impedimetric, photoelectrochemical, and electrogenerated chemiluminescence. The potentiometric sensors produce a local Nernstian equilibrium at the sensor interface when no current is allowed to flow

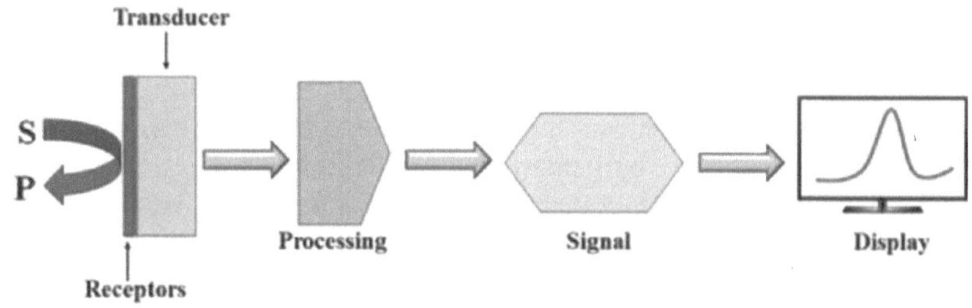

SCHEME 13.1 A schematic presentation of the sensing signal through a main part of the typical sensor.

through the system that delivers the information about the analyte's concentration. This is because of particular sensor-analyte interactions. For amperometric sensors, the electrochemical oxidation or reduction occurs upon applying a voltage between a reference electrode and a working electrode. The resulting current is then measured, providing a quantitative indicator of the analyte's concentration. On the other hand, the conductometric sensors are also known as impedimetric sensors which measure the variations of surface impedance to quantify the analyte-specific recognition. Metal-Organic Frameworks (MOFs) are an emerging class of porous materials with potential for electrochemical sensing. However, because MOFs are not always conductive, they are often embedded within a highly conductive polymer matrix to enable their use as sensors. These mixed MOF-based conductive hybrid materials can now employ for the electrochemical detection of various analytes. We are representing some MOF examples where the aforementioned electrochemical sensing techniques are exclusively used to determine the diverse analytes.

For example, Ma and co-workers developed MOF-based novel conductive electrochemical sensor, UiO-66-NH_2@polyaniline (PANI) for the detection of Cd^{2+} ions.[50] To enhance the conductivity of MOF, it is mixed with highly conducting material PANI. Therefore, the individual properties of respective components are inherited into new composite materials. The enhancement of conductivity in the composite materials compared to its pristine MOF has been confirmed by measuring with electrochemical impedance spectroscopy (EIS) which showed the electron-transfer resistance (RCT) value of UiO-66-NH_2@PANI and UiO-66-NH_2/GCE are 76 and 160 Ω, respectively. Next, the hybrid conductive MOF materials are embedded with GCE to measure differential pulse voltammetry (DPV) in the presence of Cd^{2+} ions. It is observed that 10-fold current increases in the presence of Cd^{2+} ions. The observed current enhancement is likely due to efficient electron transfer facilitated by the composite material. This efficient transfer is a result of the strong interaction between the amino group and the Cd^{2+} ion, leading to high electrochemical sensitivity. John and co-workers synthesized CAU-1 MOF for the electrochemical detection of the environmental pollutant hydroquinone (HQ) through amperometric method.[51] The MOF is coated on GCE and the composite GCE/CAU-1 is found to be very sensitive for HQ. Upon incremental addition of various amounts of HQ, the current response increased linearly with an LOD and sensitivity calculated as 0.067 μM and 1471.4 μA $mM^{-1}cm^2$, respectively. Initially, the π–π interaction between the phenyl ring of CAU-1 and the aromatic rings of HQ leads to the oxidation of HQ. Further, the conjugated aromatic rings in the HQ backbone make it simpler for the movement of electrons from HQ to modified GCE. In addition, the construction of hydrogen bonds between the hydroxyl group of HQ and $-NH_2$ functionality of 2-amino terephthalic acid leads to the high sensitivity of HQ recognition.

13.4 ELECTROCHEMICAL SENSING APPLICATION OF MOFs

In recent literature reports, MOFs and their composites are regarded as potential candidates for diverse electrochemical sensing applications. They have the ability to distinguish diverse analytes such as heavy metal ions, organic industrial wastes, agricultural waste, and it is also important to recognize DNA and larger biomolecules. The following are the examples where the involvement of MOFs has been observed as an electrochemical probe.

13.4.1 Recognition for Heavy Metal Ions

Due to the continuous dumping of industrial garbage into the environment, various heavy metal pollutants have been piled up that are particularly dangerous and have a negative impact on human health like genetic disorders and several other diseases.[52] These polluted materials swiftly get into the human body through the food supply chain. Even in tiny quantities, the majority of them are harmful.[53,54] For instance, Cd^{2+} impairs renal function and disrupts metabolism.[55] The Pb^{2+}ion can harm the neurological system.[56] Although, many natural and biological processes are accomplished with the involvement of Cu^{2+} ions, but high level of Cu^{2+} might result in cancer and some other genetic problems.[57] Likewise, mercury also badly affects our health.[58] So the monitoring of these metal ions is very imperative. It has been noticed that the ions from heavy metals produce intermediary alloys, making it difficult for their detection. The traditional sophisticated analytical techniques like atomic absorption spectroscopy (AAS), X-ray fluorescence spectroscopy, atomic emission spectroscopy, and other methods are highly effective to detect heavy metal ions.[59,60] However, these methods are intricate, expensive, and highly time-consuming. In this regard, the electrochemical procedure is superior to these conventional techniques in identifying heavy metal ions due to its cost-effectiveness, less time-consuming, and easy to handle.[61] By using this electrochemical sensing technique, several heavy metal ions, including Cd^{2+}, Pb^{2+}, Cu^{2+}, and Hg^{2+}, are possible to detect. In this respect, MOF comes out to be an excellent electrochemical sensor. However, their weak conductance and chemical stability have been the point of fundamental drawbacks.[62,63] To address these issues, MOF is combined with several other electrical conductors to produce composite materials in which the property of the individual component is inherited into the new composite. Various materials have been developed to improve the electrochemical sensing efficiency of MOFs. These are graphene oxide (GO), metal oxide, and different conducting polymeric materials.[64,65,66] The graphene aerogels (GAs) are mainly carbon-based materials which are having characteristic features such as large surface area, high porosity, low density, and customizable electrical properties. These attributes make GAs suitable for applications as absorbent, catalytic, and sensory materials.[67,68] The number of binding sites present on GAs induces strong interaction with the MOF materials resulting in the facile formation of new composite. Based on this concept, Wang and co-worker developed MOF as an electrochemical probe for the monitoring of heavy metal ions.[69] Here, GCE is integrated with GAs and solvothermally obtained UiO-66-NH_2 to construct the electrochemical probe (Figure 13.2). This probe showed the electrochemical identification of various heavy metal ions, cadmium, lead, copper, and mercury in the water phase.

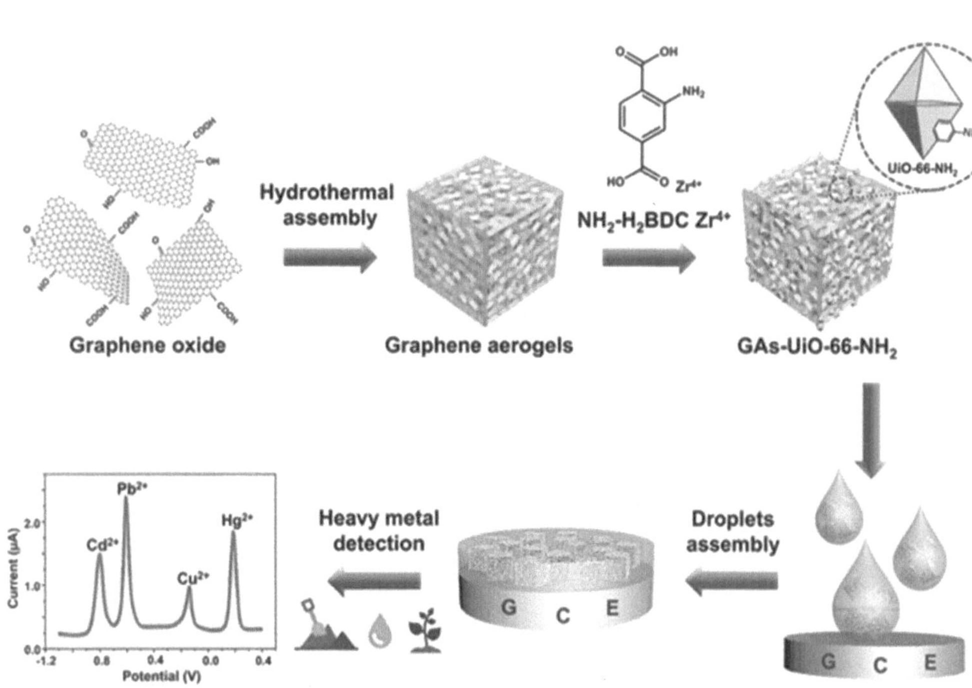

FIGURE 13.2 The synthesis of GAs-UiO-66-NH_2 through the in situ growth of UiO-66-NH_2 on GAs, followed by the detection of heavy metal ions.

Source: Permission from Ref. [69].

13.4.2 Recognition for Organic Pollutants

With an increase in the world population, industrial development is progressively increasing. As a result, the inorganic and organic contaminants are directly and indirectly pumped into the environment.[70] These contaminants are primarily found in drinking water and soil and they create negative health impacts on human living standards.[71,72] For instance, the textile and leather manufacturing industries emit colored pollutants into the water bodies, which has a significantly negative impact on aquatic life.[73] Some of the environmental pollutants like resorcinol (RS), catechol (CT), and HQ are the congener of dihydroxybenzene isomers (DBI).[74] Therefore, to safeguard public health and protect ecosystems, it is essential to detect organic contaminants. Such organic contaminants are homogeneous in terms of stereochemistry and oxidative potential. As discussed earlier, the electrochemical sensing of these pollutants is quite useful compared to the traditional analytical techniques (*vide supra*). Quite impressively, electrochemical sensing is a very convenient method because of its high sensitivity, good selectivity, and quick response to these pollutants.[75,76] For the detection of a variety of organic pollutants, diverse materials including carbon nanotubes,[77] quantum dots,[78] graphene,[79] metal sulfides,[80] etc., were utilized as electrochemical sensors. Figure 13.3 depicts the application of chitosan and GO doping to the electrode surface.[81] The GO in the CS/GO compound was reduced to reduced graphene oxide (rGO) by an electro-reduction process, leading to the prominent electrical insulator. Further, this was used as a supportive transporter again for embedding with $Cu_3(BTC)_2$. The strong covalent interaction among GO and CS in $Cu_3(BTC)_2$/ERGO/CS offered the facile formation of highly stable electrochemically active thin film. This film exhibited well-defined reductive peaks for CT, RS, and HQ, which could be distinguished from each

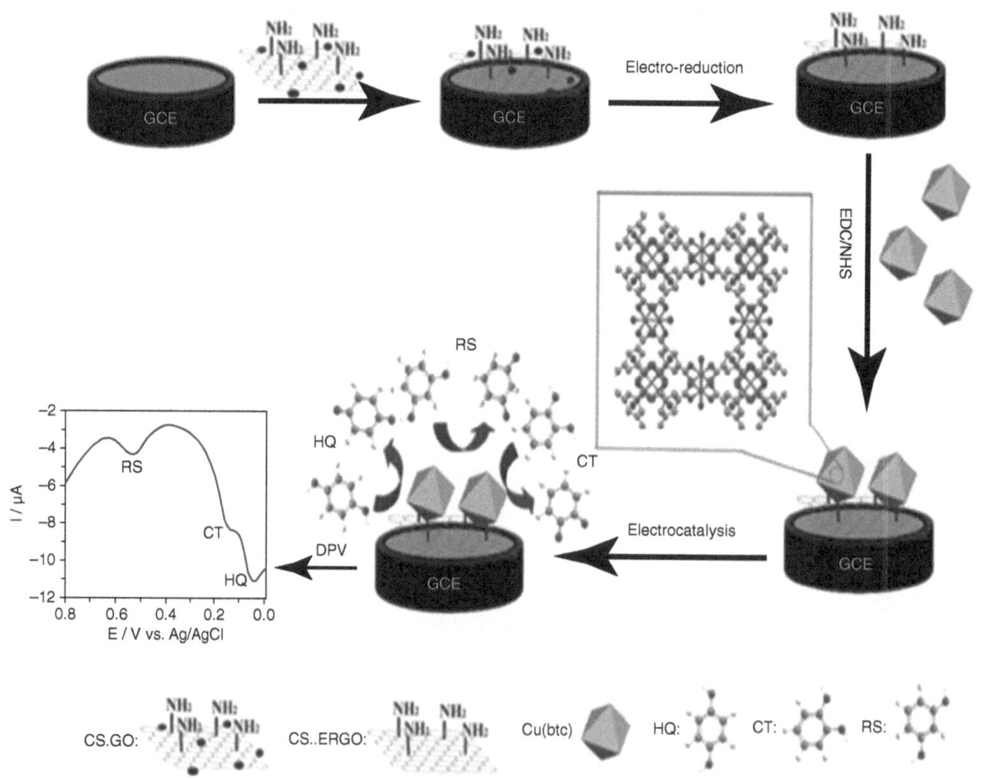

FIGURE 13.3 The fabrication of $Cu_3(BTC)_2$/CS-ERGO/GCE for the detection of DBIs.

Source: Permission from Ref. [81].

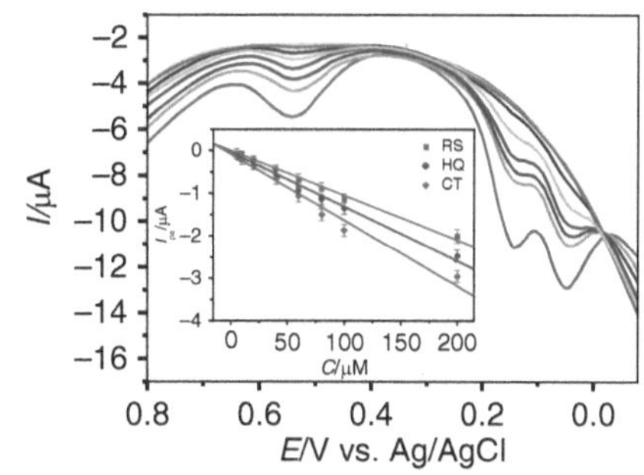

FIGURE 13.4 Differential pulse voltammetry of $Cu_3(BTC)_2$/CS-ERGO/GCE in the presence of analytes (5.0–200 μM). Inset: the linear relationship between the Ipa with the concentration of analytes.

Source: Permission from Ref. [81].

other with good selectivity. Again, $Cu_3(BTC)_2$/CS-ERGO/GCE was also used to determine HQ, CT, and RS together while continuously altering the amount of each DBI. Three distinct oxidizing peaks are shown in Figure 13.4, demonstrating the independence of reactions involving CT, HQ, with RS at $Cu_3(BTC)_2$/CS-ERGO/GCE. The good conductivity of the CS/rGO matrix considerably boosted the responsiveness, resulting in low LODs for HQ, RS, and CT around 0.44, 0.33, and 0.41 μM, respectively. Further, the sensor was applied for the identification of DBI congeners in real water samples.

13.4.3 Recognition for Pesticides

Because of their toxicity and lack of recyclability, the discharge of toxic materials into soil and water bodies caused by industrialization and other anthropogenic activities raised environmental issues. Pesticides are used in agriculture to improve crop quality and growth. These pollutants are used in one-third of all agricultural products worldwide. While pesticides are used to control pests, they leave harmful residues in agricultural products, groundwater, and soil, posing serious health risks to humans.[82] Natural problems caused by these toxins include sterility, infertility, and cancer of the lungs, brain, breast, and testicles. Due to social and economic development issues, many substances cannot be permanently outlawed. Even though pesticides have a great role in agriculture, their negative health impact is a major concern that needs to be critically considered. Therefore, it is crucial to correctly monitor these chemicals using a swift and focused strategy to protect human health and the environment.[83] The electrochemical detection of pesticide molecules by using MOF and its composite as a sensor is noticed in the published literature. For instance, Hu and co-workers developed a copper-based MOF material for the electrochemical detection of carbendazim. The pesticide carbendazim having the capacity to protect vegetables, fruits, and even plants results in large-scale production of the crops. The probe, $[Cu(adp)(BIB)(H_2O)]_n$ was synthesized by the hydrothermal reaction among the mixture of linkers 1, 4-bisimidazolebene (BIB) and adipic acid (adp) with $CuSO_4.5H_2O$.[84] This MOF was cast on the bare glassy carbon electrode (GCE) to engender active electrode components in the electrochemical sensor. This sensor showed the detection of carbendazim with very low LOD values of about 10 nM. This illustrates that the sensor is very effective for monitoring carbendazim. Qu, Lu, and co-workers screened tomato specimens by using MXene/carbon/nanohorns/cyclodextrin as an electrochemical probe.[85] The cyclodextrin-MOFs component have large pores which act as a pre-concentrator to facilitate the adsorption of large amounts of carbendazim. Quite interestingly, the suitable supramolecular interaction among MXene/carbon/nanohorns and -cyclodextrin-MOFs bestow the composite to behave as an excellent sensor. The LOD value for carbendazim is 1.0 μM obtained from the linear relationship of 3.0–10.0 nM. Further, the electrochemical sensor was found to have good repeatability, durability, and recyclability.

Dichlorophen is a class of chlorophenol antibacterial that is primarily used as a pesticide. It poses significant health risks and is very poisonous to aquatic life. To successfully detect dichlorophen, Tu et al. developed an electrochemical sensor (RGO@Ce-MOF)/GCE by utilizing a straightforward, environmentally friendly method.[86] The (RGO@Ce-MOF)/GCE was constructed in two steps. The GCE was coated with Ce-BTC MOF and GO; then, the electrode is dried at ambient temperature. After that, GO is electrochemically reduced to produce RGO. This method of manufacturing justifies the RGO coating over the MOF, stops the MOF from disintegrating in solution, and ultimately improves MOF stability. For further detection, acquired (RGO@Ce-MOF)/GCE mixture was also employed as electrode material. (RGO@Ce-MOF)/GCE hybrid detects dichlorophen with the LOD of 0.007 M upon screening the continuous concentration range of 0.02–10 M. For a high-surface-area electron conduction, Ce-MOF and RGO are combined.[86] The non-electro-active pesticide glyphosate is frequently applied in crops as a herbicide and was extremely challenging to identify with traditional electrochemical techniques. However, by using MOF materials [copper(II)-benzene-1,3,5-tricarboxylate (Cu-BTC)], the recognition of glyphosate would be accomplished.[87] For example, Cu-BTC possesses a large specific surface area that leads to an increase inside the electrode's large surface area, which improves the electrode's reaction rate and, in turn, boosts the electrode's capacity for adsorption. Synthetic sensors have a large detection limit for glyphosate, from 0.001 to 1.0 nM and 1000 to 10 μM, with an LOD of 0.14 pM. A probe has been utilized for the recognition of glyphosate from soybean specimens. To enhance the productivity of crops, the picloram is frequently used as a herbicide. It is a member of the pyridine subfamily. Because it is poisonous and easily soluble in water, it has a negative effect on the environment. Hadi and co-workers recognized picloram within environmental samples by using GC electrode modified using multi-walled carbon nanotubes and Cr-MOF.[88] Because of multi-walled channels, significant electrical conductivity, optimized pore volume, and good adsorptive properties, the composite demonstrated superior sensitivity for picloram.

13.4.4 Recognition for Antibiotics

Antibiotics are defined as antimicrobial substances that are active against bacteria. Antibiotic medicines are broadly used to treat and prevent such diseases in humans and animals.[89] Despite the antibiotics' therapeutic value, the prolonged use of antibiotics causes severe side effects such as vomiting, headaches, a rise in body temperature, and so on. More importantly, long-term uses of an antibiotic cause our bodies to develop antibiotic resistance, which is even more dangerous. On the other hand, the release of antibiotics from the pharmaceutical industry leads to pile up in the environment. As a result, because of their solubility in water, their existence is found in drinking water, triggering numerous health issues. Thus, the sensing of these pollutants by benign technique is very crucial. The electrochemical sensing approach is considered to be efficient in this case. Chloramphenicol (CAP) is normally used as an antibiotic and has great importance in curing infectious bacteria. It is necessary to recognize it so that it can be prevented from entering the body. In this regard, the development of a suitable sensor is imperative. Zhang and co-workers developed the electrochemical probe MIL-101(Cr)/XC-72 for effective recognition of CAP.[90] It was produced using a hydrothermal synthesis approach, where a Teflon-lined autoclave was used in synthesis. The GCE of MIL-101(Cr)/XC-72 gives a great positive signal for CAP as compared to GCE, MIL-101(Cr)/GCE, and XC-72/GCE. Two types of sensing techniques—differential pulse voltammetry (DPV) and cyclic voltammetry (CV)—were employed. The electrochemically active probes exhibited a very good LOD value of about $1.5 \times 10^{-9 \mathrm{M}}$ for the recognition of CAP. Furthermore, these modified sensors were also applicable in natural products like animal milk, eye drops, and honey for the recognition of CAP.

A typical drug, ciprofloxacin (Cip) is widely used for treating symptoms like acute cold infections, cystitis, and urinary tract infections. On the other side, it negatively affects the environment, contaminates local water supplies, and harms public health.[91] A nanocomposite is a good option for

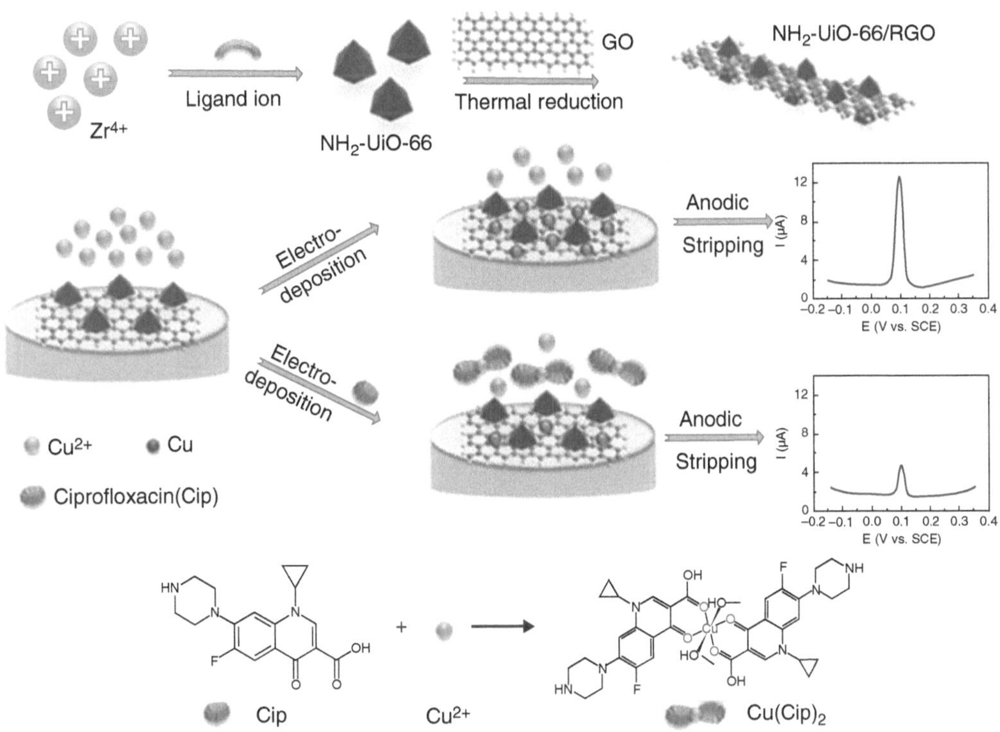

FIGURE 13.5 Synthesis of NH_2–UiO-66/RGO and its electrochemical recognition of Cip with the assistance of Cu^{2+} as anodic strips voltammetry.

Source: Permission from Ref. [92].

the detection of Cip due to its wide surface area, fast detection, high porosity, and strong electrical conductivity. In this regard, Mao and co-workers synthesized Zr-MOF (NH_2-UiO-66) and rGO composite for the identification of Cip in actual water samples (Figure 13.5).[92] To detect Cip, anodic stripping voltammetry was performed using a sensor fabricated with Cu^{2+}. In the presence of Cu^{2+} ions, Cip underwent a complex formation reaction to make a stable composite. Consequently, the oxidation potential of Cu^{2+} ions declined. The sensitivity for Cip by NH_2–UiO-66/RGO was found to be excellent with a value of 10.86 μA μM^{-1} and the LOD is 6.67 nM.

13.4.5 Recognition of DNA

In humans and animals, DNA is mostly responsible for life's functioning and many biological processes. There are two base pairs, which are essential building blocks of DNA such as adenine (A), thymine (T), cytosine (C), and guanine (G).[93] Unusual variations in the amount of DNA cause various kinds of severe diseases.[94,95,96] Therefore, the monitoring of DNA concentrations or any biomolecules is very important. There are several methods available for the detection of biomolecules, but among them, electrochemical detection techniques are superior in terms of sensitivity and selectivity.[97,98] The A and G base pairs of DNA are electrochemically active. We can measure the amount of G and A using electrochemical sensing. MOFs have exclusive structural properties, and composites can be easy to synthesize. For the electrochemical sensing of DNA, MOFs are very essential materials. In general, a single component MOF cast on the electrode surface displays less stability and sensitivity owing to the negligible electrical conductivity of MOFs. To address these issues, when MOF was coupled with high-conductivity materials (graphene), the composite's

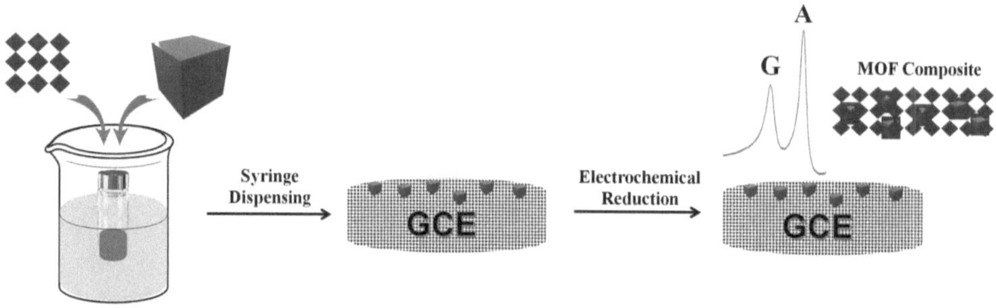

FIGURE 13.6 A schematic representation of the Cu-MOF-composites fabrication for the recognition of DNA bases.

conductivity and stability are significantly enhanced. For example, NH_2-MIL-101(Fe) was used to modify the electrode, which was then electrochemically reduced by graphene oxide. The whole modified electrode composite demonstrated high stability and good sensitivity in the recognition of purine moiety.[99] In another approach, by using the ultrasonic method a nanocomposite matrix Cu-MOF/GO was first developed and fabricated with GCE to engender metal–organic framework/electrochemically reduced graphene oxide/glassy carbon electrode (MOF/ERGO/GCE) for the recognition of adenine (A) and guanine(G) (Figure 13.6). The LOD values for these bases are 0.012 and 0.002 µM, respectively.[100]

13.5 CONCLUSION AND FUTURE PROSPECTS

MOFs are outstanding materials to act as electrochemical probes for the recognition of a wide variety of analytes. These materials are now the center of attraction to many scientific communities. In this book chapter, we have documented various synthetic techniques for the synthesis of MOFs and their electrochemical sensing applications. It has been noticed from the aforementioned discussion that generally two types of MOF materials are utilized for electrochemical sensor as synthesized MOFs and MOF-based composite materials or MOF-derived functional composite. In general, the MOF-based composite or functional materials are superior in electrochemical sensing than the pristine MOF materials. In most cases, the pristine MOF is unstable and composite MOF materials are found to be stable i.e., the individual properties of the MOF and secondary materials are inherited into the new MOF composite materials. Also, the functionalization induces a strong interaction with the analytes; therefore, the electrochemical sensing efficiency is significantly enhanced. Interestingly, MOF-based composite materials offered the optimized pore size or pore number of the overall materials that lead to the pre-enrichment of analytes, resulting in selective and sensitive recognition. Various MOFs with their crucial features for the electrochemical detection of diverse analytes are listed in Table 13.1.

Many significant achievements have taken place, but some challenges are associated with this electrochemical sensing behavior of MOFs. The two important parameters which are essential prerequisites for MOF materials to show electrochemical sensing behavior are the chemical stability and conductivity. Most of the MOFs are showing poor or no conductivity, and insufficient stability hinders their electrochemical sensing application. Sometimes, it is very problematic to construct the sensing mechanism; particularly, the problem arises in the in situ characterization and theoretical investigation of the intermediate component. Some points we may consider while considering the MOF as an electrochemical sensor are: (1) construction of 2D MOFs as a composite matrix for their good conductivity; (2) integration of MOFs with suitable materials to optimize the conductivity as well as stability; (3) in situ characterization needs to be performed in order to understand the host–guest interactions, and, in addition, the theoretical investigation is prerequisite for clear cut presentation of the sensing mechanism.

TABLE 13.1
Different MOFs and Their Electrochemical Detection of Various Analytes with Their LOD

MOF	Metal	Solvent	Analyte	LODs	References
UiO-66-NH_2	Zr wire	DMF/CH_3COOH	TNP	0.1 ppm	101
IRMOF-3	Zinc plate	DMF/CH_3CH_2OH	Fe^{3+} ions	10^{-8} mol L^{-1}	102
Cu-MOF	$Cu(NO_3)·3H_2O$	DMF: Water	–		103
HKUST-1	Cu plate	Methanol	–		104
UiO-66	$ZrCl_4$	DMF	Cd^{2+}, Pb^{2+}, Cu^{2+}, and Hg^{2+}	9, 1, 8, and 0.9 nM	105
MOF-76	$CeCl_3·7H_2O$	DMF	–	–	106
MIL-101	$Cr(NO_3)_3$·H2O	Water	Chloramphenicol	1.5×10^{-9} M	107
ZIF-67	$Co(NO_3)·6H_2O$	DMF	Nitrobenzene	0.16 μM	108
Co-MOF	$Co(AC)_2·H_2O$	Water	H_2O_2	3.76 μM	109
Co-MOF-74	$Co(NO_3)_2·6H_2O$	DMF/CH_3CH_2OH/water	–	–	110
HKUST-1	$Cu(NO_2)_3·3H_2O$	Water/CH_3CH_2OH	–	–	111
IRMOF-3	$Zn(NO_3)_2·6H_2O$	DEF/CH_3CH_2OH	–	–	112
Cu-BDC-NH_2	$Cu(NO_2)_3·3H_2O$	DMF	Determination of amaranth	0.08–360.0 μM	113
MOF-5	$Zn(NO_3)_2·6H_2O$	NMP	–	–	114
ZIF-8	$ZN(NO_3)·6H_2O$	DMF	–	–	115
Zn-MOF	$Zn(AC)_2$	DMF	Fe^{3+} ions	0.3 μM (17 ppb)	116
Bi-MOF	$Bi(NO_3)_2·5H_2O$	–			117
Cu-MOF	$Cu(OAC)_2·H_2O$	–	–	–	118
Ni-MOF	$Ni(OAC)_2·4H_2O$	–	–	–	119
HKUST-1	$Cu(OAC)_2·H_2O$	–	–	–	120

In brief, the judicial design and amalgamation of MOF materials and advanced electrochemical sensing applications would be accomplished for further real-life applications like in clinical diagnosis, biomedical sensing, and waste analysis.

LIST OF ABBREVIATIONS

AAS	Atomic Absorption Spectroscopy
Adp	Adipic acid
AES	Atomic emission spectroscopy
BIB	1, 4-bisimidazolebene
CAP	Chloramphenicol
CAU	Christian-Albrechts-University
Cip	Ciprofloxacin
CS	Chitosan
CT	Catechol
DBI	Dihydroxybenzene isomers
DNA	Deoxyribonucleic acid
Gas	Graphene aerogels
GAs	Graphene aerogels
GCE	Glassy carbon electrode
GO	Graphene oxide
HPLC	High-performance liquid chromatography
HQ	Hydroquinone
ICP-MS	Inductively coupled plasma mass spectrometry

ILAG	Ion- and Liquid-assisted grinding
LAG	Liquid-assisted grinding
LOD	Limit of detection
NH_2-BDC	2-amino terephthalic acid
NG	Neat grinding
RS	Resorcinol
RT	Room temperature
SBU	Secondary building unit

REFERENCES

1. S. Kitagawa, R. Kitaura, and S. I. Noro, *Angew.* Chem., Int. Ed., 2004, **43**, 2334–2375.
2. G. Li and Y. Han, *ACS Cent. Sci.,* 2022, **8**, 150–152.
3. R. L. Martina and M. Haranczyk, *Chem. Sci.,* 2013, **4**, 1781–1785.
4. T. K. Pal, D. De, S. Neogi, P. Pachfule, S. Senthilkumar, Q. Xu, and P. K. Bharadwaj, *Chem. Eur. J.,* 2015, 21, 19064–19070.
5. T. K. Pal, D. De, S. Neogi, and P. K. Bharadwaj, *Inorg. Chem. Front.,* 2015, **2**, 395–402.
6. T. K. Pal, *Chemistry Select*, **2019**, 4, 536–542.
7. S. Pal, T. K Pal, and P. K. Bharadwaj, *Cryst. Eng. Comm.,* 2016, **18**, 1825–1831.
8. U. Mondal, S. Bej, A. Hazra, S. Mandal, T. K. Pal, and P. Banerjee, *Dalton Trans.,* 2022, **51**, 2083–2093.
9. M. Ding, X. Cai, and H. L. Jiang, *Chem. Sci.,* 2019, **10**, 10209–10230.
10. M. Dinca and J. Long, *Chem. Rev.,* 2020, **120**, 8037–8038.
11. N. Patel, P. Shukla, P. Lama, S. Das, and T. K. Pal, *Cryst. Growth Des.,* 2022, **22**, 3518–3564.
12. O. M. Yaghi, G. Li, and H. Li, *Nature*, 1995, **378**, 703–706.
13. Q. Yin, Y. L. Li, L. Li, J. Lu¨, T.-F. Liu, and R. Cao, *ACS Appl. Mater. Interfaces*, 2019, **11**, 17823–17827.
14. P. Lama, A. Rawat, P. Sikiti, and T. K. Pal, *Inorg. Chem.,* 2022, **61**, 939–943.
15. T. K. Pal, D. De, and P. K. Bharadwaj, *Fuel*, 2022, **320**, 123904.
16. A. Rawat, S. Dhakla, P. Lama, and T. K. Pal, *J. CO_2* Utilizat., 2022, **59**, 101939.
17. T. Qiu, Z. Liang, W. Guo, H. Tabassum, S. Gao, and R. Zo, *ACS Energy Lett.,* 2020, **5**, 520–532.
18. H. Lawson, S. Walton, and C. Chan, *ACS Appl. Mater. Interfaces*, 2021, **13**, 7004–7020.
19. D. B. Kanzariya, M. Y. Chaudharya, and T. K. Pal, *Dalton Trans.,* 2023, **52**, 7383–7404.
20. T. K. Pal, Mater. *Chem. Front.,* 2023, **7**, 405–441.
21. D. B. Kanzariya, R. Goswami, D. Muthukumar, R. S. Pillai, and T. K. Pal, *ACS Appl. Mater. Interfaces*, 2022, **14**, 48658–48674.
22. A. Bavyskina, N. Kolobov, I. Khan, J. A. Bau, A. Ramirez, and J. Gascon, *Chem. Rev.,* 2020, **120**, 8468–8535.
23. D. De, T. K. Pal, and P. K. Bharadwaj, *Inorg. Chem.,* 2016, **55**, 6842–6844.
24. H. Chen, L. Fan, X. Zhang, and L. F. Ma, ACS *Appl. Nano Mater.,* 2020, **3**, 2680–2686.
25. D. B. Kanzariya, K. U. Patel, R. Desai, and T. K. Pal, *Springer Proceed. Mater.,* 2023, **22**, 135–146. https://doi.org/10.1007/978-981-99-1616-0_13
26. T. Ma, H. Li, J.-G. Ma, and P. Cheng, *Dalton Trans.,* 2020, **49**, 17121–17129.
27. M. D. Allendorf, R. Dong, X. Feng, S. Kaskel, D. Matoga, and V. Stavila, *Chem. Rev.* 2020, **120**, 8581–8640.
28. N. Kajal, V. Singh, R. Gupta, and S. Gautam, *Environ. Res.,* 2022, **204**, 112320.
29. P.M. Bersier, J. Howell, and C. Bruntlett, *Analyst*, 1994, **119**, 219–232.
30. Y. Li, Y. Jiang, and X.-P. Yan, *Anal. Chem.,* 2006, **78**, 6115–6120.
31. C. Xu, R. Fang, R. Luque, L. Chen, and Y. Li, *Coord. Chem. Rev.,* 2019, **388**, 268–292.
32. D. M. Venturi, F. Campana, F. Marmottini, F. Costantino, and L. Vaccaro, *ACS Sustainable Chem.* Eng. 2020, **8**, 17154–17164.
33. J. Gascon, U. Aktay, M.D. Hernandez-Alonso, G.P. van Klink, and F. Kapteijn, *J. Catal.,* 2009, **261**, 75–87.
34. U. Mueller, M. Schubert, F. Teich, H. Puetter, K. Schierle-Arndt, and J. Pastre, *J. Mater. Chem.,* 2006, **16**, 626–636.
35. N. Campagnol, E. R. Souza, D. E. De Vos, K. Binnemans, and J. Fransaer, *Chem. Commun.,* 2014, **50**, 12545–12547.
36. H. M. Yang, X. L. Song, T. L. Yang, Z. H. Liang, C. M. Fan, and X. G. Hao, *RSC Adv.,* 2014, **4**, 15720–15726.

37. L. L. Jiang, X. Zeng, M. Li, M. Q. Wang, T. Y. Su, X.C. Tian, and J. Tang, *RSC Adv.*, 2017, **7**, 9316–9320.
38. J.-Z. Wei, X.-L. Wang, X.-J. Sun, Y. Hou, X. Zhang, D.-D. Yang, H. Dong, and F. -M. Zhang, *Inorg. Chem.*, 2018, **57**, 3818–3824.
39. J.-Z. Wei, F.-X. Gong, X.-J. Sun, Y. Li, T. Zhang, X.-J. Zhao, and F.-M. Zhang, *Inorg. Chem.*, 2019, **58**(10), 6742–6747.
40. J. Klinowski, F. A. A. Paz, P. Silva, and J. Rocha, *Dalton Trans.*, 2011, **40**, 321–330.
41. S.-H. Jhung, J.-H. Lee, and J.-S. Chang, Bull. Kor. *Chem. Soc.*, 2005, **26**, 880–881.
42. N. A. Khan, and S. H. Jhung, Coord. *Chem. Rev.*, 2015, **285**, 11–23.
43. A. Gedanken, *Ultrason. Sonochem.*, 2004, **11**, 47–55.
44. W.-J. Son, J. Kim, J. Kim, and W.-S. Ahn, *Chem. Commun.*, 2008, **2008**, 6336–6338.
45. M. Klimakow, P. Klobes, A. F. Thünemann, K. Rademann, and F. Emmerling, *Chem. Mater.*, 2010, **22**, 5216–5221.
46. T. Stolar, and K. Uˇzarevi´c, *CrystEngComm*, 2020, **22**, 4511–4525.
47. D. Chen, J. Zhao, P. Zhang, and S. Dai, *Polyhedron*, 2019, 162, 59–64.
48. A. Pichon, A. Lazuen-Garay, and S. L. James, *CrystEngComm*, 2006, **8**, 211–214.
49. W. Yuan, A. L. Garay, A. Pichon, R. Clowes, C. D. Wood, A. I. Cooper, and S. L. James, *CrystEngComm*, 2010, **12**, 4063–4065.
50. Y. Wang, L. Wang, W. Huang, T. Zhang, X. Hu, J. A. Perman, and S. Ma, *J. Mater. Chem.* A, 2017, **5**, 8385–8393.
51. S. S. Ming, N. S. K. Gowthaman, H. N. Lim, P. Arul, E. Narayanamoorthi, I. Ibrahim, H. Jaafar, and S. A. John, *J. Electroanal. Chem.*, 2021, **883**, 115067.
52. X. Guo, S. Sang, A. Jian, S. Gao, Q. Duan, J. Ji, Q. Zhang, and W. Zhang, *Sensors and Actuators B: Chemical.*, 2018, **256**, 318–324.
53. Y.-J. Gao, H.-Y. Sun, J.-L. Li, X.-H. Qi, K.-Z. Du, Y.-Y. Liao, X.-Y. Huang, M.-L. Feng, and M.-G. Kanatzidis, *Chem. Mater.*, 2020, **32**, 1957–1963.
54. W.-Y. Zhou, J.-Y. Liu, J.-Y. Song, J.-J. Li, J.-H. Liu, and X.-J. Huang, *Anal. Chem.*, 2017, **89**, 3386–3394.
55. B. Cheng, L. Zhou, L. Lu, J. Liu, X. Dong, F. Xi, and P. Chen, *Sensor. Actuator. B Chem.*, 2018, **259**, 364–371.
56. K. Huang, B. Li, F. Zhou, S. Mei, Y. Zhou, and T. Jing, *Anal. Chem.*, 2016, **88**(13), 6820–6826.
57. S. Lee, G. Barin, C. M. Ackerman, A. Muchenditsi, J. Xu, J. A. Reimer, S. Lutsenko, J. R. Long, and C. J. Chang, *J. Am. Chem. Soc.*, 2016, **138**, 7603–7609.
58. P. Sharma, M. Mourya, D. Choudhary, M. Goswami, I. Kundu, M. P. Dobhal, C. S. P. Tripathi, and D. Guin, *Sensor. Actuator. B Chem.*, 2018, **268**, 310–318.
59. L. Zhang, D. Peng, R.-P. Liang, and J.-D. Qiu, *Trends Anal. Chem.*, 2018, **102**, 280–289.
60. G. Aragay, J. Pons, and A. Merkoçi, *Chem. Rev.*, 2011, **111**, 3433–3458.
61. J.-C. Jin, J. Wu, G.-P. Yang, Y.-L. Wu, and Y.-Y. Wang, *Chem. Commun.*, 2016, **52**, 8475–8478.
62. X. Wang, Q. Wang, Q. Wang, F. Gao, F. Gao, Y. Yang, and H. Guo, *ACS Appl. Mater. Inter.*, 2014, **6**, 11573–11580.
63. Y. Wang, L. Wang, W. Huang, T. Zhang, X. Hu, J.A. Perman, and S. Ma, *J. Mater. Chem.*, 2017, **5**, 8385–8393.
64. M. Lu, Y. Deng, Y. Luo, J. Lv, T. Li, J. Xu, Chen, S.-W., and Wang, J., *Anal. Chem.*, 2018, **91**, 888–895.
65. Y. Wang, L. Wang, H. Chen, X. Hu, and S. Ma, ACS *Appl. Mater. Inter*, 2016, **8**, 18173–18181.
66. M. Jahan, Q. Bao, and K. J. Loh, *Am. Chem. Soc.*, 2012, **134**, 6707–6713.
67. X. Hu, W. Xu, L. Zhou, Y. Tan, Y. Wang, S. Zhu, and J. Zhu, *Adv. Mater.*, 2017, **29**, 1604031.
68. C. Li, D. Jiang, H. Liang, B. Huo, C. Liu, W. Yang, and J. Liu, *Adv. Funct. Mater.*, 2018, **28**, 1704674.
69 M. Lu, Y. Deng, Y. Luo, J. Lv, T. Li, J. Xu, S.-W. Chen, and J. Wang, *Anal. Chem.*, 2018, **91**, 888–895.
70. J. Choi, D. A. Reddy, and T. K. Kim, *Ceram. Int.*, 2015, **41**, 13793–13803.
71. E.-N. Papadakis, A. Tsaboula, Z. Vryzas, A. Kotopoulou, K. Kintzikoglou, and E. Papadopoulou-Mourkidou, *Sci. Total Environ.*, 2018, **624**, 732–743.
72. Y. Zhang, and S.-J. Park, Chem. *Eng. Technol.*, 2019, **369**, 353–362.
73. R. Javaid, and U. Y. Qazi, *Int. J. Environ. Res. Publ. Health*, 2019, **16**, 2066.
74. D. A. Perry, T. M. Razer, K. M. Primm, T. Chen, J. B. Shamburger, J. W. Golden, A. R. Owen, A. S. Price, R. L. Borchers, and W. R. Parker, *J. Phys. Chem. C*, 2013, **117**, 8170–8179.
75. J.-F. He, F.-J. Yao, H. Cui, X.-J. Li, and Z.-B. Yuan, *J. Separ. Sci.*, 2012, **35**, 1003–1009.
76. A. S. Ahammad, M. M. Rahman, G.-R. Xu, S. Kim, and J.-J. Lee, *Electrochim. Acta*, 2011, **56**, 5266–5271.
77. Z. Wang, S. Li, and Q. Lv, Sensor. *Actuator. B Chem.*, 2007, **127**, 420–425.
78. H.-F. Wang, Y.-Y. Wu, and X.-P. Yan, *Anal. Chem.*, 2013, **85**, 1920–1925.
79. X. Zhou, Z. He, Q. Lian, Z. Li, H. Jiang, and X. Lu, Sensor. *Actuator. B Chem.*, 2014, **193**, 198–204.

80. S. Hu, W. Zhang, J. Zheng, J. Shi, Z. Lin, L. Zhong, G. Cai, C. Wei, H. Zhang, and A. Hao, *RSC Adv.*, 2015, **5**, 18615–18621.
81. Y. Yang, Q. Wang, W. Qiu, H. Guo, and F. Gao, *J. Phys. Chem. C.*, 2016, **120**, 9794–9803.
82. Y.-W. Hu, W. Wang, H.-E. Li, Q.-K. Li, and K.-K. Niu, *Int. J. Electrochem. Sci.*, 2018, **13**, 5031–5040.
83. J. Chen, Y. Huang, P. Kannan, L. Zhang, Z. Lin, J. Zhang, T. Chen, and L. Guo, *Anal. Chem.*, 2016, **88**, 2149–2155.
84. Y. W. Hu1, W. Wang, H. E. Li, Q. K. Li and K. K. Niu, *Int. J. Electrochem. Sci.*, 2018, **13**, 5031–5040.
85. X. Tu, F. Gao, X. Ma, J. Zou, Y. Yu, M. Li, F. Qu, X. Huang, and L. Lu, *J. Hazard Mater.*, 2020, **396**, 122776.
86. X. Tu, Y. Xie, X. Ma, F. Gao, L. Gong, D. Wang, L. Lu, G. Liu, Y. Yu, and Huang, X., *J. Electroanal. Chem.*, 2019, **848**, 113268.
87. Y. Cao, L. Wang, C. Shen, C. Wang, X. Hu, and G. Wang, *Sensor. Actuator. B Chem.*, 2019, **283**, 487–494.
88. M. Hadi, M. Bayat, H. Mostaanzadeh, A. Ehsani, and A. Yeganeh-Faal, *Int. J. Environ. Anal. Chem.*, 2018, **98**, 197–214.
89. B. Halling-Sørensen, H. C. H. Lützhøft, H. R. Andersen, and F. Ingerslev, *J. Antimicrob. Chemother.*, 2000, **46**, 53–58.
90. W. Zhang, Z. Zhang, Y. Li, J. Chen, X. Li, Y. Zhang, and Y. Zhang, *Sensor. Actuator. B Chem.*, 2017, **247**, 756–764.
91. Y. Zhao, ACS *Appl. Nano Mater.*, 2020, **3**, 4917–4919.
92. X. Fang, X. Chen, Y. Liu, Q. Li, Z. Zeng, T. Maiyalagan, and S. Mao, *ACS Appl. Nano Mater.*, 2019, **2**, 2367–2376.
93. J. D. Watson, and F. H. Crick, *Nature*, 1953, **171**, 737–738.
94. A. Wasfi, F. Awwad, and A. I. Ayesh, *Biosens. Bioelectron.*, 2018, **119**, 191–203.
95. N. Chen, J. Li, H. Song, J. Chao, Q. Huang, and C. Fan, *Acc. Chem. Res.*, 2014, **47**, 1720–1730.
96. J. Luo, L. Fang, H. Liu, Q. Zhu, H. Huang, J. Deng, F. Liu, Y. Li, and J. Zheng, *Sensor. Actuator. B Chem.*, 2020, **304**, 127266.
97. H. Li, L. Ying, J. J. Green, S. Balasubramanian, and D. Klenerman, *Anal. Chem.*, 2003, **75**, 1664–1670.
98. S. Petralia, and S. Conoci, *ACS Sens.*, 2017, **2**, 876–891.
99. J. Fu, X. Wang, T. Wang, J. Zhang, S. Guo, S. Wu, and F. Zhu, *ACS Appl. Mater. Inter.*, 2019, **11**, 33238–33244.
100. X. Wang, J. Zhang, Y. Wei, T. Xing, T. Cao, S. Wu, and F. Zhu, *Analyst*, 2020, **145**, 1933–1942.
101. J. Z.Wei, X. L. Wang, X.-J. Sun, Y. Hou, X. Zhang, D. D. Yang, H. Dong, and F. M. Zhang, *ACS. Inorganic Chem*, 2018, **57**, 3818–3824.
102. J. Z. Wei, F. X. Gong, X. J. Sun, Y. Li, T. Zhang, X. J. Zhao, and F. M. Zhang, *ACS. Inorganic Chem*, 2021, **58**, 6742–6747.
103. M. Naseri, L. Fotouhi, A. Ehsani, and S. Dehghanpour, *J. Colloid Interface Sci.*, 2016, **484**, 314–319.
104. U. Mueller, M. Schubert, F. Teich, H. Puetter, K. Schierle-Arndta, and J. Pastréa, *J. Mater. Chem*, 2006, **16**, 626–636.
105. M. Lu, Y. Deng, Y. Luo, J. Lv, T. Li, J. Xu, S.W. Chen, and J. Wang, *Anal. Chem*, 2019, **91**, 888–895.
106. P. F. Santos, and P. P. Luz, *J. Braz. Chem. Soc*, 2020, **31**, 566–573.
107. W. Zhang, Z. Zhang, Y. Li, J. Chen, X. Li, Y. Zhang, and Y. Zhang, *Sensor. Actuator. B Chem.*, 2017, **247**, 756–764.
108. X. Feng, S. Lin, M. Li, X. Bo, and L. Guo, *Anal. Chim. Acta*, 2017, **984**, 96–106.
109. L. Yang, C. Xu, W. Ye, and W. Liu, *Sensor. Actuator. B Chem*, 2015, **215**, 489–496.
110. H. Y Cho, D. A Yang, J. Kim, S. Y. Jeong, and W. S. Ahn, *Catal. Today*, 2012, **185**, 35–40.
111. Y. K. Seo, G. Hundal, I. T. Jang, Y. K. Hwang, C. H. Jun, and J. S. Chang. *Microporous Mesoporous Mater*, 2009, **119**, 331–337.
112. Z. Ni, and R. I. Masel, *J. Am. Chem. Soc*, 2006, **128**, 12394–12395.
113. S. Tajika, H. Beitollahi, H. W. Jang, and M. Shokouhimehr, *RSC Adv.*, 2021, **11**, 278–287.
114. W. J. Son, J. Kim, J. Kim, and W. S. Ahn, *Chem. Commun*, 2008, **2008**, 6336–6338.
115. H. Y. Cho, J. Kim, S. N. Kim, and W. S. Ahn, *Microporous Mesoporous Mater*, 2013, **169**, 180–184.
116. E. Moradi, R. Rahimi, and V. Safarifard, *Polyhedron*, 2019, **159**, 251–258.
117. L. Tröbs, M. Wilke, W. Szczerba, U. Reinholz, and F. Emmerling, *CrystEngComm*, 2014, **16**, 5560–5565.
118. A. Pichon, A. Lazuen-Garay, and S. L. James, *CrystEngComm*, 2006, **8**, 211–214.
119. R. Zhang, C. A. Tao, R. Chen, L. Wu, X. Zou, and J. Wang, *Nanomaterials*, 2018, **8**, 1067.
120. W. Yuan, A. Lazuen Garay, A. Pichon, R. Clowes, C. D. Wood, A. I. Cooper, and S. L. James, *CrystEngComm*, 2010, **12**, 4063–4065.

14 Porphyrin-Functionalized Hybrid Materials for Biosensing Applications

Vinod Khatri, Preeti Yadav, Vijay Bahadur, Manvendra Pratap Singh, Avinash Kumar Singh, Shatruhan Singh Rajput, and Badri Parshad

14.1 INTRODUCTION

Porphyrins are the important classes of biological ligands that are heterocyclic macrocycle organic compounds composed of four modified pyrrole subunits interconnected at α-carbon atoms via methine bridges (=CH–). Nitrogen atoms located at the interior of the large ring form a central binding cavity that can coordinate with the various metal ions with different oxidation states to form metalloporphyrin. The electronic properties of porphyrins depend on the nature of the coordinated metal, its oxidation state, and substituents on the porphyrin scaffold.[1,2] Due to the presence of a large conjugated aromatic structure, porphyrin possessed a distinctive electronic spectrum and physicochemical behavior. The suitable functionalized derivatives of porphyrins could mimic the active site of many natural enzymes and can be utilized as an effective electron transfer scaffold to catalyze various biological reactions. Heme and chlorophyll-a are two examples of naturally occurring porphyrins where heme is an iron-containing porphyrin and forms a non-protein part of hemoglobin and myoglobin, which is necessary to store and transport oxygen in the bloodstream and also involved in mitochondrial energy production, signal transduction, and antioxidant defense, whereas chlorophyll-a, a magnesium-containing porphyrin, acts as a catalyst in oxygenic photosynthesis process (Figure 14.1).

Inspired by natural porphyrins and their importance to life, researchers have developed many synthetic porphyrins which display significant electronic properties that are different from the natural porphyrins due to their structural differences. Synthetic porphyrins are useful in a diverse range of applications as their electronic properties and physicochemical behavior can easily be fine-tuned by different synthetic modifications.[3] The ability to fine-tune synthetic porphyrins for specific applications make them important scaffolds in situations where natural porphyrins would never be able to be utilized.

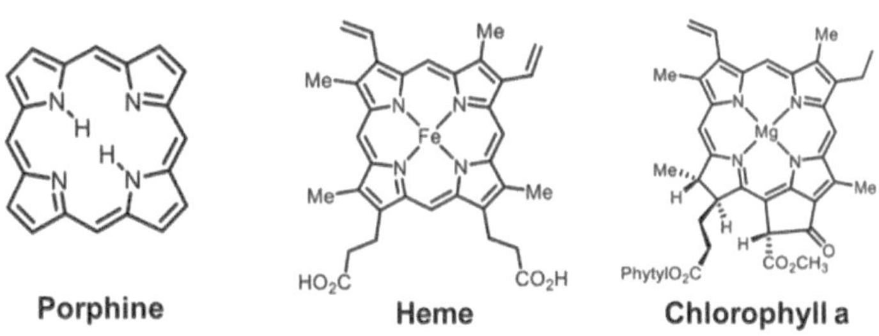

FIGURE 14.1 Chemical structures of the porphine, heme, and chlorophyll a.

DOI: 10.1201/9781032654874-14

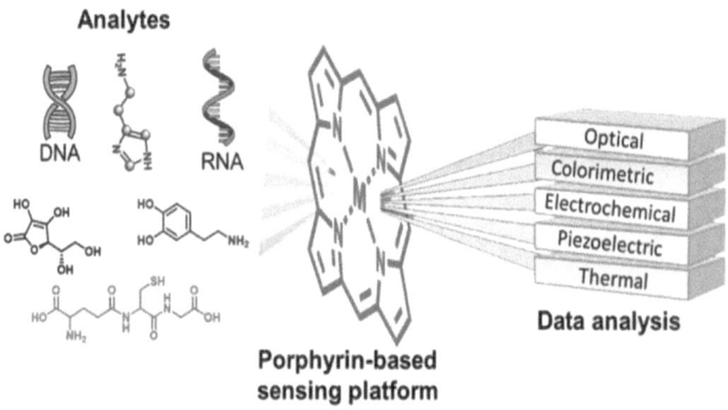

FIGURE 14.2 Schematic representation of porphyrin-based platform for biomolecule sensing.

Moreover, porphyrins can be conjugated with other useful organic or/and inorganic scaffold to get hybrid material which can lead to improved physicochemical properties with synergistic effect.[4] The hybrid material usually displays good thermal and chemical stability and provides the possibility of further modification. Typically, while incorporating properties of all the combined scaffolds, the hybrid material can demonstrate superior or even entirely new properties while incorporating the properties of the individual combining scaffolds.

Porphyrin-based hybrid materials have been utilized to sense/detect a diverse range of biomolecules such as dopamine, flavonoids, antioxidants, amino acids, carbohydrates, proteins, and nucleic acids with high specificity and sensitivity. The present chapter delineates the impact of different porphyrin conjugates on their physiochemical properties and their suitability as biosensors in a range of applications (Figure 14.2). The objective of this study is to discuss exceptional hybrid porphyrin materials designed for the electrochemical detection of biomolecules. These materials are aimed to exhibit both high selectivity and sensitivity toward biomolecules.

14.2 CLASSIFICATION OF PORPHYRIN-FUNCTIONALIZED HYBRID MATERIALS

Based on the type of formulation, porphyrin-based hybrid materials are divided into the following four categories: (1) Porphyrin-functionalized metal nanoparticles, (2) Porphyrin-functionalized organic architectures, (3) Porphyrin-functionalized graphene/carbon nanotubes, and (4) Porphyrin-based metal-organic frameworks. All these hybrid materials have been explored in terms of their sensing properties toward specific biomolecules (Figure 14.3).

14.2.1 Porphyrin-Functionalized Metal Nanoparticles

Porphyrin-functionalized nanoparticles constitute a diverse class of compounds that offer a variety of opportunities in sensing applications. By taking advantage of the unique properties of nano-scale materials and those of porphyrins, these conjugates go beyond conventional sensing materials and offer various characteristics such as high affinity, high sensitivity, and high specificity.

Sebarchievici et al. utilized manganese porphyrin and its hybrid with gold nanoparticles for electrochemical and optical detection of ascorbic acid.[5] Ascorbic acid is known for its potent antioxidant properties.[6,7] It is required for the growth and repair of the tissues in all parts of the human body. It is also used for the prevention and treatment of various illnesses such as scorbutus, mental illness, immune response activation, and infertility. The potential of bare Mn(III) tetratolylporphyrin chloride (MnTTPCl), spherical gold colloid (10–15 nm), and their hybrid (MnTTPCl-Au) was determined and compared in terms of ascorbic acid detection (Figure 14.4). Although bare gold colloid

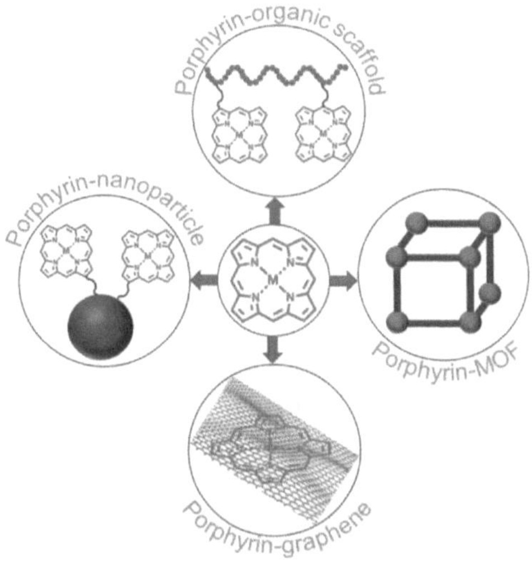

FIGURE 14.3 Classification of porphyrin-based hybrid materials explored for biosensing application in this chapter.

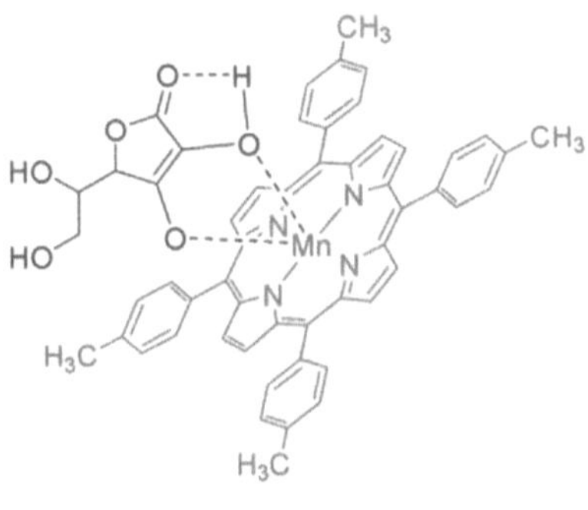

FIGURE 14.4 Chemical structure showing the host-guest interaction between the manganese core of porphyrin and ascorbic acid.

did not show any response toward ascorbic acid, its porphyrin hybrid (MnTTPCl-Au) displayed a very nice efficiency with a detection limit in the concentration range of 2.6×10^{-6} to 4.38×10^{-5} M. Moreover, the addition of ascorbic acid to MnTTPCl-Au hybrid exhibited a red shift from 664 nm for MnTTPCl-Au to 682 nm for MnTTPCl-Au@ascorbic acid host-guest complex.

Glutathione is an intracellular antioxidant tripeptide made up of glutamic acid, cysteine, and glycine. It is one of the most abundant low-molecular-weight non-protein thiol species found in most mammalian tissues including humans. It plays an important role in regulating redox homeostasis and prevents cellular damage caused by various reactive oxygen species (ROS), free radicals, and peroxides.[8,9] Various researchers have demonstrated that low or high level of glutathione is associated with various health issues such as aging, cancer, Parkinson's disease, and liver damage. In case of low glutathione level, people are likely to take exogenous glutathione to maintain their antioxidant ability; thus, it is important to develop a method to detect glutathione in real samples such as foodstuff and other biological samples. Tu et al. developed a photoelectrochemical biosensing platform for the detection of glutathione using porphyrin-functionalized titanium oxide (TiO_2) nanoparticles.[10] TiO_2 nanoparticles were functionalized with Fe(III) meso-tetrakis(4-sulfonatophenyl)porphyrin chloride (FeTPPS) through dentate binding of TiO_2 with sulfonic acid groups of FeTPPS. The TiO_2-FeTPPS-nanoparticle-coated indium tin oxide electrode was found to display

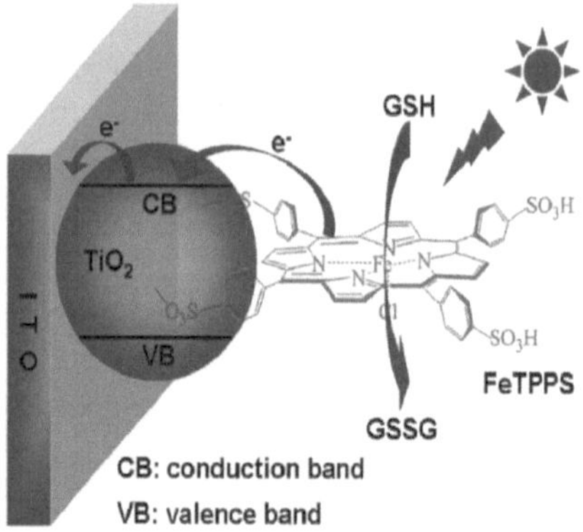

FIGURE 14.5 Schematic representation of photoelectrochemical oxidation of glutathione at indium tin oxide electrode functionalized with FeTPPS-TiO_2 nanoparticles.

Source: Reprinted with permission from Ref. [10].

a stable photochemical response that could be further sensitized via electron transfer process from glutathione to porphyrin and then to the illuminated TiO_2, thus leading to an efficient glutathione sensing platform (Figure 14.5). Under optimal conditions, the developed photochemical sensing platform exhibits glutathione detection concentration ranging from 0.05 to 2.4 mmol L^{-1} with a lower detection limit of 0.03 mmol L^{-1} at a signal-to-noise ratio of 3.

Chen et al. developed a 5, 10, 15, 20-(4-sulfonatophenyl) porphyrin (TPPS4) based turn-on fluorescence sensor for highly sensitive detection of glutathione in real samples such as fruits (red grape and cherry tomato) and fetal calf serum.[11] The bare TPPS4 porphyrin was found to display some sort of fluorescence, which was quenched by an electron transfer process with the addition of Hg^{2+}, and subsequently formed TPPS4-Hg^{2+} system. Furthermore, when a certain amount of glutathione is added, the system again displayed the fluorescence which was related to the competitive affinity of Hg^{2+} toward glutathione and porphyrin core. The TPPS4-Hg^{2+} system exhibited a lower detection limit of 16 fM for glutathione with several advantages including high sensitivity and selectivity, low cost, and quick fluorescence response. Gallo et al. developed a hybrid system containing manganese meso-tetra(pentafluorophenyl) porphyrin and gold nanoparticles for electrochemical detection of L-cysteine, a semi-essential amino acid. Cysteine deficiency/excess is associated with several disorders such as slow growth, loss of muscle and fat, liver damage, Parkinson's and Alzheimer's diseases, diabetes, and others.[12] Thus, the development of suitable detection methods is essential for health care purposes. The porphyrin-gold hybrid was deposited on low-cost fluorine-tin-oxide (FTO)-coated glass electrode, and the sensing experiment was performed at pH 7.0. The hybrid system showed a good response toward L-cysteine detection with a linear response in the range of 12.2–33.6 mmol L^{-1} with a limit of detection and limit of quantification equal to 2.40 and 8.15 mmol L^{-1}, respectively.

The specific analysis of histidine, an essential amino acid, in biological samples is of great significance in biochemical research. Histidine serves as a precursor for histamine which is an important neurotransmitter in the nervous system.[13] Ingestion of a high amount of histidine is found to result in symptoms of intoxication; thus, several histidine detection methods have been developed. Milaeva and coworkers utilized asymmetrically substituted Fe(III)-porphyrin deposited on the gold electrode for amperometric detection of histidine. Porphyrin substituted with three 2, 6-di-tertbutylphenol moieties and one palmitoyl alkyl chain was deposited on gold electrode via two different methods viz direct chemisorption and alkyl chain embedment (Figure 14.6). These methods of deposition were found to influence the system accuracy where electrode with porphyrin deposited by embedment showed better sensitivity and reproducibility as compared to electrode with

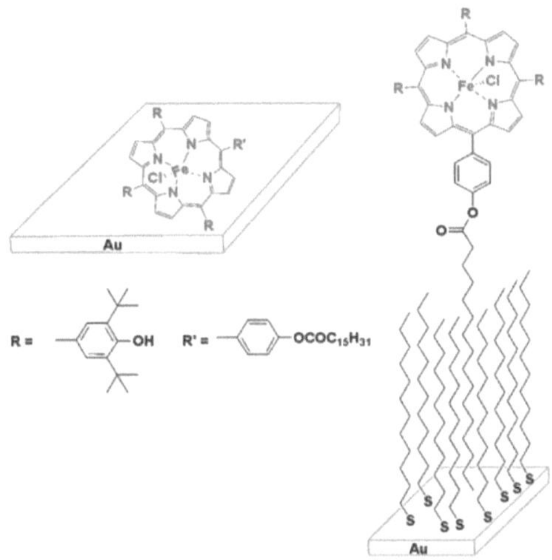

FIGURE 14.6 Chemical structure of 5-(4-palmitoyloxyphenyl)-10, 15, 20-tris[3, 5-di-tert-butyl-4-hydroxyphenyl]porphyrinato iron (III) chloride and its deposition into gold electrode via direct chemisorption and alkyl chain embedment methods.

chemisorbed porphyrin. The electrochemical properties of the system changed upon the formation of porphyrin-histidine supramolecular complex. The detection limit was found to be 2.08×10^{-7}M (signal-to-noise ratio = 3) for electrodes modified by direct chemisorption, while the electrode modified by embedment showed better detection limit with value 8.9×10^{-10}M (signal-to-noise ratio = 3).

Histamine is not only a neurotransmitter but also serves as an important biogenic amine in food, vegetables, and fruits whose concentration can increase to toxic level if food is not processed and packed in hygienic conditions. The caution level for histamine is 50 ppm, whereas the maximum acceptance level is up to 500 ppm. Highly selective and sensitive methods are required to detect such a low level of histamine. Chrisey and coworkers utilized 5-(4-carboxyphenyl)-10, 15, 20-triphenylporphyrin (CPTPP) thin film deposited by matrix-assisted pulsed laser evaporation (MAPLE) on screen printed electrode (SPE) for electrochemical detection of histamine.[14] Cyclic voltammetry studies revealed that the CPTPP-containing system is a suitable platform for histamine detection with a limit of detection within the range of 0–8 ppm.

Flavonoids are another class of important natural products which are abundant in the plant species including tea, vegetables, and fruits.[15] They are known for their remarkable antioxidant activity.[16] Research studies have demonstrated that flavonoids can modulate the activity of several enzymes and affect the behavior of many cell systems.[17] As an important member of the flavonoid family, quercetin displays biological effects that may play an important role in the prevention of various diseases. A photoelectrochemical sensor based on TiO_2 and Pt(II)-porphyrin was fabricated to determine the content of quercetin in different samples.[18] Under optimized conditions, the peak current at an applied voltage of +0.4 V was found to depend linearly on the concentration of quercetin in the 0.002–50 mg L^{-1} range with a detection limit of 0.8 μg L^{-1} (signal-to-noise ratio = 3). The system was also used to determine the content of quercetin in tea and apple juice samples.

Vaishnavi et al. utilized quantum dot-porphyrin nanohybrid for DNA sensing applications. Electrostatic interactions between the positively charged porphyrin and negatively charged thioglycolic-acid-functionalized CdTe quantum dots led to fluorescence quenching of the donor via the electron transfer process.[19] On addition of the calf thymus DNA, the porphyrin detached from the quantum dot and bound to double-stranded DNA (dsDNA) thus resulting in the restoration of fluorescence of quantum dots. Experimental results demonstrated that the increase in fluorescence intensity is directly proportional to the concentration of the DNA with detection concentration in the range of 6.5×10^{-9} M to 29.6×10^{-8} M. DNA biosensor has also been developed by employing nanohybrid of gold nanoparticle and cobalt-porphyrin. Gold nanoparticles were functionalized with cobalt-porphyrin-ssDNA complex through gold-thiol chemistry and the obtained Au-porphyrin-ssDNA was further deposited on gold electrode. In the presence of target ssDNA derived from avian influenza virus, a change in current for gold electrode was observed where the decrease in current was proportional to the concentration of the target ssDNA and sensitivity in attomolar concentration was detected.[20]

14.2.2 Porphyrin-Functionalized Organic Architectures

Porphyrin-based organic architectures such as polymers, dendrimers, and carbon-organic frameworks have also been utilized for the detection of a diverse range of biomolecules. Mukopadhyay and coworkers synthesized carbohydrate-functionalized porphyrin dendrimers for efficient detection of a mannose-binding lectin concanavalin A (ConA).[21] Two multivalent dendrimers (first and second generation) having 12 and 36 mannose units, respectively, of the periphery of Zn(II)porphyrin-based scaffold have been synthesized by employing copper-catalyzed azide-alkyne cycloaddition chemistry (Figure 14.7). Flexible linkers with a relative high degree of freedom and adaptability

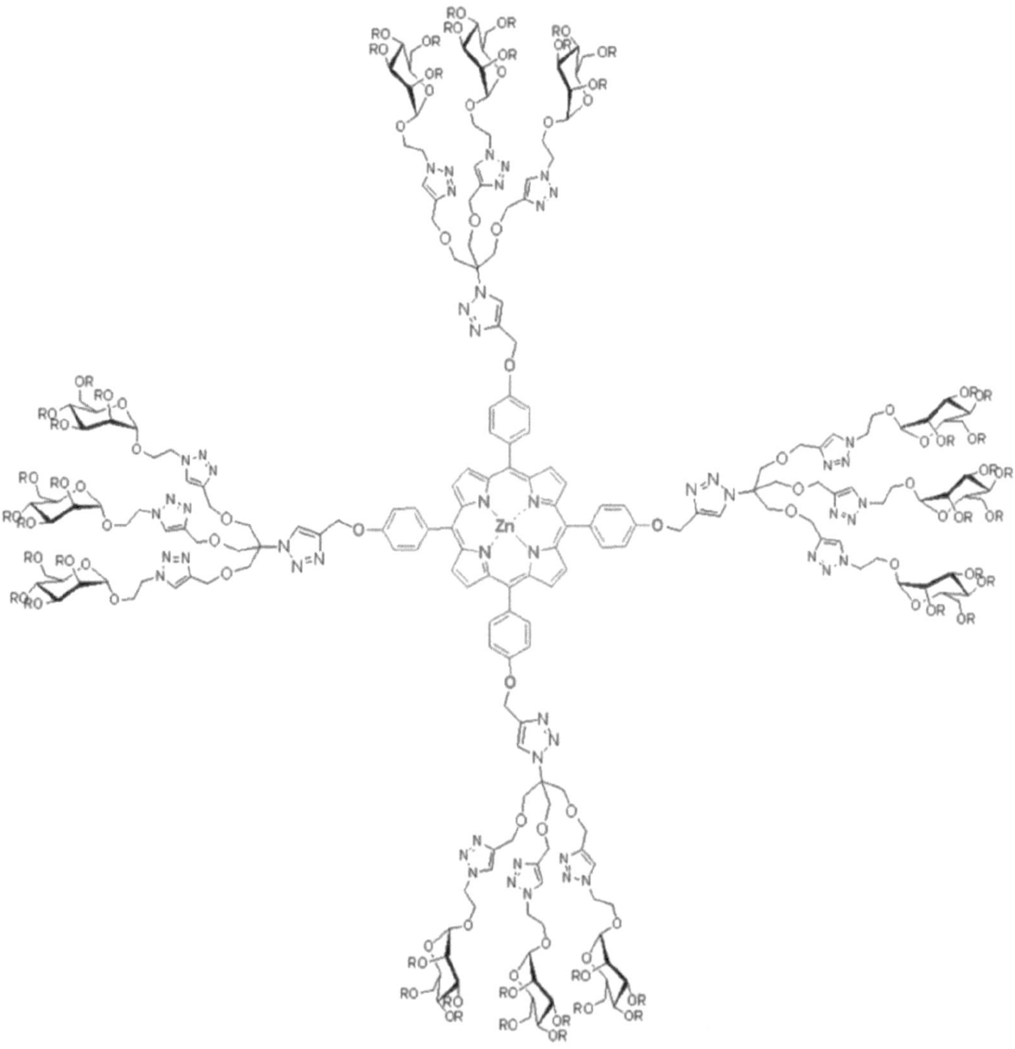

FIGURE 14.7 Chemical structure of the first-generation mannose-functionalized Zn(II)-porphyrin dendrimer utilized to sense concanavalin A lectin.

were strategically introduced between porphyrin core and mannose ligands in order to maximize the interaction with binding sites of targeted lectin. The fluorescence intensity of the water-soluble porphyrin dendrimer was found to increase in the presence of lectin which created an overall hydrophobic microenvironment around the porphyrin core owing to the formation of aggregates by the interactions of dendrimer and lectin. The extent of binding was determined by measuring the enhancement in fluorescence intensity. 1st generation dendrimer having 12 mannose units displayed dynamic linear range from 0 to 350×10^{-7}M concentration of ConA compared to that of only 0 to 50×10^{-7} for second generation dendrimer with 36 mannose units. The quicker saturation of second generation dendrimer attributed to much higher peripheral mannose density which sterically hinders the possibility of interaction with targeted concanavalin A.

Lu et al. synthesized porphyrin-glycopolymer conjugates for concanavalin A lectin binding study.[22] The porphyrin-glycopolymer conjugates showed self-assembly behavior in an aqueous medium owing to the amphiphilic nature by the presence of hydrophobic porphyrin and hydrophilic glycopolymer units and subsequently formed micelles of ~ 190 nm. The binding ability of micelles was tested using dynamic light scattering (DLS) measurements which demonstrated that the particle size of the porphyrin-glycopolymer conjugates increased steadily when lectin was attached to the particles and ultimately led to the formation of aggregates of average hydrodynamic diameter of 850 nm.

Zhang and coworkers developed a porphyrin-based porous organic polymer for colorimetric detection of antioxidant abilities of common antioxidants such as ascorbic acid, gallic acid, and tannic acid with high sensitivity.[23] Sonogashira cross-coupling reaction between bromo-functionalized iron-porphyrin and 1, 3, 5-triethynylbenzene is utilized to form desired porphyrin-based porous organic polymer (FePPOP-1). In absence of antioxidants, FePPOP-1 catalyzed H_2O_2-mediated oxidation of 3,3',5,5'-tetramethylbenzidine (TMB) to oxidized TMB (oxTMB) which resulted in the appearance of blue color, while adding any of the three studied antioxidants caused fading of blue color. UV-vis absorption measurements of TMB-H_2O_2-FePPOP-1 mixture were performed with different concentrations of all three antioxidants where absorbance intensity was observed to decrease with an increase in concentration. The detection limit of ascorbic acid, gallic acid, and tannic acid was found to be as low as 0.35, 0.75, and 0.048 μM with a linear concentration range of 1–55, 1–100, and 0.1–10 μM, respectively. Overall antioxidative abilities of these antioxidants were found to be in the order: tannic acid > ascorbic acid > gallic acid.

14.2.3 Porphyrin-Functionalized Graphene/Carbon Nanotubes

Porphyrin-functionalized graphene-based electrochemical sensor was developed and utilized for highly sensitive and selective detection of dopamine even in the presence of interfering compounds such as ascorbic acid and uric acid.[24] Even though dopamine and ascorbic acid oxidize at similar potential and ascorbic acid is present at a concentration of 100-to-1000-fold higher than dopamine *in vivo*, the graphene-porphyrin conjugate can selectively detect dopamine in the presence of a large excess of ascorbic acid.[25] Electrostatic and π-π stacking interactions between negatively charged porphyrin-graphene and positively charged dopamine enhance electron transfer, while hindering the oxidation of ascorbic acid and uric acid at the graphene-modified glassy carbon electrode. As uric acid and ascorbic acid are negatively charged in biological environments, their electrostatic repulsion with negatively charged porphyrin-graphene conjugates and weaker π–π stacking caused the lower oxidation peak on the electrode as compared to dopamine. The system was also utilized to detect dopamine in human urine and serum samples where the detection limit of dopamine was estimated to be 0.11 μM for urine sample and 0.13 μM for serum sample with signal-to-noise ratio of 3. Zhang et al. developed a novel hybrid material (PFeTPP/PSS-Gr) consisting of graphene (Gr) functionalized with poly[iron(II) tetraphenylporphyrin] (PFeTPP) and poly(sodium-p-styrenesulfonate) (PSS) for voltametric detection of dopamine.[26] When compared with bare PFeTPP or PSS-Gr, the developed composite displayed enhanced electrocatalytic behavior toward dopamine oxidation where the oxidation peak current exhibited linear response over the dopamine concentration in the range of 0.1–40 μM with a detection limit of 5.73 nM.

Wei et al. synthesized covalently functionalized porphyrin-graphene nanocomposites for the simultaneous detection of dopamine and uric acid.[27] Tetra(4-aminophenyl) porphyrin (TAPP) molecules were covalently attached on a conductive graphene oxide scaffold by a simple one-pot amide coupling reaction and subsequently deposited on glassy carbon electrode electrochemical sensing. The detection limit of as low as 38 and 47 nM for dopamine and uric acid, respectively, was observed with a wide detection range of 0.1–350 μM. Moreover, the coexistence of both dopamine and uric acid was determined simultaneously with a detection limit of 51 and 64 nM for dopamine and uric acid, respectively, which is further confirmed by accurate detection of both compounds in human urine samples.

Wang and coworkers synthesized porphyrin-graphene conjugates simultaneously both by covalent and non-covalent strategies.[28] Covalent functionalization is done onto the sp^2 carbons of π-conjugated skeleton at the edges of graphene by amide coupling as shown in Figure 14.8, while non-covalent interactions utilized π–π stacking and hydrophobic-hydrophobic interactions. As compared to bare graphene and porphyrin, the porphyrin-graphene conjugate displayed superior detection efficiency toward dopamine even in the presence of ascorbic acid and uric acid. Differential pulse voltammetry was used for quantitative detection where peak currents increased linearly with dopamine concentration in the range of 7×10^{-5} to 1×10^{-6} M with a limit of detection 9.45×10^{-9} M (signal-to-noise ratio = 3).

Porphyrin-functionalized graphene oxide nanohybrid has also been utilized for non-enzymatic electrochemical sensing of ascorbic acid.[29] The porphyrin-graphene hybrid was incorporated into cationic surfactant tetraoctylammonium bromide to improve its water dispersibility and then deposited on a glassy carbon electrode. The synergistic effect of porphyrin, graphene, and surfactant played a significant role in arousing the inner redox activity of porphyrin and improving the electron transfer between porphyrin and electrode. The system showed high sensitivity toward ascorbic acid with a detection limit of 0.28 μM with a wide linear concentration ranging from 1.33 μM to 1.46 mM.

Wang and coworkers prepared porphyrin/reduced graphene oxide nanocomposite for simultaneous detection of guanine and adenine.[30] This nanocomposite exhibited a significant synergistic effect between the porphyrin and graphene components, enabling the simultaneous detection of guanine and adenine within linear ranges of 0.05–150 μM and 0.2–100 μM, respectively. The estimated limits of detection were 0.016 μM for guanine and 0.06 μM for adenine. Guo and coworkers developed meso-tetra (4-carboxyphenyl) porphyrin (TCPP)-functionalized graphene-conjugated

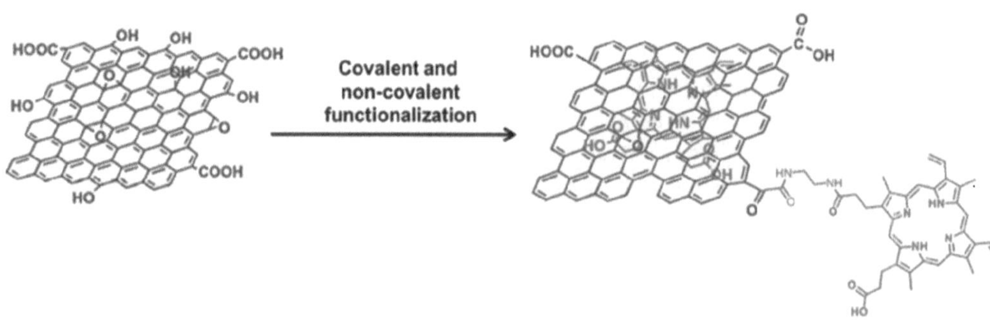

FIGURE 14.8 Simultaneous covalent and non-covalent functionalization of graphene with porphyrin. Covalent attachment utilized amide bond with edge carboxylic acid groups of graphene, while non-covalent attachment utilized π–π stacking interactions.

gold-nanoparticles-(TCPP-graphene-Au)-based aptasensor for electrochemical biosensing of myoglobin.[31] TCPP is known to attach graphene through hydrogen binding and π–π stacking interactions and improve the solubility and stability of graphene. TCPP-graphene-Au nanoconjugate provided an effective platform for the attachment of myoglobin-binding aptamer for myoglobin attachment. The electrochemical aptasensor showed high sensitivity toward myoglobin in a linear concentration range of 2.0×10^{-11} M to 7.7×10^{-7} M with a limit of detection of 6.7×10^{-12} M.

Since enzyme-linked immunosorbent tests are frequently used to create immunosensors, a novel HRP mimic trace label was created by stacking Fe-porphyrin on graphene and labeling it with streptavidin via an amidation process. The developed trace label demonstrated significantly increased peroxidase activity toward *o*-phenylenediamine oxidization in the presence of H_2O_2, improving sensitivity for electrochemical biosensing. A sensitive DNA biosensing technique was thus created by immobilizing the biotinylated molecular beacon on a modified glassy carbon electrode coated with a mixture of gold nanoparticles and single-walled carbon nanohorns (AuNPs-SWCNH).[32] The molecular beacon probe was predicted to be in the "closed" state in the absence of target DNA, which prevented the streptavidin on the trace label from specifically interacting with the biotin shielded due to the strong steric effect. On the other hand, it was anticipated that the duplex structure formed when the molecular beacon probe hybridized with the target DNA would push the biotin group away from the electrode surface, making the biotin end readily accessible to the trace label. To achieve highly sensitive and specific detection of target DNA, a 'signal-on' electrochemical biosensor was developed, utilizing a porphyrin-based HRP mimic as a trace label.

Korri-Youssoufi and coworkers have also utilized graphene-porphyrin nanoconjugates for specific electrochemical detection of DNA.[33] Authors synthesized two tetraphenylporphyrin derivatives i.e., H_2TPP-1CP and H_2TPP-4CP where only one or all four phenyl groups are substituted with carboxylic acid moiety to give mono- or tetra-carboxylic porphyrins (Figure 14.9). H_2TPP-1CP attaches to graphene oxide via π-π stacking interactions between the porphyrin's phenyl groups and the graphene surface. In contrast, H_2TPP-4CP is anchored to the periphery of graphene through hydrogen bonds formed between the hydroxyl or carboxylic groups of graphene oxide and the carboxylic groups of the porphyrin. The designed sensing platform can distinguish *M. tuberculosis* DNA from other mutated samples and displayed limit of detection in attomolar range. The detection limit of 2×10^{-18}M for H_2TPP-1CP-graphene and 7×10^{-19}M for H_2TPP-4CP-graphene conjugate, respectively, was determined toward single-stranded DNA (ssDNA) detection.

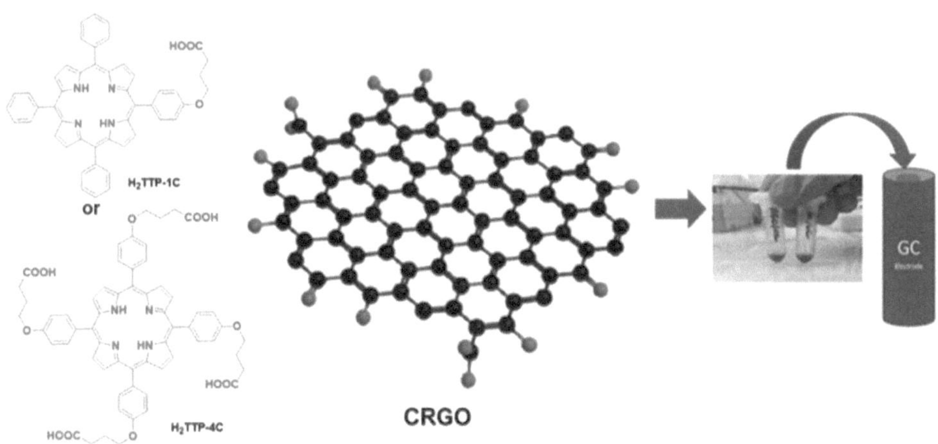

FIGURE 14.9 Chemical structures of the mono and tetracarboxylic acid substituted tetraphenylporphyrins utilized for electrochemical detection of DNA.

Source: Reprinted with permission from Ref. [33].

Benoit Piro and coworkers employed Co-porphyrin-graphene oxide nanoconjugate for electrocatalytic detection of micro RNA (miRNA).[34] Carboxylic-acid-functionalized Co-porphyrin was immobilized on reduced graphene oxide and obtained nanoconjugates were deposited on the glassy carbon electrode. The carboxylic groups were utilized to covalently attach amino-terminated oligonucleotide probes which were complementary to short microRNA targets. Co-porphyrins exhibit catalytic properties toward oxygen reduction, and it was shown that oxygen electroreduction on co-porphyrin increases upon hybridization of the probe with miRNA. Moreover, current changes were found to be amplified as compared to non-catalytic amperometric system, and detection limit in sub-nanomolar range was observed.

Fullerene-porphyrin conjugate synthesized by supramolecular interaction of fullerene and porphyrin via sonication has also been used to detect dopamine at pH 7.4.[35] The oxidation peak current was found to increase linearly with an increase in the concentration of dopamine in the range of 0–200 μM and the detection limit of 0.015 μM was observed. On studying the influence of some common organic compounds (ascorbic acid, glucose, and L-tryptophan) and ions (Na^+ and Cl^-), it was found that 5 times the concentrations of ascorbic acid, glucose, and L-tryptophan and 10 times the concentrations of Na^+ and Cl^- did not show any effect on the detection of dopamine.

Huang et al. developed Zn-porphyrin-based-multiwalled carbon nanotubes (MWCNTs) for highly sensitive electrochemical detection of ascorbic acid.[36] The porphyrin-MWCNT hybrid was first incorporated into tetraoctylammonium bromide surfactant and then deposited on a glassy carbon electrode to fabricate an electrochemical biosensor. In this hybrid conjugate, porphyrin acted as an electron intermediate; MWCNTs behaved as electron accelerators; and surfactant was utilized as an arouser to trigger porphyrin's inner redox activity. At optimal conditions, the proposed sensor exhibited outstanding electrocatalytic activity with a synergistic effect for the oxidation of ascorbic acid with a linear response in the concentration range of 18.72 μM to 1.85 nM and low detection limit of 0.18 μM. Mulchandani and coworkers developed porphyrin-functionalized single-walled-carbon-nanotube-based sensor to monitor the four volatile organic compounds (linalool, phenylacetaldehyde, tetradecene, and ethylhexanol) emitted by citrus trees when they are infected with Huanglongbing at the asymptomatic stage.[37] Porphyrin-graphene nanohybrid has also been used for the stabilization and detection of DNA G-Quadruplexes.[38] Upon binding of tetracationic 5,10,15,20-tetrakis(1-methyl-pyridinium-4-yl)porphyrin to graphene oxide, the typical absorption bands of porphyrin were found to be red shifted and quenching of fluorescence was observed. Furthermore, the recovery of fluorescence on the addition of the G-quadruplex to the porphyrin-graphene nanohybrid gave a selective "turn-off-on" fluorescence sensor for the detection of DNA G-quadruplex.

14.2.4 Porphyrin-Based Metal-Organic Frameworks

Various metal-organic frameworks are used as sensors for the detection of cancer biomarkers and food contaminant detection.[39,40,41] Porphyrin ligand has been extensively used to develop porous MOFs for biomedical applications.[42] The porphyrinic MOF have rigid structures with high surface area and porosity, which make each porphyrin accessible by substrates and also prevent the aggregation. The porphyrin-based MOFs have been reported to be used in biomimetic catalysis, photodynamic therapy, electrocatalysis, and sensing applications.[43,44,45] The incorporation of porphyrin into the framework of MOF also resulted in the enhanced stability of porphyrin. The highest occupied molecular orbital and lowest unoccupied molecular orbital (HOMO-LUMO) energy gap of porphyrinic-MOF is smaller as compared to the porphyrin which is useful for photoinduced electron transfer (PET) processes.

Ling et al. synthesized porphyrin encapsulated metal-organic framework (MOF) for the detection of DNA.[46] The sensitive electrochemical sensor was designed by one pot encapsulation of iron(III) meso-5, 10, 15, 20-tetrakis(4-carboxyphenyl)porphyrin chloride (FeTCPP) into an MOF,

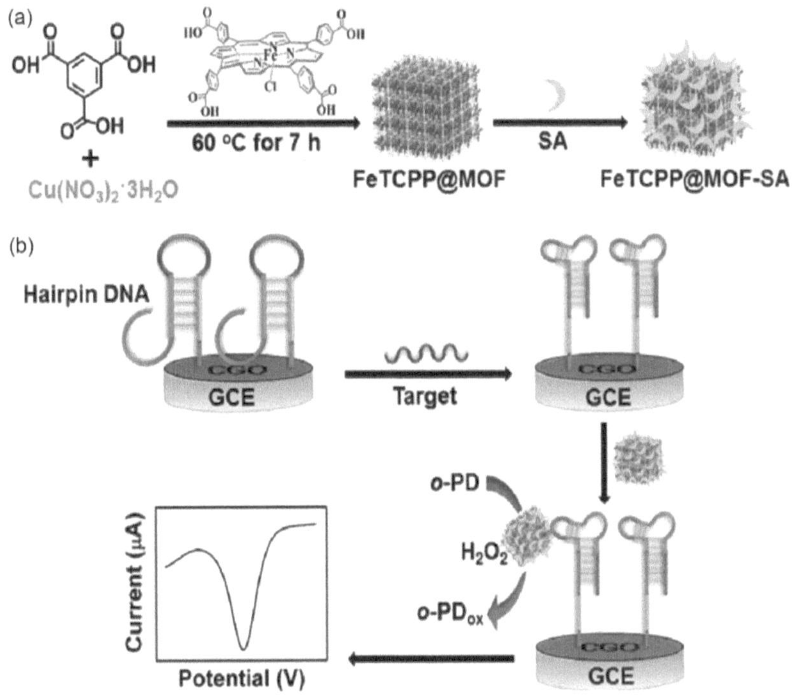

FIGURE 14.10 (a) Synthesis of FeTCPP@MOF-SA composite and (b) electrochemically DNA sensing.

Source: Reprinted with permission from Ref. [46].

HKUST-1 (Cu) [Hong Kong University of Science and Technology-1 (Copper)] and conjugated with streptavidin (SA) for recognition. The FeTCPP@MOF composite thus obtained can catalyze the oxidation of *o*-phenylenediamine (*o*-PD) to 2, 2′-diaminoazobenzene (Figure 14.10). The SEM images of FeTCPP@MOF show truncated octahedral geometry, but after modification with SA, the surface of FeTCPP@MOF-SA composite appears as agglomerates. The MOF-based sensor showed a large peak current on the addition of target DNA at –0.55V, but in the absence of target, the peak current decreases to 7% which is due to non-specific interactions.

In 2015, electrochemical probe for DNA sensing was developed which was based on streptavidin-functionalized zirconium-porphyrin MOF (PCN-222@SA).[47] The signal probe PCN-222 was synthesized by porphyrin which was further functionalized with streptavidin (SA). The sensing platform was formulated by immobilizing triple helix DNA labeled with biotin on glassy carbon electrode (GCE) modified with graphene. The addition of the PCN-222@SA nanoprobe to the sensor surface significantly amplified the electrocatalytic current for oxygen reduction, thereby enhancing electrochemical DNA sensing.. The turn-on strategy was developed for the detection of DNA with detection limit down to sub femtomolar level.

Ling et al. synthesized nanoscaled porphyrinic MOF (PorMOF) for the detection of telomerase activity through electrochemical detection.[48] Zirconium ion was used as a nod for electrochemical detection of telomerase activity. PorMOF showed excellent electrocatalytic activity for the reduction of oxygen. The surface area of nanoscaled PorMOF was found to be 1658 $m^2 g^{-1}$ determined by Brunauer–Emmitt–Teller (BET) method with porosity of 0.99 $cm^3 g^{-1}$. The PorMOF was functionalized with streptavidin (SA) which forms PorMOF@SA agglomerates and results in water-stable electrochemical tracer for telomerase activity. The functionalization of PorMOF with SA protein blocked the catalytic sites and, hence, showed a smaller current response than PorMOF in

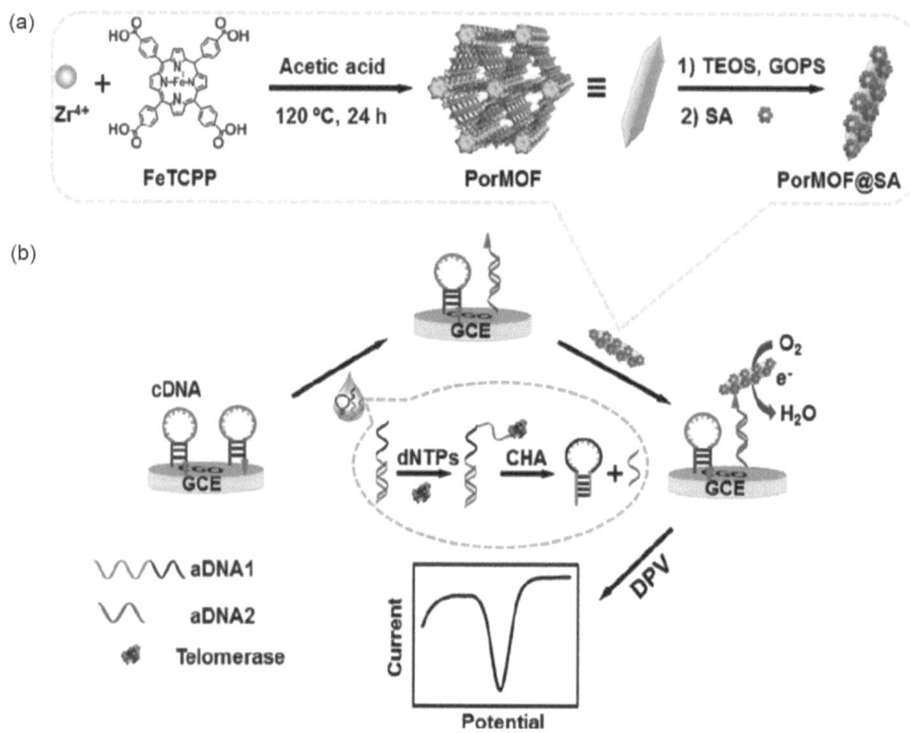

FIGURE 14.11 Schematic representation of (a) synthesis of nanoscaled PorMOF and (b) electrochemical detection of telomerase activity.

Source: Reprinted with permission from Ref. [48].

oxygen-saturated phosphate-buffered saline (PBS). The differential pulse voltammetry (DPV) was used to investigate the practicability of sensor in response to the telomerase activity (Figure 14.11). After the addition of active telomerase into the mixture of aDNA1-aDNA2 duplex and dNTP, the developed biosensor showed a strong electrochemical signal. It was demonstrated that designed assay could be used to detect the telomerase activity in 30 HeLa cells.

Zhang et al. developed a zirconium-based porphyrinic MOF (PCN-222) for label-free phosphoprotein detection.[49] The TCPP ligand having PCN-222 was working as light-harvesting unit, and the high porosity and tunable structure facilitate the oxygen enrichment. In oxygen saturation media, PCN-222 shows better photoelectronic activity for dopamine. Moreover, PCN-222 can be used for selective PEC label-free phosphoprotein assay (Figure 14.12).

In 2020, the same group explored the zirconium-based porphyrinic metal-organic framework nanoparticles (NMOFs) and luminol for the detection of polynucleotide kinase (PNK) activity.[50] The electrochemiluminescence (ECL) ratiometric strategy based on the NMOFs-luminol competitive system was used for the detection of T4 PNK activity. NMOFs worked as electron media reduces oxygen into reactive oxygen species (ROS) which led to the formation of cathodic ECL. ROS react with luminol anion radical and amplify the anodic ECL (Figure 14.13). The designed NMOFs-luminol competitive ECL biosensor was found highly specific for PNK activity and interfering proteins like bovine serum albumin (BSA), protein kinase A (PKA), and lysozyme. The detection capacity of ECL biosensors for PNK was excellent as compared to signal-driven ECL assay.

Cheng et al. synthesized MOF nanosheets for monitoring heparin activity in live rats.[51] In this work, various MOF nanosheets were successfully formulated into highly sensitive diagnostic platforms for *in vivo* monitoring of heparin activity. Hep-specific AG73 peptides were physically adsorbed on MOF nanosheets as a recognition motif which further blocks the active site

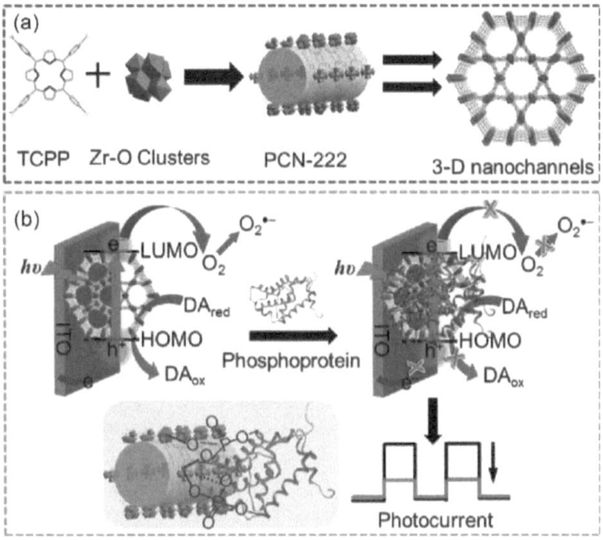

FIGURE 14.12 (a) Schematic synthesis of PCN-222 and (b) photoelectrochemical detection of phosphoprotein.

Source: Reprinted with permission from Ref. [49].

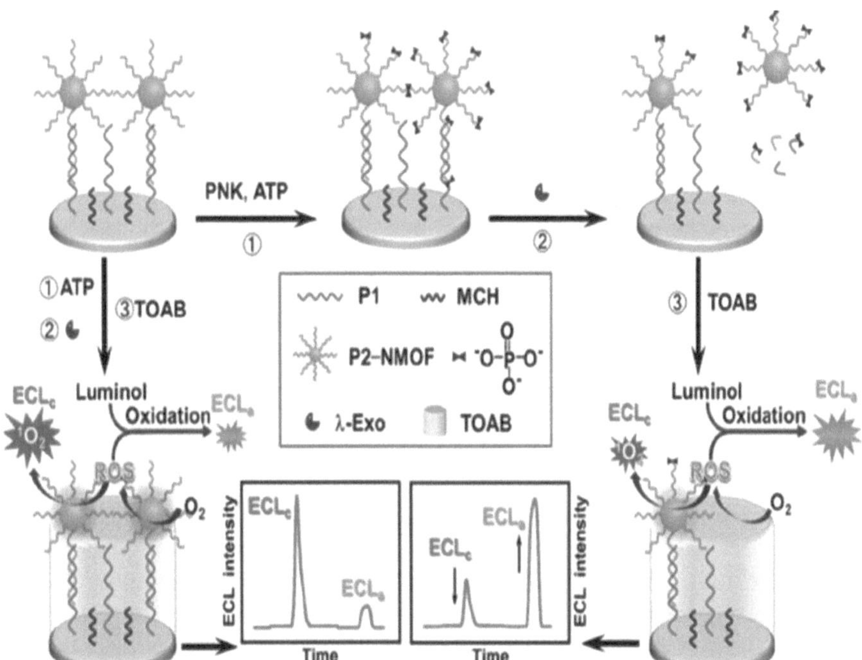

FIGURE 14.13 NMOFs-luminol-based ECL biosensor for the detection of PNK.

Source: Reprinted with permission from Ref. [50].

from non-specific interaction of substrate and catalyst (Figure 14.14). The zeta potential of 2D Zn-TCPP(Fe) nanosheet was −15.3 mV which is mainly due to the free carboxyl group on the surface of nanosheet. The addition of 10 μg mL^{-1} of AG73 peptides on MOF sheet led to a change the zeta potential to +15.1 mV which is due to the formation of complex AG73-2D Zn-TCPP(Fe).

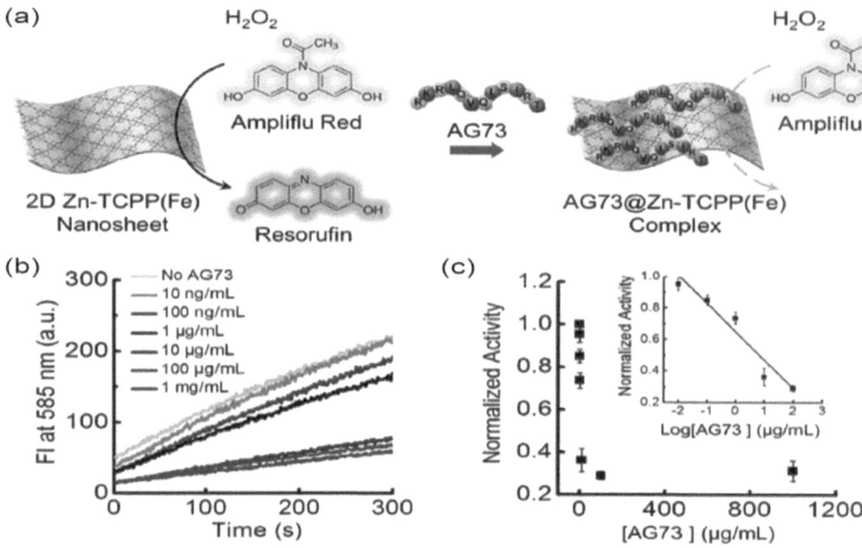

FIGURE 14.14 Schematic representation of AG73-inhibited peroxidase-like activity of 2D Zn-TCPP(Fe) nanosheet with redox substrate (Ampliflu Red).

Source: Reprinted with permission from Ref. [51].

14.3 CONCLUSION

Various hybrids of porphyrin such as porphyrin-nanoparticles, porphyrin-graphene, porphyrin-MOFs, and porphyrin-organic scaffolds have been reported for various biological applications. Various non-covalent interactions such as electrostatic interactions, π–π stacking, hydrophobic-hydrophobic interactions, metal coordination, and hydrogen bonding or stable covalent bonds have been utilized by porphyrin to interact with other useful scaffolds in order to achieve hybrid materials with enhanced chemical and thermal stability and improved dispersibility/solubility. Porphyrin-containing hybrid materials were found to display superior behavior as compared to bare porphyrin in terms of biomolecule sensing. These designed architectures led to the development of different kinds of biosensors for the detection of dopamine; ascorbic acid; glutathione histamine; and various nucleic acids, amino acids, and proteins. Hybrid materials demonstrated improved sensitivity and high specificity toward target molecules.

Hybrid-porphyrin-based biosensors have limited practical use due to their laborious and complex method of preparation. The development of affordable and sensitive biosensing material is therefore very promising. Hybrid porphyrin materials still have a lot of untapped potential for the biosensing applications. Future studies on the manufacture of hybrid porphyrins and the detection of biological molecules will take the field of nanomedicinal theranostics to new heights.

REFERENCES

1. Grinstaff, M. W., Hill, M. G., Birnbaum, E. R., Schaefer, W. P., Labinger, J. A., & Gray, H. B. (1995). Structures, electronic properties, and oxidation-reduction reactivity of halogenated iron porphyrins. *Inorganic Chemistry*, *34*(19), 4896–4902.
2. Pukhovskaya, S. G., Efimovich, V. A., & Golubchikov, O. A. (2013). Effect of structural and electronic properties of substituents on the metal porphyrin formation kinetics. *Russian Journal of Inorganic Chemistry*, *58*(4), 406–410.
3. Imran, M., Ramzan, M., Qureshi, A. K., Khan, M. A., & Tariq, M. (2018). Emerging applications of porphyrins and metalloporphyrins in biomedicine and diagnostic magnetic resonance imaging. *Biosensors*, *8*(4), 95.

4. Arora, S., Nagpal, R., Gusain, M., Singh, B., Pan, Y., Yadav, D., Ahmed, I., Kumar, V., & Parshad, B. (2023). Organic–inorganic porphyrinoid frameworks for biomolecule sensing. *ACS Sensors*, *8*(2), 443–464.
5. Sebarchievici, I., Lascu, A., Fagadar-Cosma, G., Palade, A., Fringu, I., Birdeanu, M., Taranu, B., & Fagadar-Cosma, E. (2018). Optical and electrochemical-mediated detection of ascorbic acid using manganese porphyrin and its gold hybrids. *Comptes Rendus Chimie*, *21*(3–4), 327–338.
6. Njus, D., Kelley, P. M., Tu, Y. J., & Schlegel, H. B. (2020). Ascorbic acid: The chemistry underlying its antioxidant properties. *Free Radical Biology and Medicine*, *159*, 37–43.
7. Beyer, R. E. (1994). The role of ascorbate in antioxidant protection of biomembranes: Interaction with vitamin E and coenzyme Q. *Journal of Bioenergetics and Biomembranes*, *26*(4), 349–358.
8. Aquilano, K., Baldelli, S., & Ciriolo, M. R. (2014). Glutathione: New roles in redox signaling for an old antioxidant. *Frontiers in Pharmacology*, *5*, 196.
9. Aoyama, K., & Nakaki, T. (2015). Glutathione in cellular redox homeostasis: association with the excitatory amino acid carrier 1 (EAAC1). *Molecules*, *20*(5), 8742–8758.
10. Tu, W., Dong, Y., Lei, J., & Ju, H. (2010). Low-potential photoelectrochemical biosensing using porphyrin-functionalized TiO_2 nanoparticles. *Analytical Chemistry*, *82*(20), 8711–8716.
11. Chen, J., Ma, Q., Hu, X., Gao, Y., Yan, X., Qin, D., & Lu, X. (2018). Design of a novel naked-eye and turn-on fluorescence sensor based on the 5, 10, 15, 20-(4-sulphonatophenyl) porphyrin (TPPS4)-Hg2+ system: Monitoring of glutathione (GSH) in real samples and DFT calculation. *Sensors and Actuators B: Chemical*, *254*, 475–482.
12. Goodman, M. T., McDuffie, K., Hernandez, B., Wilkens, L. R., & Selhub, J. (2000). Case-control study of plasma folate, homocysteine, vitamin B12, and cysteine as markers of cervical dysplasia. *Cancer: Interdisciplinary International Journal of the American Cancer Society*, *89*(2), 376–382.
13. Brosnan, M. E., & Brosnan, J. T. (2020). Histidine metabolism and function. *The Journal of Nutrition*, *150*(Supplement_1), 2570S–2575S.
14. Iordache, A. M., Cristescu, R., Fagadar-Cosma, E., Popescu, A. C., Ciucu, A. A., Iordache, S. M., & Chrisey, D. B. (2018). Histamine detection using functionalized porphyrin as electrochemical mediator. *Comptes Rendus Chimie*, *21*(3–4), 270–276.
15. Janabi, A. H. W., Kamboh, A. A., Saeed, M., Xiaoyu, L., BiBi, J., Majeed, F., & Lv, H. (2020). Flavonoid-rich foods (FRF): A promising nutraceutical approach against lifespan-shortening diseases. *Iranian Journal of Basic Medical Sciences*, *23*(2), 140.
16. Yadav, P., Parshad, B., Manchanda, P., & K Sharma, S. (2014). Chromones and their derivatives as radical scavengers: A remedy for cell impairment. *Current Topics in Medicinal Chemistry*, *14*(22), 2552–2575.
17. Mansuri, M. L., Parihar, P., Solanki, I., & Parihar, M. S. (2014). Flavonoids in modulation of cell survival signalling pathways. *Genes & Nutrition*, *9*(3), 1–9.
18. Tian, L., Wang, B., Chen, R., Gao, Y., Chen, Y., & Li, T. (2015). Determination of quercetin using a photo-electrochemical sensor modified with titanium dioxide and a platinum (II)-porphyrin complex. *Microchimica Acta*, *182*(3), 687–693.
19. Vaishnavi, E., & Renganathan, R. (2014). "Turn-on-off-on" fluorescence switching of quantum dots–cationic porphyrin nanohybrid: A sensor for DNA. *Analyst*, *139*(1), 225–234.
20. Kaur, B., Malecka, K., Cristaldi, D. A., Chay, C. S., Mames, I., Radecka, H., & Stulz, E. (2018). Approaching single DNA molecule detection with an ultrasensitive electrochemical genosensor based on gold nanoparticles and cobalt-porphyrin DNA conjugates. *Chemical Communications*, *54*(79), 11108–11111.
21. Das, R., & Mukhopadhyay, B. (2016). Use of 'click chemistry'for the synthesis of carbohydrate-porphyrin dendrimers and their multivalent approach toward lectin sensing. *Tetrahedron Letters*, *57*(16), 1775–1781.
22. Lu, J., Zhang, W., Yuan, L., Ma, W., Li, X., Lu, W., & Chen, G. (2014). One-pot synthesis of glycopolymer-porphyrin conjugate as photosensitizer for targeted cancer imaging and photodynamic therapy. *Macromolecular Bioscience*, *14*(3), 340–346.
23. Cui, C., Wang, Q., Liu, Q., Deng, X., Liu, T., Li, D., & Zhang, X. (2018). Porphyrin-based porous organic framework: An efficient and stable peroxidase-mimicking nanozyme for detection of H2O2 and evaluation of antioxidant. *Sensors and Actuators B: Chemical*, *277*, 86–94.
24. Sakthinathan, S., Lee, H. F., Chen, S. M., & Tamizhdurai, P. (2016). Electrocatalytic oxidation of dopamine based on non-covalent functionalization of manganese tetraphenylporphyrin/reduced graphene oxide nanocomposite. *Journal of Colloid and Interface Science*, *468*, 120–127.
25. Wu, L., Feng, L., Ren, J., & Qu, X. (2012). Electrochemical detection of dopamine using porphyrin-functionalized graphene. *Biosensors and Bioelectronics*, *34*(1), 57–62.

26. Zhang, B., Wang, Y., Li, M., Cui, L., & He, X. (2015). Graphene-supported poly [iron (II) tetraphenylporphyrin] hybrid fabricated by a solvothermally assisted π–π assembly method and its application for the detection of dopamine. *Journal of Electroanalytical Chemistry, 743*, 10–17.
27. Wei, C., Tu, J., Liu, H., Mei, Z., Gao, X., Yu, J., & Chen, Y. (2022). A porphyrin-functionalized graphene oxide nanocomposite for simultaneous detection of dopamine and uric acid. *ChemNanoMat, 8*(10), e202200229.
28. Lv, M., Mei, T., & Wang, X. (2014). Selective and sensitive electrochemical detection of dopamine based on water-soluble porphyrin functionalized graphene nanocomposites. *RSC Advances, 4*(18), 9261–9270.
29. Wu, H., Li, X., Chen, M., Wang, C., Wei, T., Zhang, H., & Fan, S. (2018). A nanohybrid based on porphyrin dye functionalized graphene oxide for the application in non-enzymatic electrochemical sensor. *Electrochimica Acta, 259*, 355–364.
30. Chen, Y., Mei, T., Chen, Y., Wang, J., Li, J., Fu, Y., & Wang, X. (2016). A sensitive porphyrin/reduced graphene oxide electrode for simultaneous detection of guanine and adenine. *Journal of Solid State Electrochemistry, 20*(7), 2055–2062.
31. Zhang, G., Liu, Z., Wang, L., & Guo, Y. (2016). Electrochemical aptasensor for myoglobin-specific recognition based on porphyrin functionalized graphene-conjugated gold nanocomposites. *Sensors, 16*(11), 1803.
32. Wang, Q., Lei, J., Deng, S., Zhang, L., & Ju, H. (2013). Graphene-supported ferric porphyrin as a peroxidase mimic for electrochemical DNA biosensing. *Chemical Communications, 49*(9), 916–918.
33. Wang, Y., Hsine, Z., Sauriat-Dorizon, H., Mlika, R., & Korri-Youssoufi, H. (2020). Structural and electrochemical studies of functionalization of reduced graphene oxide with alkoxyphenylporphyrin mono-and tetra-carboxylic acid: Application to DNA sensors. *Electrochimica Acta, 357*, 136852.
34. De Souza, C., Zrig, S., Wang, D., Pham, M. C., & Piro, B. (2014). Electrocatalytic miRNA detection using cobalt porphyrin-modified reduced graphene oxide. *Sensors, 14*(6), 9984–9994.
35. Zhang, M., & Li, J. (2020). Preparation of porphyrin derivatives and C60 supramolecular assemblies as a sensor for detection of dopamine. *Dyes and Pigments, 173*, 107966.
36. Huang, D., Li, X., Chen, M., Chen, F., Wan, Z., Rui, R., & Wu, H. (2019). An electrochemical sensor based on a porphyrin dye-functionalized multi-walled carbon nanotubes hybrid for the sensitive determination of ascorbic acid. *Journal of Electroanalytical Chemistry, 841*, 101–106.
37. Wang, H., Ramnani, P., Pham, T., Villarreal, C. C., Yu, X., Liu, G., & Mulchandani, A. (2020). Asymptomatic diagnosis of Huanglongbing disease using metalloporphyrin functionalized single-walled carbon nanotubes sensor arrays. *Frontiers in Chemistry, 8*, 362.
38. Monteiro, A. R., Ramos, C. I., Fateixa, S., Moura, N. M., Neves, M. G., & Trindade, T. (2018). Hybrids based on graphene oxide and porphyrin as tools for detection and stabilization of DNA G-quadruplexes. *ACS Omega, 3*(9), 11184–11191.
39. Mohan, B., Singh, G., Pombeiro, A. J., Solovev, A. A., Sharma, P. K., & Chen, Q. (2023). Metal-organic frameworks (MOFs) for milk safety and contaminants monitoring. *TrAC Trends in Analytical Chemistry, 159*, 116921.
40. Mohan, B., Kumar, S., Kumar, V., Jiao, T., Sharma, H. K., & Chen, Q. (2022) Electrochemiluminescene metal-organic frameworks biosensing materials for the detecting cancer biomarkers. *Trends in Analytical Chemistry, 157*, 116735.
41. Mohan, B., Kumar, S., Xi, H., Ma, S., Tao, Z., Xing, T., You, H., Zhang, Y., & Ren, P. (2022). Fabricated metal-organic frameworks (MOFs) as luminescent and electrochemical biosensors for cancer biomarkers detection. *Biosensors and Bioelectronics, 197*, 113738.
42. Chen, J., Zhu, Y., & Kaskel, S. (2021). Porphyrin-based metal–organic frameworks for biomedical applications. *Angewandte Chemie International Edition, 60*(10), 5010–5035.
43. Lin, C., Han, C., Zhang, H., Gong, L., Gao, Y., Wang, H., & Jiang, J. (2021). Porphyrin-based metal–organic frameworks for efficient photocatalytic H_2 production under visible-light irradiation. *Inorganic Chemistry, 60*(6), 3988–3995.
44. Pereira, C. F., Simões, M. M., Tomé, J. P., & Almeida Paz, F. A. (2016). Porphyrin-based metal-organic frameworks as heterogeneous catalysts in oxidation reactions. *Molecules, 21*(10), 1348.
45. Yu, W., Zhen, W., Zhang, Q., Li, Y., Luo, H., He, J., & Liu, Y. (2020). Porphyrin-based metal– organic framework compounds as promising nanomedicines in photodynamic therapy. *ChemMedChem, 15*(19), 1766–1775.
46. Ling, P., Lei, J., Zhang, L., & Ju, H. (2015). Porphyrin-encapsulated metal–organic frameworks as mimetic catalysts for electrochemical DNA sensing via allosteric switch of hairpin DNA. *Analytical Chemistry, 87*(7), 3957–3963.

47. Ling, P., Lei, J., & Ju, H. (2015). Porphyrinic metal-organic framework as electrochemical probe for DNA sensing via triple-helix molecular switch. *Biosensors and Bioelectronics*, *71*, 373–379.
48. ing, P., Lei, J., & Ju, H. (2016). Nanoscaled porphyrinic metal–organic frameworks for electrochemical detection of telomerase activity via telomerase triggered conformation switch. *Analytical Chemistry*, *88*(21), 10680–10686.
49. Zhang, G. Y., Zhuang, Y. H., Shan, D., Su, G. F., Cosnier, S., & Zhang, X. J. (2016). Zirconium-based porphyrinic metal–organic framework (PCN-222): Enhanced photoelectrochemical response and its application for label-free phosphoprotein detection. *Analytical Chemistry*, *88*(22), 11207–11212.
50. Zhang, G., Chai, H., Tian, M., Zhu, S., Qu, L., & Zhang, X. (2020). Zirconium–metalloporphyrin frameworks–luminol competitive electrochemiluminescence for ratiometric detection of polynucleotide kinase activity. *Analytical Chemistry*, *92*(10), 7354–7362.
51. Cheng, H., Liu, Y., Hu, Y., Ding, Y., Lin, S., Cao, W., Wang, Q., Wu, J., Muhammad, F., Zhao, X., Zhao, D., Li, Z., Xing, H., & Wei, H. (2017). Monitoring of heparin activity in live rats using metal–organic framework nanosheets as peroxidase mimics. *Analytical Chemistry*, *89*(21), 11552–11559.

15 Graphene-Based Metamaterial Absorbers for Electrochemical Sensing

Prince Jain, Himanshu Chhabra, and Urvashi Chauhan

15.1 INTRODUCTION TO METAMATERIAL ABSORBER (MMA)

Metamaterials (MMs) are artificial materials that are different from naturally occurring materials in a particular frequency band. These characteristics include, among others, inverse Doppler effect, negative permittivity and permeability, and backward wave propagation [1]. Numerous applications for these unusual properties of MMs have been discussed in the literature, including cloaking [2], perfect lenses [3], and metamaterial antennas [4]. Landy et al. demonstrated the first metamaterial absorber (MMA) design in 2008, which can detect nanometre-scale biomolecules like cancer cells. MMAs, classified as broadband and resonant, are a prominent application of metamaterials. The bandwidth provided by the two types of absorbers differentiates them. Resonant MMAs are narrow-band absorbers that can have single or many resonant characteristics. Because of their narrowband characteristics and polarisation sensitivity, resonant MMAs have limited practical applicability for EM absorbing applications [5,6]. Broadband MMAs exhibit absorption characteristics throughout a wide frequency range. Resonant structures are discovered to be ideal for MMA-based sensors because they provide narrow bandwidths with multiple resonance. Narrow bandwidth has a high quality factor and selectivity [7], making them ideal for sensing applications. These sensors stand out for their simple designs, low costs, great sensitivity, and high quality factor [8]. The MMAs could sense temperature, pressure, and humidity by making small modifications to the overall structure [9].

In recent years, a perfect absorption sensor has attracted many researchers due to its importance in the environment, medical treatment, defence, and other fields. Absorption sensors are currently classified into three types: physical, optical, and electrochemical sensors. The range of applications is limited by the usage of metal oxides in electrochemical sensors, as well as the significant hidden hazards associated with food quality control and biological monitoring. With the ever-expanding number of diseases we face today, it is crucial that scientists find a way to detect biomolecules rapidly and accurately. In order to assess the quality of agricultural products, they could also recognise compounds, establish their electrical characteristics like permittivity, and sense gases and biomolecules.

However, the adjustable sensitivity and flexibility requirements for sensor design are not met by these MMAs, which limits the use of sensors. MMs are dielectric and metallic structures with subwavelength periodicity that exhibit non-natural properties. During the last 15 years, this class of nanostructured artificial media has received a lot of attention and created some amazing photonic and electromagnetic (EM) phenomena. Despite its potential, the usage of MMs has been substantially limited by the high losses and utilisation of metallic top layers, as well as the challenges of producing 3D nanostructures. Planar MMs of subwavelength thickness can be created via lithography and nanoimprinting. Functional materials and optical wavefronts can be combined to achieve a variety of goals (e.g. altering amplitude, phase, and polarisation). Many optical devices have been realised using it. MMAs typically modify the wavefront of light through abrupt phase changes [10]. Refractive index (RI) biosensing is the most practical application which shifts due to biomolecular

 DOI: 10.1201/9781032654874-15

interactions in analyte layers. Due to their special capabilities for sensitive biochemical analysis, sensors can be employed in a wide range of biological and chemical sensing applications.

This resonant characteristic allows scattering output spectrum fluctuation to determine the surrounding biomolecular analytes RI. The advantages of RI sensors over traditional biosensors are manufacturing tolerance and signal stability because they sense RI change through macroscopic optical responses [10]. The other advantage of MMAs is that they have a greater quality factor, which is given by fascinating physical phenomena Fano resonances. When paired with MM, a single nanophotonic RI sensor's capability can be increased. Numerous resonances and a wide variety of slow light effects can be produced by merging multiple cells in a supercell, which are challenging to create in SPP sensors [11]. However, the name MMs are practically synonymous with optical metamaterials, which provide new optical properties. These include, for example, quantum metamaterials, transformation optic materials, invisible cloaks, chiral metamaterials, sensor metamaterials, microwave frequency selective materials, and many sophisticated optical filters and light trappers [12]. Photonics, which uses MMs or complex electrodynamic phenomena to create novel optical effects, relies on all of these [13].

Functional material has intriguing mechanical properties and can be used in interesting ways [14]. One example of such materials is polymer electrolyte electroactuators for robotics [15], as well as reverse actuators that generate electricity from mechanical motions like walking or other types of motion. Many of them have something to do with electrochemical effects. So, in this section, we'll focus on electrochemistry's potential applications to optical metamaterials. Over the past three decades, nanotechnology has made significant advancements that have made it possible to create such materials. But most of them had one thing in common. They were static; executing a fixed, given function; and were made using complicated nanotechnology (s). Following that, tunable MMs are introduced with novel scenarios such as manipulating light with light [16] or mechanically deforming MMs [17]. Electrochemistry is opening up new possibilities for electrotunability through small voltage variations (Table 15.1).

Following the discovery of surface-enhanced Raman scattering (SERS) in the 1970s, plasmonics found its way into the field of electrochemistry. This growth in nanotechnology gave birth to the second generation of plasmonics, and more broadly photonics. Indeed, the optical characteristics of common materials are derived from atoms and molecules. Nanotechnology can create structures

TABLE 15.1
Comparative Analysis of the Performance of Biosensors based on Metamaterials

References	Sensor Configuration	Target Analyte	Sensitivity	Frequency Range
[25]	Nickel (Ni), Copper (Cu), and gold (Au)	Biotin and streptavidin	–	10.64–10.84 GHz
[26]	3-nm-thick Cr followed by 97-nm-thick gold	60 nm of PRD1 virus and 30 nm of MS2 virus	6 GHz μm²/particle to 80 GHz·μm²/particle	0.5–1.5 THz
[27]	Ag nanoparticles on paper and plastic substrate	–	–	0.1–0.5 THz
[28]	Aluminium layer deposition by metal evaporation method	Bovine serum albumin	85 GHz/RIU	0.2–1.2 THz
[29]	Au nanorod on alumina matrix	Streptavidin-biotin	>30,000 nm/RIU	200–749 THz
[30]	200 nm of gold layer thickness on the silicon wafer	Glutamine transferase isozymes II (GGT-II) and alpha-fetoprotein (AFP)	3.8 GHz/(mu mL^{-1})	0.4–1.2 THz
[31]	Au (98 nm) and Cr (2 nm) metal films on the silicon wafer	Yeasts and *Aspergillus niger* and *Escherichia coli BL21 (DE3)*	~11.6 GHz/number density	0.5–3 THz

from nanoparticles, allowing larger-sized "atoms" to access new capabilities by capitalising on plasma excitations [18]. Electrochemistry hasn't fully utilised the advancement of modern plasmonics yet, but it is capable of doing so by experimenting with self-assembling nanostructures or employing electrodes that have been nanoengineered [19]. The most popular metamaterial structure is the split-ring resonator (SRR), which also includes square SRRs, edge- and broadside-linked SRRs, and SRRs with complementary symmetry. The most desirable features of a sensor are high sensitivity, selectivity, and quality factor. The modification in resonance frequency brought on by interaction with the measurement setup serves to evaluate the sensitivity of a metamaterial sensor.

The resonant frequency of a sensor is affected by several factors, including the thickness of the overlayer [20], changes in the analyte medium's permittivity [8, 21], and alterations in the refractive index (RI) due to density, temperature, or moisture [7, 22]. These factors alter the dielectric properties of the substrate or the area surrounding the resonator, leading to a shift in the resonant frequency [7, 23] The sensor's figure of merit is defined as the product of quality factor and the sensitivity. The focus of this study is on sensors that have commercial uses in biosensing; agriculture; material characterisation; detecting adulteration and authenticity of chemicals and fuels; and measuring pressure, temperature, and thickness. Tao et al. have presented a low-cost SRR sensor for quantitative blood glucose analysis, which would be beneficial to study in low-resource environments [24]. The structures were built using a photoresist-free shadow mask deposition process. SRR-based sensors are very important in viral detection.

Microorganism detection has been researched for many years. Life-threatening illnesses in humans, such as asthma [32], tuberculosis, meningitis [33], and gonorrhoea, are caused by a number of species, including bacteria and fungi. Therefore, it is essential to identify these germs quickly to stop the spread of illnesses and effectively treat them. A tetrahertz (THz) biosensor was reported by Park et al. to enable more rapid and precise detection of these organisms [34]. Photolithography can be used to create terahertz-frequency MMs [35]. Although electron beam lithography is a lengthy technique, it allows for the fabrication of high-resolution metamaterial designs [35]. Inkjet printing is environmentally friendly and significantly lowers manufacturing time [36].

15.2 NANOMATERIALS

Many sensing approaches, such as cross-resonators [37], double split ring resonator (DSRR) [38], complementary circular spiral [39], swastika-shaped resonator [40] surface-enhanced Raman scattering [41], chemiluminescence [42], electrochemistry [43], and others have been developed. The primary drawbacks of these techniques are high manufacturing costs, demanding sample processing, and challenging operation, all of which restrict the real-world applications of these sensors [44]. To meet the performance objective, traditional semiconductor-based sensors have been investigated. As a result, rather than depending on advancements in preparation technology, the development of sensing materials with good overall performance is projected to be the most effective path for the evolution of the next generation of detecting technology. Reducing the order of complexity of the sensing material and scaling up production inherently increases its response, whether optical, chemical, electrical, or all three.

The most popular zero-dimension material is quantum dots, which may be built into layered structures using various deposition mechanisms. Quantum dot materials are divided into two categories: metallic nanoparticles and organic materials. Metallic nanoparticles are on the bottom shelf, and organic materials are on the top shelf. Fullerene is a zero-dimensional (0D) nanomaterial [45]. 1D nanomaterials were suited for detection due to their high surface-to-volume ratio as well as their distinctive optical, physical, and chemical features. 1D materials such as carbon nanotubes (CNTs) [46] and nanowires (NW) [47] have shown exceptional sensing capabilities. These properties can be attributed to the material's compact size, low energy consumption, and short response times.

In 2004, scientists made history when they successfully extracted graphene, the first 2D material, from graphite via a micromechanical exfoliation procedure [48–50]. Graphene exhibits remarkable optical, thermal, mechanical, and EM properties, making it a highly promising tunable material.

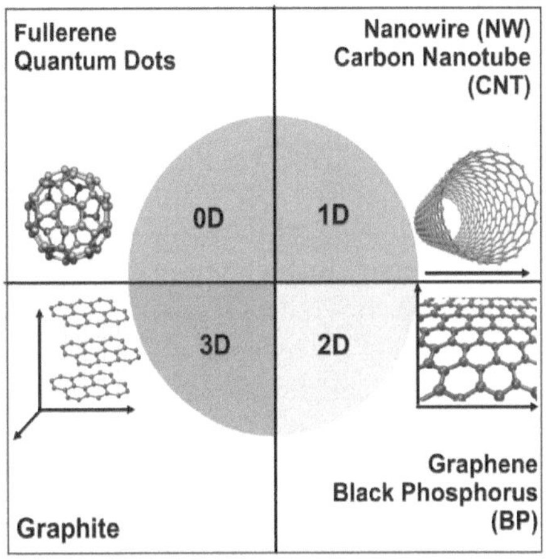

FIGURE 15.1 Various materials with different dimensionalities (0D, 1D, 2D, and 3D).

TABLE 15.2
Characteristics of 0D, 1D, 2D, and 3D Materials

Dimension	Advantages	Disadvantages
3D	• Good thermal stability • Good mechanical strength • Selectivity can be improved by good design • Strong analyte binding	• Difficulty with miniaturisation • Low surface area • Slow dynamic of analyte transport
2D	• The wide tenability of conductivity • Thickness depending on electronic properties • Large surface-to-volume ratio • Adaptable to the cutting-edge technology of ultra-thin silicon channels • Excellent mechanical flexibility • Good optical transparency	• Lack of mass production of materials with the same quality over a large area • At ambient temperatures, dome formations have limited stability • Lack of effective and dependable device integration techniques
1D	• High density of reactive sites • Excellent stability • Miniaturisation friendly • Good thermal stability	• Essential chemical modification • Hard to make reliable electrical connections • Precision control and limited structure
0D	• Atomic control of structure-property relationship • Tunable size and shape • Single molecular electronic-based sensing	• Low conductivity • Limited stability of devices • Difficulty with device integration

Because of its capacity to generate robust plasma resonance, graphene-based direct detection sensors play a crucial role in optical sensing applications. 2D materials are useful for biological diagnosis applications such as cancer detection, hyperglycemia detection, photocatalysis, etc. Sensors can detect analytes at low concentrations due to their large surface area, which permits interaction between 2D materials and analytes [51–54]. Their atomic size thickness influences the intrinsic features such as fluorescence, conductivity, magnetic permeability, and reactivity. Depending on the application, 2D nanomaterials have a wide range of conductivity. Because of their larger lateral diameters, 2D materials are more compatible with metal electrodes than other dimensional materials [55]. In this study, we will look at the advancement of 2D materials and their use in sensors for infrared radiations, antibiotics, pesticides, and bacteria that rely on 2D nanomaterials, as well as numerous detection technologies. The most important characteristics of various materials with different dimensionalities (0D, 1D, 2D, and 3D) are compiled in Table 15.2 in terms of their potential utility in detection.

15.3 APPLICATIONS OF MMA IN ELECTROCHEMICAL SENSING

15.3.1 BioSensing

Biosensing involves detecting target molecules using principles similar to those used by biological systems like the immune system. Biosensors consist of a recognition element (receptor) and a transduction element. A receptor is a biological molecule that may attach to the target molecule selectively. The binding event is transformed into a quantifiable signal by the transduction element. When it comes to biosensors, there are various receptor types that can be utilised such as antibodies, enzymes, and nucleic acids. The selection of the receptor will depend on the specific target molecule that needs to be identified. There are various devices that can serve as the transduction element. Some examples include electrochemical sensors, optical sensors, and piezoelectric sensors. The decision on which transduction element to use will depend on the biosensor's sensitivity, selectivity, and response time needs.

Because of its selective detection, MM-based sensors provide an advantage in the sensing industry and are largely employed for accurate and early diagnosis of sickness problems. By lowering detection limits, enhancing sensitivity, and enabling low-cost manufacturing, MMs give conventional plasmonic sensors new capabilities that give rise to the hybrid sensing paradigm. Localised surface plasmon resonance (LSPR) and SPP are the two main forms of plasmonic excitations. SPP sensors have been widely researched over the last few decades, yielding numerous academic articles and commercial applications [56]. Traditional SPP instruments have various disadvantages, including a lack of multiplexing capacity, which results in poor throughput, reliance on a certain binding surface, resulting in reduced sensitivity and a lack of wireless operation [57].

In 2000, an experimental microwave frequency demonstration of the MM-based RI sensor was made [58]. Due to these sensors' extraordinarily high sensitivity in comparison to traditional plasmonic biosensors, they have received a lot of interest [59]. Numerous viruses, including Zika virus [60], hepatitis B [61], and HIV DNA [62], have been detected using plasmonics and their meta structures. A review of contemporary MMA-based sensors for biosensing applications is presented in this section. In [63], a 2D photonic structure was reported for precisely sensing blood glucose levels. 3×3 air holes are engraved into the framework so that blood with various glucose levels can pass through them. The variation in output power is comparable to various blood samples. Sensing the RI of the metamaterial structure's surrounding medium is another use for MMA [7]. Teflon, a dielectric material with an RI of 1.45, is used to divide these two layers. The structure is coated with an analyte layer of constant thickness, but RI is altered (1.0–1.3). The resonance frequency shifts in accordance with the analyte layer's RI variation. The bulk of biomedical samples, as stated in the research, have RI values near to 1.3; as a result, the suggested model may function as biosensors.

In label-free biosensing, the target biomolecule is not labelled with radioactive compounds or enzymes to make measurements easier. As a result, the electric fields become more concentrated in one spot and interact with the biomolecular layer more frequently. Zhao et al. [38] suggested a metamaterial multiband sensor for detecting the dielectric characteristics of substances. If only one resonant frequency is utilised for measurements, the measurement's accuracy is limited because different compounds may produce the same shift in the resonant frequency, making it challenging to distinguish between them. When using two or more resonant frequency structures, measurement accuracy is improved and the system becomes more effective because many chemicals may be detected simultaneously. Table 15.1 compares the performances of several bio-sensor technology.

Nowadays, a lot of analytical equipment relies on biosensing as a crucial part. This is especially true for medical diagnostics, environmental monitoring, and biotechnology. Biosensing is designed to detect and analyse biological entities, like proteins, DNA, antibodies, enzymes, and even whole cells. Its main objective is to identify and measure specific biological targets with precision, speed, and sensitivity.

15.3.2 Gas Sensing

Gas sensors are used to monitor smoke and carbon monoxide emissions in homes as well as in manufacturing and industrial settings to identify gas leaks. There are now sensors that can identify gases including methane, hydrogen sulphide, ammonia, nitrogen dioxide, carbon monoxide, and ozone. These gas sensors are employed in the medical and life science sectors [64], indoor air quality/heating, ventilation, and air conditioning (HVAC) industries, fire suppression tests, transportation industries, and a variety of other applications. Electrochemical, acoustic, piezoelectric, optical, photonic crystal gas sensors, and MMA systems have all been researched for various applications and schemes [65]. Even though optical fibre-based gas sensors are very repeatable and reusable, ambient light can cause false readings [66]. MMAs are one technique for enhancing gas-sensing performance in multiple frequency ranges, from microwave to ultraviolet [67]. Plasmonics enhances performance in a metal-insulator-metal (MIM) structure by incorporating customisable optical properties [68]. To address this issue, Su et al. created a vertically oriented channel MIM (c-MIM) structure with a plasmonic molecular area for improved sensitivity [69]. Carbon dioxide and butane gas detection was shown using this c-MIM structure. Using optical sensors, mid-infrared gas detection has been the subject of extensive research [78].

Furthermore, CO_2 flow regulation must be maintained in health research and industry activities [77]. As a result, unique gas sensors must be developed that can detect CO_2 levels in this range and beyond. The sensor was made by placing a silicon nano-cylindrical meta-atom absorber on a gold substrate [76]. Table 15.3 compares the performances of several gas sensor technologies.

15.3.3 Chemical Sensing

Chemical sensing involves identifying the chemical components in a sample. Chemical sensors are usually composed of a recognition element (receptor) and a transduction element. The receptor is a molecule that selectively binds to the target chemical compound. The binding event is converted into a quantifiable signal by the transduction element. There are numerous receptors that can be used in chemical sensors. Enzymes, antibodies, and fluorescent compounds are some common examples. The selection of the receptor depends on the chemical molecule being detected. The transduction element can take different forms, such as electrochemical sensors, optical sensors, and mass spectrometry sensors. The particular transduction element chosen is based on the required sensitivity, selectivity, and response time of the sensor.

TABLE 15.3
Comparative Analysis of the Performance of Gas Sensors based on Metamaterials

References	Material on Top	Target Analyte	Sensitivity	Frequency Range (f)/ Concentration Range (C)
[70]	Aluminium-doped Zinc oxide	H_2	0.0006 a.u./%	f: 250–333 THz
[71]	A gap between two gold electrodes	CO_2 and C_4H_{10}	2.92×10^{-4}ppm^{-1}	f: 60–150 THz
[24]	Gold-coated Si spacer	CO_2	22.4±0.5 ppm·Hz$^{-0.5}$	C.: 0–5000 ppm
[72]	Polyethylenimine polymer	CO_2	500 nm/RIU	C.: 0–177 ppm
[73]	Cross-shaped top Cu resonator on dielectric spacer layer (Al_2O_3)	CO_2	1.7×10^{-4} %/ppm	C.: 0–50,000 ppm
[74]	Bimetallic Au/Pd nanorod	H_2 and N_2	–	f: 333–750 THz
[75]	Fe_3O_4 nanoparticles on two square ring-shaped slots	NO_2	0.2 MHz/ppm	f: 200–800 MHz

In 2013, Withayachumnankul et al. proposed the first metamaterial-based chemical sensor to detect ethanol and methanol at 1.9 GHz [79]. In addition to the utilisation of the microfluidic platform, the utilisation of MMA is another option that has been offered [80,81]. It's fascinating that researchers have created a paper-based wearable metamaterial sensor that can tell the difference between glycerol, methanol, and water [82]. Researchers have solved this low quality factor constraint by selecting substrates with low loss [83] and employing various resonator design methodologies [84]. In recent years, a number of MM-based chemical sensors designed for commercial application have been created [84].

The Fano model was also used to improve the performance of the sensor in carbon-nanotube-based SPR MMA in order to get the best results [85]. In order to boost the sensor's sensitivity and Q factor, certain chemical samples were inserted in a capillary glass tube that was placed parallel to the sensor surface [86]. In order to recognise the difference between branded and unbranded diesel oil, an MM-based sensor with a microstrip line was made [87]. The use of several symmetrical double SRRs led to the development of another intriguing metamaterial-based chemical sensor. A structure like that has potential for multiband chemical sensing [38]. The most important thing about this sensor is that it uses a miniaturised and non-destructive combination of MMs and microfluidics to figure out the chemical properties of liquids. Chemical sensing plays a crucial role in the identification and measurement of specific chemical compounds across a range of applications, such as environmental monitoring, industrial safety, food quality control, and healthcare.

15.3.4 Tunable Sensors

Circuit tuning can give MMs the ability to be tuned by putting discrete impedances in specific places within the unit cell [88]. It is common practise to tune the resonant frequency of resonators by including varactor diode in MMA structure and manipulating the bias voltage of these diodes [89], analogous to material tuning which includes making adjustments to the substrate properties [21]. The structure's capacity to be tuned is made possible by the use of an external gate voltage, which works to alter the conductivity of the graphene. Applications needing sensor tuning across a wide frequency range would necessitate composition and material qualities [88,90]. Due to discrepancies in manufacture and design, this would restrict the sensor's reliability. Because diverse materials create the same frequency variance, a single resonant frequency limits the sensor's reliability, leading to false positives [38]. The capability of tuning resonant frequency across a larger range to provide multiband response improves dependability and makes testing easier with the current test setup [89].

A demonstration of a permittivity sensor with the benefit of tunability can be found in Ref. [21]. The Fermi energy of graphene can be altered through the use of an external gate voltage that is applied through the electrodes. This, in turn, has an effect on the surface conductivity of the material. The structure can be modified by modifying the surface conductivity, which, in turn, alters the relative permittivity of the graphene layer. Because of the interaction between the graphene layer and the Jerusalem cross (JC) metal, an expansive absorption bandwidth with a frequency of 11.7 THz is formed as a result of the process. The analyte layer has a thickness of 0.1 micrometres and a permittivity of between two and ten. The capacity to tune the resonance frequency is a feature that is highly wanted in metamaterial sensors because it improves measurement reliability, accuracy, and simplicity. Table 15.4 displays resonance frequencies for various Fermi energies.

A tunable MMA is suggested in Ref. [89] as a temperature sensor. A varactor diode with reverse bias is installed in one of the splits. The resonance frequency of the structure is subject to a wide variety of shifts as a result of adjustments to the bias voltage. A metal plate forms the bottom layer, which blocks all input signal transmission. A sensor layer is placed between the SRR and the metal plate to perform the sensing function. Since the dielectric constant is temperature dependent, resonance frequency changes when the sensor layer is subjected to a changing temperature. Because it may be used with a variety of fixed frequency sources, the design allows for easy measurement.

TABLE 15.4
Absorption Bands, Shifts in Resonance Frequencies, and Tunability of Various MMA Tunable Sensors

Source	Absorption Band	Resonant Frequency	Resonant Frequency Shift/Remarks	Tunability
[89] (2016)	Single	4.2, 5.3 GHz	0.5 MHz/°T	Yes
[22] (2017)	Multiple	6.46, 7.68 GHz	340 MHz	No
[91] (2017)	Single	10.25 GHz	16 MHz	No
[37] (2018)	Single	7.11 GHz	410 MHz	No
[38] (2018)	Multiple	6.89, 8.21 GHz	320 MHz	No
[21] (2018)	Multiple	85.2, 93.6 THz	70–130 THz	Yes
[92] (2020)	Single	15.17 GHz	1.88 GHz	No

15.4 FUTURE WORK IN MMAS FOR SENSING APPLICATIONS

Recent advances in MMA include phase modulation, polarisation conversion, and high-resolution imaging [93]. Deep learning in combination with nanophotonics and MMAs has recently received a lot of attention. Several 3D printing systems have exhibited extraordinary capability in terms of manufacturing features with micro/nano precision throughout the previous few decades. Using biomaterials and 3D printing, the next generation of MMAs could have unique properties, thermo-optic effects and tunable electro-optic, and analyte sensing functionalities. Chalcogenide phase-change materials, on the other hand, have ushered in a new era of adaptive MMAs [94]. When these materials are incorporated in micro/nano photonic platforms, they will produce reconfigurable MMAs with desired qualities like ultra-high sensitivity, high selectivity, high resilience, and multiplexing capabilities. A nanophotonics platform seamlessly integrated with artificial intelligence and machine learning could optimise biosensor design limits like cost, performance, and power consumption [95]. Based on the above explanation, there are many chances to integrate MMAs with novel production methods and materials to create a new gas/bio/chemical sensing paradigm.

15.5 CONCLUSION

This chapter has discussed MMA for sensing applications utilised in a variety of commercial applications. Absorption sensors are currently classified into three types: physical, optical, and electrochemical sensors. The use of metal oxides in electrochemical sensors, as well as the major hidden risks involved with food quality management and biological monitoring, limits the range of applications. With the ever-increasing number of diseases, we confront today, scientists must devise a method to detect biomolecules quickly and precisely. The MMA-based gas sensor has demonstrated great repeatability and minimal regeneration time in detecting numerous gases. However, these MMAs do not meet the adjustable sensitivity and flexibility required for sensor design, limiting the application of sensors. Graphene may be a new alternative material due to its remarkable optical, thermal, mechanical, and EM capabilities. Graphene-based direct detection sensors are important for optical sensors because they feature a strong plasma resonance that increases sensitivity. Because of this property, they are valuable for biological diagnosis, such as cancer, hyperglycemia, photocatalysis, and other disorders.

REFERENCES

1. Veselago, V. G. The electrodynamics of substances with simultaneouslynegative values of ε and μ. *Sov. Phys. Uspekhi*, **1968**, 10(4), 509–514. https://doi.org/10.1070/PU1968v010n04ABEH003699.
2. Pendry, J. Metamaterials in the sunshine. *Nat. Mater.*, **2006**, 5(8), 599–600. https://doi.org/10.1038/nmat1697.

3. Ramakrishna, S. A.; Pendry, J. B.; Wiltshire, M. C. K.; Stewart, W. J. Imaging the near field. *J. Mod. Opt.*, **2003**, 50(9), 1419–1430. https://doi.org/10.1080/09500340308235215.
4. Jain, P.; Thourwal, A.; Bansal, S.; Gupta, N.; Kumar, S.; Singh, A. K.; Samanta, S.; Sardana, N. T-shaped resonator for X-band applications. In *2017 IEEE MTT-S International Microwave and RF Conference (IMaRC)*; Ahmedabad, India, 2017; pp. 1–5. https://doi.org/10.1109/IMaRC.2017.8449678.
5. Jain, P.; Singh, A. K.; Pandey, J. K.; Bansal, S.; Sardana, N.; Kumar, S.; Gupta, N.; Singh, A. K. An ultra-thin compact polarization-sensitive triple-band microwave metamaterial absorber. *J. Electron. Mater.*, **2021**, 50(3), 1506–1513. https://doi.org/10.1007/s11664-020-08680-z.
6. Jain, P.; Chhabra, H.; Chauhan, U.; Singh, D. K.; Anwer, T. M. K.; Ahammad, S. H.; Hossain, M. A.; Rashed, A. N. Z. Multiband metamaterial absorber with absorption prediction by assisted machine learning. *Mater. Chem. Phys.*, **2023**, 307, 128180. https://doi.org/10.1016/j.matchemphys.2023.128180.
7. Saadeldin, A. S.; Hameed, M. F. O.; Elkaramany, E. M. A.; Obayya, S. S. A. Highly sensitive terahertz metamaterial sensor. *IEEE Sens. J.*, **2019**, 19(18), 7993–7999. https://doi.org/10.1109/JSEN.2019.2918214.
8. Jain, P.; Prakash, K.; Sardana, N.; Kumar, S.; Gupta, N.; Singh, A. K. Design of an ultra-thin hepta-band metamaterial absorber for sensing applications. *Opt. Quantum Electron.*, **2022**, 54(9), 569. https://doi.org/10.1007/s11082-022-03917-z.
9. Jain, P.; Bansal, S.; Samant, P. Metamaterial absorbers for IoT applications. In *2022 IEEE VLSI Device Circuit and System (VLSI DCS)*; Kolkata, India, 2022; pp 300–303. https://doi.org/10.1109/VLSIDCS53788.2022.9811494.
10. Zhang, X.; Liu, Z. Superlenses to overcome the diffraction limit. *Nat. Mater.*, **2008**, 7(6), 435–441. https://doi.org/10.1038/nmat2141.
11. Wu, P. C.; Liao, C. Y.; Chen, J. W.; Tsai, D. P. Isotropic absorption and sensor of vertical split-ring resonator. *Adv. Opt. Mater.*, **2017**, 5(2), 1600581. https://doi.org/10.1002/adom.201600581.
12. Zheludev, N. I. The road ahead for metamaterials. *Science*, 2010, 328(5978), 582–583. https://doi.org/10.1126/science.1186756.
13. Liu, J.-M. *Principles of Photonics*, Cambridge University Press, Cambridge, 2016. https://doi.org/10.1017/cbo9781316687109.
14. Banerjee, S.; Tyagi, A. K. *Functional Materials*, Elsevier, Amsterdam, The Netherlands, 2012. https://doi.org/10.1016/C2010-0-65659-8.
15. Goodwin, Z. A. H.; Eikerling, M.; Löwen, H.; Kornyshev, A. A. Theory of microstructured polymer-electrolyte artificial muscles. *Smart Mater. Struct.*, **2018**, 27(7), 075056. https://doi.org/10.1088/1361-665X/aac291.
16. Zhang, J.; MacDonald, K. F.; Zheludev, N. I. Controlling light-with-light without nonlinearity. *Light Sci. Appl.*, 2012, 1(7), e18–e18. https://doi.org/10.1038/lsa.2012.18.
17. Zheludev, N. I.; Plum, E. Reconfigurable nanomechanical photonic metamaterials. *Nat. Nanotechnol.*, 2016, 11(1), 16–22. https://doi.org/10.1038/nnano.2015.302.
18. Zheludev, N. I.; Kivshar, Y. S. From metamaterials to metadevices. *Nat. Mater.*, 2012, 11(11), 917–924. https://doi.org/10.1038/nmat3431.
19. Kasani, S.; Curtin, K.; Wu, N. A review of 2D and 3D plasmonic nanostructure array patterns: Fabrication, light management and sensing applications. *Nanophotonics*, **2019**, 8(12), 2065–2089. https://doi.org/10.1515/nanoph-2019-0158.
20. Islam, M. T.; Rahman, M. N.; Samsuzzaman, M.; Mansor, M. F.; Misran, N. Resonator-inspired metamaterial sensor: Design and experimental validation for measuring thickness of multi-layered structures. *Sensors (Switzerland)*, **2018**, 18(12), 4213. https://doi.org/10.3390/s18124213.
21. Huang, H.; Xia, H.; Xie, W.; Guo, Z.; Li, H.; Xie, D. Design of broadband graphene-metamaterial absorbers for permittivity sensing at mid-infrared regions. *Sci. Rep.*, **2018**, 8(1), 1–10. https://doi.org/10.1038/s41598-018-22536-x.
22. Bakır, M.; Karaaslan, M.; Unal, E.; Akgol, O.; Sabah, C. Microwave metamaterial absorber for sensing applications. *Opto-Electronics Rev.*, **2017**, 25(4), 318–325. https://doi.org/10.1016/j.opelre.2017.10.002.
23. Jain, P.; Chhabra, H.; Chauhan, U.; Prakash, K.; Gupta, A.; Soliman, M. S.; Islam, M. S.; Islam, M. T. Machine learning assisted hepta band THz metamaterial absorber for biomedical applications. *Sci. Rep.*, **2023**, 13(1), 1792. https://doi.org/10.1038/s41598-023-29024-x.
24. Tao, H.; Chieffo, L. R.; Brenckle, M. A.; Siebert, S. M.; Liu, M.; Strikwerda, A. C.; Fan, K.; Kaplan, D. L.; Zhang, X.; Averitt, R. D.; et al. Metamaterials on paper as a sensing platform. *Adv. Mater.*, **2011**, 23(28), 3197–3201. https://doi.org/10.1002/adma.201100163.
25. Holloway, C. L.; Kuester, E. F.; Gordon, J. A.; O'Hara, J.; Booth, J.; Smith, D. R. An overview of the theory and applications of metasurfaces: The two-dimensional equivalents of metamaterials. *IEEE Antennas Propag. Mag.*, **2012**, 54(2), 10–35. https://doi.org/10.1109/MAP.2012.6230714.

26. Kildishev, A. V.; Boltasseva, A.; Shalaev, V. M. Planar photonics with metasurfaces. *Science*, **2013**, 339(6125), 1232009. https://doi.org/10.1126/science.1232009.
27. Minovich, A. E.; Miroshnichenko, A. E.; Bykov, A. Y.; Murzina, T. V.; Neshev, D. N.; Kivshar, Y. S. Functional and nonlinear optical metasurfaces. *Laser Photonics Rev.*, **2015**, 9(2), 195–213. https://doi.org/10.1002/lpor.201400402.
28. Pendry, J. B.; Holden, A. J.; Robbins, D. J.; Stewart, W. J. Low frequency plasmons in thin-wire structures. *J. Phys. Condens. Matter.*, **1998**, 10(22), 4785. https://doi.org/10.1088/0953-8984/10/22/007.
29. Su, H.; Wang, H.; Zhao, H.; Xue, T.; Zhang, J. Liquid-crystal-based electrically tuned electromagnetically induced transparency metasurface switch. *Sci. Rep.*, **2017**, 7(1), 1–7. https://doi.org/10.1038/s41598-017-17612-7.
30. Decker, M.; Staude, I.; Falkner, M.; Dominguez, J.; Neshev, D. N.; Brener, I.; Pertsch, T.; Kivshar, Y. S. High-efficiency dielectric huygens' surfaces. *Adv. Opt. Mater.*, **2015**, 3(6), 813–820. https://doi.org/10.1002/adom.201400584.
31. Lamb, H. On the reflection and transmission of electric waves by a metallic grating. *Proc. London Math. Soc.*, **1897**, 1(1), 523–546. https://doi.org/10.1112/plms/s1-29.1.523.
32. Marsland, B. J.; Gollwitzer, E. S. Host-microorganism interactions in lung diseases. *Nat. Rev. Immunol.*, **2014**, 14(12), 827–835. https://doi.org/10.1038/nri3769.
33. Kim, K. S. Pathogenesis of bacterial meningitis: From bacteraemia to neuronal injury. *Nat. Rev. Neurosci.*, **2003**, 4(5), 376–385. https://doi.org/10.1038/nrn1103.
34. Park, S. J.; Hong, J. T.; Choi, S. J.; Kim, H. S.; Park, W. K.; Han, S. T.; Park, J. Y.; Lee, S.; Kim, D. S.; Ahn, Y. H. Detection of microorganisms using terahertz metamaterials. *Sci. Rep.*, **2014**, 4(1), 1–7. https://doi.org/10.1038/srep04988.
35. Walia, S.; Shah, C. M.; Gutruf, P.; Nili, H.; Chowdhury, D. R.; Withayachumnankul, W.; Bhaskaran, M.; Sriram, S. Flexible metasurfaces and metamaterials: A review of materials and fabrication processes at micro- and nano-scales. *Appl. Phys. Rev.*, **2015**, 2(1), 011303. https://doi.org/10.1063/1.4913751.
36. Kim, H. K.; Lee, D.; Lim, S. A fluidically tunable metasurface absorber for flexible large-scale wireless ethanol sensor applications. *Sensors (Switzerland)*, **2016**, 16(8), 1246. https://doi.org/10.3390/s16081246.
37. Zhang, Y.; Zhao, J.; Cao, J.; Mao, B. Microwave metamaterial absorber for non-destructive sensing applications of grain. *Sensors (Switzerland)*, **2018**, 18(6), 912. https://doi.org/10.3390/s18061912.
38. Zhou, H.; Hu, D.; Yang, C.; Chen, C.; Ji, J.; Chen, M.; Chen, Y.; Yang, Y.; Mu, X. Multi-band sensing for dielectric property of chemicals using metamaterial integrated microfluidic sensor. *Sci. Rep.*, **2018**, 8(1), 1–11. https://doi.org/10.1038/s41598-018-32827-y.
39. ul Haq, T.; Ruan, C.; Zhang, X.; Ullah, S. Complementary metamaterial sensor for nondestructive evaluation of dielectric substrates. *Sensors*, **2019**, 19(9), 2100. https://doi.org/10.3390/s19092100.
40. Abdulkarim, Y. I.; Deng, L.; Altıntaş, O.; Ünal, E.; Karaaslan, M. Metamaterial absorber sensor design by incorporating swastika shaped resonator to determination of the liquid chemicals depending on electrical characteristics. *Phys. E Low-Dimens. Syst. Nanostruct.*, **2019**, 114, 113593. https://doi.org/10.1016/j.physe.2019.113593.
41. Ge, Y.; Chen, S.; Xu, Y.; He, Z.; Liang, Z.; Chen, Y.; Song, Y.; Fan, D.; Zhang, K.; Zhang, H. Few-layer selenium-doped black phosphorus: Synthesis, nonlinear optical properties and ultrafast photonics applications. *J. Mater. Chem. C*, **2017**, 5(25), 6129–6135. https://doi.org/10.1039/c7tc01267e.
42. Tian, Y.; Wang, Y.; Xu, Y.; Liu, Y.; Li, D.; Fan, C. A highly sensitive chemiluminescence sensor for detecting mercury (II) ions: A combination of exonuclease III-aided signal amplification and graphene oxide-assisted background reduction. *Sci. China Chem.*, **2015**, 58(3), 514–518. https://doi.org/10.1007/s11426-014-5258-9.
43. Zhao, Y.; Chen, F.; Zhang, Q.; Zhao, Y.; Zuo, X.; Fan, C. Polymerase/nicking enzyme synergetic isothermal quadratic DNA machine and its application for one-step amplified biosensing of lead (II) ions at femtomole level and DNA methyltransferase. *NPG Asia Mater.*, **2014**, 6(9), e131–e131. https://doi.org/10.1038/am.2014.84.
44. Labib, M.; Sargent, E. H.; Kelley, S. O. Electrochemical methods for the analysis of clinically relevant biomolecules. *Chem. Rev.*, **2016**, 116(16), 9001–9090. https://doi.org/10.1021/acs.chemrev.6b00220.
45. Yang, S.; Guo, F.; Kiraly, B.; Mao, X.; Lu, M.; Leong, K. W.; Huang, T. J. Microfluidic synthesis of multifunctional janus particles for biomedical applications. *Lab Chip*, 2012, 12(12), 2097–2102. https://doi.org/10.1039/c2lc90046g.
46. Chen, J. H.; Jang, C.; Xiao, S.; Ishigami, M.; Fuhrer, M. S. Intrinsic and extrinsic performance limits of graphene devices on SiO_2. *Nat. Nanotechnol.*, **2008**, 3(4), 206–209. https://doi.org/10.1038/nnano.2008.58.

47. Su, S.; He, Y.; Zhang, M.; Yang, K.; Song, S.; Zhang, X.; Fan, C.; Lee, S. T. High-sensitivity pesticide detection via silicon nanowires-supported acetylcholinesterase-based electrochemical sensors. *Appl. Phys. Lett.*, **2008**, 93(2), 023113. https://doi.org/10.1063/1.2959827.
48. Chang, H.; Wu, H. Graphene-based nanocomposites: Preparation, functionalization, and energy and environmental applications. *Energy Environ. Sci.*, **2013**, 6(12), 3483–3507. https://doi.org/10.1039/c3ee42518e.
49. Prakash, K.; Thakur, P.; Bansal, S.; Garg, S.; Jain, P.; Sharma, K.; Gupta, N.; Kasjoo, S. R.; Kumar, S.; Singh, A. K. Thermoelectric rectification in a graphene-based triangular ballistic rectifier (G-TBR). *J. Comput. Electron.*, **2021**, 20(6), 2308–2316. https://doi.org/10.1007/s10825-021-01794-5.
50. Prakash, K.; Bansal, S.; Garg, S.; Thakur, P.; Sharma, K.; Jain, P.; Gupta, N.; Kasjoo, S. R.; Kumar, S.; Singh, A. K. Thermoelectric effect in graphene-based three-terminal junction. *IEEE Trans. Nanotechnol.*, 2021, 20(733–738), 2021. https://doi.org/10.1109/TNANO.2021.3113343.
51. Xia, F.; Wang, H.; Xiao, D.; Dubey, M.; Ramasubramaniam, A. Two-dimensional material nanophotonics. *Nat. Photonics*, **2014**, 8, 899–907. https://doi.org/10.1038/nphoton.2014.271.
52. Kurapati, R.; Kostarelos, K.; Prato, M.; Bianco, A. Biomedical uses for 2D materials beyond graphene: Current advances and challenges ahead. *Adv. Mater.*, **2016**, 28, 6052–6074. https://doi.org/10.1002/adma.201506306.
53. Jain, P.; Prakash, K.; Khanal, G. M.; Sardana, N.; Kumar, S.; Gupta, N.; Singh, A. K. Quad-band polarization sensitive terahertz metamaterial absorber using gemini-shaped structure. *Results Opt.*, **2022**, 8, 100254. https://doi.org/https://doi.org/10.1016/j.rio.2022.100254.
54. Prakash, K.; Bansal, S.; Jain, P.; Garg, S.; Khanal, G. M.; Kumar, S.; Gupta, N.; Kasjoo, S. R.; Singh, A. K. Thermoelectric rectification in graphene based Y-junction. *Micro Nanostruct.*, **2022**, 167, 207242. https://doi.org/https://doi.org/10.1016/j.micrna.2022.207242.
55. Zeng, M.; Xiao, Y.; Liu, J.; Yang, K.; Fu, L. Exploring two-dimensional materials toward the next-generation circuits: From monomer design to assembly control. *Chem. Rev.*, **2018**, 118(13), 6236–6296. https://doi.org/10.1021/acs.chemrev.7b00633.
56. Jorgenson, R. C.; Yee, S. S. A fiber-optic chemical sensor based on surface plasmon resonance. *Sensors Actuators B. Chem.*, **1993**, 12, 213–220. https://doi.org/10.1016/0925-4005(93)80021-3.
57. Helmerhorst, E.; Chandler, D. J.; Nussio, M.; Mamotte, C. D. Real-time and label-free bio-sensing of molecular interactions by surface plasmon resonance: A laboratory medicine perspective. *Clin. Biochem. Rev.*, **2012**, 33(4), 161.
58. Smith, D. R.; Padilla, W. J.; Vier, D. C.; Nemat-Nasser, S. C.; Schultz, S. Composite medium with simultaneously negative permeability and permittivity. *Phys. Rev. Lett.*, **2000**, 84(18), 4184–4187. https://doi.org/10.1103/PhysRevLett.84.4184.
59. Hong, J. T.; Jun, S. W.; Cha, S. H.; Park, J. Y.; Lee, S.; Shin, G. A.; Ahn, Y. H. Enhanced sensitivity in THz plasmonic sensors with silver nanowires. *Sci. Rep.*, **2018**, 8, 1–8. https://doi.org/10.1038/s41598-018-33617-2.
60. Ahmadivand, A.; Gerislioglu, B.; Tomitaka, A.; Manickam, P.; Kaushik, A.; Bhansali, S.; Nair, M.; Pala, N. Extreme sensitive metasensor for targeted biomarkers identification using colloidal nanoparticles-integrated plasmonic unit cells. *Biomed. Opt. Express*, **2018**, 9, 373–386. https://doi.org/10.1364/boe.9.000373.
61. Wang, X.; Li, Y.; Wang, H.; Fu, Q.; Peng, J.; Wang, Y.; Du, J.; Zhou, Y.; Zhan, L. Gold nanorod-based localized surface plasmon resonance biosensor for sensitive detection of Hepatitis B virus in buffer, blood serum and plasma. *Biosens. Bioelectron.*, **2010**, 26(2), 404–410. https://doi.org/10.1016/j.bios.2010.07.121.
62. Diao, W.; Tang, M.; Ding, S.; Li, X.; Cheng, W.; Mo, F.; Yan, X.; Ma, H.; Yan, Y. Highly sensitive surface plasmon resonance biosensor for the detection of HIV-related DNA based on dynamic and structural DNA nanodevices. *Biosens. Bioelectron.*, **2018**, 100, 228–234. https://doi.org/10.1016/j.bios.2017.08.042.
63. Sethi, K. K.; Palai, G.; Sarkar, P. Realization of accurate blood glucose sensor using photonics based metamaterial. *Optik (Stuttg).*, **2018**, 168, 296–301. https://doi.org/10.1016/j.ijleo.2018.04.091.
64. Kehe, K.; Szinicz, L. Medical aspects of sulphur mustard poisoning. *Toxicology*, **2005**, 214, 198–209. https://doi.org/10.1016/j.tox.2005.06.014.
65. Nazemi, H.; Joseph, A.; Park, J.; Emadi, A. Advanced micro-and nano-gas sensor technology: A review. *Sensors (Switzerland)*, **2019**, 19, 1285. https://doi.org/10.3390/s19061285.
66. Mishra, S. K.; Gupta, B. D. Surface plasmon resonance-based fiber optic chlorine gas sensor utilizing indium-oxide-doped tin oxide film. *J. Light. Technol.*, **2015**, 33, 2770–2776. https://doi.org/10.1109/JLT.2015.2412615.

67. Zhang, Y.; Yi, Z.; Wang, X.; Chu, P.; Yao, W.; Zhou, Z.; Cheng, S.; Liu, Z.; Wu, P.; Pan, M.; et al. Dual band visible metamaterial absorbers based on four identical ring patches. *Phys. E Low-Dimens. Syst. Nanostruct.*, **2021**, 127, 114526. https://doi.org/10.1016/j.physe.2020.114526.
68. Park, J.-H.; Kodigala, A.; Ndao, A.; Kanté, B. Hybridized metamaterial platform for nano-scale sensing. *Opt. Express*, **2017**, 25, 15590–15598. https://doi.org/10.1364/oe.25.015590.
69. Su, D. S.; Tsai, D. P.; Yen, T. J.; Tanaka, T. Ultrasensitive and selective gas sensor based on a channel plasmonic structure with an enormous hot spot region. *ACS Sensors*, **2019**, 4, 2900–2907. https://doi.org/10.1021/acssensors.9b01225.
70. Bunte, G.; Hürttlen, J.; Pontius, H.; Hartlieb, K.; Krause, H. Gas phase detection of explosives such as 2,4,6-trinitrotoluene by molecularly imprinted polymers. *Anal. Chim. Acta*, **2007**, 591, 49–56. https://doi.org/10.1016/j.aca.2007.02.014.
71. Lee, H. J.; Yook, J. G. Biosensing using split-ring resonators at microwave regime. *Appl. Phys. Lett.*, **2008**, 92, 254103. https://doi.org/10.1063/1.2946656.
72. Liu, K.; Zhang, R.; Liu, Y.; Chen, X.; Li, K.; Pickwell-Macpherson, E. Gold nanoparticle enhanced detection of EGFR with a terahertz metamaterial biosensor. *Biomed. Opt. Express*, **2021**, 12, 1559–1567. https://doi.org/10.1364/boe.418859.
73. Chen, H. T.; Padilla, W. J.; Zide, J. M. O.; Gossard, A. C.; Taylor, A. J.; Averitt, R. D. Active terahertz metamaterial devices. *Nature*, **2006**, 444, 597–600. https://doi.org/10.1038/nature05343.
74. Sreekanth, K. V.; Alapan, Y.; Elkabbash, M.; Ilker, E.; Hinczewski, M.; Gurkan, U. A.; De Luca, A.; Strangi, G. Extreme sensitivity biosensing platform based on hyperbolic metamaterials. *Nat. Mater.*, **2016**, 15, 621–627. https://doi.org/10.1038/nmat4609.
75. Chen, H.; Kou, X.; Yang, Z.; Ni, W.; Wang, J. Shape- and size-dependent refractive index sensitivity of gold nanoparticles. *Langmuir*, **2008**, 24, 5233–5237. https://doi.org/10.1021/la800305j.
76. Kazanskiy, N. L.; Butt, M. A.; Khonina, S. N. Carbon dioxide gas sensor based on polyhexamethylene biguanide polymer deposited on silicon nano-cylinders metasurface. *Sensors (Switzerland)*, **2021**, 21, 378. https://doi.org/10.3390/s21020378.
77. Lashof, D. A.; Ahuja, D. R. Relative contributions of greenhouse gas emissions to global warming. *Nature*, **1990**, 344, 529–531. https://doi.org/10.1038/344529a0.
78. Abdulkarim, Y. I.; Deng, L.; Luo, H.; Huang, S.; Karaaslan, M.; Altintas, O.; Bakir, M.; Muhammadsharif, F. F.; Awl, H. N.; Sabah, C.; et al. Design and study of a metamaterial based sensor for the application of liquid chemicals detection. *J. Mater. Res. Technol.*, **2020**, 9, 10291–10304. https://doi.org/10.1016/j.jmrt.2020.07.034.
79. Withayachumnankul, W.; Jaruwongrungsee, K.; Tuantranont, A.; Fumeaux, C.; Abbott, D. Metamaterial-based microfluidic sensor for dielectric characterization. *Sensors Actuators, A Phys.*, **2013**, 189, 233–237. https://doi.org/10.1016/j.sna.2012.10.027.
80. Nguyen, T. T.; Lim, S. Wide incidence angle-insensitive metamaterial absorber for both TE and TM polarization using eight-circular-sector. *Sci. Rep.*, **2017**, 7(1), 1–11. https://doi.org/10.1038/s41598-017-03591-2.
81. Lee, D.; Jeong, H.; Lim, S. Electronically switchable broadband metamaterial absorber. *Sci. Rep.*, **2017**, 7, 1–10. https://doi.org/10.1038/s41598-017-05330-z.
82. Sadeqi, A.; Nejad, H. R.; Sonkusale, S. Low-cost metamaterial-on-paper chemical sensor. *Opt. Express*, **2017**, 25, 16092–16100. https://doi.org/10.1364/oe.25.016092.
83. Wang, W.; Zheng, L.; Xiong, L.; Qi, J.; Li, B. High Q-factor multiple fano resonances for high-sensitivity sensing in all-dielectric metamaterials. *OSA Contin.*, **2019**, 2, 2818–2825. https://doi.org/10.1364/osac.2.002818.
84. Abdulkarim, Y. I.; Deng, L.; Karaaslan, M.; Unal, E. Determination of the liquid chemicals depending on the electrical characteristics by using metamaterial absorber based sensor. *Chem. Phys. Lett.*, **2019**, 732, 136655. https://doi.org/10.1016/j.cplett.2019.136655.
85. Wang, Y.; Cui, Z.; Zhang, X.; Zhang, X.; Zhu, Y.; Chen, S.; Hu, H. Excitation of surface plasmon resonance on multiwalled carbon nanotube metasurfaces for pesticide sensors. *ACS Appl. Mater. Interfaces*, **2020**, 12, 52082–52088. https://doi.org/10.1021/acsami.0c10943.
86. Mosbah, S.; Zebiri, C.; Sayad, D.; Elfergani, I.; Bouknia, M. L.; Mekki, S.; Zegadi, R.; Palandoken, M.; Rodriguez, J.; Abd-Alhameed, R. A. Compact and highly sensitive bended microwave liquid sensor based on a metamaterial complementary split-ring resonator. *Appl. Sci.*, **2022**, 12, 2144. https://doi.org/10.3390/app12042144.
87. Tamer, A.; Karadağ, F.; Ünal, E.; Abdulkarim, Y. I.; Deng, L.; Altintas, O.; Bakır, M.; Karaaslan, M. Metamaterial based sensor integrating transmission line for detection of branded and unbranded diesel fuel. *Chem. Phys. Lett.*, **2020**, 742, 137169. https://doi.org/10.1016/j.cplett.2020.137169.

88. Turpin, J. P.; Bossard, J. A.; Morgan, K. L.; Werner, D. H.; Werner, P. L. Reconfigurable and tunable metamaterials: A review of the theory and applications. *Int. J. Antennas Propag.*, **2014**, 11, 1–18. https://doi.org/10.1155/2014/429837.
89. Bakır, M.; Karaaslan, M.; Dincer, F.; Delihacioglu, K.; Sabah, C. Tunable perfect metamaterial absorber and sensor applications. *J. Mater. Sci. Mater. Electron.*, **2016**, 27(11), 12091–12099. https://doi.org/10.1007/s10854-016-5359-7.
90. Jain, P.; Bansal, S.; Prakash, K.; Sardana, N.; Gupta, N.; Kumar, S.; Singh, A. K. Graphene-based tunable multi-band metamaterial polarization-insensitive absorber for terahertz applications. *J. Mater. Sci. Mater. Electron.*, **2020**, 31(14), 11878–11886. https://doi.org/10.1007/s10854-020-03742-8.
91. Tümkaya, M. A.; Dinçer, F.; Karaaslan, M.; Sabah, C. Sensitive metamaterial sensor for distinction of authentic and inauthentic fuel samples. *J. Electron. Mater.*, **2017**, 46(8), 4955–4962. https://doi.org/10.1007/s11664-017-5485-x.
92. Haq, T.; Ruan, C.; Zhang, X.; Ullah, S.; Fahad, A. K.; He, W. Extremely sensitive microwave sensor for evaluation of dielectric characteristics of low-permittivity materials. *Sensors*, **2020**, 20, 1916. https://doi.org/10.3390/s20071916.
93. Intaravanne, Y.; Chen, X. Recent advances in optical metasurfaces for polarization detection and engineered polarization profiles. *Nanophotonics*, **2020**, 9, 1003–1014. https://doi.org/10.1515/nanoph-2019-0479.
94. Abdollahramezani, S.; Hemmatyar, O.; Taghinejad, H.; Krasnok, A.; Kiarashinejad, Y.; Zandehshahvar, M.; Alù, A.; Alù, A.; Adibi, A. Tunable nanophotonics enabled by chalcogenide phase-change materials. *Nanophotonics*, **2020**, 9, 1189–1241. https://doi.org/10.1515/nanoph-2020-0039.
95. Singh, R.; Agarwal, A.; Anthony, B. W. Design of optical meta-structures with applications to beam engineering using deep learning. *Sci. Rep.*, **2020**, 10, 1–10. https://doi.org/10.1038/s41598-020-76225-9.

16 Electrochemical Biosensors Based on Nanomaterial Functionality

Ruchi Khare, Mohammad Yasir, Sharvari Bhat, Vivek Shrivastava, Rahul Kumar Maurya, Alok Shiomurti Tripathi, Manoj S Dikkatwar, and Rahul Shrivastava

16.1 INTRODUCTION

Biosensors are the structured assembly of a biological entity, a transducing element, and a readable output source, functionalised for the detection of an analyte. The biosensors are considered sensitive tools for onsite detection of analytes. Based on the origin or source of the analyte, biosensors are exploited for various applications like healthcare, environment monitoring, and food contamination regulation (Ruchi et al., 2020). These biosensors serve as alternatives to conventional laboratory techniques applied for the detection of analytes. It has not only increased the efficiency of detection but also takes less time and requires less skilled labour in comparison to conventional techniques. Daily examples of biosensors include pregnancy test kits, blood glucose testing kits, alcohol sensors, smart watches, onsite pollutant detecting devices, and many more (Randviir and Banks, 2013).

In spite of having sensitive conventional techniques like enzyme-linked immunosorbent assay (ELISA), liquid chromatography-mass spectrometry (LC-MS), gas chromatography-mass spectrometry (GC-MS), and many more, the need for biosensors came into existence because of their on-site detection properties. The time spent between sample collection and sample analysis is considered the "crucial hour," during which the analytes are either degraded or evaporated, which also gives false negative results. So, to detect analytes at the micro- or nanogram level within a short span of time that can be used by laymen, biosensors have emerged as an effective solution (Khare et al., 2020).

For the fabrication of a typical biosensor, there are three essential components required, namely the biorecognition element, the transducer, and the output source (Figure 16.1). The combination of the receptor of analyte immobilised on a matrix is the first component of the biorecognition element; in the case of electrochemical biosensors, matrix is made up of a semiconducting platform onto which different biological receptors are placed and which are then deposited onto the electrodes. The next component is the transducer, which is the main sensing part and is responsible for reading and recording signals produced before and after the binding of antigens to the receptors. The electrochemical biosensors provide different electrode systems that sense current and voltage fluctuations before and after the binding of antigen to the analyte. For sensing these fluctuations with respect to time, the transducers can be a potentiostat, a galvanostat, or an impedance analyser. The last component of the biosensing system is the reader output, which represents the data produced by the system (Eissa et al., 2020).

DOI: 10.1201/9781032654874-16

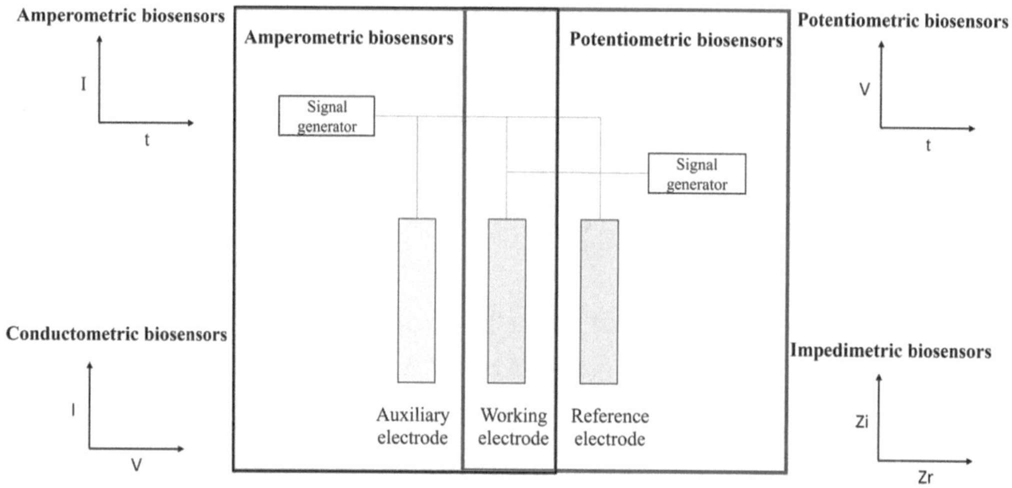

FIGURE 16.1 Diagrammatic overview of biosensors.

However, after the platform components are assembled, the platform's properties such as accuracy, sensitivity, efficacy, and limit of detection (LOD) are determined. These properties are considered key features of the biosensing system. In this chapter, detailed information about the biosensing systems is given, in which we first discuss the biomarkers.

16.2 BIOMATERIALS

With reference to the biosensing systems, biomarkers are the analytes that act as key factors in determining the nature of the biosensing platform. Biomarkers can be biological or biochemical components that have to be detected. They are proteins, hormones, enzymes, or biochemical entities whose concentration differs in the natural state and the diseased state. These are considered a mirror of the habitat. For instance, in comparison to the natural state of the body, the blood glucose concentration is higher in the altered state, so for the detection of diabetes, the blood glucose concentration serves as a biomarker of the biosensing system, which will detect the possible range of blood sugar at a particular period of time (Lee and Wong, 2009).

For electrochemical biosensors, the biomarker or its receptor should be detected by an electrochemically active species. In this case, both are electrochemically inactive, and in order to enhance the sensitivity of the systems, the receptors are conjugated with the nanomaterials. Nanomaterials are materials with a single unit size of 100 nm or less; they are also known as nano-objects. They can be of many types like nanorods, nanosheets, nanotubes, and nanocrystals (Figure 16.2). These nanomaterials can be synthesised by chemical, biological, or physical methods. These prepared nanomaterials are either tagged with receptors or coated with drugs before being used in biological applications. They serve as site-directed drug delivery agents and reducing agents in the treatment of many human diseases like cancer (Kelley et al., 2014). But for electrochemical sensing platforms, they serve as semiconducting agents, which not only enhance the sensitivity of the system but also increase the surface area and volume in order to load the maximum concentration of receptors. Although the utility and origin of biomaterials are determined by the techniques used in their production, for the preparation of biomaterials, the two basic strategies are given below:

1. Chemical synthesis of nanoparticles followed by conjugation of receptors like drugs, antibodies, etc.
2. Green synthesis of nanoparticles possessing properties like receptors, enzymes, reducing agents, etc. (Vigneshvar et al., 2016).

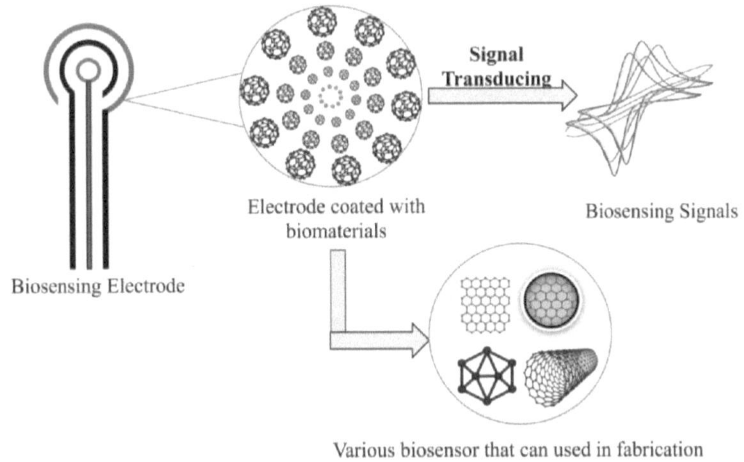

FIGURE 16.2 Diagrammatic illustration of the basic operation of an electrochemical biosensor.

However, different types of nanomaterials employed for the fabrication of electrochemical biosensing platforms are given below:

16.2.1 Nanoparticles

These are 3-dimensional spheres or materials with sizes ranging from 1 to 100 nm. These nanoparticles have been exploited in the fabrication of electrochemical biosensors because of their semiconducting nature. These nanoparticles, when attached to receptors, serve as electrochemically active species that are able to recognise analytes. Apart from receptors, these nanomaterials are synthesised by their metal precursors, where the size is controlled by variations in the concentration of reducing agents, temperature, and time duration. They are stable in lyophilised form as well as in a liquid colloidal state. They are water soluble by nature (Khan et al., 2019). These nanoparticles have localised surface plasma resonance (LSPR), due to which they possess opto-electric properties. The synthesis is primarily characterised by various microscopy techniques like atomic force microscopy (AFM), scanning electron microscopy (SEM), transmission electron microscopy (TEM), and UV-visible spectroscopy, as these nanoparticles show characteristic peaks in spectral scans (Murthy, 2007). Studies have reported the application of these nanoparticles' electrical properties in the fabrication of biosensors for medical, environmental, and healthcare devices.

16.2.2 Nanotubes

These two-dimensional nanomaterials are commonly made of graphene sheets. When the graphene sheets fold into cylindrical shapes, they are termed "nanotubes." The nanotubes can be single or double walled (Anzar et al., 2020). In electrochemical biosensors, these carbon nanotubes are used as electrochemically conductive mesh onto which the receptors are immobilised. However, the use of nanotubes is quite critical because of their very low solubility in water. In water, they are dispersed rather than soluble. As a result, conjugation methods involving the dispersion of nanotubes in water lead to an uneven distribution of receptors, which also affects the sensitivity and reproducibility of the designed electrochemical systems. For the solubility of carbon nanotubes, ultrasonication is considered an essential step. Studies have reported the use of 2% sodium dodecyl sulphate (SDS) and dimethyl formamide (DMF) as preferred solvents for the solubilisation of carbon nanotubes (Geckeler and Premkumar, 2011). Carbon nanotubes are characterised using techniques such as near-infrared-infrared spectroscopy (NIR-IR) and TEM, and their conjugation can be characterised using Raman spectroscopy (RS) and Fourier transform infrared spectroscopy (FT-IR) (Pavan Kalyan et al., 2021).

16.2.3 Nanofibres

Nanofibres are fibres with a one-dimensional structure and a size between 1 and 100 nm, prepared from different polymers. Their applications vary according to the polymer material and dimensions. They are used in the fabrication of electrochemical biosensors because of their higher receptor loading capacity, which results in high sensitivity and a faster response time of the biosensors. These are prepared using a hydrodynamic process called electrospinning, where the polymer is filled in a syringe and the droplet formed at the end of the needle is converted into fibre as the surface tension is overpowered by electric force. The stretched, elongated fibres are collected in a plate collector or rotary collector (Halicka and Cabaj, 2021).

This method is applicable to both natural polymers F (cellulose) and synthetic polymers (polyacrylnitrile). Depending upon the polymers used for fabricating nanofibres, other parameters like conductivity of nanofibres, viscosity of solutions, and solubility of polymers in solvents are standardised. After the concentrations of raw materials, the parameters for electrospinning like the applied voltage, electric field, collector type, and rotation speed are decided (Bhardwaj and Kundu, 2010). The surface and chemical characterisation is performed using spectroscopic techniques like X-ray diffraction (XRD); RS; energy dispersive spectroscopy (EDS), and FT-IR and microscopic techniques like AFM, TEM, and SEM. The morphologies of nanofibres reported till now are beaded, hollow, core shell, ribbon, porous solid, and multichannel nanofibres (Wang et al., 2021).

Nanofibres are functionalised with two major approaches: (1) new compound incorporated directly before electrospinning into polymer solution leads to the incorporation of compound with hybrid structure in material. (2) Other one is surface modification; chemical modification methods are used to conjugate the new compound on the nanoparticle surface. There are diverse molecules used for modification such as enzymes like biomolecules, carbon nanoparticles, dyes, organic compounds, nanoparticles (quantum dots), and polymers. Immobilisation capacity of biomolecules is increased with porous structure and large surface area which improves retention time and sensitivity of the sensor (Balusamy et al., 2019).

16.3 PRINCIPLE OF ELECTROCHEMICAL SENSOR

The principle of electrochemical biosensor is based on generating a signal that can be measured. These biosensors combine a biological component that can detect an analyte with a physicochemical component for signal evaluation. In 1962, a biosensor prototype was unveiled to track blood gas levels throughout surgery (Clark and Lyons, 1962). Glucose meters and home pregnancy tests are now the most popular biosensors, but there is a push to develop novel biosensors for a variety of uses, such as food testing, drug detection, and deoxyribonucleic acid (DNA) testing. These sensors could get smaller and have lower detection limits, thanks to thin coatings (Zeng et al., 2014). The following elements are typically found in biosensors:

- *Analyte*: A component that has to be found and is of interest.
- *Bioreceptor*: A molecule that precisely detects the analyte. It includes things such as antibodies, DNA, cells, aptamers, and enzymes. Biorecognition is the mechanism behind signal production, where the response occurs due to the interaction between the analyte and the bioreceptor.
- *Transducer*: Transducer is an element which transforms energy from one form to another. In biosensor transducer function, converts the signal generated by biorecognition into a measurable, quantifiable output. Energy conversion process is termed signalisation; electrical and optical signals are majorly generated by transducers, which are equal to the amount of interaction between bioreceptor and analyte.
- *Electronics*: the system which display the signals after analysing it. It is composed of electric circuit which amplifies the signals and converts into digital form. Further signals are quantified through a process in the display device (Bhalla et al., 2016).

16.4 ROLE OF NANOMATERIALS

However, the above type of biomaterials acts differently, when it comes to the measurement of properties including electrical, catalytical, magnetic, etc. These properties are often exploited to fill the gap between the transducer and biological receptors. The electrochemical properties of the nanomaterials impart signal either by exhibiting catalytic reaction or by propensity-based behaviour. The catalytic properties of the reaction are based on charge transfer and the propensity-based reactions are conducted when receptors like antibodies are coated onto the substrate surface. Although if we compare both the type of reactions on nanomaterials, the propensity-based reactions are considered catalytic reaction as later ones generally use oxidase enzymes for the reactions which are relatively less stable than proteins and their activities also vary with respect to time and temperature. Even the freeze-thaw cycles done during the preparation of platform confers to varied results, whereas other receptors provide reproducible results. The propensity-based reaction also provides better results due to specific reactivity against the concentration of antigen, whereas the catalytic reaction is highly sensitive to physiological parameters. However, the measurement methods applied for electrochemical measurements are explained in the next section.

16.5 ELECTROCHEMICAL METHODS

The analyte adheres to the biological receptors to create an analyte-receptor complex, which then causes the measurable electrical reaction. Sometimes the analyte undergoes conversion into a product that may involve the emission of heat, oxygen, electrons, or hydrogen ions. Product-linked modifications can be converted by the transducer into electrical signals on the three-electrode or two-electrode system that can be magnified and evaluated. The three-electrode systems comprise of reference electrode, counter electrode, and working electrode, whereas the two-electrode system comprises of counter electrode and working electrode only. Measurements of the natural fluctuations of current and potential around an average value in an electrochemical assembly are typically made up of two functional electrodes—these are known as electrochemical noise measurements. The electronic impulse from the transducer is often weak and superimposed on a relatively high and noisy baseline. This baseline consists of a high-frequency signal element that appears random, likely caused by electrical disturbances or generated within the electronic components of the transducer itself.. Signal processing typically entails amplifying the resulting signal differential and electronically filtering away the undesired signal noise. The reference baseline signal is produced from a transducer with a comparable design but no biocatalyst membrane. Electrical noise filtration is made much easier by the biosensor's comparatively slow reaction time. The analogue signal generated at this stage could be output directly (Figure 16.3), but it's more common to convert it to a digital signal before sending it to a microprocessor stage, where it will be processed, converted to the desired units, and then sent to a display or data store (Malhotra et al., 2017).

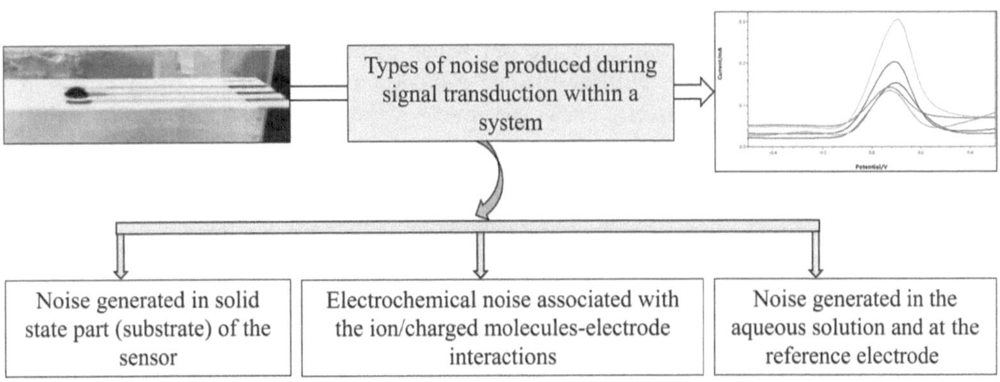

FIGURE 16.3 Schematic representation of types of noise produced during signal transduction and processing.

Biosensor are like electrochemical biosensor, which turn data of biological entities into electrical signals like impedance, current, voltage, etc. (Ahmad et al., 2020; Khosla et al., 2018). Clark invented the first electrochemical biosensor which quantifies the concentration of glucose into human serum. Biological components such as enzymes, antibodies, viruses, and proteins are quantified with electrochemical biosensors (Clark and Lyons, 1962). The most important component of it is the electrode, which regulates the electron flow and bioagents. The recently developed electrochemical sensors are as follows:

16.5.1 Amperometry Biosensors

The amperometry biosensor is a system based on the quantity of the current produced by oxidation that provides precise quantitative analytical information. Potentiometric biosensors are these kinds of biosensors in the sense of sensitivities, energy ranges, and response times. The outgoing current of the sensor is examined and used for the sensing procedure in amperometry biosensors. By contrasting the present results acquired for the various analyte concentrations, the sensitivities of the amperometry biosensor are ascertained. These biosensors only used two electrodes: one to apply the voltage and other to measure the current that was passing through the device. The amperometry biosensors rely solely on the current readings and do not need optical or electrochemical components. The first step in the creation of biosensors was glucose sensing, and the majority of this sensing is catalysed by an enzyme reaction with glucose oxidase (Gupta et al., 2019; Khan et al., 2021). Temperature and pH variations have an impact on the sensitivity of these glucose sensors. Electrochemical impedance spectroscopy (EIS) and cyclic voltammetry (CV) are the main electrochemical methods employed in the sensing process of amperometry glucose sensors. An amperometry glucose sensor's sensitivity is determined by the rise in current per mM concentration in one centimetre square area, or $\mu A\ mM^{-1}cm^{-2}$.

Purines, which are present in food and cells, are broken down metabolically to become uric acid. Blood uric acid levels range from 140 to 430 μM, and elevated uric acid levels can cause kidney stones and gout (Chen et al., 2013). To reduce the price and widen the use of uric acid measurements, smaller-scale uric acid sensing instruments have been developed. The electrodes in enzymatic or non-enzymatic sensors were often constructed utilising sophisticated materials with conductometric measurements as their main focus (Ansari et al., 2009; Castillo-Ortega et al., 2002).

Urea is eliminated through perspiration and urine and is regarded as a waste product. Blood urea levels range from 2.5 to 7.5 mM, and abnormal urea levels can cause liver and kidney issues such as renal insufficiency and uraemia (Yang et al., 2015). As urea sensors, vertically aligned ZnO nanorods were employed, although their sensitivity (41.64 $\mu A\ mM^{-1}cm^{-2}$) was rather poor for the linear range of 0.001–24.0 mM (Ahmad et al., 2014). With a tested reaction time of 4 s, a 132-μA $mM^{-1}cm^{-2}$ sensitivity value was discovered. Cross-sensitivity and durability tests, however, weren't carried out. Although the sensitivity values were lower (40 $\mu A\ mM^{-1}cm^{-2}$), electrochemically produced nanostructured ZnO films displayed a similar linear range (1.7–13.6 mM) (Ali et al., 2009).

Uric acid detection up to 0.374 mM was accomplished using 3D periodic mesoporous nickel oxide (NiO) particulates with crystalline walls and a modest sensitivity (756.26 $\mu A\ mM^{-1}cm^{-2}$) (Huang et al., 2015). However, no interference experiments were performed for this paper. NiO thin films were employed on platinum-coated glass substrates to measure uric acid levels with a sensitivity close to clinical human uric acid values (1278.48 $\mu A\ mM^{-1}$) (0.05–1.0 mM) (Arora et al., 2011).

16.5.2 Potentiometric Biosensors

On the basis of the potential difference between the working and reference electrodes, potentiometric biosensors function. Unlike in an amperometric biosensor, the detected species are not eaten. By comparing its activity to that of the reference electrode, its activity is proportionate to the analyte concentration. When a relatively stable and precise reference electrode is employed, potentiometric biosensors have a significant advantage in that they are sensitive and selective. In the realm of sensors, potentiometry is a commonly used electrochemical approach that is both affordable and effective over a broad array of ion concentrations. The signal generated by potentiometric biosensors takes the form of potential, and the biological component employed is a portion of the receptor, which are its two main benefits (Koncki, 2007). Interleukin was recently found to have an LOD of 1 pg mL^{-1} using potentiometric biosensor with a gold electrode as well as expanded field effect transistor (FET) transistor (Parrilla et al., 2019). The electrode was adsorbed with thiol that was created by the acetylcholinesterase, which was employed as an antibody.

Potentiometric-based biosensors can support early illness diagnosis and improve identification of biological component identification. These biosensors track the biomarkers in humans which are analytically portable in nature. These are used to assess human health, as they contain ion-sensitive membrane sensors which act as a biomarker to monitor the activity of chloride, sodium, and potassium ions. Biomarkers of cancer identified for early and quickly by this sensitive tool, potentiometric-based Ion-sensitive membrane devices are used independently or in combination with various techniques to detect a wide range of biological molecules. (Karimi-Maleh et al., 2021). Light-addressable potentiometric biosensors, photoelectrochemical potentiometric biomarkers, potentiometric biosensors amplified with molecularly imprinted polymer systems and wearable potentiometric biomarkers are the four primary categories.

Valproic acid is an anti-epileptic drug detected in blood samples with a potentiometric poly (vinyl chloride) (PVC)-membrane-based biosensor. Lower detection limit for the biosensor was 9.75 10^{-7}mol L^{-1}. The created valproate-selective biosensor demonstrated excellent selectivity for various ions found in human blood as well as for anti-epileptic medications like carbamazepine, levetiracetam, and phenytoin. The biosensor performed in the broad pH range of 4.0–11.0 and demonstrated a rapid response time of less than 10 seconds, strong reusability, and stability. By using SEM, the morphology of the produced PVC membrane biosensor was examined. The detection of valproic acid in patient blood samples using the newly created valproate-selective biosensor was very successful (Özbek et al., 2021).

Employing poly (2-hydroxyethyl methacrylate-glycidyl methacrylate) nanoparticles as urease enzyme nano-carriers, a highly stable novel urea biosensor was created. The urease-immobilised nanoparticles and an ammonium-selective electrode were used to create the urea biosensor. SEM was used to characterise the synthesised nanoparticles, and energy-dispersive X-ray (EDX) analysis was used to gauge how well the enzyme was attached. The potentiometric responsiveness of the biosensor was investigated in relation to various urea solution concentrations, and a linear concentration range of urea in the range of 0.01–500 mM was shown. Additionally, the LOD was examined and determined to be 0.77 M. The response time was similarly found to be 30 seconds. The urea biosensor system's repeatability was discovered to be extremely high, and it was shown that the sensor's response was protected to a degree of 95.7%. Ascorbic acid, uric acid, glucose (GL), and sodium ions (Na^+) were tested against the interference pattern of the urea biosensor, and the responses of these interfering substances were determined to be minimal (Öndeş et al., 2021).

16.5.3 Impedimetric Biosensors

By immobilising biological recognition components onto an electrode surface, an impedimetric biosensor is created. It measures and/or monitors the targeted analyte and reports the results via the generation of an electrical impedance indicator that is inversely proportional to the analyte activity. Impedance biosensors are electrochemical biosensors that employ variations in impedance to detect analytes or biological entities. This method most commonly employs electrochemical impedance spectroscopy as a methodology (EIS). EIS makes it simple to determine both the parameters of the bulk electrode and the processes occurring at the electrode interface (Maalouf et al., 2007). The Bode or Nyquist graphs, which are both dependent on frequency, are used to visualise the EIS spectrum. There is a very slight variation in the signal's amplitude in the EIS. Additionally, because the biosensor uses agents like antibodies, viruses, bacteria, enzymes, etc., to monitor a biological event, the evaluation of impedance is not dependent on the occurrence of redox couples. The goal of impedimetric biosensors is to create amino, carboxyl, and related groups on the electrode surfaces in order to trap the antibodies. The development of an impedimetric biosensor depends heavily on this step since it guarantees the sensor's durability and repeatability. Nanomaterials have additionally been applied for these goals. The resistance to electron transfer would often rise when the antibody is discovered utilising antigens, which would ultimately lead to a reduction in capacitance (Sharma et al., 2021). Due to the fact that they have no effect on the activity or functionality of antibodies, gold nanoparticles can be employed to trap them (Chullasat et al., 2011).

In order to detect the hepatitis B virus (HBV) in blood serum, a highly sensitive label-free electrochemical impedance biosensor based on gold nanocrystals (AuNCs) has been reported. It was possible to create a useful platform based on AuNCs electrodes for testing the geno-sensitivity biosensor's for detecting HBV DNA. Field electron scanning electron microscopy (FESEM) and a high-resolution transmission electron microscope (HRTEM) were used to materialise and characterise the modified morphologies of 50–100 nm beaded AuNCs on the In_2O_3 support layer (HRTEM). Under the conditions of redox reactions, EIS was used to conduct the biosensing experiments. A unique link between the HBV DNA probe single-stranded DNA (ssDNA) and gold was formed, immobilising the ssDNA on the surface of the AuNCs. ssDNA/AuNCs biosensor readings in concentration range with an LOD were used to identify the DNA target (LOD). When the biosensor's selectivity was examined, it was found that it could clearly distinguish between complementary DNA targets and non-complementary DNA, 1-, 2-, and 3-mismatch targets. Last but not least, the potential of the electrode to detect HBV in blood serum collections was investigated, and a biosensor produced intriguing findings in the detection of HBV DNA as a viable candidate for real-world use (Shariati, 2021).

Even when adequate antifungal therapy is used, fungi continue to represent a significant portion of healthcare-acquired illnesses. One of the most common microbial species among the several fungal infections found in the healthcare environment is *Candida spp.* Fungal infections like candidemia and candidiasis cause high morbidity and mortality rates among sick hospital patients. The advantages of electrochemical biosensors over traditional methods are quick analyte detection, inexpensive, and the potential for downsizing. Lectins are proteins that bind to carbohydrates and can penetrate the cell walls of microorganisms. In this work, we suggested using wheat germ agglutinin (WGA) and concanavalin A (ConA) as yeast cell identification molecules in an impedimetric biosensor for *Candida spp.* After the molecules were put together and exposed to fungal samples, AFM photos show alterations in the biosensor surface. Results from the evaluation of four *Candida* species using EIS showed a proportionate rise in charge transfer resistance (RCT) as fungal CFU increased (*Candida krusei, Candida albicans, Candida tropicalis, and Candida parapsilosis*) (Sá et al., 2020).

16.5.4 Voltametric Biosensors

Voltammetry is a type of electro-analytical technique that measures the current produced when a potential is changed in order to determine information about an analyte. Voltametric biosensors are electrochemical devices that use voltametric sensing methods such as differential pulse voltammetry (DPV), CV, square wave voltammetry (SWV), and linear sweep voltammetry (LSV). Voltametric techniques are used in sensing platforms because of their affordability, high sensitivity, and superior selectivity (Singh et al., 2021). A biosensor based on DPV and CV techniques was created by Gupta et al. to identify anti-iTG antibodies in humans. Gold nanoparticles were altered with graphene quantum dots (GQDs) and polyamidoamine and then embedded in multi-walled carbon nanotube (MWCNT) to create the biosensor. The biosensors were successful in detecting antibody concentrations as minimal as 20–50 fg mL^{-1} (Gupta et al., 2017).

For patients to survive acute myocardial infarction (AMI), a prompt and accurate diagnosis is crucial. In addition to being useful point-of-care diagnostic tools, voltammetry-based biosensors have been designed to concurrently detect multiplexed biomarkers, including cardiac troponin (cTnI, cTnT) isoforms, and offer quick and affordable methods for monitoring these biomarkers. Enzymatic and non-enzymatic processes are defined, and several voltammetry-labelled and label-free procedures are discussed, with an emphasis on the structural specifics and comparative advantages of different approaches. It is predicted that the incorporation of nanoscale advanced components and advancement in engineering through the fusion of microfluidics and voltametric techniques would lead to the development of much more reliable and sensitive voltametric biosensors (Pourali et al., 2021).

The rise in inflammatory disorders has recently piqued scientific curiosity about the development of treatments and diagnostic technologies. Acetylsalicylic acid (ASA or aspirin) is frequently used to treat several disorders when taken in therapeutic doses. However, ASA overdoses can result in harmful side effects like stomach damage and ulcers. Therefore, it would be ideal to build quick, portable, and sensitive ASA detection tools. On a carbon electrode that had been screen printed, chitosan that had been coated with gold nanoparticles (Cs+AuNPs) self-assembled to create the electrochemical sensor (SPCE). Between 1 pg mL^{-1} and 1 g mL^{-1}, it displays a logarithmic linear correlation between its reaction and the ASA concentration. To categorise urine specimens injected with various aspirin concentrations, qualitative analysis was carried out utilising the VE-Tongue based upon metal electrodes in conjunction with two chemometric techniques (Diouf et al., 2020).

16.6 APPLICATIONS OF ELECTROCHEMICAL BIOSENSORS

Impedimetric biosensors examine the charge transfer between electrodes, while biorecognition molecules like enzymes, lectins, nucleic acids, antibodies, whole cells, etc., interact with the analyte in a given sample. The charge transfer between electrodes is monitored as the generation of current which is perceived as readable signals by the transducer. These sensors are applicable in various fields like healthcare, clinical analysis and disease diagnosis, environmental monitoring, industrial processing monitoring like the food industry. Based on these applications and types of biorecognition molecules, biosensors are classified under various categories.

16.6.1 Medical Diagnostics

Various types of biosensors based on assay enable rapid diagnostics, cost-effective testing, and rapid screening of various diseases. Biosensors can easily detect different types of pathogens like HIV, mycobacteria, hepatitis virus, etc. The most important factor considered while designing biosensor is pathogenic concentration which is diagnostically relevant and it differs in different types of body fluids where they can be found. Aydin et al. (2019) developed innovative impedimetric immunosensors for detecting interleukin 6 (IL-6), a prostate cancer marker, using polypyrrole polymer-modified indium tin oxide (ITO) electrodes. The polypyrrole polymer was first electropolymerised on ITo electrodes for the formation of an immobilised matrix for IL-6 receptor. This biosensor demonstrated high selectivity and offered a cost-effective diagnostic tool for prostate cancer with a lower limit of detection (LOD) (Aydin et al., 2019). Li et al. (2020) fabricated immunosensor having glassy carbon electrode with gold nanoparticles molybdenum graphene oxide nanocomposite for the detection of porcine epidemic diarrhoea virus (PEDV). Immobilisation of monoclonal antibody PEDV-2C11 on the modified electrode in the nanocomposite was performed. The antigen-antibody interaction at hexoferrocyanate (II/III) probe which yielding the charge transfer changer was used to measure the concentration of PEDV (Li et al., 2020).

Kang et al. (2020) constructed an impedimetric biosensor for the detection of myoglobin in human serum. The myoglobin aptamer strands were immobilised on MnCo oxide nanohybrid platform which was prepared from Prussian blue analogue (PBA) nanocubes. It exhibited LOD of 0.56 fg mL^{-1} (0.034 fM) having high sensitivity (Kang et al., 2020).

16.6.2 Food and Water Assessment

Biosensors play crucial role in identification of contaminants in beverages, water, and food products. Mohato et al. (2011) demonstrated the detection of alkaline phosphatase (ALP) in human serum using AuNPs on Au nano-dendroid and graphene oxide nanocomposite on carbon-printed electrode. Label-free EIS was evaluated for assessing the performance (Mahato et al., 2020). Ochratoxin (OTA), a mycotoxin found as contaminant in beer, was detected using impedimetric aptasensor with pencil graphite electrode. OTA aptamers complex enhanced the LOD at 0.1 ng mL^{-1} (Gökçe et al., 2020).

16.6.3 Biodefence Application

Biosensors can be utilised for biological defence application. Military purpose biosensor is used to identify biological warfare agents like bacteria, viruses, etc. Besides the constructive use of biosensors, they could be applied as warfare agents for assay of bioweapons constructed. Impedimetric sensor technology is used for detection and assay of biological agents used for warfare like *Bacillus antharsis*, *Botulinum toxin*, and many more (Hobbs et al., 2019; Tournier and Rougeaux, 2020).

16.6.4 Environmental Monitoring

Detection of heavy metals and pollutants is essential in environmental monitoring. Xie et al. (2018) constructed a sensor for Hg^{+2} detection using manganese-porphyrin-modified DNA for precipitation and polymerisation. This sensor was effective in environmental monitoring of Hg^{+2} (Xie et al., 2018). Marzo et al. (2020) constructed enzyme-based biosensor by immobilisation of horse radish peroxidise (HRP) for hydrogen peroxide detection. They used 3-D printed graphene/polylactic acid (PLA) electrodes for the immobilisation of enzymes. The activated electrodes are modified by incubating them overnight with AuNPs. Further incubation of the electrode/AuNP with HRP yielded the immobilised enzyme on electrode surface. AuNPs are used as performance enhancers of the biosensor for electron transfer between enzyme and electrode (López Marzo et al., 2020).

16.7 CONCLUSION

This chapter gives a brief insight into the principle, assembly, and applications of electrochemical biosensors along with the potential role of nanomaterials. The main principle of electrochemical biosensing for the detection of analytes and their measurement is based on charge transfer within the electrode system with respect to current, voltage, and time. Not only just one molecule, but several analytes at the same time could be analysed by building a multiplexed electrode array platform because of EIS measurements with higher sensitivity and lower LOD.

Apart from all that, several other advantages like stability, cost-effectiveness, speed, low energy requirement, label-free screening, and miniaturisation ease were seen through various types of electrochemical biosensors. Construction of sensor devices by conventional methodology and materials is reliable, but the latest breakthrough of nanomaterials and polymer technology for the fabrication of these devices sheds the light on the direction for the construction of more efficient sensing devices by researchers around the globe. Among different types, nanomaterials are used at a great scale due to the ease of monitoring antigen-antibody interaction and DNA hybridisation on the immobilised electrode surface. The different types of nanomaterials were based on their applications in several fields like medical diagnosis, environmental monitoring, food industries, etc.

Sensors based on electrochemical properties are used to detect pathogenic bacterial disease biomarkers in vitro as well as in cytological behavioural studies, drug development and other applications. Apart from the applications in medical field, the main use of electrochemical sensors came in environmental monitoring of heavy metals, pollution level indicators, and in the food industry for detecting adulterants and pathogenic bacteria. Electrochemical-based sensors are remarkable sensing devices due to their properties which try to fulfil all the needs arising every time due to new problems in methodology, materials, or new analytes. But new technology and research work is also required to meet those needs efficiently, improve the quality of detection, and build a successful device for commercialisation.

REFERENCES

Ahmad, R., Khan, M., Mishra, P., Jahan, N., Ahsan, M.A., Ahmad, I., Khan, M., Watanabe, Y., Ali, S., Furukawa, H., Khosla, A., 2020. Engineered hierarchical CuO nanoleaves based electrochemical nonenzymatic biosensor for glucose detection. *J. Electrochem. Soc.* 168, 515. https://doi.org/10.1149/1945-7111/abd515

Ahmad, R., Tripathy, N., Hahn, Y.-B., 2014. Highly stable urea sensor based on ZnO nanorods directly grown on Ag/glass electrodes. *Sensors Actuators B Chem.* 194, 290–295. https://doi.org/10.1016/j.snb.2013.12.098

Ali, A., Ansari, A.A., Kaushik, A., Solanki, P.R., Barik, A., Pandey, M.K., Malhotra, B.D., 2009. Nanostructured zinc oxide film for urea sensor. *Mater. Lett.* 63, 2473–2475. https://doi.org/https://doi.org/10.1016/j.matlet.2009.08.038

Ansari, S.G., Wahab, R., Ansari, Z.A., Kim, Y.-S., Khang, G., Al-Hajry, A., Shin, H.-S., 2009. Effect of nanostructure on the urea sensing properties of sol–gel synthesized ZnO. *Sensors Actuators B Chem.* 137, 566–573. https://doi.org/https://doi.org/10.1016/j.snb.2009.01.018

Anzar, N., Hasan, R., Tyagi, M., Yadav, N., Narang, J., 2020. Carbon nanotube – A review on synthesis, properties and plethora of applications in the field of biomedical science. *Sensors Int.* 1, 100003. https://doi.org/https://doi.org/10.1016/j.sintl.2020.100003

Arora, K., Tomar, M., Gupta, V., 2011. Highly sensitive and selective uric acid biosensor based on RF sputtered NiO thin film. *Biosens. Bioelectron.* 30, 333–336. https://doi.org/10.1016/j.bios.2011.09.026

Aydın, E.B., Aydın, M., Sezgintürk, M.K., 2019. Ultrasensitive determination of cadherin-like protein 22 with a label-free electrochemical immunosensor using brush type poly(thiophene-g-glycidylmethacrylate) modified disposable ITO electrode. *Talanta* 200, 387–397. https://doi.org/https://doi.org/10.1016/j.talanta.2019.03.082

Balusamy, B., Senthamizhan, A., Uyar, T., 2019. Functionalized electrospun nanofibers as colorimetric sensory probe for mercury detection: A review. *Sensors* 19, 4763. https://doi.org/10.3390/s19214763

Bhalla, N., Jolly, P., Formisano, N., Estrela, P., 2016. Introduction to biosensors. *Essays Biochem.* 60, 1–8. https://doi.org/10.1042/EBC20150001

Bhardwaj, N., Kundu, S.C., 2010. Electrospinning: A fascinating fiber fabrication technique. *Biotechnol. Adv.* 28, 325–347. https://doi.org/https://doi.org/10.1016/j.biotechadv.2010.01.004

Castillo-Ortega, M.M., Rodriguez, D.E., Encinas, J.C., Plascencia, M., Méndez-Velarde, F.A., Olayo, R., 2002. Conductometric uric acid and urea biosensor prepared from electroconductive polyaniline–poly(n-butyl methacrylate) composites. *Sensors Actuators B Chem.* 85, 19–25. https://doi.org/https://doi.org/10.1016/S0925-4005(02)00045-X

Chen, X., Wu, G., Cai, Z., Oyama, M., Chen, Xi, 2013. Advances in enzyme-free electrochemical sensors for hydrogen peroxide, glucose, and uric acid. *Microchim. Acta* 181, 689–705. https://doi.org/10.1007/s00604-013-1098-0

Chullasat, K., Kanatharana, P., Limbut, W., Numnuam, A., Thavarungkul, P., 2011. Ultra trace analysis of small molecule by label-free impedimetric immunosensor using multilayer modified electrode. *Biosens. Bioelectron.* 26, 4571–4578. https://doi.org/10.1016/j.bios.2011.05.029

Clark, L.C.J., Lyons, C., 1962. Electrode systems for continuous monitoring in cardiovascular surgery. *Ann. N. Y. Acad. Sci.* 102, 29–45. https://doi.org/10.1111/j.1749-6632.1962.tb13623.x

Diouf, A., Moufid, M., Bouyahya, D., Österlund, L., El Bari, N., Bouchikhi, B., 2020. An electrochemical sensor based on chitosan capped with gold nanoparticles combined with a voltammetric electronic tongue for quantitative aspirin detection in human physiological fluids and tablets. *Mater. Sci. Eng. C* 110, 110665. https://doi.org/https://doi.org/10.1016/j.msec.2020.110665

Eissa, S., Alkhaldi, S., Chinnappan, R., Siddiqua, A., Abduljabbar, M., Abdel Rahman, A.M., Dasouki, M., Zourob, M., 2020. Selection, characterization, and electrochemical biosensing application of DNA aptamers for sepiapterin. *Talanta* 216, 120951. https://doi.org/https://doi.org/10.1016/j.talanta.2020.120951

Geckeler, K.E., Premkumar, T., 2011. Carbon nanotubes: are they dispersed or dissolved in liquids? *Nanoscale Res. Lett.* 6, 136. https://doi.org/10.1186/1556-276X-6-136

Gökçe, G., Ben Aissa, S., Nemčeková, K., Catanante, G., Raouafi, N., Marty, J.-L., 2020. Aptamer-modified pencil graphite electrodes for the impedimetric determination of ochratoxin A. *Food Control* 115, 107271. https://doi.org/https://doi.org/10.1016/j.foodcont.2020.107271

Gupta, J., Arya, S., Verma, S., Singh, A.P., Sharma, A., Singh, B., Prerna, Sharma, R., 2019. Performance of template-assisted electrodeposited Copper/Cobalt bilayered nanowires as an efficient glucose and Uric acid senor. *Mater. Chem. Phys.* 238, 121969.

Gupta, S., Kaushal, A., Kumar, A., Kumar, D., 2017. Ultrasensitive transglutaminase based nanosensor for early detection of celiac disease in human. *Int. J. Biol. Macromol.* 105, 905–911. https://doi.org/10.1016/j.ijbiomac.2017.07.126

Halicka, K., Cabaj, J., 2021. Electrospun nanofibers for sensing and biosensing applications: A review. *Int. J. Mol. Sci.* 22, 221226357. https://doi.org/10.3390/ijms22126357

Hobbs, R.J., Thomas, C.A., Halliwell, J., Gwenin, C.D., 2019. Rapid detection of botulinum neurotoxins-A review. *Toxins (Basel).* 11, 418. https://doi.org/10.3390/toxins11070418

Huang, W., Cao, Y., Chen, Y., Zhou, Y., Huang, Q., 2015. 3-D periodic mesoporous nickel oxide for nonenzymatic uric acid sensors with improved sensitivity. *Appl. Surf. Sci.* 359, 221–226. https://doi.org/https://doi.org/10.1016/j.apsusc.2015.10.028

Kang, M., Li, Z., Hu, M., Oderinde, O., Hu, B., He, L., Wang, M., Fu, G., Zhang, Z., Du, M., 2020. Bimetallic MnCo oxide nanohybrids prepared from Prussian blue analogue for application as impedimetric aptasensor carrier to detect myoglobin. *Chem. Eng. J.* 395, 125117. https://doi.org/https://doi.org/10.1016/j.cej.2020.125117

Karimi-Maleh, H., Orooji, Y., Karimi, F., Alizadeh, M., Baghayeri, M., Rouhi, J., Tajik, S., Beitollahi, H., Agarwal, S., Gupta, V.K., Rajendran, S., Ayati, A., Fu, L., Sanati, A.L., Tanhaei, B., Sen, F., shabani-nooshabadi, M., Asrami, P.N., Al-Othman, A., 2021. A critical review on the use of potentiometric based biosensors for biomarkers detection. *Biosens. Bioelectron.* 184, 113252. https://doi.org/https://doi.org/10.1016/j.bios.2021.113252

Kelley, S.O., Mirkin, C.A., Walt, D.R., Ismagilov, R.F., Toner, M., Sargent, E.H., 2014. Advancing the speed, sensitivity and accuracy of biomolecular detection using multi-length-scale engineering. Nat. Nanotechnol. 9, 969–980. https://doi.org/10.1038/nnano.2014.261

Khan, Ibrahim, Saeed, K., Khan, Idrees, 2019. Nanoparticles: Properties, applications and toxicities. *Arab. J. Chem.* 12, 908–931. https://doi.org/https://doi.org/10.1016/j.arabjc.2017.05.011

Khan, M., Nagal, V., Nakate, U.T., Khan, M.R., Khosla, A., Ahmad, R., 2021. Engineered CuO nanofibers with boosted non-enzymatic glucose sensing performance. *J. Electrochem. Soc.* 168, 67507. https://doi.org/10.1149/1945-7111/ac030d

Khare, R., Yasir, M., Singh, P., Shrivastava, R., 2020. Diagnostic aids for early detection of oral squamous cell carcinoma: Concepts and emerging techniques. *Minerva Biotecnol.* 32, 20–28. https://doi.org/10.23736/S1120-4826.19.02586-2

Khosla, A., Shah, S., Shiblee, M.D.N.I., Mir, S.H., Nagahara, L.A., Thundat, T., Shekar, P.K., Kawakami, M., Furukawa, H., 2018. Carbon fiber doped thermosetting elastomer for flexible sensors: Physical properties and microfabrication. *Sci. Rep.* 8, 12313. https://doi.org/10.1038/s41598-018-30846-3

Koncki, R., 2007. Recent developments in potentiometric biosensors for biomedical analysis. *Anal. Chim. Acta* 599, 7–15. https://doi.org/https://doi.org/10.1016/j.aca.2007.08.003

Lee, Y.-H., Wong, D.T., 2009. Saliva: An emerging biofluid for early detection of diseases. *Am. J. Dent.* 22, 241–248.

Li, X., Wang, Y., Zhang, X., Gao, Y., Sun, C., Ding, Y., Feng, F., Jin, W., Yang, G., 2020. An impedimetric immunosensor for determination of porcine epidemic diarrhea virus based on the nanocomposite consisting of molybdenum disulfide/reduced graphene oxide decorated with gold nanoparticles. *Mikrochim. Acta* 187, 217. https://doi.org/10.1007/s00604-020-4166-2

López Marzo, A.M., Mayorga-Martinez, C.C., Pumera, M., 2020. 3D-printed graphene direct electron transfer enzyme biosensors. *Biosens. Bioelectron.* 151, 111980. https://doi.org/10.1016/j.bios.2019.111980

Maalouf, R., Chebib, H., Saïkali, Y., Vittori, O., Sigaud, M., Jaffrezic-Renault, N., 2007. Amperometric and impedimetric characterization of a glutamate biosensor based on Nafion and a methyl viologen modified glassy carbon electrode. *Biosens. Bioelectron.* 22, 2682–2688. https://doi.org/10.1016/j.bios.2006.11.003

Mahato, K., Purohit, B., Kumar, A., Chandra, P., 2020. Clinically comparable impedimetric immunosensor for serum alkaline phosphatase detection based on electrochemically engineered Au-nano-Dendroids and graphene oxide nanocomposite. *Biosens. Bioelectron.* 148, 111815. https://doi.org/10.1016/j.bios.2019.111815

Malhotra, S., Verma, A.K., Tyagi, N., Kumar, V., 2017. Biosensors: Principle, types and applications. *Int. J. Adv. Res. Innov. Ideas Educ.* 3, 3639–3644.

Murthy, S.K., 2007. Nanoparticles in modern medicine: State of the art and future challenges. *Int. J. Nanomed.* 2, 129–141.

Öndeş, B., Akpınar, F., Uygun, M., Muti, M., Aktaş Uygun, D., 2021. High stability potentiometric urea biosensor based on enzyme attached nanoparticles. *Microchem. J.* 160, 105667. https://doi.org/https://doi.org/10.1016/j.microc.2020.105667

Özbek, O., Isildak, Ö., Isildak, I., 2021. A potentiometric biosensor for the determination of valproic acid: Human blood–based study of an anti–epileptic drug. *Biochem. Eng. J.* 176, 108181. https://doi.org/https://doi.org/10.1016/j.bej.2021.108181

Parrilla, M., Cuartero, M., Padrell Sánchez, S., Rajabi, M., Roxhed, N., Niklaus, F., Crespo, G.A., 2019. Wearable all-solid-state potentiometric microneedle patch for intradermal potassium detection. *Anal. Chem.* 91, 1578–1586. https://doi.org/10.1021/acs.analchem.8b04877

Pavan Kalyan, G. V, Sarkar, A., Aswini, C., Chakraborty, P., 2021. Characterization of CNT and It's carrier mobility dependency on diameter in Back-Gated CNT-FET. *Mater. Today Proc.* 45, 3869–3873. https://doi.org/https://doi.org/10.1016/j.matpr.2020.06.274

Pourali, A., Rashidi, M.R., Barar, J., Pavon-Djavid, G., Omidi, Y., 2021. Voltammetric biosensors for analytical detection of cardiac troponin biomarkers in acute myocardial infarction. *TrAC Trends Anal. Chem.* 134, 116123. https://doi.org/https://doi.org/10.1016/j.trac.2020.116123

Randviir, E.P., Banks, C.E., 2013. Electrochemical impedance spectroscopy: An overview of bioanalytical applications. *Anal. Methods* 5, 1098–1115. https://doi.org/10.1039/C3AY26476A

Ruchi, K., Singh, P., Yasir, M., Joseph, A., Rahul, S., 2020. Emerging trends of non-invasive nano-systems in disease detection through saliva and sweat. *Res. J. Biotechnol.* 14, 131–136.

Sá, S.R., Silva Junior, A.G., Lima-Neto, R.G., Andrade, C.A.S., Oliveira, M.D.L., 2020. Lectin-based impedimetric biosensor for differentiation of pathogenic candida species. *Talanta* 220, 121375. https://doi.org/https://doi.org/10.1016/j.talanta.2020.121375

Shariati, M., 2021. Impedimetric biosensor for monitoring complementary DNA from Hepatitis B Virus based on gold nanocrystals. *J. Electrochem. Soc.* 168, 16512. https://doi.org/10.1149/1945-7111/abdc72

Sharma, A., Ahmed, A., Singh, A., Oruganti, S.K., Khosla, A., Arya, S., 2021. Review—Recent advances in tin oxide nanomaterials as electrochemical/chemiresistive sensors. *J. Electrochem. Soc.* 168, 27505. https://doi.org/10.1149/1945-7111/abdee8

Singh, A., Sharma, A., Ahmed, A., Sundramoorthy, A.K., Furukawa, H., Arya, S., Khosla, A., 2021. Recent advances in electrochemical biosensors: Applications, challenges, and future scope. *Biosensors* 11, 336. https://doi.org/10.3390/bios11090336

Tournier, J.-N., Rougeaux, C., 2020. Anthrax toxin detection: From in vivo studies to diagnostic applications. *Microorganisms* 8, 11. https://doi.org/10.3390/microorganisms8081103

Vigneshvar, S., Sudhakumari, C.C., Senthilkumaran, B., Prakash, H., 2016. Recent advances in biosensor technology for potential applications – An overview. *Front. Bioeng. Biotechnol.* 4, 11. https://doi.org/10.3389/fbioe.2016.00011

Wang, X.-X., Yu, G.-F., Zhang, J., Yu, M., Ramakrishna, S., Long, Y.-Z., 2021. Conductive polymer ultrafine fibers via electrospinning: Preparation, physical properties and applications. *Prog. Mater. Sci.* 115, 100704. https://doi.org/https://doi.org/10.1016/j.pmatsci.2020.100704

Xie, S., Tang, Y., Tang, D., Cai, Y., 2018. Highly sensitive impedimetric biosensor for Hg^{2+} detection based on manganese porphyrin-decorated DNA network for precipitation polymerization. *Anal. Chim. Acta.* 1023, 22–28. https://doi.org/https://doi.org/10.1016/j.aca.2018.04.019

Yang, Z., Liu, X., Zhang, C., Liu, B., 2015. A high-performance nonenzymatic piezoelectric sensor based on molecularly imprinted transparent TiO_2 film for detection of urea. *Biosens. Bioelectron.* 74, 85–90. https://doi.org/10.1016/j.bios.2015.06.022

Zeng, S., Baillargeat, D., Ho, H.-P., Yong, K.-T., 2014. Nanomaterials enhanced surface plasmon resonance for biological and chemical sensing applications. *Chem. Soc. Rev.* 43, 3426–3452. https://doi.org/10.1039/C3CS60479A

17 Influence of Diffusion, Convection, and Reaction Time Scales on Cyclic Voltammetry Targeting Efficient Electrochemical Sensor Design

Shirsendu Mitra and Bharti Saini

17.1 INTRODUCTION

Electrochemistry is a branch of science, where we study the interconversion between chemical energy and electrical energy. Long back, in the year 1834, Faraday, in his article on electrochemistry,[1] gave the quantitative relation between the amount of electricity and the extent of a chemical reaction. From the fundamental knowledge of electrochemistry, the concept of cells was introduced. Typically, electrochemical cells are broadly classified into two categories: galvanic cells[2,3] and electrolytic cells.[4,5] In galvanic cells, like fuel cells and voltaic cells, chemical energy is exploited in making electrical energy. On the other hand, electrolytic cells are those where electrical energy is utilized in doing a chemical reaction. A few examples of electrolytic cells are electrolysis of water, recharging batteries, manufacturing of sodium, among others. The quantitative estimation of electrochemical potential generation for a galvanic cell, wherein an ideal reversible reaction takes place, is given by the popular Nernst equation.[6,7] Since the inception of electrochemical research to date, there have been a lot of technological and scientific advancements using the Nernst equation for various calculations. The electrochemical theory is utilized in multiverse fields, including corrosion science,[8–10] electrochemical sensing of multiple components, advanced galvanic, and electrocatalytic cell design.

Whenever a metal or other conducting material is dipped into an electrolytic solution,[11] a potential is developed around it. Quantifying the absolute potential that is formed around the dipped electrode in reference to the bulk solution is a challenging task to do. For measurement purposes, if another reference electrode[12,13] is inserted into the bulk solution, the same would also develop a potential around it. Hence, the absolute potential at both the electrodes is difficult to measure with the existing instruments available to date. Nevertheless, even the difference in potential between any two electrodes also has significant importance as far as various practical applications of electrochemistry are concerned. In relevance, it has been targeted to keep the potential of a particular electrode approximately unchanged so that any change in the other electrode can be tracked. The electrode whose potential is aimed to keep constant is called the reference electrode in electrochemistry. Whenever an active metal or some suitable conducting material is placed into an electrolytic solution, a redox reaction, i.e., transfer of electron, takes place, and the same forms a potential, and the same is quantified by Nernst equation. Various commercial electrodes are available, which are used as reference electrodes, and those are standard hydrogen electrodes (SHE),[14–16] saturated

DOI: 10.1201/9781032654874-17

calomel electrodes (SCE),[17,18] and Ag/AgCl[19,20] reference electrodes. Those electrodes are used in several instruments such as pH meter, potentiometric titration instrument, potentiostat-galvanostat, alternating current-dirrect current (AC-DC) analyzer instrument, among others.

Of late, researchers are using potentiostat-galvanostat to perform a host of electrochemical studies, which include cyclic voltammetry (CV), electrochemical impedance spectroscopy (EIS), linear sweep voltammetry (LSV), stripping voltammetry (SV), and others. In general, three different electrodes are utilized to do electrochemical studies in a potentiostat-galvanostat, and those are working electrode, reference electrode, and a counter electrode. As it is mentioned earlier, the reference electrode is used as a standard to measure the potential of the working electrode. In contrast, the working electrode is used to measure qualitative and quantitative changes happening on the electrode due to various faradic or non-faradic processes. In a faradic process, an electrochemical reaction takes place, and the same makes changes in electron generation (consumption) leading to oxidation (reduction) processes.

On the other hand, in a non-faradic process, despite a reaction, a change of potential takes place as a result of charge deposition or charge dissipation. The counter electrode[21,22] facilitates the flow of excessive charge that gets accumulated on the electrode due to faradic or non-faradic processes. High electrical conductivity materials like gold and platinum are used as counter electrodes. All these electrochemical studies are complex and can get affected due to the variation of multiple factors. Parameters like working electrode material, heterogeneous reaction kinetics, external temperature, the concentration of bulk electrolyte, diffusivity of ions, electrode polarization, presence of any local convection in the system, possibility of electrolysis, number of electron transfer in the process affect CV, EIS, and other electrochemical studies to a considerable extent. In an electrochemical cell, several changes take place when we study CV. And deciphering the same is very much essential to explain the ongoing activities, which may be a part of the inevitable research outcome, and hence, to understand the effect of all the individual parameters is significant.

Theoretical studies are the most reliable tool to understand any process in depth. In relevance, there are a few reports which studied CV and EIS incorporating an electric field and transport of species equations. Brown[23] intelligently explains the mechanism of developing CVs directly with the help of Microsoft-Excel-based calculations. In another report, Abdullah et al. tried to explain different kinds of CVs obtained from various experiments, primarily which are obtained in different kinds of working electrodes for biosensing. Also, it was categorized that protein-protein interaction shows an entirely different nature than that of other electrochemical sensing. Elgrishi et al.,[24] in their early report, explained the influence of varying reaction kinetics and reaction mechanism on CVs. Briefly, it can be concluded that both theoretical and experimental efforts have been put forth to explain several interesting observations.

Despite the availability of extensive literature on CV study, we realized that a chapter on the importance of kinetic and dynamic parameters on the characteristics of CV will be helpful for electrochemists. The current chapter gives special emphasis on the nature of CV curves as they are important for scientists working on electrochemical sensor development. The variation of reaction rate, diffusion, and local convection are inevitable for electrochemical biosensors, and hence, understanding their role is an important task. Moreover, along with the scientists working in electrochemical sensors, this chapter will guide electrochemists working in diverse areas.

17.2 PROBLEM FORMULATION

In this section, a faradic electrochemical system for studying CVs has been modeled with COMSOL Multiphysics. The rectangular space in Figure 17.1c schematically shows a part of an electrochemical cell near the working electrode. The left edge of the rectangle represents the working electrode surface, and the right edge represents the bulk solution. However, in actual situations, the electrode may be considered to be a cylindrical one, and hence, a 360° rotation of the left edge of the rectangle around a center, which is an electrode radius distance away from the electrode surface, would depict the real geometry. Furthermore, the distribution of the electrolyte and solute is considered to be symmetrical in the azimuthal direction, and the same allows taking the gradient of any dependent variable with respect to the azimuthal direction is insignificant. Different kinds of situations are virtually created near the working electrode by changing diffusivity ratios of the component ions, local convection near the working electrode, and reaction rate kinetics of the oxidation-reduction couple. For all the simulations of the present chapter, a 2D Cartesian coordinate system has been chosen. In this formulation, x and y represent two mutually perpendicular directions in the Cartesian frame. The stoichiometric equation of the reaction that takes place at the working electrode is, 'Ox', 'Red'. In this case, the concentrations of the oxidant and reductant are symbolized as cOx and cRed.

Further, a superscript 'b' to the symbols represents bulk concentration. The diffusivities of the 'Ox' and 'Red' are symbolized as D_{Ox} and D_{Red}. The density and viscosity, are represented by the symbols ρ and μ, respectively.

In this formulation, we have solved the species transport equation for the concentration of the 'Ox' and 'Red' throughout the solution space. The unsteady state transport equation is given as,

$$\frac{\partial c_i}{\partial t} + \mathbf{u} \cdot \nabla c_i = D_i \nabla^2 c_i \tag{17.1}$$

where c_i and D_i stand for concentrations and diffusivities of the oxidant and reductant, i.e. 'i' represents either 'Ox' or 'Red'. The symbol '**u**' represents a velocity vector that may appear due to any local convection near or away from the electrode surface. In this case, despite solving complicated fluid flow inside an electrochemical cell or near the electrode surface, simple flow profiles are taken where fluid either flows from the electrode surface to the bulk or it comes back from the bulk liquid to the electrode surface. In this formulation, the electrode surface is taken as $x=0$, whereas the $x=L$ is considered as the starting point of the bulk liquid. The flow profile that has been taken for the electrode to the bulk and bulk to the electrode is expressed as $u = u_0(1 - x/L)$ and $u = u_b(x/L - 1)$, respectively. Here, u_0 is the maximum convective flow.

Along with the heterogeneous reaction that takes place at the electrode surface, we also consider a case where the bulk reaction takes place, and the same influences the profile of ions in the bulk as well as at the electrode. The bulk reaction, in this case, is assumed to follow first-order irreversible rate kinetics with stoichiometry, $Ox \rightarrow P$, having rate constant k_1. Alongside bulk reaction, a competing chain reaction, i.e. Ox Red $\rightarrow P$, has been taken into account as a specific case basis, where the product (P) formation from 'Red' follows first-order irreversible rate kinetics with rate constant k_2.

17.3 BOUNDARY CONDITIONS

Physical conditions pertinent to the modeled problem, which are stated in Section 17.2, are translated into the boundary conditions. At the electrode surface, Faraday's law of electrolysis, which relates the flux of the components ('Ox' or 'Red') with current density has been imposed. This law serves as the electrode boundary condition and has the mathematical form, $-n \cdot N_i = v_i i_{loc}\, nF$. Here, the term i_{loc} represents faradic current density and is calculated from the famous Butler–Volmer[25,26] equation.

$$i_{loc} = nFk_0\left(c_{Ox}\exp\left[(n-\alpha_c)F\eta RT\right] - c_{red}\exp\left[-\alpha_c RT\right]\right) \tag{17.2}$$

In equation 17.2, the symbols n, F, k_0, α_c, R, and T represent the number of electron transfers in the oxidation or reduction reaction, the Faraday constant, the heterogeneous reaction rate constant, the cathodic transfer coefficient, the universal gas constant, and absolute temperature, respectively. At the bulk solution, the concentration of the components is kept constant, i.e. at $x=L$, $c=c^b$, as the effect of the electrode reaction would not affect the concentration away from the electrode surface. The characteristic length is calculated from the following formula, as $L = \sqrt{\frac{D_{Ox}^2(E_{max} - E_{min})}{V}}$ the equilibrium concentration is related to these parameters, and the same is reported in the literature. Here, E_{max} and E_{min} represent the maximum and minimum scan potential. The non-faradic contribution is taken as a constant capacitance near the electrode.

17.4 NUMERICAL METHODOLOGY

In this chapter, we employ COMSOL Multiphysics, a finite element-based CFD software, to solve the equation 17.1 with the appropriate boundary conditions as stated before. Triangular mesh elements are taken for the discretization and calculations at each grid point. The entire solution space has approximately 50,000 mesh elements, essential for a reliable grid-independent solution with parameters taken for simulation. The solver PARDISO has been deployed for a robust computation. In this time-dependent study, the second-order backward difference method is utilized, having ~10^{-4} seconds of each time step.

17.5 DISCUSSION

17.5.1 Introduction to Voltammetry

Voltammetry is a common characterization tool for various kinds of electrochemical cells targeting a rudimental understanding of multifarious fields like corrosion, electrochemical bio-sensing,[27,28] electrodeposition,[29,30] battery-supercapacitor modeling[31,32] and others. In general, a three-electrode electrochemical cell setup is used for CV study, and a similar arrangement is envisioned theoretically in this present work. In relevance, it is worth remembering that electrochemical systems are broadly categorized into faradic (involves a chemical reaction) and non-faradic (without a chemical reaction) processes. For the present work, we have focused on a faradic process that leads to the appearance of oxidation and reduction peaks in CV. A small change in characteristic CV curve is important to scientists and engineers working with multifaceted targets, and hence, understanding the theory of the same is very important. This theoretical journey throws light into changes that occur in CVs in the presence of unavoidable experimental parameters including component diffusion, various kinds of reaction, convection, and scan time. In a CV, the potential applied between the working electrode and the reference electrode follows a triangular waveform (refer to Figure 17.1b). To be more specific, initially, the potential sweeps from zero to a maximum positive potential ($+E_{max}$), and then it goes in the opposite direction with a linear fall to meet zero potential. After that, the potential continues up to a maximum negative potential ($-E_{max}$), and finally, it changes its direction again and reaches zero potential. The current versus potential map obtained in this electrochemical characterization process is known as CV. In an electrochemical measurement instrument, the cycles of the potential ramp can be repeated as many times as required. The conventional duck-shaped schematic of a CV is shown in Figure 17.1a. Figure also shows two different peaks in the CV, which appear due to oxidation and reduction processes and hence termed oxidation and reduction peaks, respectively. It is known that the potential of the reference electrode in a cell is fixed, and hence, any changes in the applied potential between the working and the reference electrode are sensitive to the working electrode. The rate at which the externally applied potential is varied between the working and the reference electrode is called the scan rate (υ) of the system, which is measured in volts per sec (V/s). Controlling the potential of the working electrode with respect to the reference electrode is equivalent to controlling the energy of the electrons within the working electrode. When a negative potential is applied to the working electrode (i.e. by connecting the negative terminal of the power source to the working electrode) during the negative swift, the energy of the electrons is raised. With the increase in the negative potential, a point is reached when the energy of the electrons in the working electrode reaches the highest level. At this point, the electrons from the higher energy state from the working electrode transfer to the vacant electronic states of the electrolyte. Thus, the species in the working electrode get oxidized (loses electrons to the electrolyte) and the species in the electrolyte get reduced (gain electrons from the working electrode). This flow of electrons takes place with a concurrent chemical reaction at the electrode/electrolyte interface and current flows in the external circuit. The current observed during the negative swift of the potential is known as the reducing current. Similarly, when a positive potential is applied to the working electrode (i.e. the positive terminal of the external power supply is connected to the working electrode) during the positive swift of the potential, the energy of the electrons within the working electrode is lowered. During the positive potential swift, a point is reached when the electrons on solutes in the electrolyte find more favorable energy at the working electrode and, hence, transfer there. In this process, the solute species get oxidized (transfer electrons to the working electrode) and the species on the working electrode get reduced (gain electrons from the electrolyte). The current observed in this process is termed oxidizing current. The critical potential at which the flow of electrons between the electrode and the electrolyte occurs is related to the standard potential E^0 of the specific chemical reactions taking place in the cell.

17.5.2 Effect of Diffusion of Species on the Peak Current

Diffusion controls most of the electrochemical responses on the working electrode. It is crucial to decipher the role of component diffusivities on CVs to explore the processes that go on a working electrode. Tracking the changes on the working electrode is the prime route to quantify an analyte in electrochemical processes. In this section, we talk about the role of diffusivities on CVs.

The monoelectronic system, which is considered in this work follows the redox reaction, $Ox + e \leftrightarrow Red$. In electrochemistry, for the calculation of the cell potential standard Nernst equation, $E = E^0 + RTnF_{red}^{Ox}$, is followed. Here, R is the universal gas constant, T is the temperature of the system, n is the number of electrons taking part in the chemical reaction, and F is the Faraday's constant. The standard cell potential E^0 is equal to the cell potential for a case [Ox] = [Red], which is the thermodynamic equilibrium condition. However, CV process is a bit complicated compared to a single-cell configuration.

During the measurement of CV, when a negative potential is scanned to the working electrode (refer to the path *abc* in Figure 17.1a), the 'Ox' species present in the electrolyte near the electrode gets reduced, and thus depletion of 'Ox' species appears near the electrode region. This forms a concentration gradient of the 'Ox' species between the electrode surface and the bulk solution and subsequently causes a diffusion of the species from the bulk solution toward the electrode surface. The endpoint of path *abc* in Figure 17.1a corresponds to the peak cathodic current, i_{pc}, and is controlled by the maximum delivery of 'Ox' species from the bulk solution to the electrode surface. As the 'Ox' species, which diffuses from the bulk solution, gets reduced near the electrode surface, a volume of solution containing the 'Red' species continues to grow at the electrode surface.

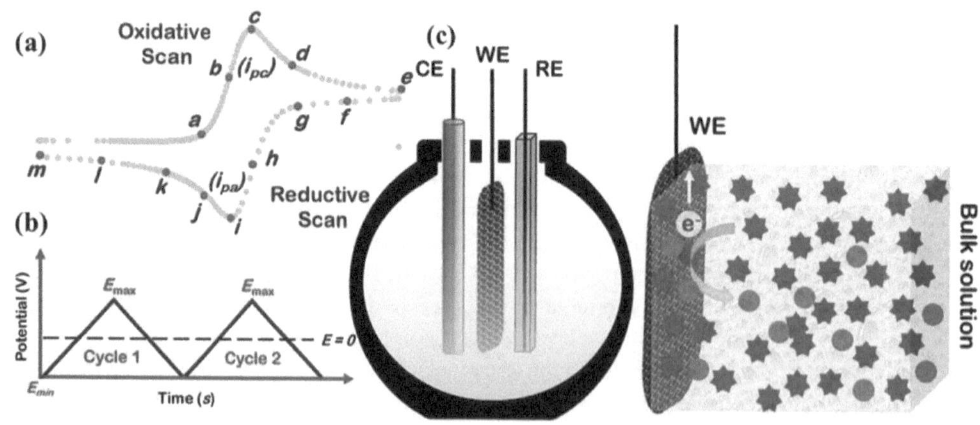

FIGURE 17.1 Plot (a) schematically shows a typical CV curve for a monoelectronic Faradic system. Plot (b) shows a triangular waveform that is fed externally during a CV study. Plot (c) depicts an experimental arrangement of an electrochemical cell and ion distribution near the working electrode.

During this negative sweep, the conversion of 'Ox' to 'Red' continues and after a critical time; this growth hinders further transport of the 'Ox' species to the electrode surface. Once the hindrance reaches its peak limit, the current starts decreasing (refer to the path *cde* in Figure 17.1a) with the increase in the negative potential, until the switching potential is reached. At the switching potential, point *e* in Figure 17.1a, the polarity of the externally applied voltage is reversed and positive potential starts continuing to the working electrode (refer to the path *efghi*, in Figure 17.1a). It is to be kept in mind that during the reduction process, the concentration of the 'Red' species was high at the electrode surface as compared to the '*Ox*' species. During the anodic sweep, the 'Red' species near the electrode surface gets re-oxidized to 'Ox'. In this case, the concentration of 'Ox' near the electrode again starts increasing, and hence diffusion of 'Ox' is observed from the electrode surface to the bulk solution (i.e. opposite direction compared to the cathodic sweep). The point *f* in Figure 17.1a corresponds to the peak anodic current, i_{pa}, and the same appears at a potential value which is collectively governed by the standard cell potential of the system and also on the diffusion kinetics of the component ions.

In an ideal situation if we consider the diffusivities of both the components are similar, then diffusion of both the species experience similar hindrance of motion, and we get oxidation and reduction peaks at a certain voltage as governed by the Nernst Equation. However, in real CVs, most of the time we see a certain distance between oxidation and reduction peaks due to the difference in diffusivities. In this work, we have done a set of systematic parametric studies, which take different ratios of diffusivities. In set 1, the diffusivity of 'Ox' is higher than 'Red', and the vice versa has been considered for the other set. For the set 1, where D_{Ox}/D_{Red} is more than one, the ratio of D_{ox}/D_{Red} is varied from 1 to 10, and for the other set, where D_{ox}/D_{Red} is less than 1, the reverse ration (D_{Red}/D_{ox}) range is taken from 1 to 10. Figure 17.2a–d show CVs of set 1 with varying υ 0.001. 0.01, 0.1, and 1 Vs^{-1}, respectively. Each line in a single subplot indicates the difference in D_{ox}/D_{Red} ratio from 1 to 10. Further, Figures 17.2e–h show CVs for set 2 having υ 0.001. 0.01, and 0.1 Vs^{-1}, respectively, and different lines in subplots again indicate D_{ox}/D_{Red} ratio from 1 to 0.1. An initial observation of the plots reveals that CVs get affected more if the D_{ox}/D_{Red} ratio is incremented, keeping the diffusivity of the reductant constant. For the set 2, where the effect of reverse ratio is shown, it is seen that CVs marginally got affected by the change of D_{Red}/D_{ox} ratio. In fact, for set 1, for a particular υ, the peak current changes gradually with the change in D_{ox}/D_{Red} ratio, whereas there was almost no change in peak current values for set 2. Hence, it can be stated that it is the diffusivity of the 'Ox' which determines peak currents in a CV curve, while other parameters are kept unchanged. A close look into the Butler–Volmer equation can also foretell this observation. Further, from the observation, it can be stated that in the Randles–Sevcik equation,[24] the diffusivity should be the diffusivity of '*Ox*' and not 'Red'. The variation of the diffusivity of 'Red' was almost non-responsive on peak current. In this regard, reading the chapter by Brown et al.[33] would guide the readers in quantitative calculations of peak current values. There are more observations worth mentioning in this context. It is observed that for both sets, with an increase in υ, the peak magnitude of peak currents overshoots. In the case of set 1, with the increase in D_{ox}/D_{Red} ratio, the voltage corresponding to peak oxidation current (i_{pc}) gradually shifts toward the right axis, whereas voltage corresponding to peak reduction current (i_{pa}) gradually shifts toward the origin. In the case of set 2, the opposite trend is observed although the magnitude of change is not as much as it is in set 1. Eventually, the time scale of a particular distance coverage for 'Ox' and 'Red' differs with the change in diffusivity ratio, which subsequently causes the distancing between two peaks. As mentioned earlier, the maximum peak current in a CV can be well explained with Randles–Sevcik equation[24] and the preliminary observation that we have obtained for set 1 matches with this equation. However, it is to be noted that the relative effect of component diffusivities on the CVs can be well understood from the present study.

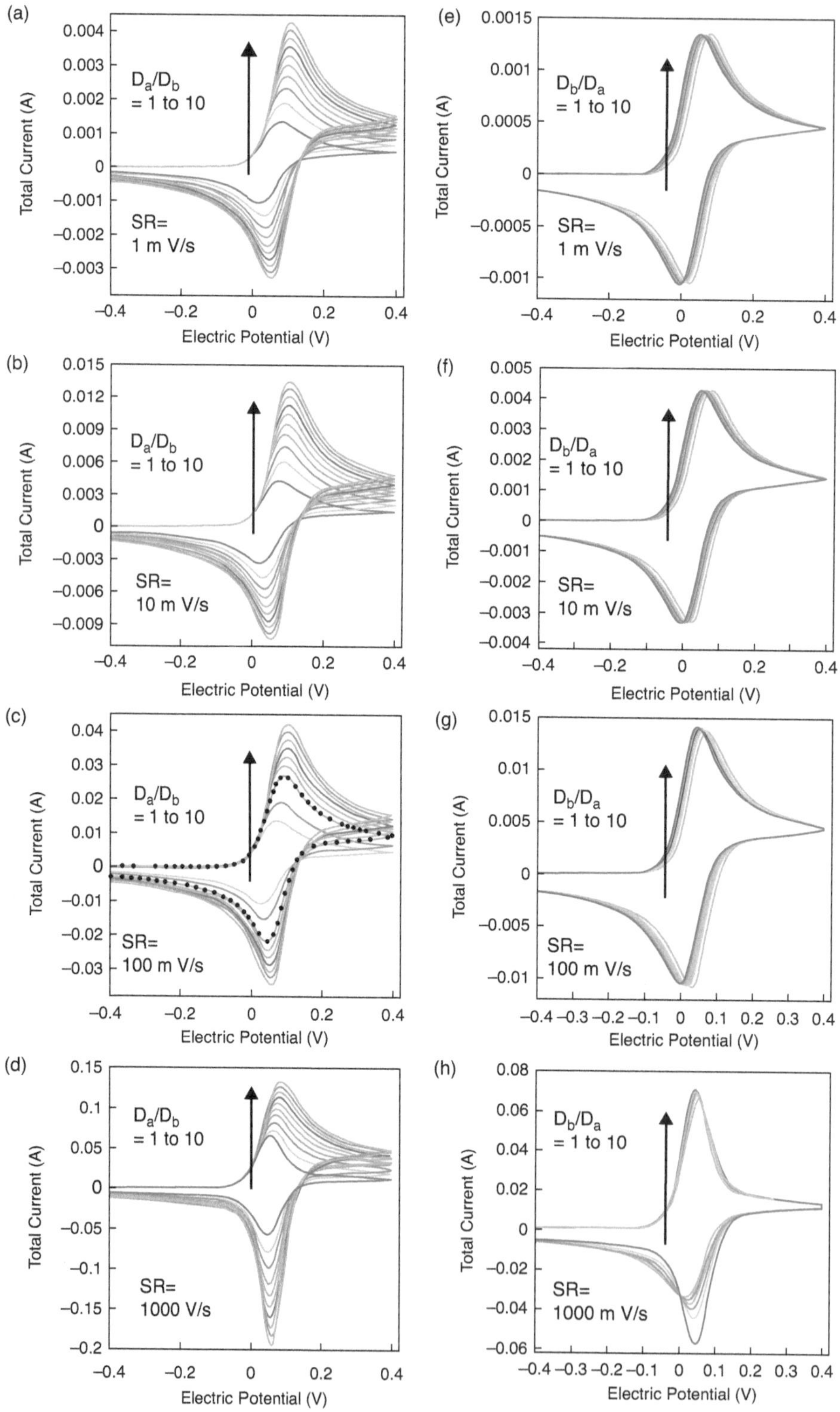

FIGURE 17.2 Plots (a–d) show CVs for scan rates 1, 10, 100, and 100 mV s^{-1}, respectively when '*Ox*' has higher diffusivity compared to '*Red*'. Different lines inside each plot indicate variations of diffusivity ratio (D_{Ox}/D_{Red}). Plots (e–h) show CVs for υ values 1, 10, 100, and 100 mV s^{-1}, respectively, when '*Ox*' has lower diffusivity compared to '*Red*'. Different lines inside each plot indicate variation of diffusivity ratio (D_{Red}/D_{Ox}) 1–10.

17.5.3 Concentration Profiles during CV

A detailed understanding of Butler–Volmer equation tells that the imbalance of the component concentrations changes the current in a CV. An exclusive analysis reveals that heterogeneous chemical reaction on the electrode surface changes the concentration of the components on the electrode surface, and the same changes with applied external potential and, hence, with the progress of scan time. Eventually, during a complete cycle, both the forward and backward reactions take place at once per cycle rate. During the forward (oxidation) reaction ('Ox' to 'Red') the concentration of the 'Ox' decreases at the electrode surface reaching a minimum before increaseing again during the backward (reduction) reaction ('Red' to 'Ox'). In addition to this, the concentration gradient that forms across the electrode and bulk solution initiate the component diffusion guided by the transient gradient and component diffusivities. In the combination of all these effects, the concentration profile varies spatiotemporally with respect to certain experimental parameters.

These parameters hold the ability to alter CVs to various possible extents and the same bears a lot of information about the electrochemical system. In this section, we show the concentration of 'Ox' and 'Red' near the electrode surface for two different diffusivity ratios, $D_{ox}/D_{Red}=5$ and $D_{Red}/D_{ox}=5$ while keeping the υ at 100 mV s^{-1}. For all these simulations, the applied voltage is scanned in the range [−0.4, 0.4], and hence, the total scan time is $4E_{max}/\upsilon=16$ s. Figure 17.3a and b show concentration profiles of 'Ox' and 'Red' near the working electrode surface for $D_{ox}/D_{Red}=5$, and $D_{Red}/D_{ox}=5$, respectively. Different solid upward lines, as shown by blue, orange, green, and pink colors, in both the plots visualize concentration profiles for 'Ox' near the electrode surface for times 4, 7, 10, and 12, respectively. Similarly, solid downward lines having colors red and black show the profile of 'Ox' at 12 and 14 seconds. The profiles shown with similar colors but in symbols indicate the concentration of 'Red' at similar time steps. A close insight into Figure 17.3a shows that during a single scan cycle, the concentration of c_{ox} (c_{Red}) reduces (increases) as we move away from the electrode surface for up to 8 seconds and after that, the concentration of c_{Red} (c_{ox}) increases (reduces). It is to be remembered that the scan starts at $E=-0.4$ V and goes for an increase in voltage (positive sweep), and during this ramp up, the oxidation reaction takes place at the electrode surface, and hence, the concentration of the 'Ox' (Red) is found increasing (diminishing) as we move away from the electrode surface. During the negative voltage sweep, the opposite reaction takes place, and hence, after 8 s, the concentration of the 'Ox' (Red) diminishes (increases) as one moves away from the electrode surface. In the case of Figure 17.3b, a similar trend is observed with a change in

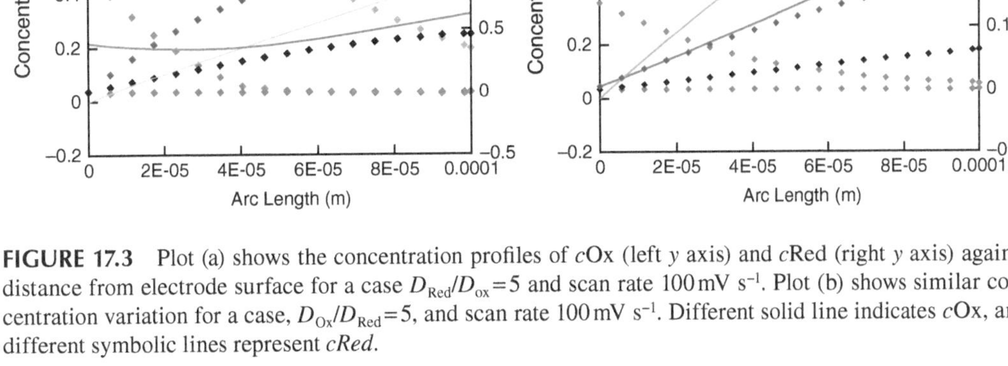

FIGURE 17.3 Plot (a) shows the concentration profiles of *c*Ox (left *y* axis) and *c*Red (right *y* axis) against distance from electrode surface for a case $D_{Red}/D_{ox}=5$ and scan rate 100 mV s^{-1}. Plot (b) shows similar concentration variation for a case, $D_{Ox}/D_{Red}=5$, and scan rate 100 mV s^{-1}. Different solid line indicates *c*Ox, and different symbolic lines represent *cRed*.

magnitude. In fact, different diffusivity ratio changes the kinetics and thereby we see a change in the concentration profile. The take-home message from this curve is once we know the concentration of both the products at the electrode surface at any point in time, we can get the current using Bulter–Volmer equation, and repeating the same process at different times gives different current values. These datasets form the CV plot. These results would certainly help readers in understanding the CV plot in a piecewise manner.

17.5.4 Effect of Reactions

In an electrochemical cell, there might be reactions, which take place apart from the desired heterogeneous catalysis reaction on the electrode surface. In this section, we discuss the possibilities of two types of reactions, which affect the CVs to a considerable extent. In general, in an ideal condition, the electrochemical reaction is expected to occur only at the electrode surface. A bulk reaction can take place due to a host of reasons including etching out of the electrode material to the bulk solution, inherent homogeneous reactions, and others.

In such scenarios, the concentration profiles of the components are disrupted, which in turn affects the CVs. In this section, to do mathematical analysis, we consider a bulk reaction that converts the 'Ox' to a product with a first-order reaction rate kinetics. Figure 17.4a shows the CVs having bulk reactions and the lines in the plot show the effect of different reactions k_1 0.01, 0.02, 0.03, 0.04, and 0.05 s^{-1}, respectively. From the figure, it is seen that with the increase in the rate of the bulk reaction, the magnitude of both oxidation and reduction peaks decreases. It is assumed that the bulk reaction which is undergoing here generates some product that does not contribute to the current. In mathematical terms, in the Butler–Volmer equation, the concentration of this product does not appear. Further, an increased reaction rate leads to the destruction of c_1 throughout the zone, and subsequently, it reduces the generation of c_2. As a combined effect of all the facts, both the oxidation and reduction peaks' intensity gradually go down.

In another case, as shown in Figure 17.4b, we consider a reaction that takes place throughout the bulk medium, including the electrode surface, as per the following stoichiometry, $Ox+e \longleftrightarrow Red \rightarrow P$. It is to be remembered that in this chain reaction, the first step is a reversible one; however, the second step is an irreversible one. Based on the rate constant of all the reaction steps, the fate of equilibrium between 'Ox' and 'Red' depends. In fact, when the 'Red' vanishes at a faster rate

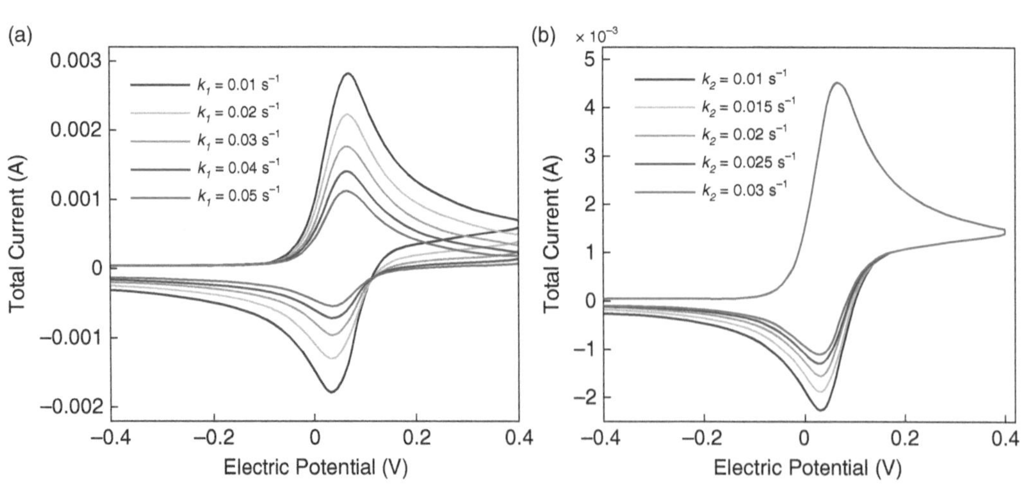

FIGURE 17.4 Plot (a) shows the CV curves in the presence of bulk reaction, $Ox+e \leftrightarrow Re$, that is decomposing the component '*Ox*' throughout the cell. Plot (b) shows similar effect in the presence of a competing reaction which follows the stoichiometry, $Ox+e \leftrightarrow Red \rightarrow P$. The simulation parameters for this simulation are given in Table 17.1.

TABLE 17.1
Values or Range of Values of Different Physical Parameters Used in Simulations

Serial No.	Symbol (s)	Description	Unit	Value/Range
1	D_i	Diffusivities of 'Ox' and 'Red'	m^2s^{-1}	$10\times^{-9}$ to $10\times^{-8}$
2	c_i	Concentrations of 'Ox' and 'Red'	mol m^{-3}	–
3	u_0	Maximum convective flow velocity	ms^{-1}	
4	F	Faraday constant	CV^{-1}	986,500
5	k_0	Heterogeneous reaction rate constant	Dimensionless	10^{10}
6	k_1	First-order rate constants for the bulk reaction	s^{-1}	
7	k_2	First-order rate constant for competing for a chain reaction	s^{-1}	
8	Y	Scan rate	Vs^{-1}	0.001–0.1
9	c_i^b	Bulk ion concentration of 'Ox' and 'Red'	mmol L^{-1}	1 and 0
10	E_{min}	Start potential	V	–0.4
11	E_{max}	Switching potential	V	0.4
12	L	$L = \sqrt{\frac{D_{Ox}^2 (E_{max} - E_{min})}{V}}$	μm	–

than that of the formation of 'Red', then we actually miss the reduction peak (the bottom one). Figure 17.4b shows that with an increase in reaction k_2 from 0.01 to 0.03 s^{-1} with an interval of 0.005 s^{-1}, the reduction peak (refer to the bottom peak in the figure) gradually diminishes, and at the rate constant 0.05 s^{-1}, the peak actually disappears. The disappearance of the peak indicates that the conversion to 'P' is so fast that no c_{Red} remains on the electrode surface for the reduction reaction to occur. During this calculation, all other simulation parameters are kept constant, and the simulation parameters for this particular case are given in Table 17.1. For this case, we only show the effect of the reaction rate maintaining constant υ, zero u_0, and other parameters constant condition.

17.5.5 Effect of Convection

It is quite relevant that in the proximity of the electrode surface, a lot of factors including reaction, bubbling, mechanical agitation of the electrode can cause local fluid flow. This local fluid flow has the strength to incur a convective mode of mass transfer and the same is potential enough to alter the CV to a considerable extent. Sometimes, during a CV experiment we come across different kinds of unexpected plots, and finding out the reason behind the same becomes difficult. In this work, we systematically explain the effects of all possible parameters to guide an experimentalist to take appropriate precautions during an experiment. Also, a proper understanding of all the effects assists in explaining a CV curve. We have done an analysis taking two different kinds of fluid flows happening near the electrode surface. In the first case, a fluid flow is taking place which originates from the electrode surface and goes to the bulk solution. Figure 17.5a shows the CVs for various convective strengths having maximum speed (u_0) 0.0002, 0.0004, and 0.0006, respectively. Here, we see with the increase in convection strength the ions irrespective of 'Ox' and 'Red' wash away from the electrode surface, and hence, both the oxidation and reduction peaks start diminishing. Moreover, due to the relatively high strength of convection, the effect of weak diffusion has vanished and the same is reflected in the distance between oxidation and reduction peaks, which shows hardly any space difference. On the other hand, in the presence of the opposite fluid flow, from the bulk solution to the electrode surface, further ion migrations are facilitated on the electrode surface and the same cause typical changes in the CV peaks. Figure 17.5b shows CVs for different convective strength which includes various maximum speeds such as 0.0002, 0.0004, 0.0006, 0.0008, and

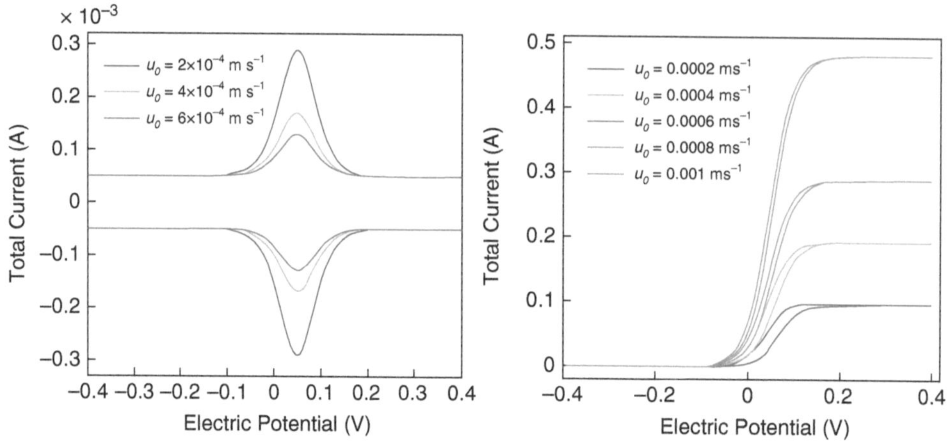

FIGURE 17.5 Plot (a) shows the CV curves in the presence of local convective flow that is taking place from the electrode surface to the bulk near the working electrode. Plot (b) shows a similar effect of the convective flow, where the flow is happening from the bulk solution to the working electrode surface. The simulation parameters for this simulation are given in Table 17.1.

0.0010, respectively. Interestingly, for this case, due to the presence of high convective strength toward the electrode surface, we do not see any distinguishable peaks in the CV. The convection disturbs the equilibrium to a large extent and equalizes the concentration of both the ions near the electrode, and hence, no prominent peak is observed. However, due to the presence of the ions on the electrode, the current increases with an increase in the convective strength.

17.5.6 Role of Different Time Scales in CV

Physical understanding of different timescales gives the best quantitative information of current change in CVs and they are reflections of the processes that undergo on a typical working electrode. Here, we explore the role of diverse timescales on CV for the assistance of the scientists working in electrochemical sensors and other diverse areas of electrochemistry. From the discussion made so far, it is well understood that multiple factors affect CVs. During the measurement of CVs, various factors can alter simultaneously to a different extent, resulting in a change in the shape of the CV. Having worked with ample numbers of parametric variation, it is understood that a game continues among different time scales, such as the diffusion time scale ($r_D = L^2/D$), reaction time scale ($r_r = 1/k_i$), convection time scale ($r_c = L/u_{avg}$), and scan period (r_{sc}). Here, $u$avg is the average convective flow velocity. The ultimate nature of a CV can be predicted via the mentioned time scales' relative strengths. With the help of 3D bar plots as shown in Figure 17.6a–f, we show the dependence of either oxidation or reduction peak currents against any of the two varying independent parameters including component diffusivity, convection strength, the reaction rate constant, and υ. In these 3D plots, we straightforward consider variables like diffusivity values, convective flow velocity, the reaction rate constant, and υ for better visualization of the readers. However, the corresponding time scale determines the ultimate fate of the dependent variables (shown in z axis) such as peak current. The relations for various time scale calculations are given in Table 17.2.

We typically work in the time scale range as mentioned in the Table 17.2. A close look into the different time scale values enables us in understanding the game among reaction, convection, scanning, and diffusion that takes place while we record a CV. For the working set of parameters, it is observed that diffusion and reaction have relatively higher time scales (signifies sluggishness) compared to the convection time scale. The scan time scale talks about the time period we allow to record the entire CV. If we do the scanning too fast (the scan time period is much smaller than

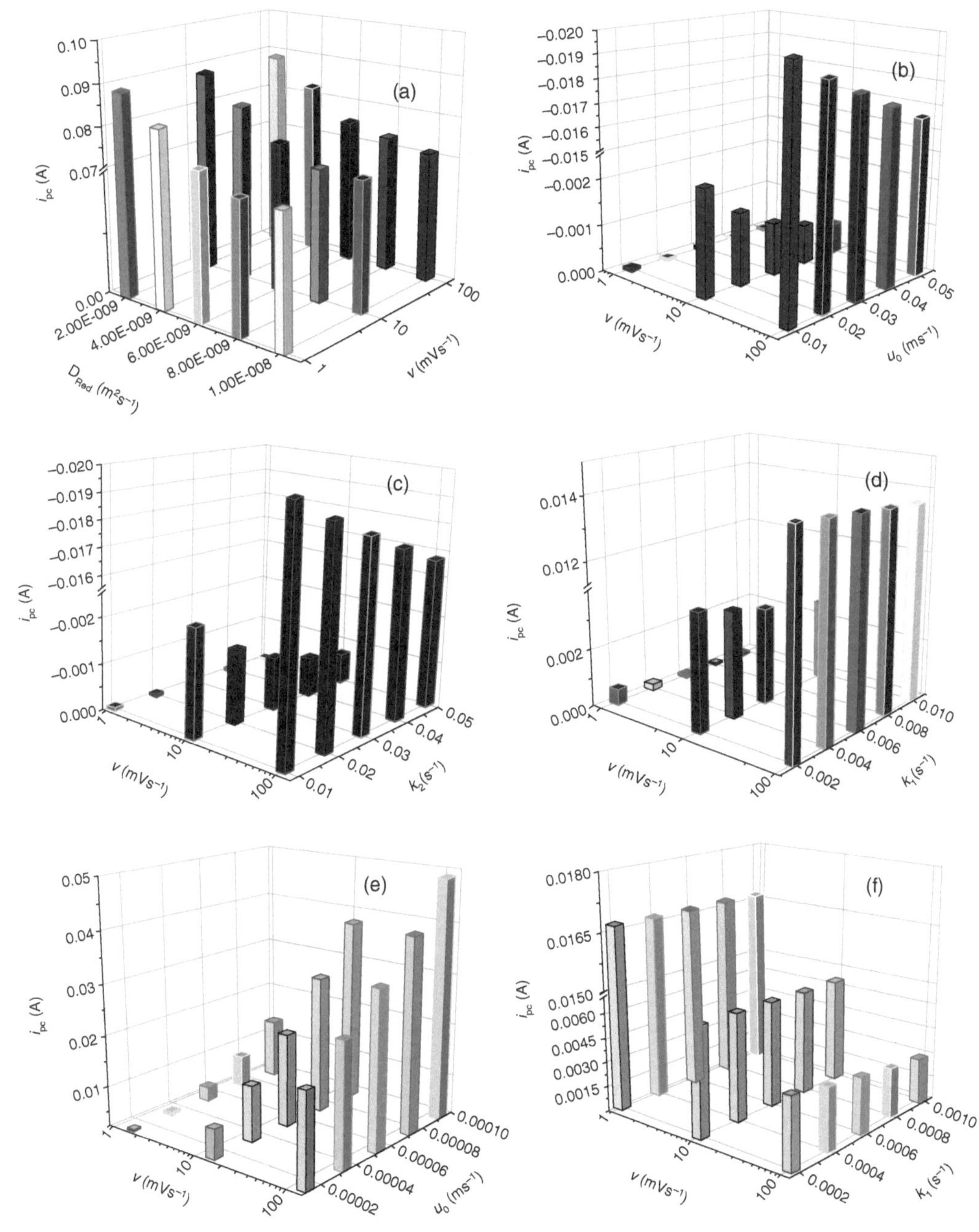

FIGURE 17.6 Plot (a) shows the variation of i_{pc} against D_{Red} and υ. Plot (b) shows the variation of i_{pc} against υ and u_0. Plot (c) shows variation of i_{pc} against υ and k_2. Plot (d) shows variation of i_{pa} against υ and k_2. Plots (e,f) show variation of i_{pc} against two sets (υ and u_0 for plot (e) and υ and k_1 for plot (f), respectively) of parameters for a situation where convection and bulk reactions are taking place in a concurrent fashion.

other time scales), then we may miss the effect of any changes with respect to diffusion, convection, and reaction. On the other hand, if we allow sufficient time (low υ) to record the CV, then we can systematically see their respective effects in the output curves. For this study, υ has been varied from 1 to 1000 mVs^{-1} and the same indicates scan time period variation from 500 to 100 seconds. Further, it is observed that the diffusion time scale approximately varies from 50 to 500 seconds. As

TABLE 17.2
Different Time Scales and Their Ranges with the Given Parameters

Serial No.	Name	Relation	Typical Range of Parameters	Typical Range of Time Scales (seconds)
1	Diffusion time scale	$r_D = L^2/D$	10^{-9} to $10^{-8}\,m^2s^{-1}$	560–56
2	Convection time scale	$r_c = L/U_{avg}$	0.0002–0.05 ms^{-1}	3.75–0.015
3	Reaction time scale	$r_r = 1/k$	0.002–0.01 s^{-1}	500–100
4	υ time scale	$r_{CV} = 2V_{range}/V$	0.001–0.1 Vs^{-1}	1600–16

we allow sufficient scanning time for the diffusion process to occur, we get the effect of diffusivity in terms of change in peak current for an entire range of υ. Contrary to Randles–Sevcik equation, an increase in the diffusivity values of the 'Red' results in the mitigation of current values, as shown in Figure 17.6a. Interestingly, it is found that the oxidation peak current (i_{pc}) changes with diffusivity values of the 'Red', and for the given set of parameters, this is eventually independent of the υ.

In an electrochemical cell, along with diffusion, there is the possibility of local convection owing to a lot of factors as mentioned in the earlier section. In a previous section, we have also mentioned that this convection may alter the CV extensively. For the set of parameters used in this work, the time scale for convection is relatively shorter (indicates faster component transfer) than the diffusion time scale. Hence, any kind of convective flow near the electrode is expected to alter the CV extensively. Further, as we mentioned earlier, the competition between the convective time scale and the scan period continues. The CV characteristics eventually change as an effect of these two-time scales. Figure 17.6b shows the i_{pc} against variations of υ and u_0 (electrode to bulk motion) near the electrode. Lower υ (higher scan period) reduces the i_{pc} drastically, which indicates that local convective flow gets sufficient time to alter the concentration profiles of both 'Ox' and 'Red' near the working electrode. On the other hand, for a higher υ (lower scan time period), we get a higher oxidation peak current. In this case, despite the presence of convection, the concentration profiles do not get the scope for a major change as the scan completes within a short time span. For a constant, υ, if we augment the convective strength (reducing the convection time scale), then i_{pc} again reduces as the increase in convective strength facilitates faster alteration of concentration profiles near the working electrode.

In general, it is expected that the reaction occurs at the working electrode surface and CV tracks the change in current owing to such heterogeneous reactions. However, for a few cases, there might be the possibility of side reactions, $Ox + e \leftrightarrow Red \rightarrow P$, at the bulk solution along with a heterogeneous electrode reaction. In such cases, the CV curve alters significantly and the same is also explained in the previous part of this chapter. In this section, we discuss how CV characteristics get affected by the reaction k_2 and υ. Figure 17.6c shows the change in i_{pa} with the change of k_2 and υ. A constant reaction rate implies a fixed time requirement for a reaction progression, the faster the reaction rate the lower the time requirement for the reaction.

In a scenario, if we complete the scan before the completion of the reaction, we miss the effect of the reaction and the CVs appear normal. However, a slow υ offers a larger time period of scan and the bulk reaction happens within this scan time and the effect of reaction starts appearing in the CVs. The plot shows that with an increase in υ (reduction of scan time scale), keeping the reaction rate constant, the i_{pa} increases, which indicates mitigation of the effect of bulk reaction. On the other hand, for a constant low υ (higher scan time scale), with the increase in reaction rate constant (alleviation of reaction time scale), the i_{pa} starts decreasing gradually, and the same indicates the dominance of reaction on CVs.

We do another relative time scale study for a case where the component 'Ox' vanishes with a particular first-order reaction rate and this reaction is considered throughout the bulk. In this section, we track the oxidation peak current (i_{pc}) against different υ and k_1. The 3D plot is shown in Figure 17.6d which shows the combined effect of k_1 and υ on *ipc*. It is seen that for a relatively slow υ, with an increase in k_1 (i.e. a decrease in r_7), the *ipc* gradually decreases. On the other hand, with a constant k_1, if we increase υ, the i_{pc} starts increasing. In a layman language, if the scanning happens so fast that the reaction does not get the scope to reach to completion, in that case, the relative change of i_{pc} is not that much significant. Evidently, for a slower υ, the reaction gets sufficient time to occur, and hence, relative change in the i_{pc} becomes significant enough. The plot shows a clear picture of how the peak current is dependent on both the time scales.

Up to this point, we have discussed the effects of convection and diffusion separately, where it is assumed that in the presence of one, the other effect becomes null and void. However, in real cases, there might be situations where both convection and reaction may happen simultaneously and both the causes become potential enough to change the CV to a large extent. Figure 17.6e shows the change in i_{pc} with two varying parameters, υ and u_0 of the electrolyte while keeping the bulk reaction at a constant value 0.010 ms^{-1}. On the other hand, Figure 17.6f shows the sensitivity of the CVs with respect to two independent variables such as k_1 and the υ while keeping u_0 0.00001 ms^{-1}. Interestingly, in the presence of bulk reaction with a constant rate constant, with an increase in u_0 (decrease in convection time scale) while keeping the υ constant, the i_{pc} increases. It is to be noted that, in the presence of concurrent bulk reaction, the observation is in contrast with the sensitivity of local convection shown in Figure 17.6b. Eventually, in this case, the strength of bulk reaction (shorter reaction time scale) dominates over the convective flow (larger convective time scale) and the same leads to such an observation. Hence, it proves our claim on multiple dependencies of a CV curve. In the other plot, we show the sensitivity of i_{pc} with the variations of υ and reaction rate constant while keeping a weak convective flow alive near the working electrode. In this case, we observe similar sensitivity with respect to a change in υ. However, for a lower υ (higher scan time period), we observe almost no change in i_{pc} with the variation of reaction rate constant. However, for a higher υ (lower scan time period), we see a steady decrease in i_{pc} with a gradual rise in bulk reaction rate constant.

17.6 SUMMARY

This present work explores CVs' salient aspects considering the effects of component diffusivities, reaction rates, scan time, and convection strength. This is a comprehensive theoretical study that quantitatively targets all parameters that alter a CV curve. This work is compiled for straightforward clarification of a CV curve, and the knowledge would help decipher many complicated CV plots. The key findings of the present study are given pointwise in the following section.

1. Ion diffusivities play a significant role in deciding the oxidation and reduction peak of a CV. More the deviations in diffusivities of the component more deviations in CVs are observed. Further, diffusivities of 'Ox' and 'Red' have contrasting effects on CVs and thereby diffusivity ratios play an important role. Hence, similar diffusivities make both the peaks appear at a specific potential. Deciphering these concepts from the characterization studies of electrochemical sensors helps designing the sensor better.
2. A slow υ can capture the effect of diffusion on CVs, whereas a fast υ may miss the impact of diffusion. Hence, it is recommended to know about the system before setting a particular system υ. Scan rate plays a vital role to control bio-receptor and target analyte interactions, and hence, a close look into the effects of scan rates is needed for chemists, biochemists, engineers who work on electrochemical sensor design.
3. A bulk reaction or side reaction affects CVs significantly. It is shown that a faster bulk reaction changes the CV curve significantly. A chain reaction with stoichiometry, $Ox+e \leftrightarrow Red \rightarrow P$, can either mitigate or remove the reduction peak based on the reaction rate constant values.

4. Any sort of local convection, i.e. fluid flow, near the working electrode surface during the CV operation process may hamper the nature of the CV. The local convection may arise in the system.

   Due to many reasons, including mechanical stirring, bubbling near the electrode's surface, and chemical gradient-driven flow near the electrode. Working electrode of an electrochemical sensor also involves reactions of different types, and hence, this understanding maps with the required know how for electrochemical sensor interaction kinetics.
5. Different time scales associated with diffusion, reaction, scan speed in combination determine the ultimate fate of a CV curve. It is to be remembered that to understand an electrochemical system, the knowledge about the relative strengths of those time scales is very much relevant.

   Finally, this is claimed that this chapter will be helpful for electrochemists, engineers, researchers working in the field of electrochemistry to know a lot about CVs in a concise place. It is also to be mentioned that the chapter contains the most important points that are of interest to scientists working in electrochemical sensors. We also attempted to reveal how the contents of this chapter would help scientists and technologists working with design of a novel electrochemical biosensor.

REFERENCES

1. Faraday, M., Experimental researches in electricity. Seventh Series. *Philosophical Transactions of the Royal Society of London* **1834**, *124*, 77–122.
2. Sanger, M. J.; Greenbowe, T. J., Common student misconceptions in electrochemistry: Galvanic, electrolytic, and concentration cells. *Journal of Research in Science Teaching: The Official Journal of the National Association for Research in Science Teaching* **1997**, *34*, 377–398.
3. Özkaya, A. R., Conceptual difficulties experienced by prospective teachers in electrochemistry: Half-cell potential, cell potential, and chemical and electrochemical equilibrium in galvanic cells. *Journal of Chemical Education* **2002**, *79*, 735.
4. Van Berkel, G. J.; Zhou, F., Characterization of an electrospray ion source as a controlled- current electrolytic cell. *Analytical Chemistry* **1995**, *67*, 2916–2923.
5. Branca, P. A.; Hubis, D. E.; Mallouk, R. S.; Perry, R. L., *Electrolytic Cell with Composite, Porous Diaphragm*. WL Gore and Associates Inc: Canada, **1993**.
6. Feiner, A.-S.; McEvoy, A., The Nernst equation. *Journal of Chemical Education* **1994**, *71*, 493.
7. Walczak, M. M.; Dryer, D. A.; Jacobson, D. D.; Foss, M. G.; Flynn, N. T., pH dependent redox couple: An illustration of the Nernst equation. Journal of *Chemical Education* 1997, 74, 1195.
8. Pourbaix, M., Applications of electrochemistry in corrosion science and in practice. Corrosion Science **1974**, *14*, 25–82.
9. Kelly, R. G.; Scully, J. R.; Shoesmith, D.; Buchheit, R. G., *Electrochemical Techniques in Corrosion Science and Engineering*. CRC Press: Boca Raton, FL, **2002**.
10. Perez, N., *Electrochemistry and Corrosion Science*. Springer: New York, **2004**; Vol. 412.
11. Hansen, W. N.; Hansen, G. J., Absolute half-cell potential: A simple direct measurement. Physical Review A **1987**, 36, 1396.
12. Jin, M.; Xu, J.; Jiang, L.; Gao, G.; Chu, H.; Xiong, C.; Gao, H.; Jiang, P., Electrochemical characterization of a solid embeddable Ag/AgCl reference electrode for corrosion monitoring in reinforced concrete. Electrochemistry 2014, 82, 1040–1046.
13. Schmidt, B.; King, D.; Kariuki, J., Designing and using 3D-printed components that allow students to fabricate low-cost, adaptable, disposable, and reliable Ag/AgCl reference electrodes. *Journal of Chemical Education* 2018**,** 95, 2076–2080.
14. Reiss, H.; Heller, A., The absolute potential of the standard hydrogen electrode: A new estimate.*The Journal of Physical Chemistry* **1985**, *89*, 4207–4213.
15. Tripkovic, V.; Björketun, M. E.; Skúlason, E.; Rossmeisl, J., Standard hydrogen electrode and potential of zero charge in density functional calculations. *Physical Review B* 2011, 84, 115452.
16. Biegler, T.; Woods, R., The standard hydrogen electrode. A misrepresented concept. *Journal of Chemical Education* **1973**, *50*, 604.
17. Kolthoff, I.; Thomas, F., Electrode potentials in acetonitrile. Estimation of the liquid junction potential between acetonitrile solutions and the aqueous saturated calomel electrodel. *The Journal of Physical Chemistry* **1965**, *69*, 3049–3058.

18. El Din, A. S.; Kamel, L., The effect of amalgam formation on the potential of the saturated calomel electrode. *Journal of Electroanalytical Chemistry* **1966**, *11*, 111–121.
19. Sophocleous, M.; Atkinson, J. K., A review of screen-printed silver/silver chloride (Ag/AgCl) reference electrodes potentially suitable for environmental potentiometric sensors. Sensors Actuators A: Physical 2017, *267*, 106–120.
20. East, G. A.; Del Valle, M., Easy-to-make Ag/AgCl reference electrode. *Journal of Chemical Education* **2000**, *77*, 97.
21. Yoon, C. H.; Vittal, R.; Lee, J.; Chae, W.-S.; Kim, K.-J., Enhanced performance of a dye- sensitized solar cell with an electrodeposited-platinum counter electrode. *Electrochimica Acta* **2008**, *53*, 2890–2896.
22. Huo, J.; Zheng, M.; Tu, Y.; Wu, J.; Hu, L.; Dai, S., A high performance cobalt sulfide counter electrode for dye-sensitized solar cells. *Electrochimica Acta* **2015**, *159*, 166–173.
23. Brown, J. H., Development and use of a cyclic voltammetry simulator to introduce undergraduate students to electrochemical simulations. *Journal of Chemical Education* **2015**, *92*, 1490–1496.
24. Elgrishi, N.; Rountree, K. J.; McCarthy, B. D.; Rountree, E. S.; Eisenhart, T. T.; Dempsey, J. L., A practical beginner's guide to cyclic voltammetry. Journal of Chemical Education 2018, 95, 197–206.
25. Noren, D.; Hoffman, M., Clarifying the Butler–Volmer equation and related approximations for calculating activation losses in solid oxide fuel cell models. *Journal of Power Sources* **2005**, *152*, 175–181.
26. Mann, R.; Amphlett, J.; Peppley, B.; Thurgood, C., Application of Butler–Volmer equations in the modelling of activation polarization for PEM fuel cells. *Journal of Power Sources* **2006**, *161*, 775–781.
27. Opoku, F.; Osikoya, A. O.; Dikio, E. D.; Govender, P. P., One-step synthesized 2D heteroatom doped graphene for high throughput electrochemical biosensing: A combined experimental and computational studies. Diamond Related Materials 2019, 100, 107592.
28. Lephalala, M.; Kanchi, S.; Sabela, M. I.; Bisetty, K., Electrochemical enzymatic biosensing of neotame supported by computational methods. Electroanalysis 2020, *32*, 2669–2680.
29. Schwarzacher, W., Electrodeposition: A technology for the future. *Electrochemical Society Interface* **2006**, *15*, 32–33.
30. Landolt, D., Electrodeposition science and technology in the last quarter of the twentieth century. *Journal of the Electrochemical Society* 2002**,** 149, S9.
31. Dixit, A.; Middya, S.; Mitra, S.; Maity, S.; Bhattacharjee, M.; Bandyopadhyay, D., Unexplored pathways to charge storage in supercapacitors. *The Journal of Physical Chemistry* C **2018**, *123*, 195–204.
32. Raza, W.; Ali, F.; Raza, N.; Luo, Y.; Kim, K.-H.; Yang, J.; Kumar, S.; Mehmood, A.; Kwon, E. E., Recent advancements in supercapacitor technology. *Nano Energy* **2018**, *52*, 441–473.
33. Brown, J. H., Development and use of a cyclic voltammetry simulator to introduce undergraduate students to electrochemical simulations. *Journal of Chemical Education* **2015**, *92*(9), 1490–1496.

18 Advances, Opportunities, and Challenges in Development of 2-D MXene-Based Aptasensors for Early Cancer Detection

Bushra Rashid, Syed Shahabuddin, and Nanthini Sridewi

18.1 INTRODUCTION

Cancer is one of the major threats to the life and leading cause of death. As per the WHO estimations, cancer is the first or second leading cause of death for people aged below 70 in 112 countries out of 183 countries and is the third or fourth leading cause in 23 other countries (Bray et al., 2018). As per the estimation provided by The International Agency for Research on Cancer (IARC), during the lifetime of a person, 1 in 5 people develop cancer. One woman out of 11 women dies with cancer, whereas 1 out of 8 men dies. About 50 million people are living with a 5-year history of cancer. Breast, colorectal, lung, cervical, and thyroid cancers are common cancers among women. Lung and prostate cancer are common cancers in men. It was predicted in GLOBOCAN 2020 report that the countries which are identified as low or medium human development index would have the most increase in cancer cases by 2040 (Are et al., 2020). The estimated number of incident cases and mortality associated with various types of cancer is represented in the bar graph in Figure 18.1.

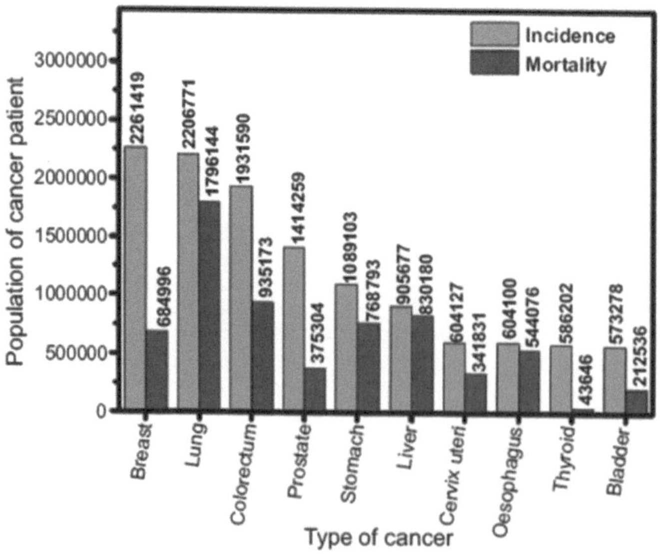

FIGURE 18.1 Number of incident cases and mortality associated with various types of cancer (Parihar et al., 2022).

DOI: 10.1201/9781032654874-18

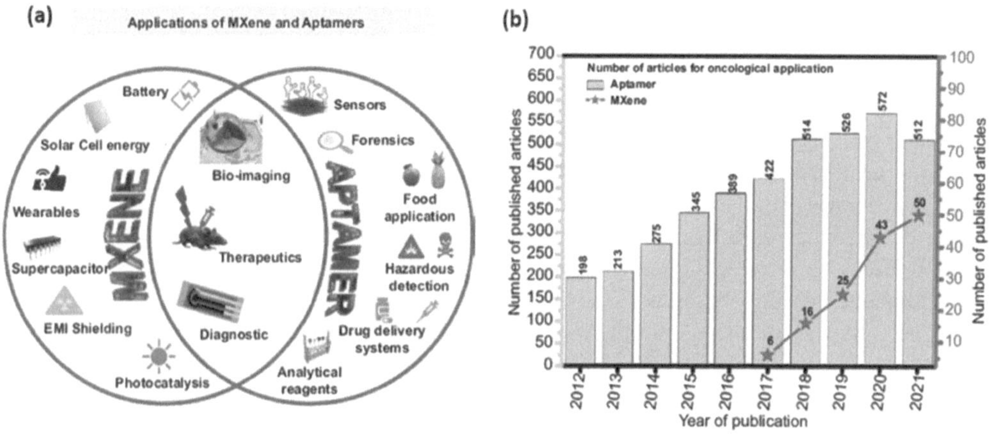

FIGURE 18.2 (a) Utilization of MXenes in various applications and (b) year-wise publications (Parihar et al., 2022).

Recently, biosensor-based advanced diagnostic approaches have shown potential for the early diagnosis of cancer and other deadly diseases (Khan, Parihar, & Sanghi, 2021). Several techniques employed for the detections are optical (Bernard & Wittwer, 2002), electrochemical (EC) (Kumar et al., 2021), and piezoelectric (Kiio, Mbui, Ndangili, Onam, & Oloo, 2021). The electrochemical-based detection of several biomarkers such as epithelial cell adhesion molecule (EpCAM), CD44, vascular endothelial growth factor (VEGF), Mucin 1, carcinoembryonic antigen (CEA) have fetched great attention (Ranjan et al., 2020). The EC techniques (amperometry, impedemetric, or potentiometric) deliver a highly sensitive rapid and cost-effective platform for early detection of cancer biomarkers. The miniaturization of EC devices helps in their handy usage. Moreover, the EC techniques can reach to attomole detection level and offer high selectivity (Lakshmanakumar, Nesakumar, JBB, & Rayappan, 2021). Recent studies have shown the crucial role of nanomaterials in the enhancement of the performance of EC devices for the early detection of cancer. The excellent electrical, mechanical, electrochemical, and optical features of 2D materials have fetched much attention over the past decade in terms of their wide applicability in various fields including disease diagnosis and therapeutics (Mas-Balleste, Gomez-Navarro, Gomez-Herrero, & Zamora, 2011). Recently, MXene has been efficiently used for various biosensing applications (Xu et al., 2021), targeted drug delivery, energy storage (Cao et al., 2021), heat resistance material synthesis (Ye, Jiang, Li, Wu, & Chen, 2022), catalysis (Ahmad et al., 2021), and many other. Figure 18.2 shows the use of MXene in various applications along with year-wise papers published (Parihar et al., 2022).

18.2 SELECTION OF APTAMERS

Biosensors have emerged as an essential component of laboratory medicine's state-of-the-art devices, notably in point-of-care testing (Razmi et al., 2018). Biosensing platforms are frequently used for the quantification, detection, and analysis of biomolecules with substantial health implications (Khalilzadeh, Shadjou, Charoudeh, & Rashidi, 2017). Analytical devices like biosensors have two key components: a bio-recognition component like enzymes, antibodies, or nucleic acids and a transduction part for transmitting signals via EC or optical means to display quantitative analytical data and measurable signals (Zahra et al., 2021). Signal transduction and bio-recognition element loading are critical in biosensor configuration. Integrating bio-recognition features into biosensors increases their specificity and sensitivity. Interfaces with maximum loading capacities of recognition components and strong biocompatibility are therefore essential (Li, Lu, & Liu, 2021). When transforming from bio-signals to optical or electrical signals, a transducing element with superior electrical conductivity and surface area is recommended since it yields higher output signals while producing less noise. 2D materials are frequently used as photon-electron or electrode interface materials in optical and EC biosensors.

Aptasensors recognize biomolecules using ssDNA or RNA aptamers (Bayat, Taghdisi, Rafatpanah, Abnous, & Ramezani, 2019). These oligonucleotide sequences exhibit high specificity and affinity against various biomolecules and are selected *in vitro* using SELEX from randomized nucleic acid libraries (Robertson & Joyce, 1990). Tuerk and Gold (1990) were the first to select an RNA-based oligonucleotide against bacteriophage T4 DNA polymerase, followed by Ellington and Szostak (1990) who selected an RNA oligonucleotide the same year against an organic dye. They were subsequently termed "aptamers", the name derived from the Latin word "aptus", which means "to fit" (Lee, Stovall, & Ellington, 2006). Conventional SELEX, based on the preferred aptamer requirements, involves several steps. To screen DNA aptamers, target molecules are incubated with a specially designed DNA library. After incubation, the DNA-target complexes or the bound sequences are retained while removing the unbound sequences. A polymerase chain reaction (PCR) is used to amplify the bound sequences (Figure 18.3). This process is repeated for several selection cycles to get enriched sequences, which are then sequenced and characterized to analyze their binding kinetics in various ways (Lee et al., 2006), for example, surface plasmon resonance (SPR), circular dichroism (CD), and isothermal titration calorimetry (ITC) etc. Several other techniques can also be used to precisely investigate the binding kinetics between a target and its aptamer. Even though the traditional SELEX method was not meant to be used for aptamer selection, it was changed repeatedly to meet the needs of the desired aptamers and eventually became the main way to screen aptamers. The SELEX technique has been continuously modified and improved over the last three decades to fulfil a variety of requirements. Recently, aptamers can be selected using various techniques in just a few hours, rather than weeks or months as compared to traditional SELEX. Modern strategies assisted by computer software systems enable the prediction of the binding affinity of an aptamer and structure prior to its selection, significantly reducing the cost of the selection process. Structure-switching particle display, molecular dynamics simulation, and a few other methods have also been used for aptamer selection (Ullah et al., 2022).

To date, the literature has documented over 1000 well-characterized high-affinity oligonucleotide aptamers to detect or quantify a variety of analytes (from small molecules to whole cells or tissues). In the last few years, the number of papers on aptamer technologies has undergone

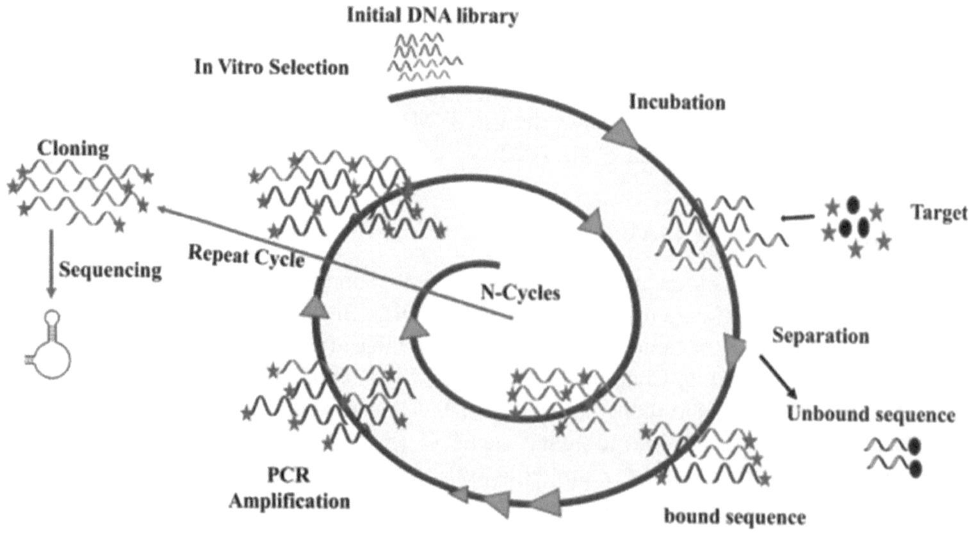

FIGURE 18.3 Schematic diagram to show the basic steps such as incubation, separation, and PCR amplification involved in a generalized conventional SELEX for DNA aptamer selection (Ullah et al., 2022).

exponential growth. Biosensors are the primary focus of most aptamer technology investigations. Drolet, Moon-McDermott, and Romig (1996) and Davis et al. published the first biosensor based on aptamer in 1996. Following that, aptamers were coupled to a variety of sensing platforms (electrochemical, surface-enhanced Raman scattering, fluorescent, and colorimetric) for the purpose of analyzing the target molecules in the sample (Alkhamis, Canoura, Yu, Liu, & Xiao, 2019). In a recent review of aptasensors, Zahra et al. (2022) have discussed the fundamentals and the design strategies used to develop the most common types of aptamer-based biosensors. These include pre-programmed techniques to convert aptamer identification actions into readable signals (Billet et al., 2021).

Aptamers act as molecular bio-recognition elements when they fold into distinct tertiary or secondary structures, allowing them to detect and bind targets with greater affinity (the dissociation constant Kd value ranges from micromolar to picomolar concentrations) (Lubin & Plaxco, 2010). Proteins, metal ions, small organic molecules, cells, tissues, bacteria, viruses, etc. are among some of the possible targets (Ping et al., 2015). To put it another way, because of the peculiar features of aptamers e.g. high chemical stability, cost-effectiveness, and in vitro synthesis (Zhao et al., 2019), they can be used as beneficial molecular recognition markers in the development of high-efficiency biosensors (Vinkenborg, Karnowski, & Famulok, 2011) and have garnered a huge interest in the last 30 years from a range of research disciplines, including disease diagnostics (Cella et al., 2010), food safety (Mei et al., 2013), etc. The fundamental step in developing an aptasensor is immobilizing the aptamer on a substrate. A variety of aptamer immobilization techniques have been documented in the literature. Controlling the immobilization process is critical for ensuring the high stability, accessibility, orientation, enhanced target binding, and reactivity of the aptamer as well as avoiding non-specific adsorption processes. To prevent the non-specific binding of interfering molecules, an affinity reaction must be conducted with the target. The immobilization method chosen is determined by the detection principle and assay format, and it is believed to have a direct effect on the capability of an aptamer to bind the target. The size of the target molecules determines the format of the assay and the mechanism for aptasensing. Once the target is captured on the sensor surface, multiple approaches for transducing the bio- recognition event can be employed (Ravalli, Voccia, Palchetti, & Marrazza, 2016). Since the aptamer immobilization and binding occur at the surface of the transducer, effective transducer utilization is vital for developing promising aptasensors (Zhao, Xu, & Chen, 2016). As previously stated, MXene has shown to be an excellent substrate for transducer surface applications.

Aptamers are compatible with a wide range of diagnostic and biosensor platforms, allowing them to contribute significantly to the global diagnostic sector (Dhiman, Kalra, Bansal, Bruno, & Sharma, 2017). Even though aptamer-based sensors are still in their early phase of development in comparison to enzyme-based biosensors and immunoassays, nevertheless when combined with nanomaterials (NMs), they have the potential to become powerful biosensing tools in the future (Kim, Raston, & Gu, 2016). To facilitate bio-functionalization, 2D-NMs that combine the advantages of an abundance of active sites, stable dispersity, and superior biocompatibility are extremely competitive. While aptamers, enzymes, and antibodies are frequently exploited as recognition components in the construction of biosensors, MXenes typically provide an excellent transducing platform and circumvent some of the constraints associated with graphene or other related materials (Bostan et al., 2017). MXenes' fundamental features make them an excellent candidate for use as a biosensor material. Biosensing technology based on MXene has advanced at a breakneck pace over the last several years. Because of their exceptional combination of properties, MXene materials together with an effective bio-recognition platform have been demonstrated to be a highly selective and sensitive detection system for biosensor applications despite their extremely short journey (Sinha et al., 2018). Nanotechnology has enabled the coupling of various nanomaterials with aptamers, resulting in extremely selective and sensitive aptasensors that have added great value to the diagnostics and analytical fields. The unique features of NMs, including shape- and size-dependent optical properties, catalytic activity, and flexible tuning of surface chemistry, make them extremely valuable for signal induction and amplification (Burda, Chen, Narayanan, & El-Sayed, 2005). The combination of NMs and aptamers has been extensively investigated in order to build sensitive and unique sensing devices (Hao Wang, Yang, Yang, & Tan, 2009). It is essential to synthesize reliable, stable, and reproducible nanostructures that could offer an appropriate environment for the immobilization of recognition components (i.e. aptamer or other probes), before moving on to building biosensor techniques for use in complicated bio-environments (Wongkaew, Simsek, Griesche, & Baeumner, 2018). Composite NMs have now been widely used in EC biosensors for labelling and signal amplification (Ping, Fan, Sindoro, Ying, & Zhang, 2017). Even though MXene was originated in late 2011 and possessed excellent intrinsic physicochemical features, the very first MXene-based biosensing device was introduced in 2015. MXenes were introduced by Xu and colleagues for probing neural function (Xu et al., 2016). They developed a very sensitive field-effect transistor (FET) biosensor based on extremely thin MXene micropatterns for label-free probing of small molecules under ordinary biological settings and rapid identification of action potential in primary neurons. This discovery provided a new pathway for biosensing applications based on MXene, allowing more scientists to discover the significance of this broad family of 2D NMs known for their superior electrical behaviour (Xu, Zhi, & Shi, 2020).

18.3 ROLE OF MXenes IN APTAMER-BASED BIOSENSORS

MXenes have the ability to interact with various biomolecules via coordination bonds, electrostatic interactions, van der Waals forces, and hydrogen bonds (Chen et al., 2018), making them an ideal interface for biosensor development (Zhang et al., 2018). The long-term durability of the constructed MXene biosensors ensures great reproducibility of data over time, which might be immensely beneficial when using these biosensors for a variety of diagnostic applications. MXenes' high biocompatibility and increased surface area provide an impetus for incorporating innovative nano(bio) hybrid frameworks with bio-receptors such as antigen-antibody (Ag-Ab), enzymes, proteins, whole cells, and aptamers [DNA/microRNA (miRNA)] that enable suitable integration of biomolecules onto their surface to behave as a sensitive recognition interface. It is anticipated that these advances in the construction of MXene-based biosensors will further contribute to the emergence of modern detection methodologies utilizing techniques such as EC, colorimetric, and fluorescent luminescence for the highly selective identification of multiple biomarkers for use in point-of care diagnostics and early detection of diseases (Sinha et al., 2018). Recently, the development of 2D NM-based biosensors for ultrasensitive disease diagnostic testing applications, such as the identification of tumour biomarkers existing in ultra-low amounts in various clinical samples (e.g. urine, serum, blood, and tissues), has been driven by emerging concepts in bio-sensing systems and their translation into a clinical trial (Wang et al., 2015). Thus, in the sensing arena, combining various functionalized nanomaterials (including MXenes) with aptamers is expected to open new opportunities in next-generation biosensors. Numerous advanced aptasensors have been employed exclusively to fabricate field-deployable and clinical analytical devices in view of their excellent optical, thermal, mechanical, and electronic features (Mahmoudpour et al., 2021).

Recently, MXenes have been employed for the detection of exosomes. Exosomes are extracellular vesicles, ranging from 30 to 150 nm in diameter, present in human circulatory systems, that transfer molecular proteins and/or nucleic acids between cells for cell-to-cell communication. Exosomes are considered one of the most important cancer biomarkers because they play a key role in the modulation of tumour microenvironments (Raimondo, Morosi, Chinello, Magni, & Pitto, 2011). Recently, many fluorescence- or colorimetric-based aptasensors for exosome detection have been developed (Zhu, Li, Wang, Irfan, & Qu, 2020). Yu et al. used fluorescence-based aptasensors to detect an exosome transmembrane protein, cluster of differentiation 63 (CD63), on the surface of exosomes via a competitive strategy. 180 CD63-binding aptamers immobilized on magnetic beads were hybridized with a short, Cy3-labelled complementary strand at the binding site. Upon the addition of exosomes, the CD63 on the exosome surface bound to the aptamers, resulting in dissociation of the short DNA strand into the supernatant. The number of exosomes could then be determined simply by measuring the fluorescence intensity of the supernatant. This magnetic-bead-based aptasensor could detect targets at a concentration of 1.0×10^5 particles μL^{-1}. Yu et al. (2019) also made use of the same aptamer for exosome detection in human plasma by fluorescence polarization with a detection limit of 500 particles μL^{-1}.

To improve the sensitivity, nanomaterials have been combined with fluorescence moieties. Zhang et al. (2018) made use of the transition metal carbides- and carbonitrides-based 2D material, Ti_3C_2 MXene, as a platform for immobilizing the Cy3-labelled CD63-binding aptamers by metal chelate interaction and hydrogen bonds. Initially, the Cy3 signal was quenched owing to fluorescence resonance energy transfer (FRET) between Cy3 and MXene. After binding of CD63 protein on the exosome surface to the aptamers, the exosome–aptamer complexes dissociated from the MXene's surface, causing the Cy3 fluorescence signal to recover (Figure 18.4a). Graphene oxide is also considered a biocompatible platform for effective adsorption of RNA or DNA via π–π interactions. GO has an excellent distance-dependent fluorescence quenching capability to reduce the signal-to-noise ratio. Jin et al. (2018) and Hui Wang et al. (2018) took advantage of this quenching property of graphene oxide (GO) to develop turn-on aptasensors for accurate detection of exosomes. To amplify the fluorescence signal after target binding, deoxyribonuclease I (DNase I) was used to digest the bound aptamers from the exosome surface, enabling the free exosomes to interact with more fluorescence aptamers. Particularly, GO-based aptasensors are useful for sensitive detection of colorectal cancer

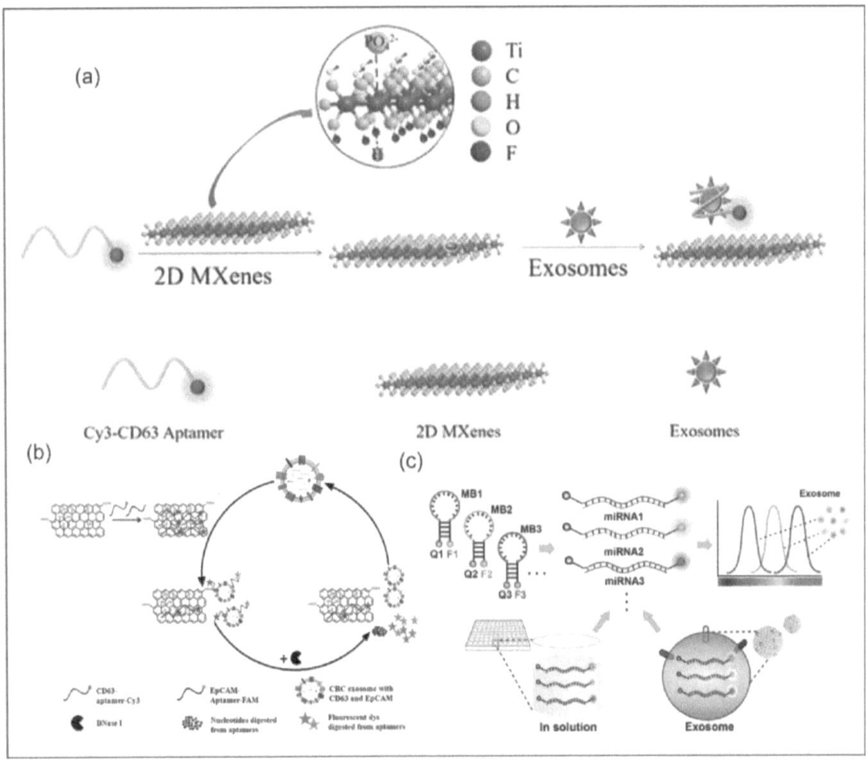

FIGURE 18.4 (a) FRET biosensing platform based on the Cy3-labelled aptamer/Ti_3C_2 MXenes nanocomplex for quantitative detection of exosomes. (b) GO-based aptasensor for enzyme-aided exosome detection. (c) In situ simultaneous and multiplexed detection of miRNAs in exosomes based on Mxene biosensors (MBs).

Source (a): Reproduced with permission from Zhang et al. (2018). (b): Reproduced with permission from Hui Wang et al. (2018). (c): Reproduced with permission from Hui Zhang et al. (2020).

exosomes with a detection limit of 2.1×10^4 particles μL^{-1} (Figure 18.4b). Besides the CD63 protein on the surface of exosomes, miRNAs inside exosomes are also excellent biomarkers for cancer diagnosis. Lee, Kim, Jeong, and Rhee (2016) demonstrated this for the first time using molecular aptamer beacons for in situ simultaneous detection of three targets microRNA (miR) miR-21, miR 375, and miR-27a in exosomes from MCF-7 cells (Figure 18.4c). In general, this is a time- and labour-saving strategy for high-throughput analysis for disease diagnosis and treatment. In addition, a label-free aptasensor linked with protein tyrosine kinase-7 (PTK-7) aptamers was developed to recognize and capture exosomes containing PTK-7 (Hui Zhang et al., 2020).

18.4 MXene-ENABLED ADVANCED EC APTASENSORS FOR CANCER DIAGNOSTICS

MXene, a 2D nanomaterial with plausible electroconductive properties, has been identified as a viable molecule for the fabrication of EC biosensors due to its simple manufacturing process. Aptamers, on the other hand, have proven to be a boon for manufacturing low-cost sensing devices due to their great selectivity and specificity, as well as their mass-production ability. The advantageous electroconductive properties of MXene-enabled with selective and specific aptamers against cancer-specific biomarkers can be potentially employed for early and efficient diagnostics of cancer which is the need of an hour. In this section, we have discussed the studies on MXene-enabled EC aptasensors for the detection of cancer-specific biomarkers. For the synthesis of various types of MXene composite, Naguib et al. added the Ti_2C, Ta_4C_3, TiNbC, (V0.5, Cr0.5) $3C_2$, and Ti_3CN_x, where $x < 1$ which expands the MXenes family (Naguib, Mashtalir, et al., 2012). In further work, Naguib, Come, et al. (2012) proved that MXenes can also be used as electrodes when intercalated with Li ions in lithium-ion batteries). Aptamer-enabled MXene-based EC biosensors have been used in recent studies to detect a variety of cancer biomarkers, as stated in Table 18.1.

In a recent study, the breast cancer marker Mucin 1 was detected using a competitive EC aptasensor fabricated on a complementary DNA (cDNA)-ferrocene/MXene probe [Mucin 1(MUC1)]. Herein, MXene (Ti_3C_2) nanosheets with high specific surface area and excellent electrical conductivity were chosen as aptamer-probe carriers. To make a cDNA-Fc/MXene probe, ferrocene-labelled complementary DNA (cDNA-Fc) was coupled to the surface of MXene followed by attachment of MUC1 aptamer on the electrode via Au-S bonds. A cDNA-Fc/MXene/Apt/Au/glassy carbon electrode (GCE) aptasensor was made and utilized to detect MUC1 through a competitive process that occurs between the cDNA-ferrocene/MXene probe and MUC1. The reduction in an electrical signal happens due to the detachment of the cDNA-Fc/MXene probe from the sensing electrode. This aptasensor has a broad linear range of 1.0–10 μM and a limit of detection (LOD) of 0.33 pM, making it suitable for clinical diagnostics (Bray et al., 2018). An electrochemiluminescent (ECL) aptasensor for the detection of exosomes in breast cancer cells is described by Qiao et al. ECL emitters and reactants used in this study were H_2O_2 and mercaptopropionic acid (MPA)-modified Eu3+-doped CdS nanocrystals (MPA-CdS: Eu NCs), respectively. The CD63 aptamer recognizes and captures exosomes, which subsequently create a G-quadruplex/hemin DNAzyme that competently causes the breakdown of H_2O_2, and thereby reduces the ECL signal in MPA-CdS: Eu NCs. The exosomes from breast cancer cells (MCF-7cells) can be found in concentrations ranging from 3.4×10^5 to 1.7×10^8 particles mL^{-1}. The LOD and signal-to-noise ratio was determined to be 7.41×10^4 particles mL^{-1}. Exosomes in the serum have been effectively detected using these aptasensors (Qiao et al., 2019). In a similar study, CD63-aptamer-modified poly(amidoamine) (PAMAM)-Au NP electrode interface which has a high binding affinity for CD63 protein on exosomes generated from OVCAR cells has been fabricated for the detection of exosomes. Furthermore, the CD63-modified Ti_3C_2 MXene was employed as a nanocarrier for several aptamers and was adsorbed to exosomes. The Ti_3C_2 MXene is generated in situ and loads it efficiently as well as magnifies the EC signal at a low potential, minimizing interference from the electrochemically active species. This aptasensor shows a linear range

TABLE 18.1
Recent Studies to Detect a Variety of Cancer Biomarkers

MXene-Aptamer Probe	Analytes	Detection Methods	Detection Limits	Linear Range	Applications in	References
Cy3-CD63-Apt/Ti_3C_2	Exosomes	FRET	1.4×103 particles mL^{-1}	104–109 particles mL^{-1}	Cancer	Zhang et al. (2018)
$Ti_3C_2T_x$/Au/SiO_2/Apt	STX	EC	0.03 nM	1.0–200 nM	Food safety	Ullah et al. (2021)
Fe_2O_4@Ag-HB5Apt/BSA/Ti_3C_2/GE	HER2-positive circulating tumor cells (CTCs)	EC	47 cells mL^{-1}	102–106 cells mL^{-1}	Cancer	Vajhadin et al. (2022)
AuNPs-Ti_3C2-CD63Apt/GCE/SA-PAM	Exosomes	ECL	30 particles μL^{-1}	102–105 particles μL^{-1}	Cancer	Huixin Zhang et al. (2020)
MCH/Apt/ILs/SiO_2NUs/GCE	Exosomes	ECL and photo-thermal	1.1×10^2 particle μL^{-1} exosomes	–	Cancer	Fang et al. (2020)
Apt2/Apt1/PNIPAM-AuNPs/GCE/Ti_3C_2	Exosomes (on MCF-7 cells)	ECL	125 particles μL^{-1}1	–	Cancer	Zhang, Wang, Zhang, Wang, and Liu (2019)
Apt/He@CCNT/Ti_3C_2/SPCE	CEA	M-EC	2.88 pg mL^{-1}	10 to 100	Cancer	Zhao et al. (2021)
Au-Pd-Pt/$Ti_3C_2T_x$/HP1/HP2	CEA	EC	0.32 fg mL^{-1}	–	Cancer	Song, Gao, Yuan, and Xiang (2022)
cDNA-Fc/MXene/Apt/Au/GCE	MUC1	EC	0.33 pM	1.0 pM to 10 µM	Cancer	Haiyan Wang et al. (2020)
PPy@ $Ti_3C_2T_x$/PMo12/Apt	OPN	EC	0.98 fg mL^{-1}	0.05 pg mL^{-1} to 10.0 ng mL^{-1}	Cancer	Zhou et al. (2019)
Ti_3C_2/PAMAM-AuNPs/CD63-	Exosomes	EC	229 particles	5×10^2 particles μL^{-1} to 5×10^5	Cancer	Zhang et al. (2021)
$Ti_3C_2T_x$ @FePcQDs hybrid/Apt	miRNA-155	EC	0.0043 fM	0.01 to 1.0×10^4 fM	Cancer	Duan et al. (2020)
$Ti_3C_2T_x$ MXene carbon dots or CoCu-ZIF@CDs/Apt	B16-F10 cells	EC	33 cells mL^{-1}	1×10^2 to 1×10^5 cells mL^{-1}	Cancer	Liu et al. (2021)
AuNP dimers/ $Ti_3C_2T_x$ NSs/Apt	Aflatoxin B1 (AFB1)	SERS	0.6 pg mL^{-1}	0.001–100 ng mL^{-1}	Food safety	Wu, Sun, Pu, Wei, and Lin (2022)
GCE/P4VP/Apt/ $Ti_3C_2T_x$ /GO-	AFB1	EC	3 pg mL^{-1}	0.01–50 ng mL^{-1}	Food safety	Guo et al. (2021)

(Continued)

TABLE 18.1 (*Continued*)
Recent Studies to Detect a Variety of Cancer Biomarkers

MXene-Aptamer Probe	Analytes	Detection Methods	Detection Limits	Linear Range	Applications in	References
AuAg Janus NPs/MXene NSs/Aptamer	Ochratoxin A (OTA)	SERS	1.28 pM	–	Food safety	Zheng, Ke, Shi, Liu, and Zhao (2019)
Ti_3C_2/Apt/Al	Endotoxin (DON)	EC	1 fg mL^{-1}	1 fg mL^{-1} to 1 ng mL^{-1}	Food safety	Sangu, Illias, Ong, Gopinath, and Saheed (2021)
Ti_3C_2NSs-TDNs/GCE/Apt	Gliotoxin	EC	5 pM	5 pM to 10 nM	Food safety	Wang et al. (2019)
Ti_3C_2/HD22Apt/HD1Apt/NU172Apt@FAM	Tuberculosis (TB)	FRET	5.27 pM	–	Cardiovascular disease	Cui, Fu, Yang, Xing, and Wang (2021)
MoS_2@$Ti_3C_2T_x$MXene NH/	T4	EC	0.39 pg mL^{-1}	7.8×10 to 7.8×10^6 pg mL^{-1}	Thyroid disorders	Kashefi-Kheyrabadi, Koyappayil, Kim, Cheon, and Lee (2021)
Ti_3C_2 NSs@Apt/OPD/H_2O_2	TB	Colorimetric COL	1.0×10^{-11} M	1.0×10^{-11} to 1.0×10^{-8} M	Blood thrombin levels	Li et al. (2021)
POSS/PQDs/Ti_3C_2/Apt	*Vibrio parahaemolyticus V.P*	FRET	30 cfu mL^{-1}	102–106 cfu mL^{-1}	Water-borne pathogens	Hong et al. (2021)
PBA-Fc@Pt@Ti_3C_2/Apt	*V.P*	ECL-COL	5 and 30 CFU mL^{-1}	–	Food-borne pathogens	Wang et al. (2022)
$Ti_3C_2T_x$ /ZIF-8/GCE/Apt	HIV protein	ECL	0.3 fM	1 fM to 1 nM	HIV/AIDS	Wang et al. (2021)
TDNA@Au/Ti_3C_2-MXene/Apt-H1/Apt-H2	Cardiac troponin 1 (cTnI)	ECL	0.04 or 0.1 fM	0.1 fM^{-1}pM or 0.1–500 fM	MI/heart attack	Mi, Li, Tan, Feng, and Tu (2021)
Apt/MXene	Tumour necrosis factor-α (TNF-α) Interferon-γ (IFN-γ)	EC	0.25 pg mL^{-1} 0.26 pg mL^{-1}	1 pg mL^{-1} to 10 ng mL^{-1}	Cytokine storm	Noh et al. (2022)
CRISPR-Cas12a/gRNA-treated AuNPs@Ti_3C_2/GCE/Apt	Sialic-acid-binding immunoglobulin-type lectins (Siglec-5)	ECL	20.22 fM	–	AML	Zhang et al. (2021)
CRISPR-Cas12a/crRNA/MXene/FAM-Apt	LPS in Gram-negative bacteria	FLR	11 pg mL^{-1} (LPS) & 23 CFU mL^{-1}	–	Food safety	Sheng et al. (2021)

(Continued)

TABLE 18.1 (*Continued*)
Recent Studies to Detect a Variety of Cancer Biomarkers

MXene-Aptamer Probe	Analytes	Detection Methods	Detection Limits	Linear Range	Applications in	References
ZnO/N-Ti_3C_2/GCE/Apt	Catabolic activator protein (CAP)	ECL	0.019 ng mL^{-1}	0.1–100 ng mL^{-1}	Food safety	Jiang, Wei, Du, Qin, Shan, and Chen (2022)
Ti_3C_2–AgBrNCs/GCE/Apt	Enroflaxacin (ENR)	ECL	5.97×10^{13} mol L^{-1}	1.0×10^{12} to 1.0×10^{6} mol L^{-1}	Food safety	Jiang, Wei, Du, Qin, Shan, Wang, et al. (2022)
PMXF-5/HP2/ssDNA/AuNR/HP1	TB	EC	0.67 pM	2 pM–10 nM	Blood thrombin levels	Yang et al. (2022)
PMXF5/HP2/ssDNA/HCuPtA/HP1			16.67 pM	50 pM–50 nM		
GOPS-Ti_3C_2-GCE/Apt	Programmed cell death ligand 1 (PD-L1) protein	EC	0.01–100 ng mL^{-1}	7.8 pg mL^{-1}	Cancer	Li et al. (2022)
MXene-Au/PEI-UPNPs/Cas12a-	DON	Luminescence resonance energy transfer (LRET)	0.64 ng mL^{-1}	1–500 ng mL^{-1}	Food safety	Lin et al. (2022)
Au@$Nb_4C_3T_x$/GCE/Thiol-Apt	Pb2+	EC	4 nM	10 nM–5 μM	Environmental monitoring	Rasheed, Pandey, Jabbar, and Mahmoud (2022)
Apt/MXene-AuNPs/GCE	CAP	EC	0.03 pM	0.0001–10 nM	Food safety	Yang et al. (2022)
Nb_2C-SH QDs/N-58Apt	Nucleocapsid of SARS-CoV-2	SPR	4.9 pg mL^{-1}	0.05–100 ng mL^{-1}	COVID-19 diagnosis	Chen et al. (2021)

of 5×10^2 to 5×10^5 particles μL^{-1}, and the LOD was 229 particles μL^{-1}. This EC aptasensor can detect exosomes from a variety of cancer cells, including OVCAR, HeLa, and BT474, and in serum samples with high specificity suggesting its clinical diagnostic potential for early cancer detection (Zhang et al., 2021). The nanohybrid of $Ti_3C_2T_x$ MXene and phosphomolybdic acid (PMo12) embedded with polypyrrole (denoted as PPy@ $Ti_3C_2T_x$/PMo12) was synthesized by Zhou et al. Further, it was attached to osteopontin (OPN) aptamer to build an impedimetric aptasensor for the detection of OPN. The fabricated sensor PPy@$Ti_3C_2T_x$/PMo12 hybrid is endowed with outstanding stability, great biocompatibility, and a significant binding affinity for OPN aptamer. Thus, compared to previously developed bicomponent aptasensors, the PPy@$Ti_3C_2T_x$/PMo12 hybrid demonstrated improved EC sensing. The aptasensor based on PPy@ $Ti_3C_2T_x$/PMo12 had a detection limit of 0.98 fg mL^{-1}, as well as high selectivity and stability, better repeatability, reasonable regenerability, and can be used to detect OPN in human serum samples (Zhou et al., 2019).

The direct laser patterning of various coplanar of MXene on the paper device and multiple devices with series and parallel connections can be fabricated using hydrofluoric acid (HF)-etched and clay-like Ti_3C_2 MXene slurries in just 17 s using additive manufacturing technique (Figure 18.5a) (Kurra, Ahmed, Gogotsi, & Alshareef, 2016). The fabrication steps of on-chip MXene solid-state micro-supercapacitors (MSCs) by employing a spray-coating method for deposition of highly conductive $Ti_3C_2T_x$ (L-$Ti_3C_2T_x$) flakes on a glass substrate were demonstrated by Peng et al. (Figure 18.5b). MXene-based composite meets the requirement of the bandgap value which should be between 1.55 and 3.0 eV (Figure 18.5c) (Zhan, Si, Zhou, & Sun, 2020). They used three steps which include spray coating of small-sized $Ti_3C_2T_x$ flakes (s-$Ti_3C_2T_x$) on top as an electroactive layer, interdigital pattern carved by direct laser cutting of a specific center area ($8\times6\,mm^2$) on the stacked MXene film, and a PVA/H_2SO_4 gel electrolyte was carefully dropped onto the interdigital pattern area, respectively (Peng et al., 2016). Employing this process would enable the sensor fabrication with ease and in a cost-effective manner. Fang et al. used black phosphorous quantum dots (BPQDs) and MXenes as a signal amplifier for the fabrication of ECL and photothermal dual-mode aptasensor for detection

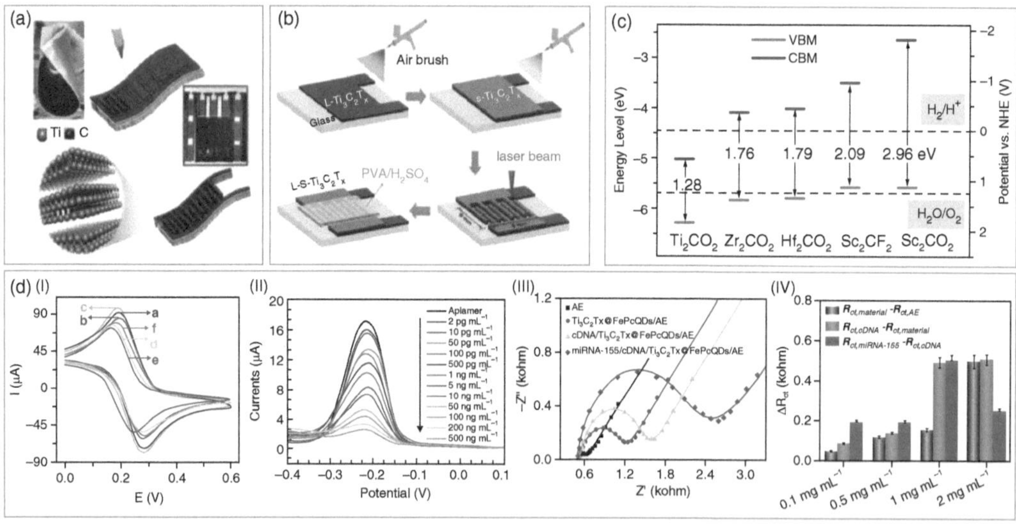

FIGURE 18.5 (a) Representation of the laser patterning of paper coated with MXene to form an interdigitated electrode for MSC devices (Kurra et al., 2016). (b) Schematic illustrating the fabrication process of the all-$Ti_3C_2T_x$ MXene MSC (Peng et al., 2016). (c) The electronic band gap for various MXenes sheets (Zhan et al., 2020). (d) (I) CV curves of different modified electrodes (Fang et al., 2020), (II) DPV responses to different concentrations (2–500 pg mL^{-1}) of CEA antigens (Wang et al., 2019), (III) EIS Nyquist plots of the miRNA-155 detection in 5.0 mM [Fe(CN)6]3−/4− containing 0.01 M PBS (Duan et al., 2020). (IV) The corresponding variations of the Rct values with different concentrations (0.1, 0.5, 1, and 2 mg mL^{-1}) for detecting miRNA-155 detection procedures ($n=3$) (Fang et al., 2020).

of the cancer-associated exosome. Herein, BPQDs catalyze the oxidation of Ru(dcbpy)32+ and are utilized as a co-reactant. The self-enhanced Ru(dcbpy)32+@BPQDs ECL system generates a strong ECL signal by shortening electron transfer distance and minimizing energy loss. MXenes provide large specific surface area and excellent conductivity and act as a supporter to enhance the number of Ru(dcbpy)32+ and BPQDs are immobilized, which improves the ECL signal.

They studied the cyclic voltammetry (CV) behaviours of ECL biosensors over electrodes in 5 mM [Fe(CN)6]3⁻/4⁻ including 0.1 M KCl. As shown in Figure 18.5d(I), bare GCE displayed a pair of well-defined redox peak currents (curve a); however, when SiO_2 nano-urchins (NUs) were placed over the electrode, the redox peak current diminished (curve b), which was linked to SiO_2 NUs with weak conductivity and hence blocking electron transport. The redox peak currents dramatically increased when interleukins (ILs) were coated onto the electrode due to the promotion impact of ILs for electronic transmission (curve c). The redox peak currents reduced dramatically (curve d) after Apt was incubated on the electrode, which was attributed to the aptamer with negatively charged phosphate backbone inhibiting the diffusion of the redox probe [Fe(CN)6]3⁻/4⁻ to the electrode surface. In exosome-modified electrodes, the redox peak currents are reduced much further (curve e). In contrast to (curve f), when MXenes- BPQDs@Ru(dcbpy)3 2+-PEI-AbCD63 (curve f) was deposited onto the electrode, the redox peak currents increased dramatically, indicating that MXenes and BPQDs have a synergistic promotion effect for electron transfer (Figure 18.5d(I)). Additionally, both BPQDs and MXenes have a good photothermal effect, which was deftly exploited as a thermal converter device in the development of a photothermal biosensor for exosome characterization. The developed dual-modality MXenes-BPQDs probe aptasensor in this study not only enhanced the signal while detection but also provided an effective and reliable approach for exosome detection in cancer patients (Fang et al., 2020). A label-free, ultrasensitive, and multiplexed microfluidic paper-based EC aptasensor for simultaneous detection of carcinoembryonic antigen (CEA) and neuron-specific enolase (NSE) was developed by Wang et al. which shows LOD 2 pg mL^{-1} for CEA and 10 pg mL^{-1} for NSE. The degree of decreased peak currents in differential pulse voltammetry (DPV) responses, which was attributable to the formation of aptamer antigen complex on the electrode surface, was used to detect the two analytes. The currents steadily reduced when CEA (3 A) and NSE increased, as shown in Figure 18.5d(II). In the ranges of 0.01 ~ 500 ng mL^{-1} for CEA (R^2=0.989) and 0.05 ~ 500 ng mL^{-1} for NSE (R^2=0.944), the calibration plots revealed a satisfactory linear detection relationship between peak currents and analyte concentrations (Wang et al., 2019). Recently, iron phthalocyanine quantum dots (FePcQDs) decorated MXene nanosheets (denoted as $Ti_3C_2T_x$@FePcQDs) and employed as promising nanocarrier of cDNA toward miRNA-155. This construct was used as a novel ultrasensitive impedimetric autosensing system for the detection of microRNA-155 (miRNA-155). The $Ti_3C_2T_x$@FePcQDs-based aptasensor demonstrated ultrahigh sensitivity with LOD of 4.3 aM (S/N=3) within the miRNA-155 concentration range of 0.01 fM to 10 pM for miRNA-155 detection; the suggested impedimetric autosensing system outperformed other published miRNA-155 aptasensors in terms of ease of fabrication, lack of labels, fast reaction time, and better sensing performance for detecting miRNA-155. This technique for determining cancer-associated miRNAs holds a lot of potential for early cancer biomarker detection (Fang et al., 2020).

Electrochemical impedance spectroscopy (EIS) Nyquist plots of miRNA-155 detection techniques employing an EC aptasensor based on a $Ti_3C_2T_x$@FePcQDs nanohybrid in 5.0 mM [Fe(CN)6]3⁻/4⁻ in 0.01 M PBS are shown in Figure 18.5d(III) (Duan et al., 2020). The experimental parameters specifically the use of $Ti_3C_2T_x$@FePcQD nanohybrid, cDNA concentration, and miRNA-155 binding time onto the $Ti_3C_2T_x$@ FePcQDs-based aptasensor were tuned to achieve excellent sensing performance for miRNA-155 detection. EIS was used for the entire fabrication and detecting process.

The charge transfer resistance (Rct) values generated by each step for miRNA-155 detection (fM) utilizing aptasensors based on the $Ti_3C_2T_x$@FePcQD nanohybrid with varied usages (0.1, 0.5, 1.0, and 2.0 mg^{-1}) under the same conditions are shown in Figure 18.5d(IV). The results showed that as the concentration of the $Ti_3C_2T_x$@FePcQDs dispersion grew from 0.1 to 1 mg mL^{-1}, the Rct values originating from the detection of miRNA-155 increased. When the $Ti_3C_2T_x$@FePcQDs dispersion concentration was more than 1 mg mL^{-1}, the Rct value for detecting miRNA-155 dropped dramatically (Duan et al., 2020).

18.5 CURRENT CHALLENGES

Since the discovery of MXene in 2011, MXenes are widely used in different fields including diagnostics and therapeutics. Although MXene is used widely, some lacunae need to be fulfilled to harvest its full potential. Although their 70+MXene are known, the number is growing rapidly. There is the possibility to explore many more compounds which are waiting to be included in the MXene family. The synthesis of new MAX phases and other layered carbide and nitride precursors is a hot research direction. The prediction of the various electronic, magnetic, thermal, and chemical properties to get the tunable size, ordered structures, strong surface terminations, and high yielding is needed. This motive can be achieved using computational strategies which can further increase the feasibility of MXene synthesis.

With advanced knowledge, the ion dynamics between the sheets of MXene can be controlled or altered to obtain the desired electronic conductivity of the MXene. This can be utilized to fabricate sensors with higher conductivity and the lowest resistance to obtain low LOD. Understanding the electrical properties of the MXene can pave new paths in the research field.

Very recently, the ionic-gel-based highly durable electronic skins were fabricated utilizing ionic gel and MXenes embedded into the polymer matrix. The electronic skin showed excellent mechanical properties, super adhesion, high sensitivity to strain and pressure and could tolerate harsh environments. Owing to the various excellent properties, the fabricated skin can be applied for multifunctional sensing purposes (Lu et al., 2022). Similarly, for serving the healthcare monitoring purpose, the strain sensor was fabricated. The sensor was based on the carbon nanotubes and MXenes in polydimethylsiloxane matrix. The sensor was found to be highly stable; durable; and, moreover, washable and could be used for the real-time monitoring of the electrocardiogram (ECG) and joints movements (Xu et al., 2021). These types of wearable sensors need to be explored more and utilized for healthcare monitoring purposes.

The conventional synthetic approach of MXene from the MAX phase includes the use of fluoride-containing compounds. The fluoride-containing compounds pose safety hazards and also limit the yield of MXene as they can alter the synthesis of MXene from the Al-containing MAX phase. The fluoride-free approach using hydrochloric acid was also used for the EC etching of the MXene from the MAX phase, but the over-etching and scaling up remain the challenge.

The search for new etching methods has become a frequent topic of research among scientists working in the area of synthesis, so major developments can be expected soon. The main challenge is to develop a sensitive, easy-to-use, cost-effective point-of-care test (POCT) to eradicate the false-negative and false-positive results. Certain points limit the use of MXene practically. Firstly, strong acids are used for the exfoliation as the MAX phase layers have strong interactions. The use of these strong acids restricts the use of MXenes in various fields, especially in green chemistry. The other harsh chemicals and use of sophisticated instruments for the synthesis are also limiting points. This point raises the demand for new etching methods and experimental conditions which use alternative chemicals. Despite the excellent features, several issues limit the practical applications of MXenes. The majority of the problems arise due to the synthesis process that requires the use of toxic chemicals and harsh experimental conditions. Hence, new experimental routes and conditions should be explored to synthesize novel MXenes with unique properties.

In this regard, recently, MXene quantum dots were synthesized using watermelon peel extract. The synthesized material was utilized for the biocompatible sensing purposes and also integrated with the smartphone. More similar work in this field is expected in the future for the sensing of various analytes and fabrication of wearable sensor (Sun, Du, Chen, Wang, & Shen, 2021). The use of M in the MAX phase is only limited to some elements Ti, V, Nb, Mo, Ta, Hf. However, many more elements can be used as M in MAX which needs to explore. Instead of metal carbides, metal nitrides can also be used. Till now, there are no more studies available on the different morphology of the MXenes; only the sheet is broadly studied even though other morphology like tube spheres also exist. The complex structure of MXenes during synthesis is still a challenge for researchers (Liu et al., 2021). To understand MXene better, due to its multivariate structure, further studies are needed.

Computational approaches can pave the way for designing new MXenes structures. MXenes family can be further expanded by designing and discovering new heterostructures. The search for new etching methods has become a frequent topic of research among scientists working in the area of synthesis, so major developments can be expected soon. Work has been already done on the MXenes, but still, there is a long way to go. Researchers are working on the greener synthesis of MXenes for replacing the harsh chemical. By exploring the various properties of MXenes like superconductivity, thermal transport, biocompatibility, etc., it can be exploited to the various multidisciplinary fields. The MXenes displayed tunable magnetic and electric properties and used as an excellent modifier in EC studies and hence showed adaptable applications in various fields such as energy storage, nanomedicine, diagnosis (Shukla, 2020).

MXene-enabled aptasensors have shown considerable promise for the early detection of cancer biomarkers. Aptamers have gained much attention in the last few decades and are explored for diagnostic and even therapeutic purposes. The ease of synthesis, low cost, and stability, make aptamer a preferable candidate over other B recognition elements (BREs). Aptamers also have many pitfalls including difficult multianalyte detection, cross-reactivity, poor precision, etc. The limiting factor in the designing of aptamers is the library design. It usually depends on the oligonucleotide's quality and length, its structural stability, specific binding sequences. The quality of the oligonucleotides depends on the nucleotide's ratio and the level of complexity of an aptamer library. More and more research in the field to make aptamer preferable and cost effective has improved its demand. More research is in demand for better integration and detection of cancer biomarkers. Advanced bioinformatics techniques and genetic-algorithm-based methods are required for studying multiple cancer cells and studying the binding affinities of biomarkers with specific aptamers.

Major challenges which need to be addressed include the methods to screen aptamer rapidly with high specificity, maintaining stability in the biological environment, reducing the toxicity of nanomaterials used in conjugation with aptamers. The MXene and aptamer-based devices can be commercialized as POCT devices. Current approaches for the diagnosis of various diseases have been using a lateral-flow-based system either in the form of electrical- or optical-based devices. The major challenge is device miniaturizations and multiplexing. POCT devices should be affordable, sensitive, specific, user friendly, robust and can always be performed outside a laboratory or hospital by a non-technical person or by a layman as per the assured criteria. Other than cost cutting, the POC test devices should give a signal in a lesser sample volume (microliter). The device that can differentiate between different biomolecules and can be used for multiplex analyte sensing is the need of the hour. MXene-enabled EC aptasensors have the potential to fulfil the criteria for effective POCT devices for early disease diagnosis.

There is an enormous interest in AI-integrated POCT devices to improve treatment efficacy and healthcare. A convolutional neural network (CNN) is one of the important AI-based tools that can be integrated with devices for the prediction of disease in resource-limited settings. These devices can also be interlinked with top-class portable healthcare interfaces such as smartphones, smartwatches, and other wearable devices. These devices can also be used to monitor the healthy as well as the patient's conditions. By integrating these devices with the Internet of Things (IoT) and Internet of medical Things (IoMT) devices, the past and present records of the individual's health condition can be assessed by medical practitioners for better disease prognosis and decision making (Ranjan et al., 2022). These devices can be used to forecast clinical conditions digitally. This will be a rapid, cost-effective, intelligent approach to deal with the disease condition even in resource-limited settings. With the interdisciplinary amalgamation of various fields such as material chemistry, biomedical engineering, computational chemistry, the MXene-enabled EC aptasensor will have the potential to tackle the challenges associated with the fabrication and development of POCT devices for early cancer diagnostics.

18.6 CONCLUSIONS

Despite the various advancements in the field of oncology, cancer still poses an enormous risk and challenges to tackle. There are several challenging factors which are needed to be addressed and resolved. This review emphasizes MXene, MXene-based materials, and their conjugation with aptamers for the fabrication of EC aptasensor for cancer diagnostics with a focus on the current trends, important hurdles, and their future insights. The role of nanotechnology in dealing with deadly diseases such as cancer is important.

Many portable biomedical devices for imagining, sensing, and other purposes were successfully developed using nanomaterials. MXene is a new class of nanomaterial emerging at a high rate. MXene is significantly used for diagnostic and other biomedical purposes. Many advancements have been made globally by many investigators to explore the properties of MXene. On the other hand, aptasensors as POCT have achieved significant attention in the past few decades. The aptamers have remarkable advantages over the conventional bio-recognition elements.

The aptasensors are economical, stable, sensitive, selective, and show negligible batch-to-batch variation. This review focussed on MXene-enabled EC aptasensors for the detection of cancer biomarkers. However, the validation of these aptasensors in a larger patient cohort using a real sample is still needed. Further, challenges associated with device miniaturization, multiplexing, and its integration with IoT-enabled smartphones need to be resolved by using an amalgamation of interdisciplinary fields to optimize the POCT sensing systems.

The development of MXene-enabled EC aptasensor for the real-time, cost-effective, and early diagnosis of cancer still has a long way to go. Nevertheless, there are many hurdles in the commercial utilization of the MXene-based aptasensors; however, further research in this field will prepare the base for the commercialization of the MXene-based aptasensors for cancer diagnostics.

REFERENCES

Ahmad, M. S., Pandey, A., Abd Rahim, N., Aslfattahi, N., Mishra, Y. K., Rashid, B., & Saidur, R. (2021). 2-D Mxene flakes as potential replacement for both TCO and Pt layers for Dye-Sensitized Solar cell. *Ceramics International*, *47*(19), 27942–27947.

Alkhamis, O., Canoura, J., Yu, H., Liu, Y., & Xiao, Y. (2019). Innovative engineering and sensing strategies for aptamer-based small-molecule detection. *TrAC Trends in Analytical Chemistry*, *121*, 115699.

Are, C., Bartlett, D. L., Nissan, A., Dov, Z., Gupta, A., Savant, D., & de Castro Ribeiro, H. S. (2020). Global Forum of Cancer Surgeons: position statement to promote cancer surgery globally. *Annals of Surgical Oncology*, *27*, 2573–2576.

Bayat, P., Taghdisi, S. M., Rafatpanah, H., Abnous, K., & Ramezani, M. (2019). In vitro selection of CD70 binding aptamer and its application in a biosensor design for sensitive detection of SKOV-3 ovarian cells. *Talanta*, *194*, 399–405.

Bernard, P. S., & Wittwer, C. T. (2002). Real-time PCR technology for cancer diagnostics. *Clinical Chemistry*, *48*(8), 1178–1185.

Billet, B., Chovelon, B., Fiore, E., Oukacine, F., Petrillo, M. A., Faure, P., & Peyrin, E. (2021). Aptamer switches regulated by post-transition/transition metal ions. *Angewandte Chemie*, *133*(22), 12454–12458.

Bostan, H. B., Danesh, N. M., Karimi, G., Ramezani, M., Shaegh, S. A. M., Youssefi, K., & Taghdisi, S. M. (2017). Ultrasensitive detection of ochratoxin A using aptasensors. *Biosensors and Bioelectronics*, *98*, 168–179.

Bray, F., Ferlay, J., Soerjomataram, I., Siegel, R. L., Torre, L. A., & Jemal, A. (2018). Global cancer statistics 2018: GLOBOCAN estimates of incidence and mortality worldwide for 36 cancers in 185 countries. *CA: A Cancer Journal for Clinicians*, *68*(6), 394–424.

Burda, C., Chen, X., Narayanan, R., & El-Sayed, M. A. (2005). Chemistry and properties of nanocrystals of different shapes. *Chemical Reviews*, *105*(4), 1025–1102.

Cao, B., Liu, H., Zhang, X., Zhang, P., Zhu, Q., Du, H., & Xu, B. (2021). MOF-derived ZnS nanodots/$Ti_3C_2T_x$ MXene hybrids boosting superior lithium storage performance. *Nano-Micro Letters*, *13*, 1–17.

Cella, L. N., Sanchez, P., Zhong, W., Myung, N. V., Chen, W., & Mulchandani, A. (2010). Nano aptasensor for protective antigen toxin of anthrax. *Analytical Chemistry*, *82*(5), 2042–2047.

Chen, R., Kan, L., Duan, F., He, L., Wang, M., Cui, J., & Zhang, Z. (2021). Surface plasmon resonance aptasensor based on niobium carbide MXene quantum dots for nucleocapsid of SARS-CoV-2 detection. *Microchimica Acta, 188*(10), 316.

Chen, X., Sun, X., Xu, W., Pan, G., Zhou, D., Zhu, J., & Song, H. (2018). Ratiometric photoluminescence sensing based on Ti_3C_2 MXene quantum dots as an intracellular pH sensor. *Nanoscale, 10*(3), 1111–1118.

Cui, H., Fu, X., Yang, L., Xing, S., & Wang, X.-F. (2021). 2D titanium carbide nanosheets based fluorescent aptasensor for sensitive detection of thrombin. *Talanta, 228*, 122219.

Dhiman, A., Kalra, P., Bansal, V., Bruno, J. G., & Sharma, T. K. (2017). Aptamer-based point-of-care diagnostic platforms. *Sensors and Actuators B: Chemical, 246*, 535–553.

Drolet, D. W., Moon-McDermott, L., & Romig, T. S. (1996). An enzyme-linked oligonucleotide assay. *Nature biotechnology, 14*(8), 1021–1025.

Duan, F., Guo, C., Hu, M., Song, Y., Wang, M., He, L., & Zhou, L. (2020). Construction of the 0D/2D heterojunction of $Ti_3C_2T_x$ MXene nanosheets and iron phthalocyanine quantum dots for the impedimetric aptasensing of microRNA-155. *Sensors and Actuators B: Chemical, 310*, 127844.

Ellington, A. D., & Szostak, J. W. (1990). In vitro selection of RNA molecules that bind specific ligands. *Nature, 346*(6287), 818–822.

Fang, D., Zhao, D., Zhang, S., Huang, Y., Dai, H., & Lin, Y. (2020). Black phosphorus quantum dots functionalized MXenes as the enhanced dual-mode probe for exosomes sensing. *Sensors and Actuators B: Chemical, 305*, 127544.

Guo, W., Umar, A., Algadi, H., Albargi, H., Ibrahim, A. A., Cui, K., & Wang, Y. (2021). Design of a unique "ON/OFF" switch electrochemical aptasensor driven by the pH for the detection of Aflatoxin B1 in acid solutions based on titanium carbide/carboxylated graphene oxide-poly (4-vinyl pyridine)/Aptamer composite. *Microchemical Journal, 169*, 106548.

Hong, J., Wang, W., Wang, J., Wang, X., Xie, H., Li, T., & Gan, N. (2021). A turn-on–type fluorescence resonance energy transfer aptasensor for vibrio detection using aptamer-modified polyhedral oligomeric silsesquioxane-perovskite quantum dots/Ti_3C_2 MXenes composite probes. *Microchimica Acta, 188*, 1–12.

Jiang, D., Wei, M., Du, X., Qin, M., Shan, X., & Chen, Z. (2022). One-pot synthesis of ZnO quantum dots/N-doped Ti_3C_2 MXene: Tunable nitrogen-doping properties and efficient electrochemiluminescence sensing. *Chemical Engineering Journal, 430*, 132771.

Jiang, D., Wei, M., Du, X., Qin, M., Shan, X., Wang, W., & Chen, Z. (2022). Ultrasensitive near-infrared aptasensor for enrofloxacin detection based on wavelength tunable AgBr nanocrystals electrochemiluminescence emission triggered by O-terminated Ti_3C_2 MXene. *Biosensors and Bioelectronics, 200*, 113917.

Jin, D., Yang, F., Zhang, Y., Liu, L., Zhou, Y., Wang, F., & Zhang, G.-J. (2018). ExoAPP: Exosome-oriented, aptamer nanoprobe-enabled surface proteins profiling and detection. *Analytical Chemistry, 90*(24), 14402–14411.

Kashefi-Kheyrabadi, L., Koyappayil, A., Kim, T., Cheon, Y.-P., & Lee, M.-H. (2021). A MoS_2@ $Ti_3C_2T_x$ MXene hybrid-based electrochemical aptasensor (MEA) for sensitive and rapid detection of Thyroxine. *Bioelectrochemistry, 137*, 107674.

Khalilzadeh, B., Shadjou, N., Charoudeh, H. N., & Rashidi, M.-R. (2017). Recent advances in electrochemical and electrochemiluminescence based determination of the activity of caspase-3. *Microchimica Acta, 184*, 3651–3662.

Khan, R., Parihar, A., & Sanghi, S. K. (2021). *Biosensor Based Advanced Cancer Diagnostics: From Lab to Clinics*. Academic Press, Cambridge, MA.

Kiio, L. K., Mbui, D. N., Ndangili, P. M., Onam, O. J., & Oloo, F. (2021). Current biosensors used for early detection of lung cancer biomarkers. *Journal of Cancer Research Reviews & Reports, 3*(3), 1–8.

Kim, Y. S., Raston, N. H. A., & Gu, M. B. (2016). Aptamer-based nanobiosensors. *Biosensors and Bioelectronics, 76*, 2–19.

Kumar, N., Sorna Gowri, V., Ranjan, P., Sadique, A., Yadav, S., Singhal, A., & Khan, R. (2021). Efficiency of nanomaterials for electrochemical diagnostics based point-of-care detection of non-Invasive oral cancer biomarkers. *Advanced Materials Letters, 12*(8), 1–20.

Kurra, N., Ahmed, B., Gogotsi, Y., & Alshareef, H. N. (2016). MXene-on-paper coplanar microsupercapacitors. *Advanced Energy Materials, 6*(24), 1601372.

Lakshmanakumar, M., Nesakumar, N., JBB, A. J., & Rayappan, J. B. B. (2021). Electrochemical DNA biosensors for cervical cancers. *Biomarkers and Biosensors for Cervical Cancer Diagnosis, 6*, 57–69.

Lee, J. F., Stovall, G. M., & Ellington, A. D. (2006). Aptamer therapeutics advance. *Current Opinion in Chemical Biology, 10*(3), 282–289.

Lee, J. H., Kim, J. A., Jeong, S., & Rhee, W. J. (2016). Simultaneous and multiplexed detection of exosome microRNAs using molecular beacons. *Biosensors and Bioelectronics, 86*, 202–210.

Li, B., Pu, W., Weng, L., Lyu, P., Xu, H., Zhang, W., & Wu, Q. (2022). Aptamer-functionalized Ti_3C_2-MXene nanosheets with one-step potentiometric detection of programmed death-ligand 1. *Electroanalysis, 34*(1), 2–7.

Li, M., Peng, X., Han, Y., Fan, L., Liu, Z., & Guo, Y. (2021). Ti_3C_2 MXenes with intrinsic peroxidase-like activity for label-free and colorimetric sensing of proteins. *Microchemical Journal, 166*, 106238.

Li, X., Lu, Y., & Liu, Q. (2021). Electrochemical and optical biosensors based on multifunctional MXene nanoplatforms: Progress and prospects. *Talanta, 235*, 122726.

Lin, X., Li, C., Meng, X., Yu, W., Duan, N., Wang, Z., & Wu, S. (2022). CRISPR-Cas12a-mediated luminescence resonance energy transfer aptasensing platform for deoxynivalenol using gold nanoparticle-decorated $Ti_3C_2T_x$ MXene as the enhanced quencher. *Journal of Hazardous Materials, 433*, 128750.

Liu, Y., Huang, S., Li, J., Wang, M., Wang, C., Hu, B., & Zhang, Z. (2021). 0D/2D heteronanostructure–integrated bimetallic CoCu-ZIF nanosheets and MXene-derived carbon dots for impedimetric cytosensing of melanoma B16-F10 cells. *Microchimica Acta, 188*, 1–12.

Lu, Y., Qu, X., Wang, S., Zhao, Y., Ren, Y., Zhao, W., & Dong, X. (2022). Ultradurable, freeze-resistant, and healable MXene-based ionic gels for multi-functional electronic skin. *Nano Research, 15*(5), 4421–4430.

Lubin, A. A., & Plaxco, K. W. (2010). Folding-based electrochemical biosensors: The case for responsive nucleic acid architectures. *Accounts of Chemical Research, 43*(4), 496–505.

Mahmoudpour, M., Ding, S., Lyu, Z., Ebrahimi, G., Du, D., Dolatabadi, J. E. N., & Lin, Y. (2021). Aptamer functionalized nanomaterials for biomedical applications: Recent advances and new horizons. *Nano Today, 39*, 101177.

Mas-Balleste, R., Gomez-Navarro, C., Gomez-Herrero, J., & Zamora, F. (2011). 2D materials: To graphene and beyond. *Nanoscale, 3*(1), 20–30.

Mei, Z., Chu, H., Chen, W., Xue, F., Liu, J., Xu, H., & Zheng, L. (2013). Ultrasensitive one-step rapid visual detection of bisphenol A in water samples by label-free aptasensor. *Biosensors and Bioelectronics, 39*(1), 26–30.

Mi, X., Li, H., Tan, R., Feng, B., & Tu, Y. (2021). The TDs/aptamer cTnI biosensors based on HCR and Au/Ti_3C_2-MXene amplification for screening serious patient in COVID-19 pandemic. *Biosensors and Bioelectronics, 192*, 113482.

Naguib, M., Come, J., Dyatkin, B., Presser, V., Taberna, P.-L., Simon, P., & Gogotsi, Y. (2012). MXene: A promising transition metal carbide anode for lithium-ion batteries. *Electrochemistry Communications, 16*(1), 61–64.

Naguib, M., Mashtalir, O., Carle, J., Presser, V., Lu, J., Hultman, L., & Barsoum, M. W. (2012). Two-dimensional transition metal carbides. *ACS Nano, 6*(2), 1322–1331.

Noh, S., Lee, H., Kim, J., Jang, H., An, J., Park, C., & Lee, T. (2022). Rapid electrochemical dual-target biosensor composed of an Aptamer/MXene hybrid on Au microgap electrodes for cytokines detection. *Biosensors and Bioelectronics, 207*, 114159.

Parihar, A., Singhal, A., Kumar, N., Khan, R., Khan, M. A., & Srivastava, A. K. (2022). Next-generation intelligent MXene-based electrochemical aptasensors for point-of-care cancer diagnostics. *Nano-Micro Letters, 14*(1), 100.

Peng, Y.-Y., Akuzum, B., Kurra, N., Zhao, M.-Q., Alhabeb, M., Anasori, B., & Gogotsi, Y. (2016). All-MXene (2D titanium carbide) solid-state microsupercapacitors for on-chip energy storage. *Energy & Environmental Science, 9*(9), 2847–2854.

Ping, J., Fan, Z., Sindoro, M., Ying, Y., & Zhang, H. (2017). Recent advances in sensing applications of two-dimensional transition metal dichalcogenide nanosheets and their composites. *Advanced Functional Materials, 27*(19), 1605817.

Ping, J., Zhou, Y., Wu, Y., Papper, V., Boujday, S., Marks, R. S., & Steele, T. W. (2015). Recent advances in aptasensors based on graphene and graphene-like nanomaterials. *Biosensors and Bioelectronics, 64*, 373–385.

Qiao, B., Guo, Q., Jiang, J., Qi, Y., Zhang, H., He, B., & Shen, J. (2019). An electrochemiluminescent aptasensor for amplified detection of exosomes from breast tumor cells (MCF-7 cells) based on G-quadruplex/hemin DNAzymes. *Analyst, 144*(11), 3668–3675.

Raimondo, F., Morosi, L., Chinello, C., Magni, F., & Pitto, M. (2011). Advances in membranous vesicle and exosome proteomics improving biological understanding and biomarker discovery. *Proteomics, 11*(4), 709–720.

Ranjan, P., Parihar, A., Jain, S., Kumar, N., Dhand, C., Murali, S., & Srivastava, A. K. (2020). Biosensor-based diagnostic approaches for various cellular biomarkers of breast cancer: A comprehensive review. *Analytical Biochemistry, 610*, 113996.

Ranjan, P., Singhal, A., Sadique, M. A., Yadav, S., Parihar, A., & Khan, R. (2022). Scope of biosensors, commercial aspects, and miniaturized devices for point-of-care testing from lab to clinics applications. *Biosensor Based Advanced Cancer Diagnostics, 41*, 395–410.

Rasheed, P. A., Pandey, R. P., Jabbar, K. A., & Mahmoud, K. A. (2022). $Nb_4C_3T_x$ (MXene)/Au/DNA aptasensor for the ultraselective electrochemical detection of lead in water samples. *Electroanalysis, 34*(10), 1540–1546.

Ravalli, A., Voccia, D., Palchetti, I., & Marrazza, G. (2016). Electrochemical, electrochemiluminescence, and photoelectrochemical aptamer-based nanostructured sensors for biomarker analysis. *Biosensors, 6*(3), 39.

Razmi, N., Baradaran, B., Hejazi, M., Hasanzadeh, M., Mosafer, J., Mokhtarzadeh, A., & de la Guardia, M. (2018). Recent advances on aptamer-based biosensors to detection of platelet-derived growth factor. *Biosensors and Bioelectronics, 113*, 58–71.

Robertson, D. L., & Joyce, G. F. (1990). Selection in vitro of an RNA enzyme that specifically cleaves single-stranded DNA. *Nature, 344*(6265), 467–468.

Sangu, S. S., Illias, N. M., Ong, C. C., Gopinath, S. C. B., & Saheed, M. S. M. (2021). MXene-based aptasensor: Characterization and high-performance voltammetry detection of deoxynivalenol. *BioNanoScience, 11*, 314–323.

Sheng, A., Wang, P., Yang, J., Tang, L., Chen, F., & Zhang, J. (2021). MXene coupled with CRISPR-Cas12a for analysis of endotoxin and bacteria. *Analytical Chemistry, 93*(10), 4676–4681.

Shukla, V. (2020). The tunable electric and magnetic properties of 2D MXenes and their potential applications. *Materials Advances, 1*(9), 3104–3121.

Sinha, A., Zhao, H., Huang, Y., Lu, X., Chen, J., & Jain, R. (2018). MXene: An emerging material for sensing and biosensing. *TrAC Trends in Analytical Chemistry, 105*, 424–435.

Song, X., Gao, H., Yuan, R., & Xiang, Y. (2022). Trimetallic nanoparticle-decorated MXene nanosheets for catalytic electrochemical detection of carcinoembryonic antigen via Exo III-aided dual recycling amplifications. *Sensors and Actuators B: Chemical, 359*, 131617.

Sun, J., Du, H., Chen, Z., Wang, L., & Shen, G. (2021). MXene quantum dot within natural 3D watermelon peel matrix for biocompatible flexible sensing platform. *Nano Research, 53*(7), 1–7.

Tuerk, C., & Gold, L. (1990). Systematic evolution of ligands by exponential enrichment: RNA ligands to bacteriophage T4 DNA polymerase. *Science, 249*(4968), 505–510.

Ullah, N., Chen, W., Noureen, B., Tian, Y., Du, L., Wu, C., & Ma, J. (2021). An electrochemical $Ti_3C_2T_x$ aptasensor for sensitive and label-free detection of marine biological toxins. *Sensors, 21*(14), 4938.

Ullah, S., Shahzad, F., Qiu, B., Fang, X., Ammar, A., Luo, Z., & Zaidi, S. A. (2022). MXene-based aptasensors: Advances, challenges, and prospects. *Progress in Materials Science, 129*, 100967.

Vajhadin, F., Mazloum-Ardakani, M., Shahidi, M., Moshtaghioun, S. M., Haghiralsadat, F., Ebadi, A., & Amini, A. (2022). MXene-based cytosensor for the detection of HER2-positive cancer cells using $CoFe_2O_4$@Ag magnetic nanohybrids conjugated to the HB5 aptamer. *Biosensors and Bioelectronics, 195*, 113626.

Vinkenborg, J. L., Karnowski, N., & Famulok, M. (2011). Aptamers for allosteric regulation. *Nature Chemical Biology, 7*(8), 519–527.

Wang, H., Chen, H., Huang, Z., Li, T., Deng, A., & Kong, J. (2018). DNase I enzyme-aided fluorescence signal amplification based on graphene oxide-DNA aptamer interactions for colorectal cancer exosome detection. *Talanta, 184*, 219–226.

Wang, H., Li, H., Huang, Y., Xiong, M., Wang, F., & Li, C. (2019). A label-free electrochemical biosensor for highly sensitive detection of gliotoxin based on DNA nanostructure/MXene nanocomplexes. *Biosensors and Bioelectronics, 142*, 111531.

Wang, H., Sun, J., Lu, L., Yang, X., Xia, J., Zhang, F., & Wang, Z. (2020). Competitive electrochemical aptasensor based on a cDNA-ferrocene/MXene probe for detection of breast cancer marker Mucin1. *Analytica Chimica Acta, 1094*, 18–25.

Wang, H., Yang, R., Yang, L., & Tan, W. (2009). Nucleic acid conjugated nanomaterials for enhanced molecular recognition. *ACS Nano, 3*(9), 2451–2460.

Wang, W., Xiao, S., Jia, Z., Xie, H., Li, T., Wang, Q., & Gan, N. (2022). A dual-mode aptasensor for foodborne pathogens detection using Pt, phenylboric acid and ferrocene modified Ti_3C_2 MXenes nanoprobe. *Sensors and Actuators B: Chemical, 351*, 130839.

Wang, X., Chu, C., Shen, L., Deng, W., Yan, M., Ge, S., & Song, X. (2015). An ultrasensitive electrochemical immunosensor based on the catalytical activity of MoS_2-Au composite using Ag nanospheres as labels. *Sensors and Actuators B: Chemical, 206*, 30–36.

Wang, Y., Luo, J., Liu, J., Sun, S., Xiong, Y., Ma, Y., & Cai, X. (2019). Label-free microfluidic paper-based electrochemical aptasensor for ultrasensitive and simultaneous multiplexed detection of cancer biomarkers. *Biosensors and Bioelectronics, 136*, 84–90.

Wang, Y., Sun, W., Li, Y., Zhuang, X., Tian, C., Luan, F., & Fu, X. (2021). Imidazole metal-organic frameworks embedded in layered $Ti_3C_2T_x$ Mxene as a high-performance electrochemiluminescence biosensor for sensitive detection of HIV-1 protein. *Microchemical Journal, 167*, 106332.

Wongkaew, N., Simsek, M., Griesche, C., & Baeumner, A. J. (2018). Functional nanomaterials and nanostructures enhancing electrochemical biosensors and lab-on-a-chip performances: Recent progress, applications, and future perspective. *Chemical Reviews*, *119*(1), 120–194.

Wu, Z., Sun, D.-W., Pu, H., Wei, Q., & Lin, X. (2022). $Ti_3C_2T_x$ MXenes loaded with Au nanoparticle dimers as a surface-enhanced Raman scattering aptasensor for AFB1 detection. *Food Chemistry*, *372*, 131293.

Xu, B., Zhi, C., & Shi, P. (2020). Latest advances in MXene biosensors. *Journal of Physics: Materials*, *3*(3), 031001.

Xu, B., Zhu, M., Zhang, W., Zhen, X., Pei, Z., Xue, Q., & Shi, P. (2016). Ultrathin MXene-micropattern-based field-effect transistor for probing neural activity. *Advanced Materials*, *28*(17), 3333–3339. doi: https://doi.org/10.1002/adma.201504657.

Xu, X., Chen, Y., He, P., Wang, S., Ling, K., Liu, L., & Cao, J. (2021). Wearable CNT/$Ti_3C_2T_x$ MXene/PDMS composite strain sensor with enhanced stability for real-time human healthcare monitoring. *Nano Research*, *14*(8), 2875–2883.

Yang, J., Zhong, W., Yu, Q., Zou, J., Gao, Y., Liu, S., & Lu, L. (2022). MXene–AuNP-based electrochemical aptasensor for ultra-sensitive detection of chloramphenicol in honey. *Molecules*, *27*(6), 1871.

Yang, P., Guo, X., Zhang, J., Chen, C., Gan, Y., Xie, W., & Wu, Z. (2022). Picomolar thrombin detection by orchestration of triple signal amplification strategy with hierarchically porous $Ti_3C_2T_x$ MXene electrode material-catalytic hairpin assembly reaction-metallic nanoprobes. *Biosensors and Bioelectronics*, *208*, 114228.

Ye, Z., Jiang, Y., Li, L., Wu, F., & Chen, R. (2022). Enhanced catalytic conversion of polysulfide using 1D CoTe and 2D MXene for heat-resistant and lean-electrolyte Li–S batteries. *Chemical Engineering Journal*, *430*, 132734.

Yu, X., He, L., Pentok, M., Yang, H., Yang, Y., Li, Z., & Liu, T. (2019). An aptamer-based new method for competitive fluorescence detection of exosomes. *Nanoscale*, *11*(33), 15589–15595.

Zahra, Q. U. A., Fang, X., Luo, Z., Ullah, S., Fatima, S., Batool, S., & Shahzad, F. (2022). Graphene based nanohybrid aptasensors in environmental monitoring: Concepts, design and future outlook. *Critical Reviews in Analytical Chemistry*, *53*(7), 1–22.

Zahra, Q. U. A., Luo, Z., Ali, R., Khan, M. I., Li, F., & Qiu, B. (2021). Advances in gold nanoparticles-based colorimetric aptasensors for the detection of antibiotics: An overview of the past decade. *Nanomaterials*, *11*(4), 840.

Zhan, X., Si, C., Zhou, J., & Sun, Z. (2020). MXene and MXene-based composites: Synthesis, properties and environment-related applications. *Nanoscale Horizons*, *5*(2), 235–258.

Zhang, H., Qiao, B., Guo, Q., Jiang, J., Cai, C., & Shen, J. (2020). A facile and label-free electrochemical aptasensor for tumour-derived extracellular vesicle detection based on the target-induced proximity hybridization of split aptamers. *Analyst*, *145*(10), 3557–3563.

Zhang, H., Wang, Z., Wang, F., Zhang, Y., Wang, H., & Liu, Y. (2020). In situ formation of gold nanoparticles decorated Ti_3C_2 MXenes nanoprobe for highly sensitive electrogenerated chemiluminescence detection of exosomes and their surface proteins. *Analytical Chemistry*, *92*(7), 5546–5553.

Zhang, H., Wang, Z., Wang, F., Zhang, Y., Wang, H., & Liu, Y. (2021). Ti_3C_2 MXene mediated Prussian blue in situ hybridization and electrochemical signal amplification for the detection of exosomes. *Talanta*, *224*, 121879.

Zhang, H., Wang, Z., Zhang, Q., Wang, F., & Liu, Y. (2019). Ti_3C_2 MXenes nanosheets catalyzed highly efficient electrogenerated chemiluminescence biosensor for the detection of exosomes. *Biosensors and Bioelectronics*, *124*, 184–190.

Zhang, K., Fan, Z., Yao, B., Ding, Y., Zhao, J., Xie, M., & Pan, J. (2021). Exploring the trans-cleavage activity of CRISPR-Cas12a for the development of a Mxene based electrochemiluminescence biosensor for the detection of Siglec-5. *Biosensors and Bioelectronics*, *178*, 113019.

Zhang, Q., Wang, F., Zhang, H., Zhang, Y., Liu, M., & Liu, Y. (2018). Universal Ti_3C_2 MXenes based self-standard ratiometric fluorescence resonance energy transfer platform for highly sensitive detection of exosomes. *Analytical Chemistry*, *90*(21), 12737–12744.

Zhao, L., Qi, X., Yan, X., Huang, Y., Liang, X., Zhang, L., & Tan, W. (2019). Engineering aptamer with enhanced affinity by triple helix-based terminal fixation. *Journal of the American Chemical Society*, *141*(44), 17493–17497.

Zhao, P., Zheng, J., Liang, Y., Tian, F., Peng, L., Huo, D., & Hou, C. (2021). Functionalized carbon nanotube-decorated MXene nanosheet-enabled microfluidic electrochemical aptasensor for carcinoembryonic antigen determination. *ACS Sustainable Chemistry & Engineering*, *9*(46), 15386–15393.

Zhao, W.-W., Xu, J.-J., & Chen, H.-Y. (2016). Photoelectrochemical aptasensing. *TrAC Trends in Analytical Chemistry*, *82*, 307–315.

Zheng, F., Ke, W., Shi, L., Liu, H., & Zhao, Y. (2019). Plasmonic Au–Ag janus nanoparticle engineered ratiometric surface-enhanced raman scattering aptasensor for Ochratoxin A detection. *Analytical Chemistry*, *91*(18), 11812–11820.

Zhou, S., Gu, C., Li, Z., Yang, L., He, L., Wang, M., & Zhang, Z. (2019). $Ti_3C_2T_x$ MXene and polyoxometalate nanohybrid embedded with polypyrrole: Ultra-sensitive platform for the detection of osteopontin. *Applied Surface Science*, *498*, 143889.

Zhu, C., Li, L., Wang, Z., Irfan, M., & Qu, F. (2020). Recent advances of aptasensors for exosomes detection. *Biosensors and Bioelectronics*, *160*, 112213.

19 Process Design of Electrochemical Sensors

Subhankar Roy, Md. Aurangzeb, and Ravi Tejasvi

19.1 INTRODUCTION

Electrochemical sensors can detect chemical responses as electrical potentials or currents and convert them into electrical signals [1–6]. Electrochemical sensors are pivotal instruments that bring together the worlds of chemistry and electronics in a harmonious and functional manner. At their core, these sensors can detect and measure chemical reactions and almost instantly convert these reactions into electric signals. This rapid transformation, which facilitates the direct transmission of electric signals, stands out as one of their most significant advantages, particularly as it enables real-time monitoring of various chemical processes. The operational foundation of these sensors is rooted in their capability to measure key electrical parameters. Typically, these parameters are current or potential (often referred to as voltage). These measurements come to fruition primarily through reduction-oxidation (redox) reactions. Such reactions are initiated when specific target chemicals come into contact and interact with specially prepared reactive substances on electrodes. Central to these sensors is the component known as the transducer. It plays a vital role by acting as a bridge that translates or transforms intricate chemical information into discernible electrical data. This conversion is essential, as it permits the accurate quantification and analysis of the target chemical or substance in various applications.

Figure 19.1 shows a general scheme of electrochemical sensing.

The versatility and wide applicability of electrochemical sensors are significantly enhanced by a suite of specialised techniques. Techniques like amperometry, potentiometry, voltammetry, and impedance spectroscopy not only provide a broader framework for these sensors to operate but also are instrumental in ensuring their heightened sensitivity and specificity. These attributes are of paramount importance in analytical chemistry and various real-world applications. Modern technological advancements have further augmented the capabilities of these sensors. One notable progress is the miniaturisation of these sensors, which allows for their seamless integration into portable

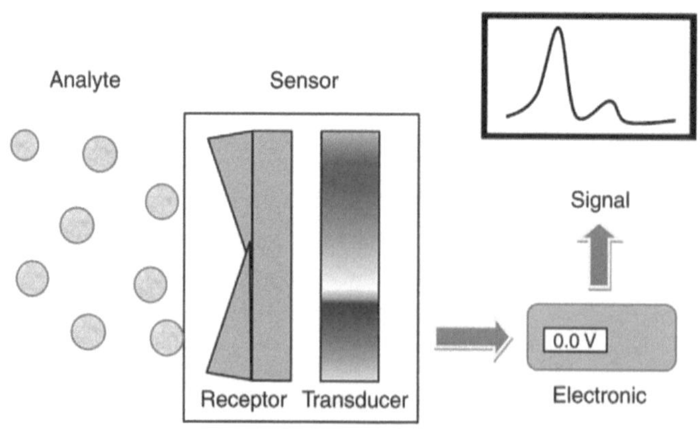

FIGURE 19.1 A general scheme of electrochemical sensing [7].

DOI: 10.1201/9781032654874-19

devices. This feature has revolutionised fields such as point-of-care diagnostics, where immediate results are crucial. Furthermore, their adaptability extends to environmental monitoring, where real-time data can be critical in decision-making processes, and to industrial control systems that require constant and accurate monitoring.

The range of applications for electrochemical sensors is impressively vast. For instance, pH electrodes, commonly found in laboratories and industrial setups, are designed to measure the potential difference between a pH-sensitive electrode and a reference point. Ion-selective electrodes take this a step further, being meticulously calibrated to detect specific ions like potassium or calcium. In the medical arena, Clark electrodes have become indispensable tools that measure oxygen concentration in blood. Additionally, diabetic patients globally rely on specialised glucose sensors, which harness the enzyme glucose oxidase (GO_x) to monitor blood glucose levels through specific oxidation processes. The realm of water treatment employs conductivity sensors to evaluate water quality, while redox probes provide crucial insights into the redox status of various solutions. Other specialised sensors, like amperometric gas sensors, are tailored to detect gases like chlorine, and biosensors employ biological molecules, such as enzymes, to detect and quantify specific substances. The importance of heavy metal detection in environmental conservation has led to the creation of dedicated sensors, with the mercury drop electrode being a classic example. The landscape of electrochemical sensors is both diverse and expansive. They offer bespoke solutions catering to an array of applications across environmental, health, and industrial sectors, marking their profound impact on modern science and technology.

Different people disagree on whether a device is an electrochemical sensor. Some scientists reject conductance probes because these rely on the physical responses generated by mobile ionic species, instead of the chemical reaction aspect of the electrochemical processes. When conductivity measurements are made across the interfaces, marking the interfaces in a cell may resolve the ambiguity because that marks individual cell compartments where reactions can take place. All these consecutive cells offer different types of impedances, and these can be combined to create the total or overall impedance. Both the bulk solution and electrode contact contribute to this complex quantity. Indeed, conductance and impedimetric sensors are often interchangeable. Based on these broad concepts, electrochemical sensors are mostly potentiometric or amperometric. All electrochemical sensors detect the analyte by its unique reaction with the sensing platform. Potentiometric sensors detect potential drop through the analyte at a fixed current in the electrochemical circuit. Amperometric sensors measure current responses that may occur due to the flow of electrons in the electrochemical circuit when the analyte reacts within the sensor platform bounds. Coulometric sensors measure the charge involved in an electrochemical reaction to determine concentration. Since the charge is the current multiplied by the elapsed time, coulometric sensors are considered a subtype of amperometric sensors and are not studied separately. Impedimetric electrochemical sensors measure produced impedance as an on-board-calculated ratio of generated potential to the generated current. Capacitive electrochemical sensors measure selective layer capacitive resistance which is a special case of impedance where the impedance contributors, other than capacitive, have minimal impedance contributions. Therefore, capacitive electrochemical sensors are not studied individually.

19.2 POTENTIOMETRIC SENSORS

Potentiometry experimental setup is straightforward and the reason behind its cherished use. There are typically three electrodes constituting an electrochemical cell. The working electrode (WE) is the electrode where the reaction locale under investigation is situated. The counter electrode (CE) is the electrode used to complete the electrical circuit and is chose to not be a rate-limiting factor in the study. The reference electrode (RE) is the electrode with stable standardised potential and the potentials at WE and CE are measured against it. For the role of RE, the second type is beneficial. The second type of RE contains a metallic M/M^+ electrode within an M^+ salt solution. The Ag/AgCl electrode with a KCl solution is a popular choice among the RE of this class. The equipment used for measuring potential must have a very high input resistance. Since their input resistances are often in the giga-ohms region, so-called pH metres are ideal for this application. When addressing potentiometric sensors, however, it is useful to begin with traditional electrode shapes. The only difference between traditional potentiometric electrodes and potentiometric sensors is their geometric dimensions.

All potentiometric measurements are based on Nernst's equation, which is typically expressed in terms of the species' activity because fundamentally, measured potentials are the responses that are generated out of the electron activities in the reaction system. The Nernst equation also infers that the responses generated by the activities of single-charged ions can be measured with the greatest accuracy vis-à-vis other higher-charged ions. Diverse thermodynamic deductions of the Nernst equation sometimes disregard the fact that the potential difference value recorded at the WE and RE is composed of many interfacing potentials. The differences between two such consecutive potentials are known as the Galvani potential differences. Such interface potential changes are not, in principle, detectable. Therefore, the cell's Electromotive Force (EMF) is measured at the electrode/electrolyte interfaces by measuring the difference between the electric potentials of two phases present at the interface.

In the matter of electrochemical sensors, the sensing platform becomes the WE. It is only the interface between the analyte and the sensing where the measurement of Galvani potential g_i is of utmost importance. In fact, the result of a potentiometric-sensor-based experiment may be summarised by this single value. The care necessary to fulfil this condition is to maintain the status quo for all prospective contributions save the one of interest. One can take advantage of the fact that, despite not being quantifiable, each Galvani potential difference follows the Nernst equation.

The design of the RE presents a challenging technological issue in the manufacture of small potentiometric sensors. All established forms necessitate liquid-liquid contact junctions between the RE and the sample solution. Typically, an $M|M^+$ wire is immersed in a tube containing the M^+ salt solution. For example, in the Ag/AgCl in KCl-based RE, an Ag wire is coated with AgCl and kept immersed in the KCl solution within a tube. The native potential of the RE originates from the equilibria of Cl^- mobility at the Ag/AgCl and the AgCl/KCl interfaces. There is a plug, also called *the frit*, at the end of the tube where the tube gets in the contact with the sample solution containing analyte. The plug is a semipervious barrier designed to allow only the motion of electrons and specified ionic species. Various porous plugs of ceramics and other materials have been proposed, along with labyrinth seals, gel plugs. When not in use, the RE is stored to keep the plug immersed in the salt solution to maintain the plug's structure, porosity, and physicochemical behaviour. Absent such care, the plug dries which causes significant, and sometimes irreparable, damages to the RE. Therefore, to avoid the complicated construction of a tiny RE, a simpler alternative, a pseudo-RE is constructed. The pseudo-RE is more of a phenomenon than a structure. It is created by coating the sensing platform with an active material containing M^+, providing metallic contact of metal M, and adulterating the analyte solution with the counter-ion of M^+ in the coated active material. For example, let's say that the conventional RE having the Ag/AgCl in KCl is to be replicated as a pseudo-RE for sensors. The sensing platform will need to be coated with AgCl, and the analyte solution will need to be adulterated with constant Cl^- concentration. Then, the presence of a thin Ag wire as metallic contact will not only provide an electrical contact but also cause the locale to act as a pseudo-RE due to the combined presence of the three intended components (Ag, AgCl, and Ag^+ at the sensor surface) of the original RE.

Selectivity is one of the potentiometric sensor's most significant advantages. Ideal would be a sensor that exclusively responds to a single ion type. The ideal cannot be realised; however, certain types of sensors approximate it. Typically, a certain number of interfering chemicals exist for each analyte. The interfering ions simulate the presence of the intended ions. It presents a virtual, unquantified, and inflated population of the intended ions in the analyte, which, in the lean analyte solutions and at times in rich solutions too, may subsume the actual activities. Various strategies to solve this issue have been proposed including the popular Nikolskiy–Eisenman modification of the Nernst equation. The Nikolskiy–Eisenman modification considers the competition between the sample ion *i* and the interfering ion *j* while quantifying the interference level through the selectivity coefficient. The selectivity coefficient represents the ratio of interfering ion sensitivity to sample ion sensitivity. For instance, a selectivity coefficient value of 0.01 indicates that the interferent must be 100 times more abundant than the sample to produce the same effect on the sensor. The Nikolskiy–Eisenman adaptation of the Nernst equation further demonstrates that the interference may reduce for increased sample activities. However, it is to be noted that the selectivity coefficient can indicate only the possibility of an interference and is not conclusive proof.

19.2.1 Case Study-1: The Lambda Probe

Solid electrolytes, as opposed to electrolytic solutions, can be used to assemble Galvanic cells. The lambda probe is a well-known potentiometric gas sensor used to assess the air-to-fuel ratio (AFR). The AFR is basically an indirect index of the oxygen available in the combustible mixture in the combustion chamber. The Lambda probe, one of the most used chemical sensors, is based on the solid-electrolyte-based Galvanic cells. The combustion process in the engine is aimed to reduce carbon monoxide emissions. The Lambda probe is utilised in all modern automobiles to monitor and help regulate the input oxygen level to optimise the degree of combustion. The probe can operate in both modes, potentiometric as well as amperometric modes, depending upon the requirements and the design of the sensing system.

In the Lambda probe, the electrolyte is solid, and the analyte sample is in the gas phase at an operating temperature of around 900°C. At such high temperatures, a platinum surface can easily obtain the redox couple O_2/O^{2-} from a supply of pure O_2 and constitute a redox electrode. Next, the yttria-stabilised-zirconia (specified doping of YO_3 in ZrO_2, also called YSZ)-based solid electrolyte provides ionic mobility. The YSZ electrolyte has inherent oxygen vacancy and utilises these vacancies to cause the ionic mobility of O^{2-} for a gaseous mixture with constant partial pressure of O_2, e.g. air. Finally, the other side of the electrolyte is attached with platinum surface with the supply of air, making the cell scheme O_2(pure)|Pt|ZrO_2|Pt|O_2(air). Since an electrolyte has different partial pressures of O_2 on either of its sides, it forms a concentration cell with a symmetrical form but varying concentrations of the active component. The mobile O^{2-} ions moving through the solid electrolyte carry an electrolytic current and it results in the O_2/O^{2-} redox reaction equilibrium on each electrode. An EMF is developed between electrodes and can be used to characterise the oxygen concentration through the Nernst equation.

Figure 19.2 depicts a lambda probe in schematic form. A finger-shaped, zirconium dioxide ceramic hollow component protrudes into the examined gas compartment. Its interior volume is exposed to ambient air. The EMF is measured between the inside and the outside surface of the cylinder, at the gas-permeable Pt surfaces available on both sides.

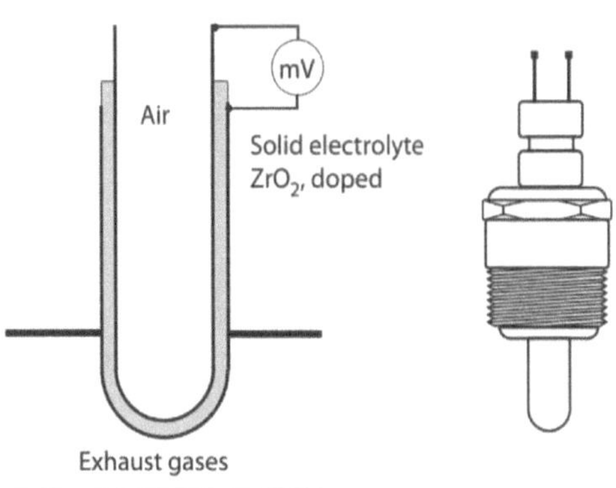

FIGURE 19.2 A Lambda probe in schematic form [8].

Fuel is converted into CO_2 and H_2O through the full oxidation of the fuel by oxygen taken from the air during combustion. The *stoichiometric* AFR means that oxygen is available in an amount just enough to completely combust the fuel. The ratio of actual AFR to stochiometric AFR determines the lambda number for the fuel-air mixture about to be combusted. The signal from the lambda probe is transmitted to an electronic control circuit that maintains the actual lambda value close to 1, so achieving nearly full combustion. In addition to engines, the lambda probe is also utilised to regulate and monitor other combustion processes, such as in power plants, metallurgical plants, and in chemical analysers. The lambda probe can function in two distinct modes. The potentiometric mode of operation follows the transducer principle of energy conversion. Alternately, the galvanic cell can be shorted briefly to measure the current instead of the potential, making the cell function as an amperometric sensor. It is an example of limited transduction by current. Amperometric lambda probes are easier to miniaturise than potentiometric probes, and the linear relationship between the signal (a limiting current) and the oxygen strengths can be harnessed to widen the sensing limits of concentrations.

19.3 AMPEROMETRIC SENSORS

Amperometric sensors measure electrolysis current by polarising the electrode. Under certain conditions, saturation occurs, and the limiting current is proportional to concentration. Contrary to energy conversion transduction in the potentiometric sensors, here current-limiting transduction is used. Under these conditions, the response rate is represented by an electric current. Now, the electrode and the analyte are not always in thermodynamic equilibrium. So, the waiting for the equilibrium is avoided. Therefore, the response times of amperometric sensors are much shorter than that of potentiometric sensors. The strongly linear proportionality of concentration to the limiting current is an additional advantageous characteristic here. Also, the amperometric sensors are easier to miniaturise than potentiometric ones and mass production generates inexpensive prices. Consequently, there is a trend towards amperometry, particularly with biosensors.

The amperometric sensors have no inherent selectivity; rather, the selectivity must be manufactured through extra effort. Without such efforts, amperometric sensors are very unlikely to be selective. Every electrochemically active material has a unique value of half-wave potentials assigned to its electrode reaction. For two compounds to be distinguished, their half-wave potentials must differ by at least 200 mV. However, the potential range of a common electrode is limited to approximately 1.5 V. Consequently, only a few compounds can be determined concurrently in a single voltammogram. Typically, at an electrode working in amperometric mode, many processes are occurring simultaneously. Multiple chemicals are typically concurrently reduced or oxidised. Only a few exceptions exist to this rule. Among these is the previously described lambda probe. Because only a specific ion is mobile between the electrodes through the solid electrolyte, it has inherent selectivity.

Amperometric procedures involve analyte molecules to reach the active site at the electrode to react chemically. Selectivity can be improved by filtering ion flow before a reaction. Utilisable filter layers allow only certain chemicals. Semipermeable membranes or nanostructured membranes with intramolecular or intermolecular routes can form such layers. Alternatively, positioning the selective catalyst molecules before the electrode can also do the trick. Catalysts favour specific electrode reactions, so the electrolysis current is only due to the analyte concentration. Catalytic admixtures in carbon-paste-based electrodes include copper, nickel, cobalt, and ruthenium in metallic and oxide forms. Semipermeable membranes block the molecules based on size or other characteristics. Dialysis membranes rely on the size of molecules for separation. The ion-selective membranes such as Nafion® (Du Pont) allow only ions with a specific charge sign to pass through. Similarly, lipophilic membranes do not allow polar molecules to permeate. The membranes with molecular recognition of the analyte also own similar features. The various techniques for enhancing selectivity are not exactly distinct from one another. For instance, we can consider several enzymes. In terms of steric (molecular) recognition, they are like selective ligands-based membranes when they pick and get attached to specific molecules. They also catalyse a specific process, like other selective catalysts.

Commonly, amperometric sensors take the form of microelectrodes in which a steady current, limited by molecular diffusion, is created without external convection. These electrodes are ideally suited for acquiring concentration-proportional signals in situ, for instance, when a biological material is pierced by the electrode. Microdisks, interdigitated lead structures, and narrow bands on support are the most typical morphologies of microelectrodes. Macroscopic forms with a diffusion barrier or a protective layer overlaying the active membrane containing immobilised molecules are also commonly employed.

19.3.1 Case Study-2 Clark Sensor

The Clark sensor is a crucial amperometric sensor for the detection of dissolved oxygen. Widespread use is made of the classical macroscopic design, but tiny variants are also frequent. Figure 19.3 presents a schematic diagram of the Clark sensor.

The Clark sensor has a gas-permeable membrane in direct contact with the WE surface. The membrane separates the sensor from the sample solution. Thus, the electrode surface is safeguarded. Selectivity is automatically enhanced because only gaseous components can reach the active area. Preferably, macroporous polytetrafluoroethylene (PTFE) membranes are utilised. The substance is highly hydrophobic. Liquid water is incapable of passing through the membrane pores. In contrast to liquid water, gases spread easily via pores. Some models of the Clark sensor also have silicone membranes which are oxygen permeable because oxygen is dissolved uniformly in the matrix.

The membrane allows the diffusion of O_2 from the analyte solution. Near the membrane, a fine electrolyte coating separates the WE. Platinum surfaces are usually tiny and flat for active electrodes. The enforced polarisation potential reduces each oxygen molecule at the electrode. Behind the membrane, oxygen concentration is near zero. Thus, only the sample's oxygen content determines the concentration profile between the sample solution and the electrode surface. In this case, the diffusion layer and gas-permeable membrane are equal. The constant thickness gives the current-voltage curve a sigmoidal shape if the scan rate is slow. Applying a constant polarisation potential and measuring the restricted current to create an amperometric sensor. Regions where diffusion limits current should have a negative imposed potential.

19.4 IMPEDIMETRIC AND OTHER ELECTROCHEMICAL SENSORS

The three methods for analysing electrochemical sensors are potentiometry, amperometry, and impedance. Electrochemical impedance studies focus on electrode surface processes [9,10]. Charge carrier properties affect a chemical sensor's total impedance as well as those far from the electrode.

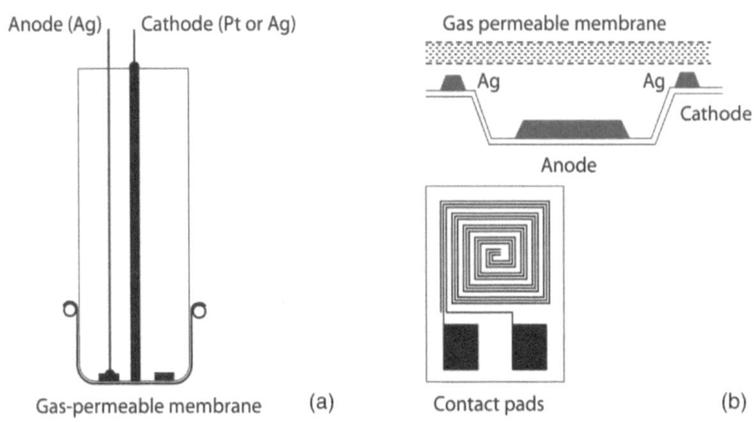

FIGURE 19.3 A Clark sensor in schematic diagram (a) classic and (b) miniature form [8].

The experimental design can focus on electrode surface processes or ion properties in homogenous solution. Variations in the dielectric constant within a layer affect the overall impedance. Electro-biochemical sensors, impedimetric sensors, use biological reactions as their information source [11–13].

The impedimetric glucometers are examples of the class of impedimetric electrochemical sensors. An impedimetric glucometer measures blood glucose levels based on changes in impedance. Typically using a pair of electrodes, these sensors might be modified with enzymes like GO_x. When glucose is present, GO_x catalyses its conversion to gluconolactone, producing hydrogen peroxide (H_2O_2). This production of H_2O_2 or glucose consumption alters the impedance of the electrode interface. Applying an AC voltage, which is generated by an onboard inverter circuit, the sensor measures the resulting current, calculating impedance changes that correlate to glucose concentration. Impedimetric sensors may offer advantages like reduced interference from other substances and longer electrode lifetimes. However, challenges include ensuring that impedance changes are exclusively due to glucose and no other interfering substances.

Biosensors can be based on biocatalysts or bioaffinity. Biocatalytic sensors use enzymes as selective catalysts on their electrodes. Enzymes speed up slow responses. Under ideal conditions, the reaction rate quantifies substrate concentration. Chemical sensors can be developed if the sample is a reactant, such as a substrate. Electrolysis current can be used to measure reaction rate with amperometric biosensors. Alternative methods include potentiometric biosensors that specifically indicate the catalysed reaction product. Bioaffinity sensors, the second type of sensor, form stable complexes with sample molecules that are strongly bonded to the sensor surface. This complex formation quantifies sample concentration. Complex creation changes many electrode characteristics, allowing indirect detection. The antibody-antigen reaction dominates bioaffinity sensors.

Immobilising bioactive receptor molecules on an electrode is necessary. Adsorption is the easiest way to attach bioactive compounds to electrodes. Adsorbed molecules are bound by weak forces, making sensors less durable. Adsorption immobilises enzymes, antibodies, and nucleic acids. Contact between an electrode's surface and a solution can often generate adsorptive layers without chemicals. This applies only to carbon-based surfaces. However, an adsorptive bond with a carbon surface may deactivate an enzyme by affecting its activity. Occasionally, denaturation occurs. Most molecules are poorly bonded and gradually desorb. Electrodes with adsorptive layers are best for initial tests. They are best for rapid tests.

19.5 CONCLUSIONS

In conclusion, the process design of electrochemical sensors, as outlined in the chapter, represents a multifaceted and dynamic field that bridges the realms of chemistry, physics, and engineering. This comprehensive overview of electrochemical sensors, with a focus on potentiometric and amperometric sensors, reveals their fundamental principles, applications, and challenges. Potentiometric sensors and amperometric sensors serve as the backbone of electrochemical sensing. They are highly versatile and find utility in a myriad of applications, from monitoring air-fuel ratios in combustion engines to detecting dissolved oxygen in various contexts. Various strategies, including the Nikolskiy–Eisenman modification, semipermeable membranes, and selective catalysts, are practiced in design to enhance the selectivity of the sensors. In addition, the emergence of biosensors, whether biocatalytic or bioaffinity, signifies the integration of biological elements into sensor design. This innovation expands the horizons of electrochemical sensing, enabling highly specific and sensitive detection of biological molecules. As technology advances, the miniaturisation of sensors becomes increasingly feasible, widening the scope for their deployment. The insights provided in this text shed light on the critical considerations and methodologies that underpin the process design of electrochemical sensors. Future developments in this field hold the promise of even more sophisticated and specialised sensor designs, revolutionising our ability to monitor and understand the chemical world.

REFERENCES

1. Skoog, D. A.; Holler, F. J.; Crouch, S. R. *Principles of Instrumental Analysis*, 7th ed.; Brooks Cole, Pacific Grove, CA, 2018.
2. Ronkainen, N. J.; Halsall, H. B.; Heineman, W. R. Electrochemical biosensors. *Chem. Soc. Rev.*, **2010**, 39(5), 1747–1763. https://doi.org/10.1039/B714449K.
3. Lu, S. M.; Peng, Y. Y.; Ying, Y. L.; Long, Y. T. Electrochemical sensing at a confined space. *Anal. Chem.*, **2020**, 92(8), 5621–5644. https://doi.org/10.1021/ACS.ANALCHEM.0C00931.
4. Shukla, A. K.; Mitra, S.; Dhakar, S.; Chakraborty, R.; Sharma, S.; Dey, K. K. Non-faradaic electrochemical energy transduction using catalytic motors. *ACS Appl. Energy Mater.*, **2022**, 10, 12916–12924. https://doi.org/10.1021/ACSAEM.2C02546.
5. Bakker, E.; Telting-Diaz, M. Electrochemical sensors. *Anal. Chem.*, **2002**, 74(12), 2781–2800. https://doi.org/10.1021/AC0202278.
6. del Valle, M. Electronic tongues employing electrochemical sensors. *Electroanalysis*, **2010**, 22(14), 1539–1555. https://doi.org/10.1002/ELAN.201000013.
7. Simões, F. R.; Xavier, M. G. Electrochemical sensors. *Nanosci. its Appl.*, **2017**, 5, 155–178. https://doi.org/10.1016/B978-0-323-49780-0.00006-5.
8. Gründler, P. Chemical sensors: An introduction for scientists and engineers. *Chem. Sensors An Introd. Sci. Eng.*, **2007**, 5, 1–273. https://doi.org/10.1007/978-3-540-45743-5/Cover.
9. Wang, S.; Zhang, J.; Gharbi, O.; Vivier, V.; Gao, M.; Orazem, M. E. Electrochemical impedance spectroscopy. *Nat. Rev. Methods Prim.* **2021**, 11(1), 1–21. https://doi.org/10.1038/s43586-021-00039-w.
10. Chang, B. Y.; Park, S. M. Electrochemical impedance spectroscopy. *Annu. Rev. Anal. Chem.*, **2010**, 3(1), 207–229. https://doi.org/10.1146/annurev.anchem.012809.102211.
11. Lisdat, F.; Schäfer, D. The use of electrochemical impedance spectroscopy for biosensing. *Anal. Bioanal. Chem.*, **2008**, 391(5), 1555–1567. https://doi.org/10.1007/S00216-008-1970-7.
12. Heine, V.; Kremers, T.; Menzel, N.; Schnakenberg, U.; Elling, L. Electrochemical impedance spectroscopy biosensor enabling kinetic monitoring of fucosyltransferase activity. *ACS Sens.*, **2023**, 25, 39. https://doi.org/10.1021/acssensors.0c02206.
13. Shukla, A. K.; Mitra, S.; Dhakar, S.; Maiti, A.; Sharma, S.; Dey, K. K. Electrochemical energy harvesting using microbial active matter. *ACS Appl. Bio Mater.*, **2023**, 6(1), 117–125. https://doi.org/10.1021/acsabm.2c00785.

20 Challenges and Future Prospects in Electrochemical Sensing

Rama Gaur, Nandini Mukherjee, and Syed Shahabuddin

20.1 INTRODUCTION

Due to its many benefits, including its high sensitivity, selectivity, and low detection limits, the electrochemical sensing platform has been thoroughly investigated in recent years as a sensitive and quick instrument in analytical chemistry [1]. Due to their distinctive characteristics, such as high surface-area-to-volume ratios, variable porosity, and high sensitivity, nanomaterials, in particular, have drawn attention as a promising sensing material in recent years [2]. Electrochemical sensors that are highly sensitive, stable, and selective with the ability to detect a variety of analytes at low concentrations have been developed as a result of the inclusion of nanomaterials. Due to their higher sensitivity and expanded surface area, nanoparticles have improved detection limits in electrochemical sensors [3].

Nanomaterials have also shown to be extremely stable and highly selective to particular analytes due to the above-mentioned unique features, making them perfect for use in electrochemical sensors. Over conventional sensing methods, the use of nanotechnology in sensor development has many benefits. Nanotechnology has experienced substantial growth in recent years, and its uses have spread into many industries, including environmental monitoring, healthcare, and food safety. Electrochemical sensors can be used in a variety of fields and are not just limited to one.

Nanomaterial electrochemical sensors have significant real-world applications in various fields such as environmental monitoring, healthcare, and food safety. In the environmental monitoring sector, nanomaterials have found utility in the detection of various pollutants. For example, carbon nanotubes have been used to detect volatile organic compounds (VOCs) in the atmosphere [4]. In the medical field, the use of nanomaterial electrochemical sensors has facilitated the detection of biomarkers, disease diagnosis, and drug monitoring [5]. In the food safety sector, nanomaterial-based electrochemical sensors have been used to detect food-borne pathogens, such as *Salmonella* and *E. coli*, and also in detecting contaminants such as pesticides and antibiotics in food products [6]. Figure 20.1 shows the nanomaterial used in the development of electrochemical sensors and their applications in various fields.

However, these sensors face several challenges which limit their performance that need to be discussed and addressed. At the same time, the recent development points towards the prospects and opportunities for researchers to expand the horizon and scope of applications of the electrochemical sensors. This chapter will discuss the challenges and prospects of developing nanomaterials as electrochemical sensors. Our aim is to provide an overview of the current state-of-the-art technology and to identify research priorities and challenges in the field.

20.2 CHALLENGES OF NANOMATERIALS AS ELECTROCHEMICAL SENSORS

Despite the great potential of nanomaterials in electrochemical sensors, there are several challenges that researchers are still grappling with. One such challenge is the lack of a universal fabrication method for nanomaterials. The synthesis of nanomaterials is highly dependent on the desired properties and application of the sensor, which makes it difficult to develop a single protocol applicable to all sensors. Another main challenge is achieving high sensitivity, especially for trace analytes.

DOI: 10.1201/9781032654874-20

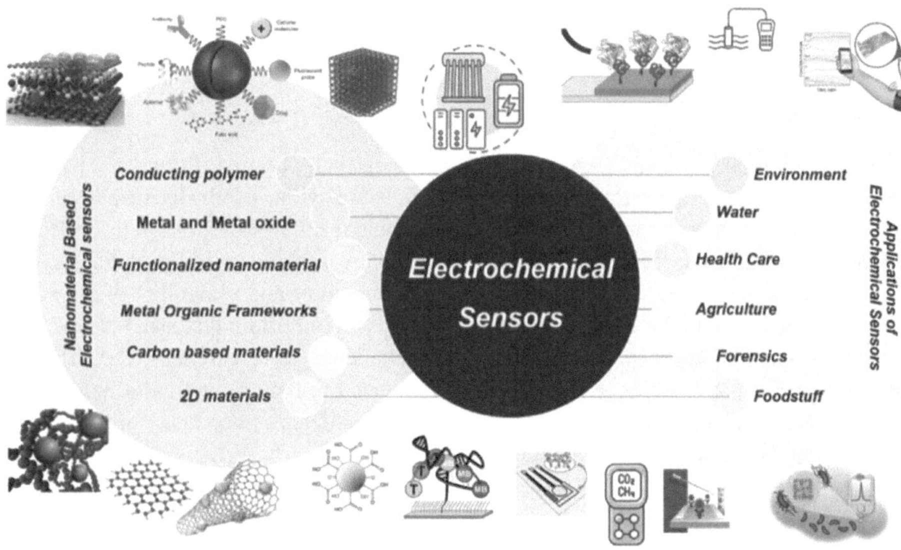

FIGURE 20.1 Schematic for applications of electrochemical sensors.

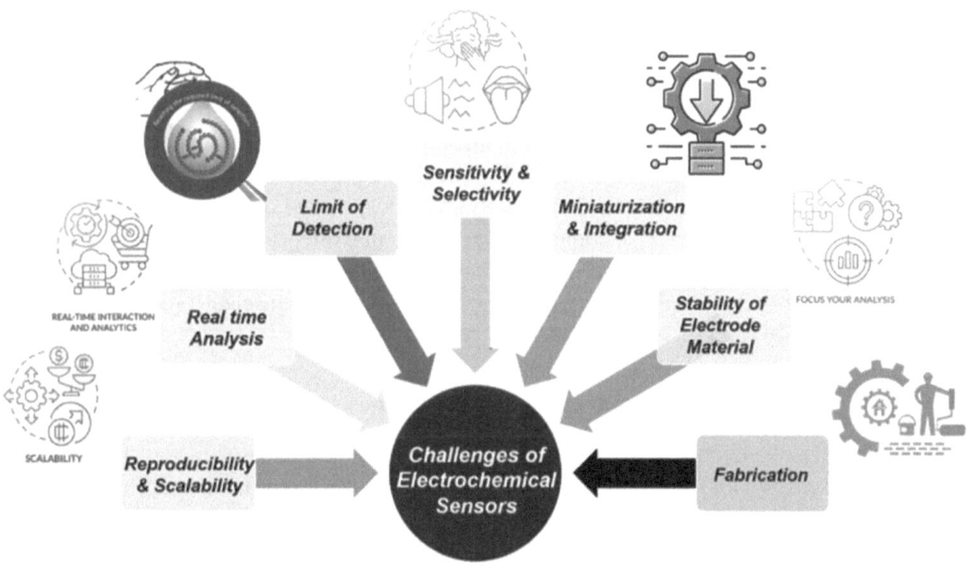

FIGURE 20.2 Schematic representation of the challenges associated with electrochemical sensors.

Improving the detection limit of electrochemical sensors will enable accurate measurement of low concentrations of analytes. Achievement of high selectivity in complex sample matrices, stability, reliability, and repeatability are a few concerns to list. Figure 20.2 depicts the major challenges associated with the use of electrochemical sensors.

The development of new materials and sensor design strategies is necessary to improve selectivity for specific analytes, minimising interference from other components. The stability of the electrochemical sensors is another major challenge faced by researchers in this domain. The degradation of electrodes over time can affect the reliability and repeatability of measurements. Developing more stable electrode materials is crucial to ensure long-term sensor performance. Furthermore, miniaturisation and integration of electrochemical sensors into portable devices pose engineering and scalability challenges. Overcoming these obstacles will enable on-site and real-time analysis. The challenges in electrochemical sensing have been discussed in detail in the following subsections.

20.2.1 Fabrication

The absence of a common fabrication process for nanomaterials is the first difficulty. The intended qualities and use of the sensor determine how nanomaterials are synthesized. Because of this, scientists must create unique techniques for each sensor, which could result in variations in the repeatability of created nanomaterials. A major challenge faced by researchers working in the development of nanomaterials for electrochemical sensors is ensuring the reproducibility and scalability of the synthesis methods The reproducibility of nanomaterials has emerged as an important factor in the development of electrochemical sensors, and it presents a significant challenge to researchers. The difficulty in reproducibility is due to the complex synthesis process involved in the production of nanomaterials. In addition, the scalability of the synthesis method is another challenge, since many of the fabrication methods are incompatible with large-scale production.

20.2.2 Electrode Material

The demand for real-time, highly sensitive, and selective analyte detection has fuelled the development of electrochemical sensors. It has been demonstrated that adding nanoparticles to electrochemical sensors improves their performance by enhancing sensitivity and selectivity and offering a high surface-area-to-volume ratio. Quantum dots, metal nanoparticles, metal oxide nanoparticles, and carbon-based materials are the most often researched nanomaterials in the creation of electrochemical sensors. Since they have a high surface area and conductivity, carbon-based materials like graphene and carbon nanotubes have been extensively used because they are perfect for electrochemical sensing. Due to their distinctive optical and electrical characteristics, metal nanoparticles like gold (Au) and silver (Ag) have been utilised in the development of electrochemical biosensors. Due to their large surface area, stability, and catalytic activity, metal oxide nanoparticles like zinc oxide (ZnO), titanium dioxide (TiO_2), copper oxide (CuO), tin oxide, (SnO_2), tungsten oxide (WO_3), and iron oxide (Fe_3O_4) have also found use in sensing [7]. Quantum dots have been employed as sensing materials due to their high sensitivity and specificity and unique electronic features, such as size-dependent and wavelength-tunable fluorescence.

20.2.3 Reproducibility and Scalability of the Synthesis Method

The reproducibility of nanomaterials has emerged as an important factor in the development of electrochemical sensors, and it presents a significant challenge to researchers. The difficulty in reproducibility is due to the complexity involved in the production of nanomaterials. As a result, the synthesized nanomaterials may differ in size, shape, and surface properties.

The scalability of the synthesis method is another challenge, as many of the fabrication methods are incompatible with large-scale production. Synthesis methods such as chemical vapour deposition (CVD) and the sol-gel method are examples of processes that are time consuming, expensive, and labour intensive. There is, therefore, a need to develop scalable and cost-effective methods for the synthesis of nanomaterials for use in electrochemical sensing.

20.2.4 Improving the Detection Limit

20.2.4.1 Sensitivity and Selectivity

One of the main challenges in electrochemical sensing is achieving high sensitivity, especially for trace analytes. The detection limit of electrochemical sensors needs to be improved to accurately measure low concentrations of analytes. Electrochemical sensors often face challenges in achieving high selectivity, particularly in complex samples containing interfering species. The development of new materials and sensor design strategies is required to improve selectivity for specific analytes.

20.2.5 Stability and Degradation of Electrochemical Sensors

Electrodes used in electrochemical sensing frequently corrode over time as a result of fouling, chemical corrosion, or degradation of the electrode material. For electrochemical sensors to be long-term reliable and repeatable, more stable electrode materials must be developed to prevent the sensor contamination. When doing measurements in a biological medium, biofouling can happen. Proteins can drastically alter the surface chemistry when they come into contact with a surface during the fouling process due to the interaction of several forces (hydrophobic effect, hydrogen bond, coulombic, and other polar interactions). The current and sensitivity will decrease if there are any substances in complex samples that bind the electrodes in an unspecific way. Bovine serum albumin (BSA) and poly(ethylene glycol) self-assembled monolayers (PEG-SAMs)3 are two examples of conventional antifouling coatings that achieve the same thing by preventing electron transfer. This creates a two-pronged antifouling technique [8].

20.2.6 Miniaturisation and Integration

Engineering and scalability difficulties arise when electrochemical sensing is integrated into compact, portable devices and combined with other parts like microfluidic systems or wireless data transmission [9]. On-site and real-time analysis is made possible by miniaturisation and integration, but this requires overcoming technical challenges. Numerous improvements in microfabrication methods, materials, and electronics have been realized in recent years. These advancements aided in the creation of more affordable, portable, and compact sensors. The creation of portable sensing devices required the integration and miniaturisation of electrochemical cells. This has been discussed in Section 20.3.6 [10].

20.3 PROSPECTS OF NANO MATERIALS AS ELECTROCHEMICAL SENSORS

Despite these challenges, electrochemical sensing holds great prospects for various applications. Electrochemical sensors' quick and inexpensive analysis can be used for point-of-care testing to enable real-time monitoring of disease biomarkers. Finding pollutants and toxins can help with environmental monitoring and food safety. The development of energy storage and conversion systems also benefits from the use of electrochemical sensors. Personalized health monitoring is also made possible by the incorporation of electrochemical sensors into wearable and implantable devices. In conclusion, electrochemical sensing has tremendous potential in a variety of sectors, despite some issues that need to be resolved. Nanomaterials have a lot of potential as electrochemical sensors in a variety of fields. The potential for nanoparticles to detect many analytes from a single sample is one such possibility. This is made feasible by the distinctive characteristics of various nanomaterials, which may be altered to react to a variety of analytes, enabling multi-parameter analysis. The possibility of functionalising nanomaterials to increase their selectivity and sensitivity is another application for them as electrochemical sensors. In order to increase the nanomaterial's selectivity toward the desired analyte, functionalisation entails altering the surface of the nanomaterial. The surface modification can ultimately result in an increase in sensitivity and aids in the removal of interference from other molecules present in the sample.

Additionally, the detection of biomolecules, the search for new drugs, and the monitoring of environmental pollutants are all areas in which nanomaterials as electrochemical sensors can be used in addition to more conventional analytical applications. These novel uses have created new avenues for the creation of electrochemical sensors based on nanomaterials. Figure 20.3 summarises the prospects of nanomaterials as electrochemical sensors.

The prospects for the use of nanoscale materials in electrochemical sensing giving it a direction by enhancing the performance in terms of selectivity and sensitivity and real-time analysis has been discussed in detail as follows.

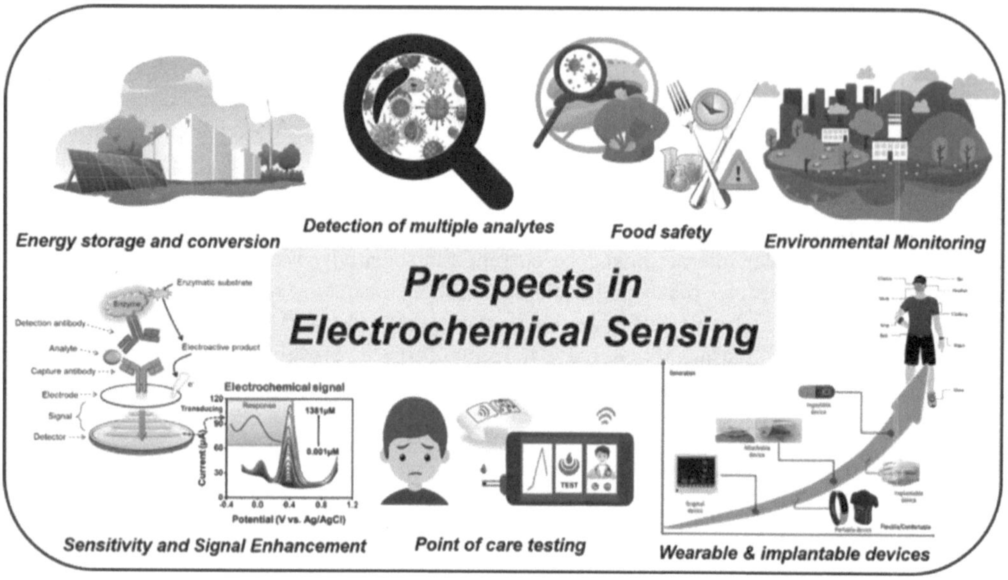

FIGURE 20.3 Schematic for the prospects of nanomaterials as electrochemical sensors.

20.3.1 Detection of Multiple Analytes

Due to its low cost, simplicity, great selectivity, and sensitivity, electrochemical sensing is a potential tool for therapeutic applications. However, biofouling frequently happens when the electrodes are exposed to biological fluids like plasma or blood because of the non-specific absorption of biological macromolecules, particularly proteins, which obstructs electron transport and consequently causes a rapid loss of sensitivity [8,11]. Small molecule glucose could diffuse through the semi-permeable membrane and reach the electrode in the glucose meter configuration with a semi-permeable membrane, but not for macromolecule proteins [12]. Nonenzymatic electrochemical sensors combined with antibiofouling techniques would be a promising commercial tool used in clinical analysis for small and electroactive biological molecules related to human health, such as uric acid (UA), ascorbic acid (AA), dopamine (DA), guanine (G), and adenine (A). Yuan et al. have developed a microelectrochemical sensor for multi-analyte detection based on oxygenated graphene-modified screen-printed electrode (Figure 20.4) [13].

The electrochemical multi-analyte point-of-care sweat sensors designed by Bauer et al. utilize on-chip three-dimensional graphene electrodes. With a detection limit of 13.7±0.5 M for glucose and 28.3 M for lactate, respectively; enzyme-based biosensors for glucose and lactate; an ion-selective chemosensor for potassium ions; and a straightforward conductometric electrolyte setup were developed through laser-induced graphene (LIG) surface modifications. The sensor performed effectively at various pH levels, with tissues that absorb sweat, on a model skin system, as well as in synthetic sweat and artificial tear fluid. Since the response time for each analytical cycle is 75 s, all analytes can be monitored simultaneously and almost continuously. Thus, for point-of-care applications, this multi-analyte all-LIG system is a useful, adaptable, and the most straightforward method that has the potential to outperform conventional screen-printed electrodes. Figure 20.5 demonstrates the multiple analyte detection using electrochemical multi-analyte point-of-care perspiration sensors using on-chip three-dimensional graphene electrodes [14].

The method used PS@PDA-metal nanocomposites based on graphene nanoribbon (GNR)-modified hot screen-printed carbon electrode (HSPCE) to detect interleukin-6 (IL-6) and matrix metallopeptidase-9 (MMP-9) quickly. Shi et al. created the ultrasensitive electrochemical immunoassay. The GNRs were employed to immobilize antibodies (Ab) and increase the electrochemical

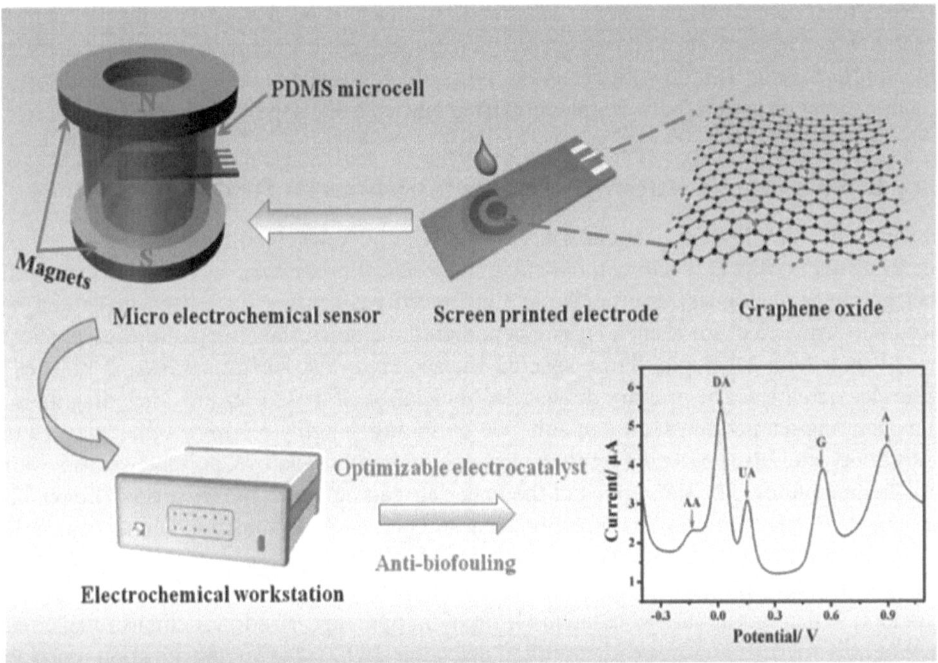

FIGURE 20.4 Schematic representation of electrochemical sensing of multi-analyte based on electrochemically tuned GO electrode [13].

Source: Reproduced with permission from Ref. [13] under creative common license copyright @2022.

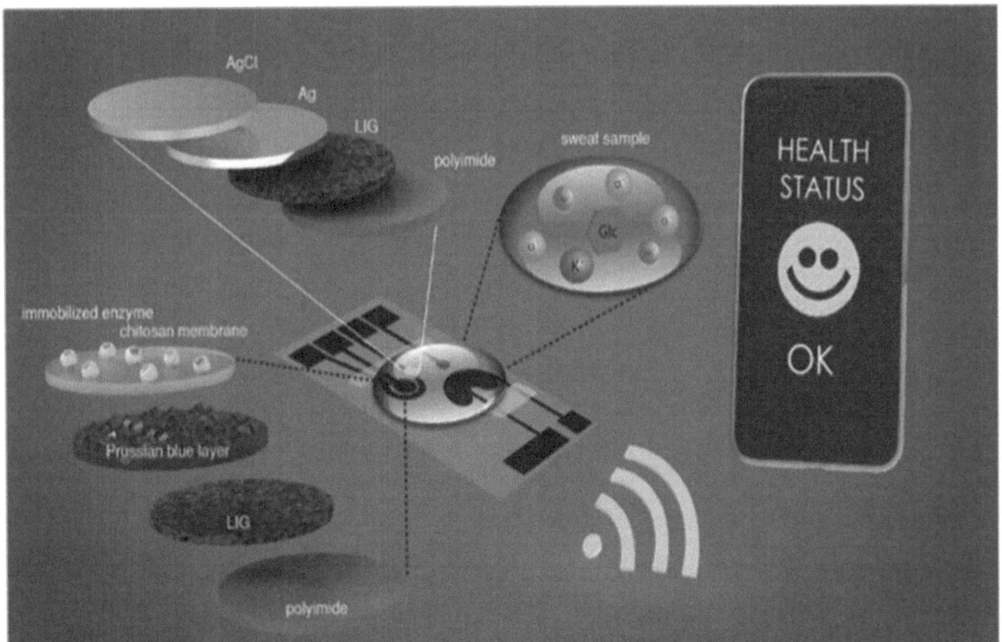

FIGURE 20.5 Schematic representation of the multiple analyte detection using electrochemical multi-analyte point-of-care perspiration sensors using on-chip three-dimensional graphene electrodes [14]. *Open Access.*

signal due to their good hydrophilicity and low toxicity. In acetic buffer, Ab were labelled with PS@PDA-metal, which also produced a potent electrochemical signal. In order to simultaneously detect MMP-9 and IL-6 using HSPCE with detection limits of 5 fg mL^{-1} and 0.1 pg mL^{-1} ($S/N=3$), respectively, and without cross-talk between neighbouring electrodes, a sandwich method was used [15].

20.3.2 Enhancement of Selectivity and Sensitivity through Functionalisation

Two-dimensional (2D) layered nanomaterials currently provide fascinating opportunities for sensor use due to distinct physical, chemical, optical, and electrical properties. Increased research efforts on other graphene analogues for effective and enhanced gas sensing have been motivated by the enhanced performance of graphene and graphene-based nanomaterials for gas sensing applications because of their high sensitivity, large specific surface area, fast electron transport kinetics, and strong surface activities. The primary drawbacks of metal-oxide-based sensors, including their need for high operating temperatures, ease of sulphide poisoning, lengthy recovery times, limited maximum sensitivity, etc., also sparked a significant search for highly sensitive, portable sensors with low operating temperatures. The limitations of the materials can be overcome by surface functionalisation or composite formation. In this direction Chacko et al. have demonstrated enhancement in the selectivity and sensitivity of Ni- and Pd-functionalized MoS_2 toxic gas sensors [16]. The results demonstrated that the Pd-MoS_2 layers exhibited a very high relative response to NO gas (700%) at 2 ppm concentration with a minimum NO detection limit of 0.1 ppm, and Ni-MoS_2 demonstrated a relative response of 80% towards H_2S gas with a limit of detection (LOD) of 0.3 ppm with good repeatability and selectivity, owing to the increased adsorption energy of NO on Pd-MoS_2 and H_2S on Ni-MoS_2 through the formation of $PdNO_x$ and NiS_2 complexes, respectively. The improved sensing capabilities of this MoS_2-based sensor also highlight the great potential and likelihood of combining 2D materials related to MoS_2 to produce highly sensitive, ultra-futuristic nanosized gas sensors that are appropriate for anti-pollution automotive systems and as volatile biomarkers.

Kim et al. reported the synthesis of small organic oligomer-based indandione derivative, 2-(5″-hexyl-[2,2′:5′2″ terthiophen]-5-yl) methylene)-1H-indene-1,3(2H) dione oligomer (HTD), for functionalising graphene oxide to form HTD@GO functionalized nanocomposites as shown in Figure 20.6 [17]. The exceptional simultaneous trace detection of heavy metals like Cr^{2+} and Cu^{2+} ions was achieved by the HTD@GO/GCE sensor probe. When compared to other metal ions (Al^{3+}, Zn^{2+}, K^{+}, Mn^{2+}), the HTD@GO/GCE sensor probe showed a strong preference towards Cr^{2+} and

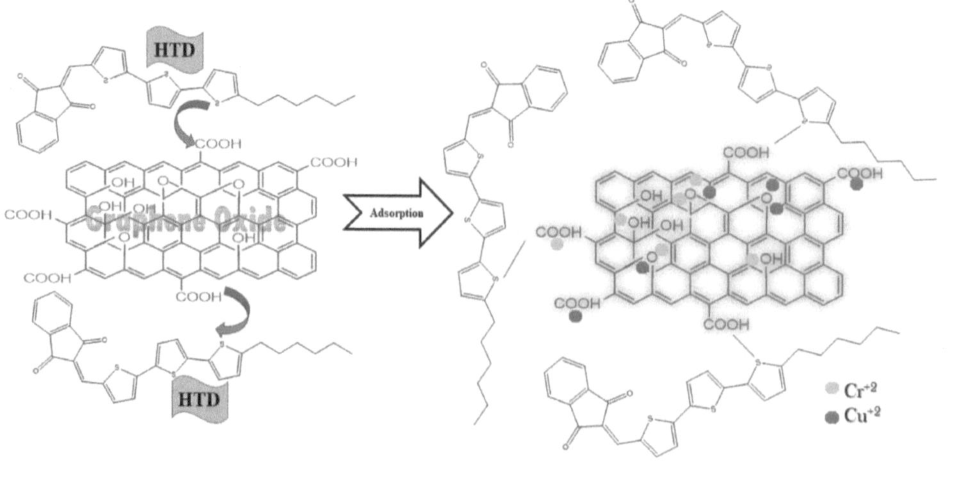

FIGURE 20.6 Graphical representation of Indandione oligomer@graphene-oxide-functionalised nanocomposites for enhanced and selective detection of trace Cr^{2+} and Cu^{2+} ions [adapted from Ref. 17].

Cu^{2+} ions. The HTD@GO/GCE-based sensor, which is significant, demonstrated quite acceptable dynamic linear ranges of 1–100 M and detection limit values of 3.65 and 2.25 M, respectively, for trace Cr^{2+} and Cu^{2+} ions [17].

Another functionalisation strategy is the use of ionic liquids to improve the selectivity and sensitivity of electrochemical sensors. The production of ionic-liquid-functionalized graphene (IL-graphene) by adding surface charges can successfully avoid graphene aggregation. Ionic liquids are typically made up of inorganic anions and organic cations. Ionic liquid is widely soluble; hence, IL-graphene compounds exhibit good dispersion and long-term stability in a variety of solvents. The enhanced electrochemical sensor for sensitive detection of bisphenol A in environmental water has been described by Wang et al. It is based on gold nanoparticles, and MoS_2 nanoflowers adorned ionic-liquid-functionalized graphene. AuNPs/MoS_2-NFs/IL-graphene nanocomposites were used to build an electrochemical sensor, and they demonstrated excellent electrocatalytic activity and selectivity for bisphenol A as well as low detection limits, high sensitivity, and good anti-interference performance [18].

Since macrocyclic compounds can be utilized as modifiers to improve the sensitivity and selectivity of electrodes, their superior supramolecular recognition properties have sparked research attention. Crown ethers, cyclodextrins, calixarenes, cucurbiturils, and pillararenes are the five classic macrocyclic compounds that have all been used as electrochemical sensor receptors. These macrocycles have led to a wider detection range, a lower detection limit, and a better ability of anti-interferences against coexisting ions or molecules. Greater sensitivity and selectivity in detection have been demonstrated by the macrocycles modified electrodes [19].

20.3.3 Point-of-Care Testing

Electrochemical sensors have the potential to revolutionize healthcare by providing rapid and low-cost point-of-care testing. They can be easily integrated into handheld devices and enable real-time monitoring of biomarkers for diseases such as diabetes, cardiovascular diseases, and infectious diseases. Blood and urine analysis has only been performed clinically in centralised clinical laboratories. Large sample volumes are needed for these tests, which are then examined by skilled individuals using heavy equipment. Samples must be transported from the sampling location to the testing location and may need to be stored before analysis, which causes delays in the results and increases the risk of contamination or losing traceability. These tests are insufficient for an early and quick diagnosis due to their limited throughput and lengthy turnover times. Point-of-care testing (POCT) completely disproves this theory. In order to do decentralised analysis whenever it is needed (such as in schools, nursing homes, and workplaces), POCT devices are low-cost, miniaturised equipment. They provide quick replies and have a significant positive impact, especially in low-income nations. The World Health Organization (WHO) has identified the development of POCT as a priority and established some requirements that must be satisfied, including being equipment free, affordable, sensitive, specific, user friendly, rapid, and resilient. Recently, further criteria were added: connectivity in real time and sample collection simplicity [20]. The POCT's capabilities will be enhanced by the combination of connectivity and sample collection, enabling quick analysis times and nearly real-time result sharing.

One of the most significant objectives of POCT devices is the utilisation of low-cost consumables and portable equipment. Numerous original and inventive solutions have been and are still being presented in this situation [21]. As a pertinent illustration, it is important to note that paper is used as a substrate for electrochemical paper-based analytical devices (ePADs), which are used in this field. Paper has other functions like screening and immobilising chemicals [22,23]. The extensive fabrication and testing would be made possible by the materials' low cost.

These devices must, analytically speaking, fulfil ultralow detection limits (LODs) and quick analysis times in order to enable the early identification and monitoring of diseases, which is essential for patients' survival. Due to the need for cognitive testing and magnetic resonance imaging of the brain in addition to chemical information, POCT may not be sufficient to identify complicated disorders like Alzheimer's disease (AD). However, because of their quickness, they can act as an initial screening that expedites diagnosis, cutting down on the quantity of doctor visits and the expense of the procedure. Here, long-term performance of the electrodes is not necessary because electrochemical sensors are designed to be single-use devices [24].

The decentralised diagnosis of a number of biomarkers using POCT devices will be crucial for the ultimate early detection and surveillance of a number of diseases. By boosting the ubiquity and prompt release of data through Internet of Things (IoT), the combination of cutting-edge electrochemical sensors with great sensitivity and specificity, low cost, high-throughput microfabrication processes, and smartphone-based platforms would expand the applicability of POCT. All of these attributes will result in the development of more approachable preventative medicine tools that everyone may use. Nevertheless, even if microfluidics can aid in reagent management, the creation of cartridges or preloading chemicals onto paper substrates can aid in preventing mistakes made by amateurs.

20.3.4 Environmental Monitoring

Electrochemical sensors offer great prospects for environmental monitoring and pollution control. They can detect various pollutants in air, water, and soil, enabling early detection and prevention of pollution-related problems. Because they enable the instrument to be brought to the sample (instead of the conventional method of taking the sample to the laboratory), electrochemical sensing devices have a significant impact on the monitoring of priority pollutants. These tools can assess a range of inorganic and organic contaminants quickly, reliably, and affordably while performing automated chemical analysis in complicated matrices [25].

Joshi et al. have reviewed the use and prospects of boron-doped diamond (BDD) towards the environmental remediation and sensing applications [26]. Due to the BDD electrode's low background current, wide potential window, and resistance to fouling, it is frequently utilised in electrochemical applications. For the detection of inorganic chemicals like H_2O_2 and arsenic as well as organic substances like pharmaceuticals and pesticides, BDD-based sensing is helpful. The ability of BDD electrodes to produce hydroxyl radicals (OH) enables the breakdown of organic micropollutants during wastewater treatment. In addition to detecting atrazine in spiked river water samples, BDD electrodes have been used to quantify the cytostatic chemical flutamide in pharmaceutical formulations, spiked human urine, and water samples. Additionally, they are helpful for processes like soil restoration, CO_2 reduction, and electrogenerated chemiluminescence [26].

Similarly Amina Rhouati et al. have discussed in detail the use of novel materials like MXenes and their composites towards the sensing of environmental contaminants [27]. MXenes are attracting interest from researchers in a variety of fields, including environmental water engineering, such as desalination and wastewater treatment, as well as designing and manufacturing effective sensors to detect dangerous pollutants in water. These characteristics include high conductivity, hydrophobicity, and large specific surface areas. A new path towards the sensing and detection of environmental contaminants is opened up by the incorporation of these 2D materials in electrochemical enzyme- and affinity-based biosensors for environmental pollutants.

The use of novel 2D materials like BDD and MXenes paves a new road towards the sensing and detection of environmental pollutants. 2D material-based sensors produce data that can be effectively used for environmental and health monitoring since they are extremely effective and compatible with current manufacturing technologies. Due to their distinctive material properties resulting from their innate structures, graphene and its oxides, black phosphorus (BP), transition metal dichalcogenides (TMDCs), metal oxides, and other 2D nanomaterials have demonstrated properties that have been appealing for the production of highly sensitive sensors [27,28].

As we move into the 21st century, we would prefer to have quick, small, simple-to-use, portable, affordable, and disposable electrode systems rather than relying on heavy electrochemical cells and large electrodes. In recent years, a wide variety of electrodes for in situ and on-site environmental monitoring have been created [29]. The use of molecularly imprinted polymers (MIPs) combined with nanomaterials and disposable screen-printed-electrochemical sensors are examples that demonstrate the breadth, strength, adaptability, and use of such miniature electrodes for environmental monitoring [29–31].

20.3.5 Food Safety

Electrochemical sensors can play a crucial role in ensuring food safety by detecting contaminants, such as pesticides, heavy metals, and pathogens, in food products. They offer rapid, sensitive, and portable solutions for food quality control and traceability.

Concern over food safety as a matter of public health has spread around the globe. Every step of the food supply chain may be subject to food safety risks brought on by the presence of food safety hazards in food. As a result, sensitive, picky, and practical analytical methods of different food safety concerns in food are highly valued. Nanomaterials are currently leading the way in many electrochemical sensing applications. Nanomaterials are regarded as ideal options for the construction of electrode surfaces to create electrochemical sensors with outstanding analytical performance because of their distinctive physical, chemical, and biological properties. Electrochemical sensors are a potential replacement for current approaches to analysing food safety with the use of nanomaterials [32].

Electrochemical sensors for food analytes have two main categories of analyte detection mechanisms: affinity-based and non-affinity-based detection. Affinity-based detection relies on the recognition of analytes by antibodies, functional nucleic acids, enzymes, or MIPs as opposed to non-affinity-based detection, which depends on the electroactivity of the analytes. Various multi-signal output, multi-target detection, and common nanomaterial-assisted signal amplification methodologies are discussed in relation to flexible sensing systems based on nanomaterials. To aid in the signal amplification of electrochemical sensors, nanomaterials play three primary functions: electrode modifiers, nanocarriers, and signal tags. The roles that different nanomaterials play depend on the individual goals and sensing techniques. The development of portable electrochemical tools for identifying risks to food safety is also mentioned. In order to investigate small electrochemical sensors, potable electrode devices, microfluidic analytical systems, and smartphone-based CPUs are crucial components.

The development of portable electrochemical sensors for on-site food safety analysis was aided by the spread of cellphones. Researchers are increasingly using smartphones as electrochemical sensors. Smartphones can drastically save hardware and software expenses when used as handheld processors. Smartphones with customised application software loaded, a wireless transmitter, and the ability to connect the system to the Cloud can be used to conduct data collecting and analysis as well as display detection findings. Smartphone-based electrochemical sensors that are more affordable, lightweight, flexible, and durable will result from the development of these cutting-edge capabilities. A smartphone-based electrochemical system made up of a smartphone, a handheld detector, and a modified screen-printed carbon electrode was described by Xu et al. [33] for quick determination of nitrite in water (Figure 20.7a). In a linear range of 1–500 mol L^{-1}, the smartphone-based system exhibits outstanding analytical performance for the detection of nitrite with a low LOD of 0.2 mol L^{-1}. Additionally, some researchers created and modified unique instruments for the extraction, collection, and detection of analytes. Lin et al. [34] created integrated exogenous antigen testing (iEAT), a portable, point-of-use detection system, as depicted in Figure 20.7b. For quick sensing and communication, the system combines a disposable antigen extraction tool with an electronic keychain reader. Simple extraction, capture, labelling, and detection were used to test for allergens.

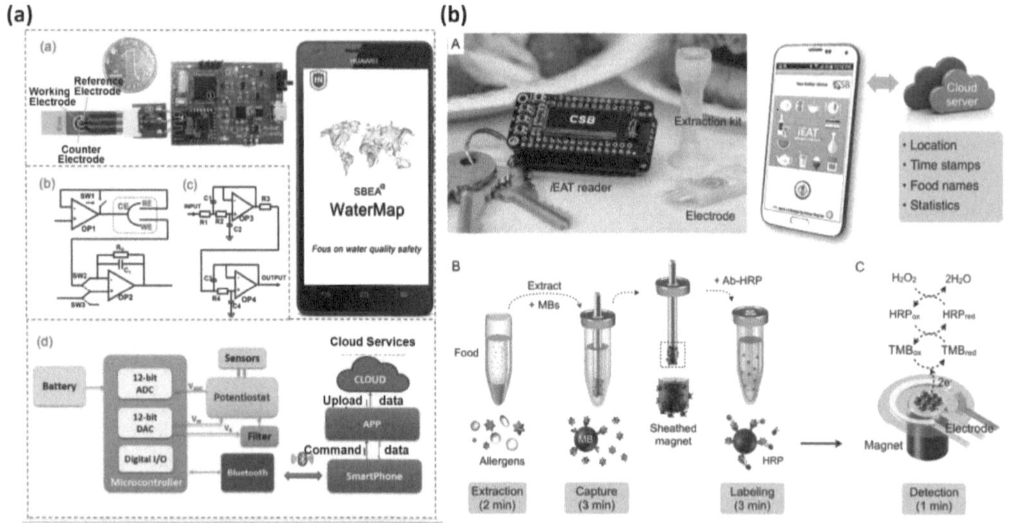

FIGURE 20.7 (a) (i) The image of the smartphone-based electrochemical system and the key circuit module of the detector; (ii) and (iii) the circuit design of potentiostat and hardware filter; (iv) schematic diagram of smartphone-based electrochemical (b) iEAT system consisting of a pocket-sized detector, an electrode chip, and a disposable kit for extraction for on-site allergen detection.

Source: Reproduced with permission from Xu et al. [33], copyright 2020 Elsevier. Reproduced with permission from H. Y. Lin et al. [34], copyright 2017 The American Chemical Society.

For a small fee (less than $4 per assay), iEAT can give analysis findings in a timely manner (less than 10 minutes for the complete experiment).

The suggested sensor outperformed the industry standard enzyme-linked immunosorbent assay (ELISA) by a factor of ten and was five times faster. In summary, the practical application and commercialisation of electrochemical sensors for food safety analysis depend on integrated smartphone-based sensors that do not require expensive equipment or a high degree of knowledge [32].

20.3.6 Wearable and Implantable Devices

Electrochemical sensors can be integrated into wearable and implantable devices for personalised health monitoring. They enable continuous and non-invasive monitoring of biomarkers in sweat, tears, or other body fluids, providing valuable real-time data for personalised healthcare applications. Wearable technology is expected to transform how we live. In fact, the majority of readers likely wear a smart watch right now to track their fitness activities or monitor their daily steps, heart rate, oxygen saturation, and sleep cycle. By implementing wireless data transmission systems, the benefits of the Internet of Things can be easily shared using the pervasive cellphones. A more individualised, patient-controlled healthcare system will be possible thanks to the data mining that they can extract for wearers. A wearable device is one that can be applied without the help of a doctor and is worn on the body with minimal or no invasiveness. Examples include watches, bracelets, skin patches with needles, mouthguards, contact lenses, and skin patches. A wearable device that is fully integrated consists of three components: sensors that interface with body fluids; a data communication module that can use Bluetooth or near-field communication (NFC); and a power source that can come from batteries, supercapacitors, or energy harvesters. Figure 20.8 shows the schematic representation of the path that electrochemical biosensing is following in terms of future market trends.

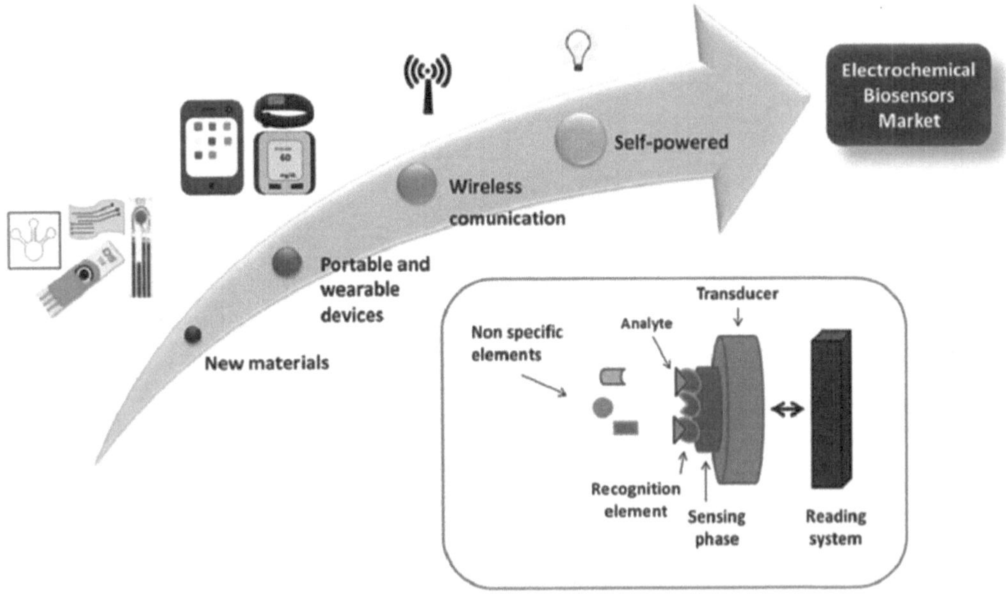

FIGURE 20.8 Schematic representation of the path that electrochemical biosensing is following in terms of future market trends [35].

Source: Reprinted with permission from Neves, M., M.B. Gonzalez-Garcia, D. Hernandez-Santos, and P. Fanjul-Bolado, Future trends in the market for electrochemical biosensing. *Current Opinion in Electrochemistry*, 2018. 10: pp. 107–111. Copyright 2021 American Chemical Society.

The most common sensors that are currently used in commercially available gadgets rely on actual readings of physiological parameters. Monitoring chemical targets in biofluids, on the other hand, is a very potent source of clinical data that will let to gain a full picture of the health condition. As a result, a lot of research is being done to create chemical sensors that can conduct continuous on-body analysis [36,37]. Since the measuring conditions are substantially different from those in a traditional laboratory, this type of measurement is highly difficult. Wearable sensors must be able to analyse very complex biofluids under uncontrolled conditions of temperature, pH, or humidity with analytical performance comparable to laboratory-based approaches in terms of sensitivity and selectivity [38]. The absence of standard materials for the analytical validation of the sensors, which is often done using blood-based gold standard laboratory assays, is another significant problem. Figure 20.9 depicts various wearable devices for the real-time electrochemical sensing of analytes [24].

One of the finest illustrations of how analytical chemistry is shifting to decentralised settings is wearable sensors. These technologies enable a fundamental paradigm shift towards precision medicine by being able to remotely gather priceless data from patients, transfer it to the cloud, and provide access to doctors in real time. This problem might be solved by creating novel self-healing materials based on polymers, conductive polymers, and bioinspired materials. Data dependability is a major challenge due to the lack of standards, especially in such dynamic on-body circumstances.

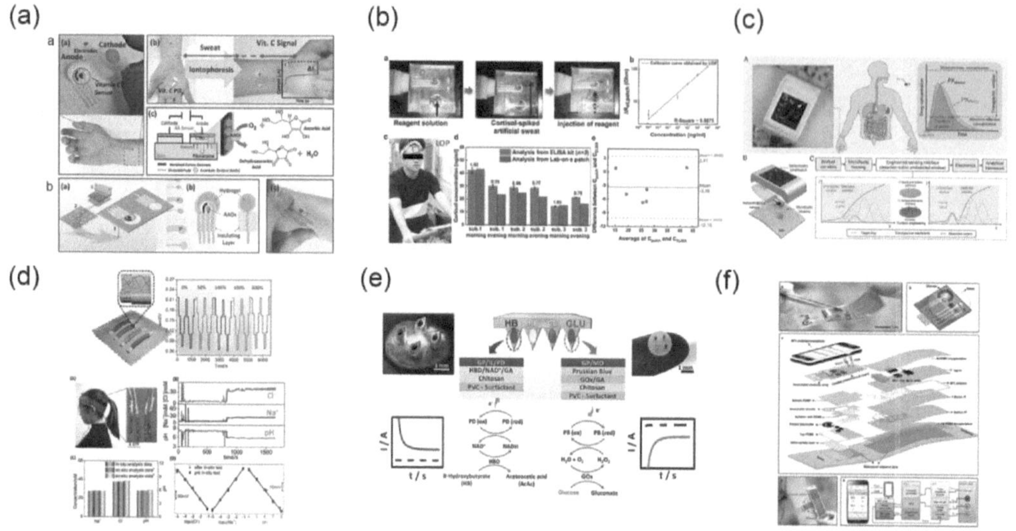

FIGURE 20.9 (a) Wearable device with integrated iontophoresis electrodes for sweat vitamin C determination. (b) Microfluidic wearable device performing different steps of an immunoassay. (c) General scheme of the operation of the fully integrated wearable for tracking acetaminophen concentrations. (d) General structure of thread-based potentiometric sensors and on-body operation. (e) Microneedle device for simultaneous monitoring of glucose and ketone bodies. (f) General design of the battery-free and wireless electrochemical system for sweat biomarkers analysis [24].

Wiorek et al. provided a good practice example in this area [39]. The addition of pH and temperature sensors made it possible to dynamically correct the data for those parameters. Additionally, Lin et al. recently showed how the addition of intelligent hydrogel-based microfluidic valves can provide new capabilities for wearable technology by actively managing fluid sampling and measurement [40]. Figure 20.10 shows the concept and operational principle of a fully integrated wearable valving system.

A barrier to widespread use of the technology is the integration of the sensing unit with wearable instruments, power sources, and/or energy harvesters. The development of fully working prototypes and possibly even commercially available gadgets will be made possible in large part by developments in flexible hybrid electronics [41]. Energy supply is frequently a significant issue, particularly when there are strict restrictions on size and weight. Batteries are now the most adaptable alternative, although additional improvements in capacity and compactness are required. Energy harvesters often produce an oscillatory energy supply, which is insufficient for electrochemical instrumentation; thus, these problems can be resolved by combining them with flexible and lightweight energy storage systems. Nanomaterials will help to enhance the performance of energy harvesters and energy storage devices in addition to wearable sensors. It is also required to include additional, less-studied indicators, such as proteins or DNA, in the wearable sensor panel.

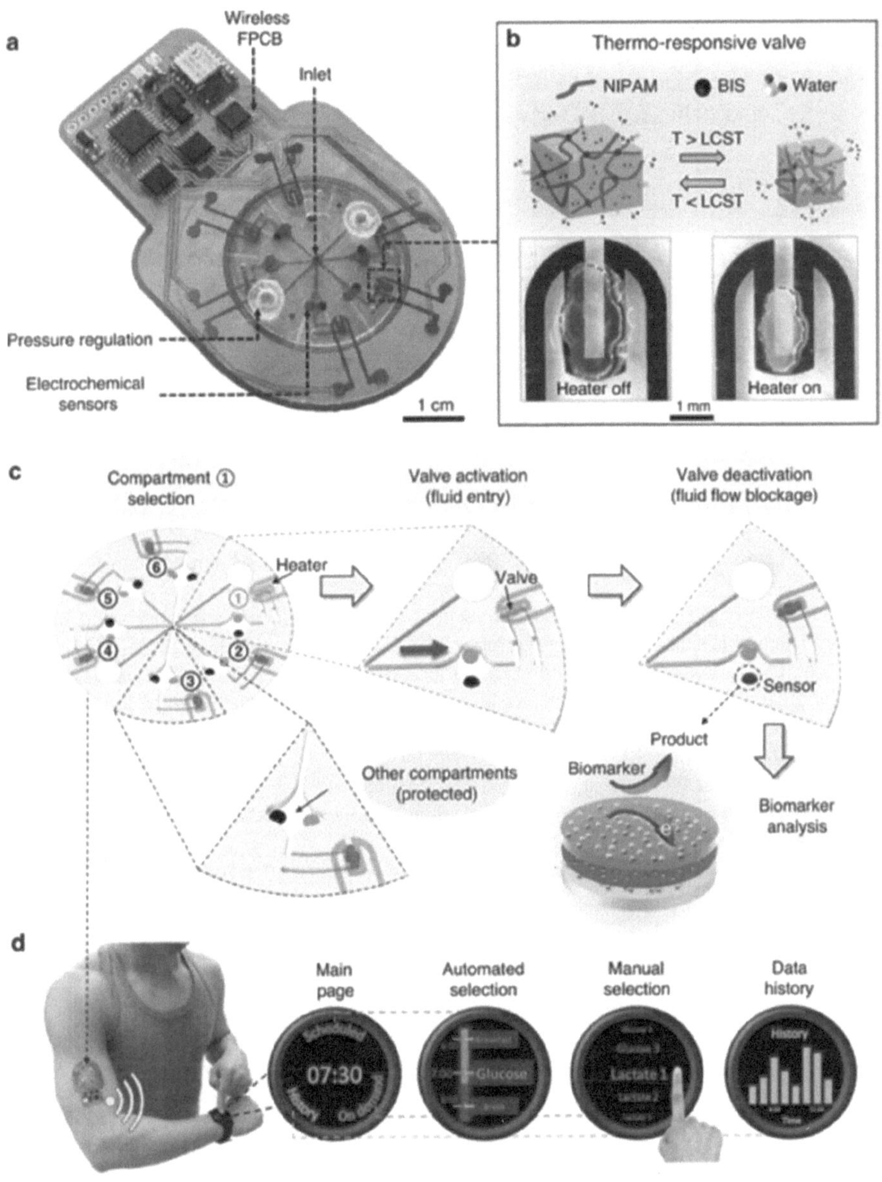

FIGURE 20.10 A fully integrated wearable valving system (concept and operational principle). From: A programmable epidermal microfluidic valving system for wearable biofluid management and contextual biomarker analysis [40].

Source: Reproduced with permission from Ref. [40] under creative common license copyright @2022.

20.4 CONCLUSION

In conclusion, the creation of highly sensitive, selective, and stable electrochemical sensors is a promising application for nanomaterials. Electrochemical sensors that use nanomaterials may detect many analytes from a single sample with higher sensitivity and good selectivity, among other benefits. The development of electrochemical sensors based on nanomaterials faces significant difficulties. Researchers in the field are focussing on problems such as the lack of a common fabrication technique, difficulties with repeatability, and scalability of the synthesis processes. However, with numerous existing real-world applications, the prospects for the development of nanomaterials as electrochemical sensors remain positive. Further development of electrochemical sensors based on nanomaterials is set to revolutionise the field of analytical chemistry, leading to the development of sensitive, selective, and stable sensors for various applications.

REFERENCES

1. Negahdary, M., et al., Recent electrochemical sensors and biosensors for toxic agents based on screen-printed electrodes equipped with nanomaterials. *Microchemical Journal*, 2023. 185: p. 18.
2. Baig, N., I. Kammakakam, and W. Falath, Nanomaterials: A review of synthesis methods, properties, recent progress, and challenges. *Materials Advances*, 2021. 2(6): pp. 1821–1871.
3. Curulli, A., Nanomaterials in electrochemical sensing area: Applications and challenges in food analysis. *Molecules*, 2020. 25(23): p. 35.
4. Nath, N., et al., Carbon nanostructure embedded novel sensor implementation for detection of aromatic volatile organic compounds: An organized review. *ACS Omega*, 2023. 8(5): pp. 4436–4452.
5. Macovei, D.G., et al., Point-of-care electrochemical testing of biomarkers involved in inflammatory and inflammatory-associated medical conditions. *Analytical and Bioanalytical Chemistry*, 2023. 415(6): pp. 1033–1063.
6. Curulli, A., Electrochemical biosensors in food safety: Challenges and perspectives. *Molecules*, 2021. 26(10): p. 62.
7. Yoon, Y., P.L. Truong, D.H. Lee, and S.H. Ko, Metal-oxide nanomaterials synthesis and applications in flexible and wearable sensors. *ACS Nanoscience Au*, 2021. 2(2): pp. 64–92.
8. del Rio, J.S., O.Y.F. Henry, P. Jolly, and D.E. Ingber, An antifouling coating that enables affinity-based electrochemical biosensing in complex biological fluids. *Nature Nanotechnology*, 2019. 14(12): p. 1143.
9. He, Q.G., et al., Research on the construction of portable electrochemical sensors for environmental. *Materials Today Advances*, 2023. 17: p. 17.
10. Neves, M., M.B. Gonzalez-Garcia, D. Hernandez-Santos, and P. Fanjul-Bolado, A miniaturized flow injection analysis system for electrogenerated chemiluminescence-based assays. *Chemelectrochem*, 2017. 4(7): pp. 1686–1689.
11. Jiang, C., et al., Antifouling strategies for selective in vitro and in vivo sensing. *Chemical Reviews*, 2020. 120(8): pp. 3852–3889.
12. Barfidokht, A. and J.J. Gooding, Approaches toward allowing electroanalytical devices to be used in biological fluids. *Electroanalysis*, 2014. 26(6): pp. 1182–1196.
13. Yuan, B.Q., et al., A micro electrochemical sensor for multi-analyte detection based on oxygenated graphene modified screen-printed electrode. *Nanomaterials*, 2022. 12(4): p. 17.
14. Bauer, M., et al., Electrochemical multi-analyte point-of-care perspiration sensors using on-chip three-dimensional graphene electrodes. *Analytical and Bioanalytical Chemistry*, 2021. 413(3): pp. 763–777.
15. Shi, J.J., et al., Ultrasensitive multi-analyte electrochemical immunoassay based on GNR-modified heated screen-printed carbon electrodes and PS@PDA-metal labels for rapid detection of MMP-9 and IL-6. *Biosensors & Bioelectronics*, 2014. 55: pp. 51–56.
16. Chacko, L., E. Massera, and P.M. Aneesh, Enhancement in the selectivity and sensitivity of Ni and Pd functionalized MoS_2 toxic gas sensors. *Journal of the Electrochemical Society*, 2020. 167(10): p. 9.
17. Kim, E.B., et al., Indandione oligomer@graphene oxide functionalized nanocomposites for enhanced and selective detection of trace Cr^{2+} and Cu^{2+} ions. *Advanced Composites and Hybrid Materials*, 2022. 5(2): pp. 1582–1594.
18. Wang, Y.F., et al., Enhanced electrochemical sensor based on gold nanoparticles and MoS_2 nanoflowers decorated ionic liquid-functionalized graphene for sensitive detection of bisphenol A in environmental water. *Microchemical Journal*, 2021. 161: p. 8.
19. Luo, H., et al., Applications of macrocyclic compounds for electrochemical sensors to improve selectivity and sensitivity. *Journal of Inclusion Phenomena and Macrocyclic Chemistry*, 2019. 95(3–4): pp. 171–198.
20. Land, K.J., et al., Reassured diagnostics to inform disease control strategies, strengthen health systems and improve patient outcomes. *Nature Microbiology*, 2019. 4(1): pp. 46–54.
21. Gonzalez-Lopez, A., M.C. Blanco-Lopez, and M.T. Fernandez-Abedul, Micropipette tip-based immunoassay with electrochemical detection of antitissue transglutaminase to diagnose celiac disease using staples and a paper-based platform. *ACS Sensors*, 2019. 4(10): pp. 2679–2687.

22. Noviana, E., et al., Electrochemical paper-based devices: Sensing approaches and progress toward practical applications. *Lab on a Chip*, 2020. 20(1): pp. 9–34.
23. Caratelli, V., et al., Precision medicine in Alzheimer's disease: An origami paper-based electrochemical device for cholinesterase inhibitors. *Biosensors & Bioelectronics*, 2020. 165: p. 10.
24. Hernandez-Rodriguez, J.F., D. Rojas, and A. Escarpa, Electrochemical sensing directions for next-generation healthcare: Trends, challenges, and frontiers. *Analytical Chemistry*, 2021. 93(1): pp. 167–183.
25. Hanrahan, G., D.G. Patil, and J. Wang, Electrochemical sensors for environmental monitoring: Design, development and applications. *Journal of Environmental Monitoring*, 2004. 6(8): pp. 657–664.
26. Joshi, P., et al., Recent advances of boron-doped diamond electrochemical sensors toward environmental applications. *Current Opinion in Electrochemistry*, 2022. 32: p. 9.
27. Rhouati, A., M. Berkani, Y. Vasseghian, and N. Golzadeh, MXene-based electrochemical sensors for detection of environmental pollutants: A comprehensive review. *Chemosphere*, 2022. 291: p. 10.
28. Tyagi, D., et al., Recent advances in two-dimensional-material-based sensing technology toward health and environmental monitoring applications. *Nanoscale*, 2020. 12(6): pp. 3535–3559.
29. Hayat, A. and J.L. Marty, Disposable screen printed electrochemical sensors: Tools for environmental monitoring. *Sensors*, 2014. 14(6): pp. 10432–10453.
30. Rahman, S., et al., Molecularly imprinted polymers (MIPs) combined with nanomaterials as electrochemical sensing applications for environmental pollutants. *Trends in Environmental Analytical Chemistry*, 2022. 36: p. 22.
31. Choudhari, U., et al., Screen-printed electrochemical sensors for environmental monitoring of heavy metal ion detection. *Reviews in Chemical Engineering*, 2022. 39: p. 42.
32. Gu, Y., et al., Recent advances in nanomaterial-assisted electrochemical sensors for food safety analysis. *Food Frontiers*, 2022. 3(3): pp. 453–479.
33. Xu, K., et al., Cost-effective, wireless, and portable smartphone-based electrochemical system for on-site monitoring and spatial mapping of the nitrite contamination in water. *Sensors and Actuators B-Chemical*, 2020. 319: p. 10.
34. Lin, H.Y., et al., Integrated magneto-chemical sensor for on-site food allergen detection. *ACS Nano*, 2017. 11(10): pp. 10062–10069.
35. Neves, M., M.B. Gonzalez-Garcia, D. Hernandez-Santos, and P. Fanjul-Bolado, Future trends in the market for electrochemical biosensing. *Current Opinion in Electrochemistry*, 2018. 10: pp. 107–111.
36. Yu, Y., H.Y.Y. Nyein, W. Gao, and A. Javey, Flexible electrochemical bioelectronics: The rise of in situ bioanalysis. *Advanced Materials*, 2020. 32(15): p. 25.
37. Bocchetta, P., et al., Soft materials for wearable/flexible electrochemical energy conversion, storage, and biosensor devices. *Materials*, 2020. 13(12): p. 35.
38. Sempionatto, J.R., I. Jeerapan, S. Krishnan, and J. Wang, Wearable chemical sensors: Emerging systems for on-body analytical chemistry. *Analytical Chemistry*, 2020. 92(1): pp. 378–396.
39. Wiorek, A., M. Parrilla, M. Cuartero, and G.A. Crespo, Epidermal patch with glucose biosensor: pH and temperature correction toward more accurate sweat analysis during sport practice. *Analytical Chemistry*, 2020. 92(14): pp. 10153–10161.
40. Lin, H.S., et al., A programmable epidermal microfluidic valving system for wearable biofluid management and contextual biomarker analysis. *Nature Communications*, 2020. 11(1): p. 12.
41. Ray, T.R., et al., Bio-integrated wearable systems: A comprehensive review. *Chemical Reviews*, 2019. 119(8): pp. 5461–5533.

Index

Note: **Bold** page numbers refer to tables and *italic* page numbers refer to figures.